Frank Austin Gooch

Methods in Chemical Analysis

Originated or Developed
in the Kent Chemical Laboratory
of Yale University

Elibron Classics
www.elibron.com

Elibron Classics series.

ISBN 0-543-69308-2 (paperback)
ISBN 0-543-69307-4 (hardcover)

This Elibron Classics Replica Edition is an unabridged facsimile
of the edition published in 1912 by John Wiley & Sons, New York.

METHODS IN CHEMICAL ANALYSIS

ORIGINATED OR DEVELOPED
IN THE KENT CHEMICAL LABORATORY
OF YALE UNIVERSITY

COMPILED BY

FRANK AUSTIN GOOCH

PROFESSOR OF CHEMISTRY AND DIRECTOR OF THE KENT CHEMICAL
LABORATORY IN YALE UNIVERSITY

FIRST EDITION

NEW YORK
JOHN WILEY & SONS
LONDON: CHAPMAN & HALL, LIMITED

Stanhope Press
F. H. GILSON COMPANY
BOSTON, U.S.A.

PREFATORY NOTE

THE object of this volume is to present concisely the principal results reached by workers in the Kent Chemical Laboratory of Yale University in the investigation and development of methods in chemical analysis. In the account of processes, modified or original, only proved procedure and immediately related experimental data are, as a rule, given. For further details in respect to the elaboration of processes, the discussion of difficulties, and the experimental illustration of the effects of varying the prescribed procedure, references are given to the original sources from which this summary has been compiled. To his colleagues, Professors Philip E. Browning, R. G. Van Name, and W. A. Drushel, the compiler is much indebted for valuable criticism of the proof sheets.

CONTENTS.

CHAPTER I.

APPLIANCES AND GENERAL PROCEDURE.

CHAPTER II.

THE ALKALI METALS.

CHAPTER III.

COPPER; SILVER; GOLD.

CHAPTER IV.

BERYLLIUM; MAGNESIUM; CALCIUM; STRONTIUM; BARIUM.

CHAPTER V.

ZINC; CADMIUM; MERCURY.

CHAPTER VI.

BORON; ALUMINIUM; LANTHANUM; THALLIUM.

CHAPTER VII.

CARBON; SILICON; TITANIUM; ZIRCONIUM; CERIUM; TIN; LEAD.

CHAPTER VIII.

NITROGEN; PHOSPHORUS; ARSENIC; ANTIMONY; BISMUTH; VANADIUM.

CHAPTER IX.

OXYGEN; SULPHUR; SELENIUM; TELLURIUM.

CHAPTER X.

CHROMIUM; MOLYBDENUM; URANIUM.

CHAPTER XI.

FLUORINE; CHLORINE; BROMINE; IODINE.

CHAPTER XII.

MANGANESE; NICKEL; COBALT; IRON.

METHODS IN CHEMICAL ANALYSIS

CHAPTER I.

APPLIANCES AND GENERAL PROCEDURE.

Mechanical Processes.

The Determination of Products Gaseous at Ordinary Temperatures by Loss of Weight.

Various forms of apparatus have been designed for determining, by loss of weight, reaction products which are gaseous at ordinary temperatures, but many of these are cumbersome or require skill in glass blowing for their construction. A form of apparatus described by Kreider* is light and easily made from three test tubes, modified and fitted as shown in the figure. The test tube, A, serves as the reaction chamber. B is perforated with a hole about 1 cm. in diameter and fits tightly within A; and C, so selected that it fits loosely within B, is drawn out to a small capillary tube.

Fig. 1.

When the apparatus is to be used, the capillary of C, which has been fitted as described, is pushed through the hole of B, packed loosely with cotton; B is filled to the depth of from 6 cm. to 8 cm. (about two-thirds of its contents) with granular calcium chloride; and B and C are adjusted as shown.

To the test tube, C, is fitted a one-holed stopper, through which passes a short glass tube which is to be closed by a rubber cap and plug. Upon removing the plug, and applying suction to the short tube, the reagent employed to liberate the volatile product to be determined is drawn up through this capillary until C is sufficiently filled. Upon replacing the plug the reagent remains within C, held by atmospheric pressure. Gentle pressure upon the cap

* J. Lehn Kreider, Am. Jour. Sci., [4], xix, 188.

expels a drop of liquid from the capillary, and upon the release of the cap a little air is drawn in to allow for expansion of air in the large tube without loss of liquid during subsequent handling.

The tubes A and B are so selected that very little of the product evolved can escape between them, and, in case they fit very loosely, a ring of paraffin melted into the mouth of A, about B, by means of a hot wire, seals the joint securely. A very convenient way to attach the paraffin is to melt it between A and another tube, which fits A, as does B, and may be removed by a turning motion, leaving the ring into which B will fit. Very little heating is then required to make a tight joint. If care be used in taking apart A and B, at the close of an experiment, such a ring of paraffin remains in place and may be used many times without replacement, being remelted by a touch of the hot wire before every new experiment.

In making a determination, the substance under examination is weighed and placed in the bottom of A. The reagent to be employed, 10 $cm.^3$ to 15 $cm.^3$, is drawn into C, and held there in the manner described. The test tube A is slipped over B, and the joint is sealed with paraffin, as has been shown. The apparatus is wiped, placed on the balance and weighed.

Upon removing the cap from the small tube in C, the reagent runs from C into A. The volatile product, forced upward through the drying column of calcium chloride, escapes through the annular space between B and C. When action ceases, a current of dry air is forced through C, to remove all the volatile product, the cap is replaced, and the apparatus is weighed. The loss of weight represents the volatile product.

Hydrogen by Loss.

	Metal taken. grm.	Hydrogen found. grm.	Error. grm.
Magnesium	0 1000	0 0087	+0 0003
	0 1000	0 0085	+0 0001
	0 1000	0 0084	0 0000
	0 1000	0 0084	0 0000
	0 1000	0 0083	−0.0001
Zinc	0 2000	0 0061	0 0000
	0 2000	0 0062	+0.0001
	0 2000	0.0062	+0.0001
	0 2000	0.0060	−0.0001
	0 2000	0.0061	0.0000

Tests of this apparatus in the determination of carbon dioxide in carbonates, and of nitrogen in urea and in ammonium salts, are described later.

In the preceding table are given results of experiments made to determine thus the weights of hydrogen liberated by the action of magnesium and zinc upon dilute hydrochloric acid.

The Distillation and Condensation of Volatile Products.

The rapid evaporation of liquid charged with soluble or insoluble matter is apt to carry mechanically to the distillate some material which should remain in the residue. A form

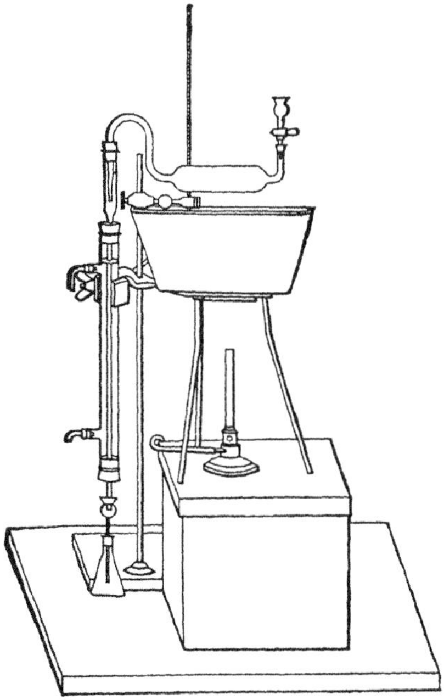

Fig. 2.

of apparatus elsewhere described* and shown in the accompanying figure (Fig. 2) solves the problem successfully. The retort, made of a pipette, bent as shown, with stoppered funnel

* Gooch, Am. Chem. Jour., ix, 28.

sealed on or attached by a rubber joint, is fitted to an upright condenser which, in turn, is connected by a stopper to a thistle tube, fitted tightly to the receiver by means of a stopper perforated or grooved to permit the passage of air. For work to be described the apparatus has been modified by substituting for the perforated or grooved stopper a tight stopper carrying a bulbed trap.*

In making a distillation, the liquid is introduced by the funnel, the glass cock is closed, the water started through the condenser, and the retort, not more than half filled and inclined backward, is carefully heated. For the heating a paraffin bath is in many cases most convenient, and it is advantageous to lower the retort into the paraffin, already heated to a temperature considerably above the boiling point of the liquid, so that evaporation may take place rapidly and often without actual boiling. The diameter of the gooseneck should be at least 0.7 cm. to prevent the formation of bubbles within it.

The use of this apparatus in the determination of boric acid is described elsewhere.

The Distillation and Absorption of Volatile Products.

Apparatus with Ground Joints. A convenient apparatus for the distillation and absorption of volatile products † is easily constructed, with glass joints throughout, by sealing together a separating funnel A, a Voit flask B, a Drexel wash bottle C, and a bulbed trap *g*, as shown in the figure. Upon the side of the distillation flask B is pasted or etched a graduated scale, by means of which the volume of liquid within the flask may be known at any time. The separating funnel is connected with a Kipp generator set up for the delivery of carbon dioxide, hydrogen, hydrogen chloride or other suitable gas. The flask serves as the retort, the wash bottle properly charged as

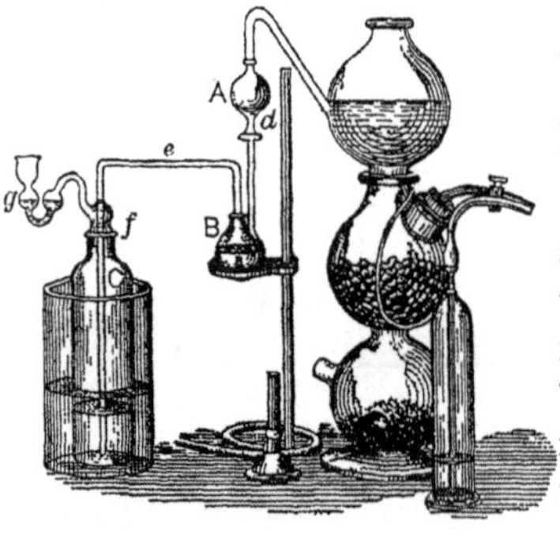

Fig. 3.

* See trap of Fig. 7, p. 6.

† F. A. Gooch and John T. Norton, Jr., Am. Jour. Sci., [4], vi, 168.

the receiver, and the products of distillation are swept forward by the generator gas, which may serve as a reagent or simply as a medium for aiding the transfer of products from the retort to the receiver.

This apparatus has served a useful purpose in processes to be described for the determination of molybdenum, vanadium, and iodine liberated from the iodide-iodate mixture by acids free or evolved.

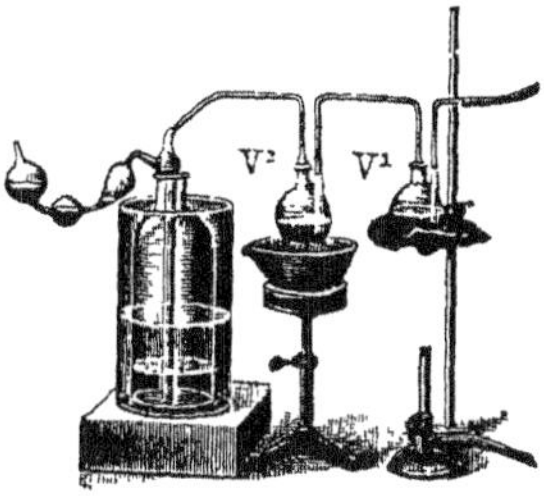

Fig. 4.

A similar device adapted to double distillation is shown in Fig. 4. Application of this apparatus will be given later.*

Fig. 5.

Apparatus with Sealed Joints. Fig. 5 shows a convenient device by Edgar,† put together without ground joints, for the distillation and absorption of volatile products. The distillation retort, similar in design to that of Fig. 2, consists of a modified pipette,

* F. A. Gooch and A. W. Peirce, Am. Jour. Sci., [4], i, 181.
† Graham Edgar, Am. Jour. Sci., [4], xxvii, 174.

with the inlet tube bent upward and sealed to a separatory funnel while the outlet tube, expanded to a small bulb, is bent upward and then downward to enter the absorption flask.

A slow current of hydrogen, or other suitable gas, is made to enter at the bottom of the retort to stir the liquid so that a very small volume may be distilled without danger of "bumping."

The Removal of Volatile Products without Loss of Non-volatile Material Reserved for Treatment.

In processes which involve the elimination of a volatile reagent or product of reaction from a boiling solution, it is often essential to prevent losses by spattering or by mechanical transfer of non-volatile material in the steam. In many such processes the simple device shown in the figure is effective in preventing appreciable error by loss.* A flask, preferably of the Erlenmeyer shape, with a broad bottom, permits boiling of the liquid in a shallow layer favorable to the checking of explosive ebullition. A two-bulbed trap, made by cutting short an ordinary calcium chloride drying tube and hung with the large opening downward, obstructs the steam while permitting sufficient relief of pressure and thus serves to catch and return to the liquid particles of the non-volatile matter thrown upward.

Fig. 6.

The Prevention of Mechanical Loss of Solution in Reactions Evolving Gaseous Products.

The danger of mechanical loss in reactions accompanied by effervescence (as in the neutralization of carbonates by strong acids) or by formation of spontaneously volatile product (as in the liberation of iodine to be subsequently titrated) may be minimized by making use of a trapped reaction chamber. For this purpose the apparatus shown in the figure is serviceable.† It consists of a Drexel washing bottle with a separatory funnel sealed to the inlet tube, and a Will and Varrentrapp absorption apparatus joined to the outlet tube. The reaction is brought

Fig. 7.

* F. A. Gooch and P. E. Browning, Am. Jour. Sci., [3], xxxix, 197.

† F. A. Gooch and C. F. Walker, Am. Jour. Sci., [4], iii, 293.

about by admitting the appropriate reagents through the funnel tube to the solution to be acted upon in the cylinder, so that all volatile products must escape through the properly charged absorption bulbs.

The Transfer of Liquids and Gases under Pressure.

A simple form of force pump, with Bunsen valves of special construction, has been described by Kreider.*

Valve. — In forcing a liquid or gas indifferent to rubber from one vessel to another, the ordinary Bunsen valve is apt to collapse in such a way as to permit a back flow. Kreider finds that a stout glass tube of desirable size, sealed at one end and drawn out with an opening in the constriction, as indicated in the accompanying figure, and a piece of rubber tubing containing a smooth slit placed over it, makes a valve in which collapse is impossible. A valve similar in appearance to the one here described has been previously used; but the similarity is confined to the appearance, as will be evident from the following description: The constriction should not be greater than is necessary to leave a small space between the glass and the rubber when the latter is loosely drawn over it; but it should be long enough to permit a slit of about a centimeter's length in the rubber to close tightly, or about twice the length of the slit. A slit one centimeter long will be found to open under very slight pressure, and, to accomplish its purpose, it is only required to close sufficiently for the external pressure to force the rubber against the opening in the tube. This opening should be carefully rounded and a little higher rather than any lower than the surrounding glass, and is better made before sealing the end, in order to keep the tube perfectly straight. The rubber should fit tightly about the larger parts of the glass tube and be put on with care to have the smoothly cut slit straight, and loose enough to close tightly. If the slit is placed about 90° from the opening in the tube, sufficient space will remain to permit the escape of the gas or liquid, but the moment the pressure outside becomes greater than that within, the rubber will be pressed tightly over this opening and thus a return made impossible. When dry the valve does not resist high pressure perfectly; but when wet, or better, when both glass and rubber,

Fig. 8.

* D. A. Kreider, Am. Jour. Sci., [3], l, 132.

including the slit, are moistened with glycerin, a nearly perfect vacuum may be retained for several days. The valve thus lubricated with glycerin, when used as a protection in an ammonia wash bottle, will prevent absolutely the access of ammonia to the mouth, and if made according to the directions will act with very little pressure. Placed in the connection between the vacuum flask and water pump ordinarily used in filtration, it has been found a valuable check on the valve of the pump, and when the latter fails this device prevents the back flow of water into the filtrate. In processes which necessitate the use of a partial vacuum, this valve may be employed to hold the vacuum in continual readiness.

Force Pump. — By adjusting two of the valves just described to the opposite extremities of a T-tube, with the horizontal limb enlarged or sealed to a larger tube so as to permit the attachment of a large and stout piece of rubber tubing closed at one end, as shown in Fig. 9, a convenient and powerful little force pump is obtained. A stout T-tube of small-bore is cut off short at the two ends at right angles to one another; to one is sealed a tube just large enough to permit the insertion of a valve; to the other, a large tapering tube, slightly lipped so as to hold a piece of rubber tubing firmly and allow of tying the latter if necessary. Of the third end of the tube, a valve like that shown in Fig. 8 is made. The compressing rubber should not be of greater length than the hand is able to cover completely, and may be closed with a glass stopper selected to fit tightly.

Fig. 9.

Providing the space through the T-part is kept at a minimum compared with that of the compressing rubber, rapid pumping will be found possible and the power limited only by the strength of the user's grip. The apparatus may be quickly constructed of materials always at hand. Originally it was made, in about fifteen minutes, of a T-tube to which the necessary enlargements were connected by rubber tubing and the unused space filled by a glass rod. The valves may be inserted directly into the ends of the compressing rubber, but the form shown in the figure is more serviceable. By attaching the lower end to a tapering tube as shown, the pump is easily inserted into a perforated stopper of any size.

The pump has been found serviceable in various applications. For filling burettes it is better than a siphon, the stoppers of the standard solution bottles being provided with two holes, through one of which the delivery tube passes, while to the other the pump is applied by the adapter shown in the figure. It may be applied to a Kipp generator in which higher pressure is momentarily required. In various other ways it has been found to be a useful piece of apparatus.

A Convenient Form of Rotary Shaker.

An apparatus designed by Perkins* serves admirably for putting a liquid into rotary motion for the purpose of securing gentle but thorough agitation. The container is an Erlenmeyer

Fig. 10.

flask. This is suspended in a retort clamp held loosely in another clamp, which in turn is also loosely held by another clamp firmly attached to the upright rod, the whole forming a system of loose joints at right angles, which permits oscillatory movement of the flask. Motion is given to the flask by a wire crank attached eccentrically to the rotating table driven by the motor. The use of this apparatus in the absorption of iodine, free or liberated in reaction, by metallic silver will be described later.

* Claude C. Perkins, Am. Jour. Sci., [4], xxviii, 33.

The Purification of Precipitates by Solution and Reprecipitation.

In many processes of analytical chemistry, the preparation of substances in pure condition is brought about by precipitation, solution and reprecipitation; and sometimes this cycle of operations must be repeated. When a precipitate, gathered upon a filter, is easily acted upon by the appropriate solvent, the process of dissolving the precipitate from the filter is simple; but when the precipitate is refractory toward solvents or difficult to attack on account of its physical condition, as is the case with many gelatinous precipitates, the proper handling of the precipitate involves some inconvenience and delay.

In meeting such difficulties, it is advantageous to place within the ordinary paper filter, before filtering, a movable lining of platinum gauze upon which the precipitate rests for the most part and with which it may be removed.* The simplest form of this device is easily made by cutting platinum gauze to the shape shown in the accompanying figure. In ordinary use, this piece of gauze, folded to make a cone of a little less than 60°, and held by pincers at the point of overlapping, is placed within this filter and allowed to fit itself closely by the natural spring of the gauze when released.

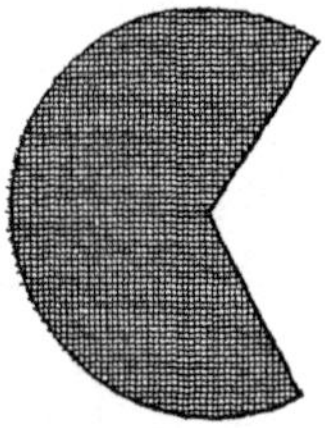

Fig. 11.

Upon filters so prepared a precipitate may be collected and washed as usual; and, at the end of the operation, the cone with nearly all the precipitate may be transferred (conveniently by means of ivory-pointed pincers) to a dish or beaker for suitable treatment. The small amounts of the precipitate which have passed through the gauze, being somewhat protected by the gauze against the compacting action of filtration and washing, are generally removable with ease from the filter by a jet of the washing liquid. After washing, the gauze may be replaced within the same filter and serve for a second collection of the precipitate, to be subsequently dissolved, in case double precipitation and solution are desirable. The final collection of the precipitate is, of course, made upon paper without the gauze lining, when precipitate and filter are to be ignited.

* Gooch, Am. Jour. Sci., [4], xx, 11.

This device has proved to be very serviceable in handling such precipitates as ferric hydroxide, aluminium hydroxide and the basic acetates.

Precipitates collected upon asbestos in the perforated crucible are frequently removable without difficulty by allowing a suitable solvent to percolate precipitate and felt; but in case the precipitate is pasty solution in this manner may be unpleasantly slow. In such cases, it is convenient to remove the greater part of the precipitate, collected and washed in the usual manner, upon a disk of platinum foil, perforated, fitted with a wire handle, as shown in the figure, and placed upon the asbestos felt before the transfer of the precipitate to the crucible. To make such a disk, shown in Fig. 12, is the work of a few moments only; and by its use pasty precipitates, such as cuprous sulphocyanate or the sulphides of the metals, are easily handled for solution.

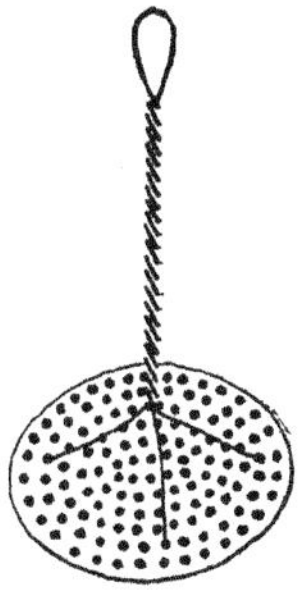

Fig. 12.

ELECTROLYTIC PROCESSES.

The Rotating Cathode.

The rotating cathode, previously utilized in the arts in the manufacture of seamless copper tubing (by electroplating with currents of low electromotive force and continuous replenishment of the bath by the use of a soluble copper anode), and by von Klobukow* for slow stirring of the electrolytic bath in analysis, has been applied in rapid motion by Gooch and Medway† to analytical processes in which the object is to remove metals completely and expeditiously from solution. An ordinary 20 $cm.^3$ platinum crucible is used as the cathode, and this is rotated at a speed of from 600 to 800 revolutions by means of a small, inexpensive electric motor fastened so that its shaft is vertical. Upon this shaft the crucible is fixed by pressing it over a rubber stopper bored centrally and fitted tightly on the end of the shaft. To secure electrical connection between crucible and shaft, a narrow strip of sheet platinum is soldered to the shaft and then bent

* Jour. prakt. Chem. (N. F.), xxxiii, 473.

† F. A. Gooch and H. E. Medway, Am. Jour. Sci., [4], xv, 320.

upward along the sides of the stopper, thus putting the shaft in contact with the inside of the crucible when the last is pressed over the stopper. The shaft is made in two parts as a matter of convenience in removing the crucible, and is joined, with care to make a good contact between the two pieces of shafting, by a rubber connector of sufficient thickness to prevent the crucible from wabbling when rotated.

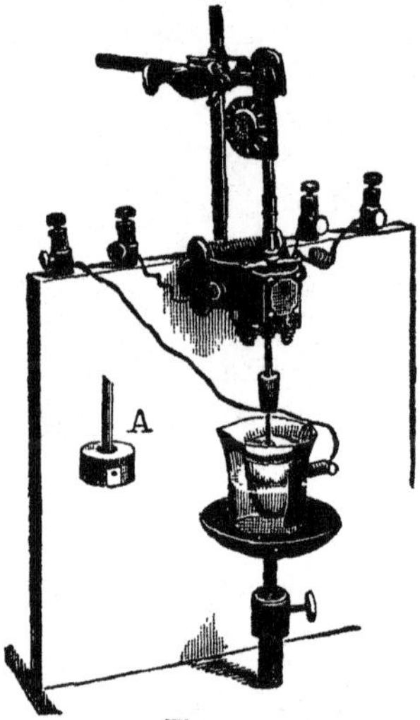

Fig. 13.

The solution to be electrolyzed is placed in a beaker upon a small adjustable stand, so that the crucible may be dipped into the liquid to any desired depth. A platinum plate is employed as an anode, and this is connected to the positive pole of a series of storage batteries, while the negative pole of this series is connected to the bearing in which the shaft rotates, thus allowing the current to go from the platinum plate through the solution to the crucible, up the shaft of the motor, and back to the batteries. The power to run the motor may be conveniently taken from the incandescent light circuit.

The stand carrying the beaker is raised until the liquid covers about two-thirds of the crucible adjusted to the shaft, thus giving a cathode surface of about 30 $cm.^2$. The anode is introduced and the motor started. The wires from the storage batteries are connected and the current allowed to pass through the solution. The duration of the electrolysis is varied according to the strength of current used, but in each case, after the deposit is nearly complete, the current from the batteries is shut off, the motor stopped, the beaker, platinum anode and crucible carefully washed with a fine jet of water, the motor again started, and the current allowed to pass for the remaining time.

When the deposition is complete the crucible is removed and washed, first with water, then with alcohol, and finally is dried by passing it over a flame.

In subsequent study of the material and shape of the rotating cathode, Medway* has shown that a silver crucible may, with some economy and without sacrifice of accuracy, be substituted for the platinum crucible, at least in the determination of copper; that neither nickel nor aluminium is a suitable metal for use as the cathode; and that a rotating disk of platinum is inferior to the crucible for use as the cathode.

The results of experimental tests of the rotating cathode in the determination of various metals — copper, silver, nickel, cadmium, tin, gold, zinc — are given under the headings of these metals.

The Filtering Crucible in Electrolytic Analysis.

The rapidity with which a metal or oxide may be thrown upon the electrode and thereafter handled successfully in the ordinary processes of electrolytic analysis depends upon keeping to conditions under which deposits are compact and adherent. It is for the purpose of getting adherent deposits that in modern rapid processes use is made of rotating electrodes,† of apparatus so arranged that gases evolved or introduced shall stir the liquid,‡ and of the agitating action of a magnetic field.§

The use of these methods is, however, limited to those cases in which attainable conditions and the nature of the processes are such that the deposits may be handled and washed without loss of material from the electrode. Plainly, the range of conditions and processes may be very much extended by the adoption of means for handling easily and safely electrolytic deposits more or less loose. Gooch and Beyer ‖ have made use of devices for this purpose, in which the filtering crucible of platinum or of porcelain is adapted to use as an electrolytic cell.

Electrolysis and Subsequent Filtration. Fig. 14 shows a convenient form of apparatus for such use in electrolytic analysis. The crucible (A), fitted in the usual manner with an asbestos felt (*a*), serves as an electrode (*e*), the surface of which is very much

* Am. Jour. Sci., [4], xviii, 180.

† V. Klobukow, Jour. prakt. Chem. (N. F.), xxxiii, 473. Gooch and Medway, Am. Jour. Sci., [4], xv, 320. (See page 11.) Exner, Jour. Am. Chem. Soc., xxv, 896.

‡ Levoir, Zeit. anal. Chem., xxviii, 63. Richards, Jour. Am. Chem. Soc., xxvi, 530.

§ Frary, Zeit. Elektrochem., xiii, 308. Jour. Am. Chem. Soc., xxix, 1592.

‖ F. A. Gooch and F. B. Beyer, Am. Jour. Sci., [4], xxv, 249.

increased by a layer of pieces of platinum foil (*b*) within the crucible and in contact with its walls. The joint between cap and crucible is made water-tight by a thin rubber band (F). The capacity of the cell is made conveniently ample by attaching to the crucible, by means of a close-fitting, thin rubber band (E), a glass chamber (C) easily made from a wide, short test tube. The second electrode (*f*) is introduced from above through the glass funnel (D), which serves to prevent spattering of the liquid during the electrolysis, and hangs within the glass chamber. The cell, held by a clamp, may be kept cool during action by immersing it in water contained in a cooler, as indicated in Fig. 15.

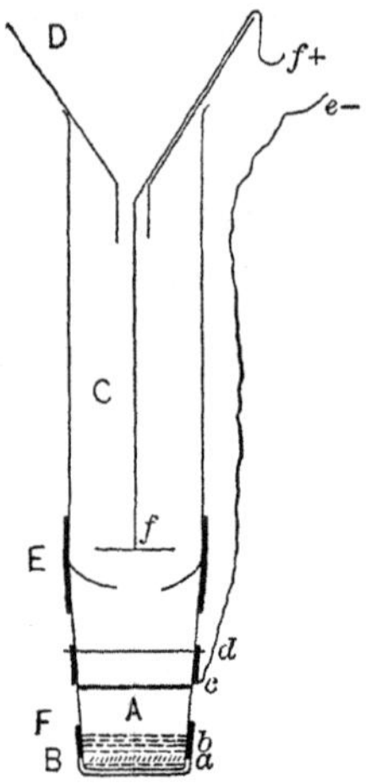

Fig. 14.

Electrical connection is made with the crucible by means of a platinum triangle (*c*), bent as shown and held tightly against the outer wall of the crucible by a rubber band (*d*). Fig. 15 shows, on the left, the apparatus adjusted for work.

In using the apparatus, the crucible, fitted with asbestos and containing clippings of platinum foil, is capped, ignited and weighed. The glass chamber with the wide rubber band folded back against itself is set upon the crucible and the band is snapped into place. The other adjustments are made in the manner shown. The electrolyte is introduced and the current turned on. After the expiration of time enough to complete the electrolysis, the cooler is lowered and arrangements are made to draw off the liquid in the cell. If the process is such that no harm can follow the stopping of the current before removing the liquid, the upper electrode and funnel are washed and removed, the cap and band are slipped off, and the apparatus is set in the holder of the filtering flask as for an ordinary filtration. The liquid is drawn through the felt to the flask, the chamber washed down, and removed from the crucible, and the deposit is well washed. The crucible and contents are dried and weighed, the increase over the original weight being, of course, the weight of the deposit.

The details of experiments made to test this form and use of the apparatus are given in the table. Copper sulphate strongly acidulated with sulphuric acid was the electrolyte. Deposition was completed in the times given, and the ferrocyanide test applied to the whole filtrate showed the absence of copper in every case. The apparatus and deposit were washed first with water and finally with alcohol. It was noticed that, though the filtrate contained no copper, the washings did sometimes contain a bare trace. When the filtrate was allowed to stand after treatment with potassium ferrocyanide it turned blue rapidly, and this

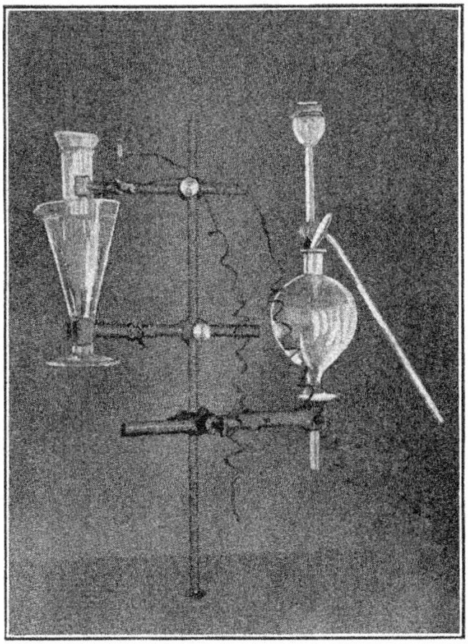

Fig. 15.

action, which indicated probably the presence of hydrogen dioxide or of persulphuric acid produced in the electrolysis of the sulphuric acid, suggests that the liquid should be drawn from the deposit as quickly as may be after the current is cut off. In the first two experiments no special care was taken in this respect, and in these experiments the results are a trifle higher than those of the other experiments, in which the manipulation was quickly made.

Obviously this process of electrolytic analysis is fairly rapid, easily executed, and accurate; but the desirability of quickly removing the liquid from the deposit after stopping the current is evident.

Electrolysis with Filtration after Interruption of the Current.

$CuSO_4.5H_2O$ taken.	Volume of liquid.	H_2SO_4 (1 : 1).	Current.		Time.	Theory for copper.	Copper found.	Error.
grm.	cm.³	cm.³	Amp.	Volt.	min.	grm	grm.	grm.
0.5038	50	5	2 4	5 7	5 40	0.1283	0.1290*	+0.0007
0.5010	50	5	2 4	5 7	5 40	0.1276	0.1282*	+0.0006
0.5009	50	5	2 4	5 7	5 40	0.1276	0.1279*	+0.0003
0.5005	50	5	2 4	5 7	5 40	0.1275	0.1277*	+0.0002
0.5047	50	5	2 4	5 7	5 40	0.1285	0.1286*	+0.0001
0.5039	50	5	2 4	5 7	5 40	0.1283	0.1285*	+0.0002
0.5030	50	5	2 4	5 7	5 40	0.1281	0.1282†	+0.0001

* No copper in filtrate or in washings. † Trace of copper in washings.

Electrolysis with Filtration. Following are results obtained as in the preceding process excepting the single point that the liquid was drawn off while the current was still running. In these experiments the filtration was effected by removing the cooler, taking off the cap and band from the crucible, and quickly swinging into place the filtration apparatus shown at the right in Fig. 15. The liquid was then drawn through the crucible and replaced by wash

Electrolysis and Filtration without Interruption of the Current.

$CuSO_4.5H_2O$ taken.	Volume of liquid.	H_2SO_4 (1 : 1).	Current.		Time.	Theory for copper.	Copper found.	Error,
grm.	cm.³	cm.³	Amp.	Volt.	min.	grm.	grm.	grm.
0.5030	50	5	2 4	5 7	5 30	0.1281	0.1278†	−0.0003
0.5008	50	5	2 4	5 7	5 40	0.1275	0.1275‡	0.0000
0.5024	50	5	2 4	5 7	5 40	0.1280	0.1277†	−0.0003
0.5014	50	5	2 4	5 7	5 40	0.1277	0.1276*	−0.0001
0.5018	50	5	2 4	5 7	5 40	0.1278	0.1278*	0.0000

* No copper in filtrate or in washings. † Trace of copper in filtrate.
‡ Trace of copper in washings.

water until the current ceased to flow because there was no electrolyte to carry it. The apparatus was washed with water and finally with alcohol, and the crucible and contents were dried for periods of ten minutes at 100°–110°, to constant weight.

Electrolysis with Continuous Filtration. I. When a deposit is so loosely adherent as to be moved by the liquid, it may be compacted upon the filtering felt by keeping the liquid in process of filtration and constant motion through the cell to the receiver. The adjustment of apparatus for this purpose is shown in Fig. 16.

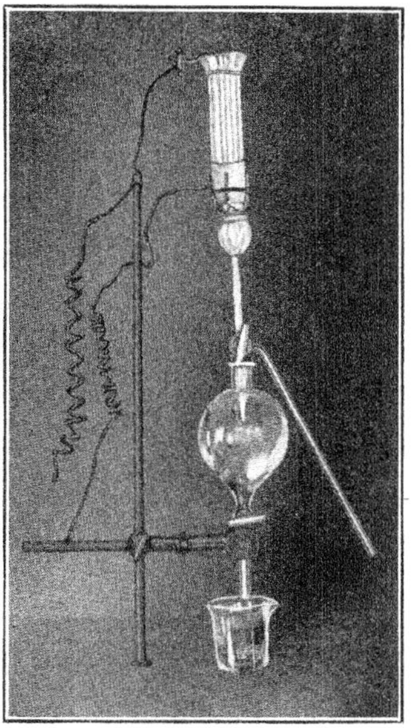

Fig. 16.

Here the electrolytic cell rests in the crucible holder fitted to a separating funnel used as a receiver and connected into the vacuum pump. A stopcock in the tube of the crucible holder is convenient but not necessary.

The manner of using the apparatus is simple. First, the weighed crucible, fitted in the usual manner with an asbestos felt and containing the platinum clippings, is adjusted to the glass chamber. The cell is pressed into the platinum triangle and set into the holder. The funnel which carries the wire electrode is put in place. The cell is charged with the electrolyte and the current is turned on. The electrolysis begins and, under regulated action of the vacuum pump, the liquid is drawn through to the receiver at a convenient rate. Usually, before the upper electrode is uncovered the stopcock is closed, the suction pump disconnected, and the liquid drawn off from the receiver and returned to the electrolytic cell. The pump is again connected, the stopcock is opened and filtration begins again.

Should the deposit be noticeably loose, it may be compacted by allowing the cell to drain completely under the action of the suction pump. The electrolyte is thus kept in circulation, and loose particles of the deposit are held upon the filtering layer. From time to time, the process of emptying the receiver and filling the cell is repeated. When the electrolysis is complete, as shown by proper testing of the filtrate, the liquid is drawn through the crucible and replaced by water from above until the current no longer flows. The electrodes are disconnected, the extension chamber easily slipped off, and the washing of the crucible and its contents continued sufficiently, with care, should the deposit be spongy, to give time enough in the washing to properly soak out absorbed material. The crucible and contents are dried, ignited, and weighed as usual. This method of manipulation was also put to the test in the electrolysis of copper sulphate. Experimental details are given in the table.

Electrolysis with Continuous Filtration.

$CuSO_4.5H_2O$ taken.	Volume of liquid.	H_2SO_4 (1 : 1).	Current.		Time.	Theory for copper.	Copper found.	Error.
grm.	cm.³	cm.³	Amp.	Volt.	min.	grm.	grm.	grm.
0.5013	50	5	2 4	5 7	5 20	0.1277	0.1280†	+0.0003
0.5003	50	5	2 4	5 7	5 20	0.1274	0.1276‡	+0.0002
0.5015	50	5	2 4	5 7	5 20	0.1277	0.1279*	+0.0002
0.5001	50	5	2 4	5 7	5 20	0.1274	0.1274*	0.0000
0.5041	50	5	2 4	5 7	5 20	0.1284	0.1285‡	+0.0001

* No copper in filtrate or washings. † Trace of copper in filtrate.
‡ Trace of copper in washings.

The results show that there is no difficulty in getting accurate results while maintaining continuous filtration during the process, and that the time needed to complete the action is somewhat shortened when the liquid is kept in circulation by filtering.

II. Another form of apparatus for electrolysis with continuous filtration, in which a porcelain filtering crucible replaces the platinum filter crucible, is shown in Fig. 17. In this apparatus, it is necessary to make the connection from above with the electrode inside the crucible, and this is accomplished by a linked platinum wire, as shown. In putting together and using this apparatus, a finely perforated disk of platinum foil (*c*) is laid upon the more coarsely perforated bottom of the porcelain crucible (A). Upon this disk the asbestos felt (*a*) is deposited in the usual manner. Platinum clippings (*b*) form a layer of suitable thickness above the asbestos, and upon this layer, and in contact with it, is placed another perforated disk of platinum foil to which is attached a twisted wire (*e*) so linked that it may be folded within the crucible. This apparatus is ignited and weighed, and to it is adjusted, as shown, a chamber to hold the electrolyte. The other electrode (*f*), inclosed within a funnel (D) made from a thistle tube, is introduced in the manner indicated. This apparatus is adapted only to use in the method of continuous filtration, and it is used exactly as in Process I. Experimental details are given in the table.

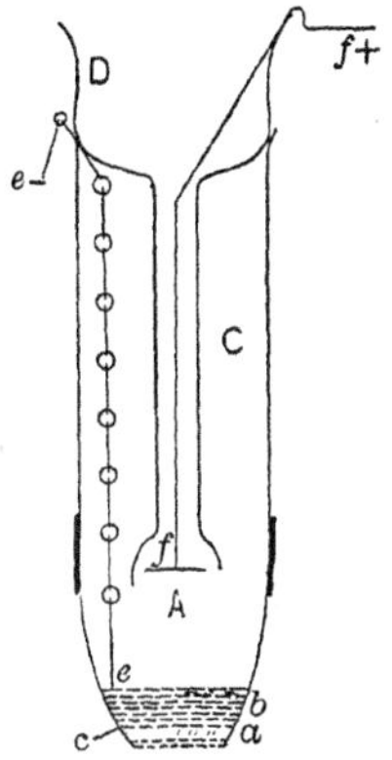

Fig. 17.

By either of the processes described, reasonably rapid and accurate electrolytic determinations may be made without the use of rotating motors or special stirring apparatus, and without large and expensive apparatus of platinum. The use of the filtering crucible as a part of the electrolytic cell makes possible the utilization of operations and conditions in which the deposit may lack the degree of adhesiveness necessary in ordinary electrolytic processes. The application of the processes to the more difficult determinations of manganese and lead as the dioxides formed upon the anode in very imperfectly adherent condition will be described later.

Electrolysis with Continuous Filtration: the Use of the Porcelain Crucible.

$CuSO_4.5H_2O$ taken.	Volume of liquid	H_2SO_4 (1 : 1).	Current.		Time.	Theory for copper	Copper found.	Error.
			Amp.	Volt.				
grm.	cm.³	cm ³			min.	grm.	grm.	grm.
0 5025	50	5	2 3 4	6 8 10	5 15 10	0.1280	0.1277†	−0 0003
0 5009	50	5	2 3 4	6 8 10	5 15 15	0 1276	0.1279*	+0.0003
0 5025	50	6	2 4	6 10	5 25	0 1280	0.1278*	−0 0002
0 5011	50	5	2 4	6 10	5 25	0.1276	0.1273†	−0.0003
0 5013	50	5	2 4	6 10	5 25	0 1277	0 1276‡	−0 0001

* No copper in filtrate or in washings. † Trace of copper in filtrate.
‡ Trace of copper in washings.

The Fixation of Chlorine on the Silver Anode.

Hydrochloric Acid, Silver Anode and Platinum Cathode. From a consideration of the apparently very exact results obtained by many investigators* in the determination of the chlorine in chlorides by fixation of that element upon an anode plated with silver, it would seem that nothing could be simpler than the accurate determination of the chlorine in hydrochloric acid by procedure advocated for the treatment of metallic chlorides. That such is not the case, however, has been shown by Gooch and Read.† In an experimental study of the electrolysis of hydrochloric acid with use of the silver anode and platinum cathode, it is shown that silver oxide is formed at the anode and must be decomposed by heating to a high temperature (incipient redness, at the tip of the Bunsen flame), not simply dried over a steam radiator‡; that silver deposited from the double cyanide solution upon platinum gauze to make the silver anode always includes more or less alkali salt, which is lost from the anode surface attacked by chlorine during electrolysis; that to avoid contamination of the silver anode by nonvolatile material it should be plated from a solution of silver

* Smith, Jour. Chem. Soc., xxv, 890. Myers, ibid., xxvi, 1124. Withrow, ibid., xxviii, 1350. Hildebrand, ibid., xxix, 447. McCutcheon, ibid., xxix, 1445. Lukens and Smith, ibid., xxix, 1455. Lukens and McCutcheon, ibid., xxix, 1460.

† F. A. Gooch and H. L. Read, Am. Jour. Sci., [4], xxviii, 544.

‡ Smith's Electro-analysis (1907), page 305.

oxalate in ammonium hydroxide; and that the silver anode is attacked and dissolved by oxygen-chlorine acids produced chiefly toward the end of the electrolysis.

The Electrolysis of Hydrochloric Acid with a Silver Anode Plated in the Oxalate Solution.

	Chlorine taken in HCl.	Time.	Current.		Increase of anode dried in air bath at 105°–110°	Increase of anode ignited.	Loss of dried anode on ignition.	Apparent error in chlorine when anode was dried.	Apparent error in chlorine when anode was ignited
	grm.	min.	Amp.	Volt	grm.	grm.	grm.	grm.	grm.
1*a*	0.0533	25	0.5 –0.00	2–4	0.0513	0.0502	0.0011	–0.0020	–0.0031
2*b*	0.0533	30	0.45–0.01	2–4	0.0529	0.0520	0.0009	–0.0004	–0.0013
3*c*	0.0533	25	0.5 –0.01	3–4	0.0513	0.0501	0.0012	–0.0020	–0.0032
4*d*	0.0533	30	0.5 –0.01	3–4	0.0523	0.0517	0.0006	–0.0010	–0.0016
5*e*	0.0533	25	0.5 –0.01	3.8–4	0.0517	0.0499	0.0018	–0.0016	–0.0034
6*f*	0.0533	25	0.4 –0.03	3.8–4	0.0502	0.0493	0.0009	–0.0031	–0.0040
7*g*	0.0533	25	0.4 –0.02	3.8–4	0.0506	0.0493	0.0013	–0.0027	–0.0040

a. The solution became suddenly opalescent and soon thereafter the current practically ceased, the liquid being neutral to litmus paper. Silver chloride (0.0017 grm.) was recovered from the liquid, and silver was found upon the cathode.

b. The electrolysis was interrupted at the first appearance of opalescence, the liquid being neutral. No silver was found in solution and none upon the cathode. Iodine, indicated by starch, was liberated when potassium iodide was added to a portion of the solution.

c. At the end of the electrolysis the liquid was slightly opalescent and neutral to litmus. Upon the addition of potassium iodide to a portion of it a trace of iodine was set free. In another portion, silver nitrate was without immediate effect. A trace of silver was found upon the cathode.

d. At the end of the electrolysis the liquid was slightly opalescent. It was neutral to litmus, but slowly bleached the color. Upon standing it developed acidity. From potassium iodide it liberated iodine equivalent to 0.0010 grm. of chlorine, as was determined by sodium thiosulphate. A trace of silver was found upon the cathode.

e. At the end of the electrolysis the liquid was slightly opalescent. It was neutral to litmus but developed acidity in the course of a half-hour. From potassium iodide a portion set free iodine. In the remainder silver nitrate gave, after standing two days, an amount of silver chloride equivalent to 0.0016 grm. of chlorine. Silver was found upon the cathode.

f. At the end of the electrolysis, the liquid was slightly opalescent. It was neutral to litmus, but developed acidity on standing a half-hour. From potassium iodide a small portion set free iodine. From the remainder silver nitrate precipitated silver chloride, which, when filtered off after five days, was found to be equivalent to 0.0035 grm. of chlorine. A trace of silver was found on the cathode

g At the end of the electrolysis, the liquid was slightly opalescent. It was neutral to litmus, but developed acidity in a half-hour. A small portion of it set free iodine from potassium iodide, Silver nitrate produced in the remainder, after four days, a precipitate of silver chloride equivalent to 0.0036 grm. of chlorine. No silver was found upon the cathode.

In experiments made with the rotating anode of gauze plated with silver from a solution of silver oxalate in ammonium hydroxide, it was noted that near the end of electrolysis, when neutrality to litmus indicated the exhaustion of hydrochloric acid, the solution suddenly became opalescent and soon afterward the current practically ceased to flow. Upon standing, the liquid, which at the end of the electrolysis had slowly bleached blue litmus paper without reddening it, developed distinct acidity, and when tested in separate portions gave further opalescence with silver nitrate and set free iodine from potassium iodide. All these phenomena point to the formation of hypochlorous acid in the process of electrolysis and its attack upon the anode to form silver hypochlorite and derived silver salts. It appears further that soluble silver hypochlorite, apparently formed chiefly when the hydrochloric acid approaches the point of exhaustion, is thrown into solution, to be partially decomposed with production of opalescent silver chloride. The details of experiments made with the silver anode plated in the oxalate solution are given in the accompanying table. In the first experiment the electrolysis was continued until the current practically ceased to pass. In the other experiments the operation was ended when the diffusion of the opalescent silver chloride indicated that the silver anode was being attacked, dissolved and partially reprecipitated in the liquid. In all the liquid was neutral at the end of the electrolysis.

The results of these experiments show the fixing of oxygen as well as chlorine upon the anode, the removal of silver from anode to cathode, and the formation of hypochlorous acid.

It is plain, therefore, that the electrolytic determination of the chlorine of hydrochloric acid with the use of the silver anode and platinum cathode is by no means an exact process.

Sodium Chloride, Silver Anode and Mercury Cathode. The electrolysis of sodium chloride with the use of a silver or silver-plated anode and the mercury cathode has been subjected by Peters* to a very careful and critical examination. The cell used was similar to that described and figured by Hildebrand,† consisting of a bottomless beaker 6.3 cm. in diameter and 6.3 cm. tall set into a crystallizing dish 11.3 cm. in diameter and 5.7 cm. high. Midway between the cells, above the mercury, was a coil of 6 turns of nickel wire 1 mm. in diameter. At three places on the nickel wire a single wire ran down forming legs upon the feet of which rested the ends of a Y. This Y, made of glass rod 3 mm. in diameter, formed the support for the inner cell and also held the nickel wire in position when the whole apparatus was inverted in

* Am. Jour. Sci, [4], xxxii, 365.
† Jour. Am. Chem. Soc., xxix, 451.

emptying. Three rubber stoppers placed radially held the inner cell in position. Mercury sealed the two compartments. The contact with the cathode was made through mercury in an S-shaped tube, hung on the edge of the outer cell with a platinum wire sealed in one end and dipping under the cathode mercury.

The anode of platinum gauze or of pure silver gauze was rotated in the inner cylinder.

In every case the liquid of the inner cell, charged with sodium chloride, showed alkalinity to indicators whenever tested. In a number of experiments the alkalinity of the inner cell was determined by titration with sulphuric acid using methyl orange. The amounts of acids used, together with the initial and final current conditions, are given in the table.

Alkalinity of Inner Cell.

Sulphuric acid, 0 08996 N. cm.³	Time of electrolysis. min.	Ampere, initial and final.	Voltage, initial and final.	Remarks.
13.69	140	0.24–0.035	4 cells	
8.66	70	0.28–0.063	4–3	Changed during experiment.
2 39	14	1 4 –0.1	7.2–8	
2 38	18	1 4 –0.1	7.2–8	
2.15	19	1.3 –0.09	7.2–8	
2 09	18	1.17–0.1	7.4–8	
1.31	42	0.77–0.03	4 cells	
1.01	38	0 6, 1 0 –0 03	6–8	0.6 raised to 1.0 near the start.
0.92	13	0.63–0.03	7.4–8	
0 62	31	1 3 –0.03	7–8	

There is no rapid decrease in the ammeter reading at any time to indicate the complete decomposition of sodium chloride, the accumulation of alkali in the inner cell being sufficient to carry a considerable amount of current.

Silver was found in the mercury cathode in every electrolysis with anode of silver or silver-plated, though under some conditions the amount of silver transferred to the cathode was small enough to be negligible in analysis. Upon the supposition that the transfer of silver to the cathode takes place in greatest degree toward the end of the electrolysis of chloride, the experiment of adding the sodium chloride in small portions was made in order to keep the concentration low and to produce repeatedly in one experimental process the conditions which exist toward the end of the ordinary electrolysis. In this manner the maximum transfer of silver for a given amount of chloride electrolyzed should be obtained. The details of experiments made in this manner, and of other experiments in which the entire amount of chloride to be electrolyzed was introduced at once, are shown in the table.

The results show the transfer of silver from the plated anode to the mercury to be a regular feature of the electrolysis, under the conditions described, and that the amount of silver so transferred is considerably increased if the electrolysis is continued beyond the point where the sodium chloride is all decomposed; but that if the electrolysis is interrupted as soon as the salt is all decomposed, very little silver is transferred to the mercury.

Transfer of Silver to the Cathode.

Total NaCl sol., 0 1 N, added.	Number of additions.	Number of times anode was heated during process to free from oxide.	Total time of electrolysis.	Initial and final current conditions.		Amount of silver recovered by distillation
cm.³			min.	Amperes.	Volts	grm
Silver-plated anode.						
47 98	24	3	305	0 06 } -0 03	8	0 0075
48 12	24	3	232	0 09 }		0 0083
Silver anode.						
17±	12	4	1077	. .		0 0162
Silver-plated anode.						
50 00	1	0	38	0 6 -0 3	7 4-8	0 0006
50 00	1	0	31	1 0 -0 3	7 4-8	0 0005
Silver-plated anode.						
50 02	1	0	14	1 4 -0.1	7 2-8	0 0005*
50 00	1	0	18	1 4 -0.1	7.2-8	
50 00	1	0	19	1 3 -0 09	7.2-8	
50 00	1	0	18	1 17-0 1	7.4-8	
Silver-plated anode.						
50	1	0	14	1.2 -0 2	7.4-8	0 0001†
50	1	0	16	1.2 -0 2	7.4-8	
50	1	0	13	1 5 -0 19	7 4-8	
50	1	0	11	1.2 -0 19	7.4-8	

* Examination of the liquid of the inner cell showed that the decomposition of sodium chloride was complete.

† Examination of the liquid of the inner cell showed that the decomposition of sodium chloride was not complete. Amounts varying from 8 mgrm. to 30 mgrm. of sodium chloride were found at the end of the operation.

The experimental data were utilized by Peters to determine the best conditions for the electrolysis of 50 cm.^{3} of $n/10$ sodium chloride with the apparatus at hand. The procedure prescribed for the quantitative electrolysis of .2923 grm. of sodium chloride in 50 cm.^{3} of water is as follows: Introduce about 2 kg. of mercury into the apparatus or sufficient mercury to rise 6 mm.–8 mm. above the bottom of the inner cell. Cover the nickel wire with water (70 cm.^{3}–80 cm.^{3}) and add to it 1 cm.^{3} of saturated salt solution. Introduce the 50 cm.^{3} of salt solution (.2923 grm. NaCl) into the inner cell and electrolyze with the bottom of the anode 6 mm.–10 mm. above the surface of the mercury. The current, 1.2 amp.–1.5 amp. at the start, should be stopped at 0.1 amp. Take out the anode, hold it over the inner cell and wash with the jet of the wash bottle. Introduce into the heating apparatus (the electric furnace or the heating crucible); heat at 400°–450°, and raise the temperature during 5–10 minutes until the silver chloride is fused, avoiding temperatures over 550°. Cool and weigh. Titrate the liquid over the mercury to near the end-point, using methyl orange with $n/10$ hydrochloric acid, then transfer to a separatory funnel and shake well to decompose the remaining sodium amalgam. Draw off the mercury and finish the titration, using sodium hydroxide to determine the end-point. In case violent shaking introduces suspended mercury into the liquid, making the end-point difficult to determine, the mercury may be allowed to settle out, or the mixture may be filtered, or more methyl orange may be added.

The results of experiments made in accordance with the procedure outlined are given below. The time required for a single determination was one hour. The temperatures given are in each case those of the electric oven at the beginning and end of heating.

Under the most favorable conditions, determined empirically for one definite amount of sodium chloride, results show an average of fair analytical accuracy. The chlorine tends to be low, averaging -0.0005 grm. between the extremes of $+0.0008$ grm. and -0.0030 grm.; and the sodium high, averaging $+0.0005$ grm. between the extremes of $+0.0016$ grm. and -0.0018 grm. What the proper conditions would be with a sodium chloride solution of different strength, an apparatus of slightly different dimensions, or different current conditions, the author feels would have to be determined by preliminary experimentation. The ammeter cannot be relied upon, without previous experimentation, to determine the end of the electrolysis, as the current, during the operation, falls gradually from the start to an indefinite end. There is no rapid decrease in the ammeter reading at any point to indicate the complete decomposition of the sodium chloride, the accumulation of the alkali in the inner cell being sufficient to carry a considerable current.

Analysis of Sodium Chloride.

Chlorine found. grm.	Sodium found. grm.	Time of electrolysis. min.	Time of heating anode. min.	Temperature of heating anode. C°.	Error, chlorine. grm.	Error, sodium. grm.
0.1755	0.1152	19	10	390–455	−0.0018	+0.0002
0.1771	0.1159	21	23	410–460	−0.0002	+0.0009
0.1770	0.1158	18	14	550–458	−0.0003	+0.0008
0.1781	0.1166	18	13	400–470	+0.0008	+0.0016
0.1775	0.1157	20	19	400–450	+0.0002	+0.0007
0.1743	0.1132	20	72	450–480	−0.0030	−0.0018
0.1780	0.1161	19	7	405–440	+0.0007	+0.0011
0.1767	0.1155	19	6	405–440	−0.0006	+0.0005
Aver. 0.1768	0.1155	19			−0.0005	+0.0005

The work shows that in the electrolysis of sodium chloride, under the conditions described, with the anode of silver, or silver-plated, and the mercury cathode, silver is always transferred from the anode to the cathode mercury; although, under conditions determined by experiment to be most favorable for a given amount of chloride, the amount of silver transferred may, for analytical purposes, be neglected.

With the apparatus described, the electrolysis of 50 $cm.^3$ of $n/10$ sodium chloride, 0.2923 grm. of the salt, is most favorably accomplished with a current of 1.2–1.5 amp. at the start, which is allowed to fall off to 0.1 amp., the operation requiring 18–20 minutes.

It is recommended that the anode covered at the end of the electrolysis with silver chloride be first heated below the fusing point of the chloride to decompose all of the silver oxide, and then for five or ten minutes at a temperature of 400°–500° to fuse the chloride.

Sodium hydroxide is always present in the inner cell after the beginning of the electrolysis.

The best method of treating the anode covered with fused silver chloride to prepare it to be used in a subsequent electrolysis is by heating about 20 minutes at about 500°, or a little over, in a current of hydrogen.

Iodometric Processes.

The Standardization of Iodine Solutions by the Action of Metallic Silver.

The affinity between silver and iodine has been made by Gooch and Perkins* the basis of a gravimetric method for the determination of iodine in general, and, incidentally, for the gravimetric standardization of iodine solutions to be used in volumetric analysis. In studying the absorption of iodine by silver, experiments were made with silver reduced in the wet way by the action of zinc upon silver chloride, nitrate or iodide; in the dry way, by the action of hydrogen upon silver sulphide or oxide; or, electrolytically, from a solution of silver nitrate, with anode inclosed in a porous cell and with an oscillating cathode of platinum, experience having shown that, while the bright and crystalline deposit formed upon a stationary cathode is lacking in absorptive power, the broken and dark product obtained by continually oscillating and scraping the cathode during the deposition of the metal is sensitive to iodine. Silver reduced from a silver salt by zinc or from silver sulphide by hydrogen may serve the purpose, provided it is subjected to a preliminary treatment with potassium iodide, and silver reduced from the oxide by hydrogen is also serviceable; but the best form of silver, and the one most easily prepared in the pure state, is that deposited electrolytically as a black mass upon a small oscillating cathode of platinum from a solution of silver nitrate, the platinum anode being inclosed in a porous cell. As long as the mass adheres to the electrode it

* F. A. Gooch and Claude C. Perkins, Am. Jour. Sci., [4], xxviii, 33.

remains perfectly intact, but as soon as it is shaken off into the solution it turns to a dull-gray color and settles to the bottom in a fine flourlike powder. It should not be allowed to remain upon the electrode after it begins to change color, as the silver then collects in a crystalline form and does not absorb iodine readily.

It was shown that when finely divided silver is shaken with a solution of iodine in potassium iodide, the iodine is absorbed by the metal, but that more iodine is absorbed than was originally free if the mixture is exposed to air in the shaking. Experiments in which the silver was shaken with 50 $cm.^3$ of a solution of potassium iodide, 20 grm. to the liter, proved fully that, either in neutral or alkaline solution, the action of air must be prevented during the agitation of the iodide with silver; for in these experiments it was found that, from the solution of potassium iodide shaken in contact with air, finely divided electrolytic silver absorbed 0.0010 grm. of iodine in fifteen minutes, silver reduced by zinc from silver iodide 0.0012 grm. in fifteen minutes, silver reduced from the sulphide by hydrogen 0.0032 grm. in one hour, and crystalline electrolytic silver 0.0051 grm. in one hour and forty-five minutes.

The procedure finally shown to be best is the following: Into a 250-$cm.^3$ Erlenmeyer flask are put the iodine, in solution in potassium iodide, and a weighed amount of the finely divided silver. The glass stopper of a Drexel bottle is fitted into the neck of the flask and held in a tight joint made by slipping over neck and stopper a broad thin rubber band. The air in the flask is replaced by hydrogen and the tubes of the stopper are capped. The flask thus trapped is shaken with a rotary motion by hand, or preferably by a mechanical shaker, until the iodine color vanishes. The liquid, usually 50 $cm.^3$ in volume, is diluted to about 100 $cm.^3$ and the residue of silver and silver iodide is collected in a perforated crucible fitted with asbestos felt, washed, dried at 130° to 140° and weighed. The difference between the weight of silver taken and that of the residue of silver and silver iodide is the measure of the free iodine. Fig. 10 shows the adjustment of the apparatus.

Figures to show the accuracy of the procedure are given in connection with processes for the determination of iodine.

Arsenic Trioxide as an Iodometric Standard.

Carefully sublimed and anhydrous arsenic trioxide serves as the most exact standard in iodometric processes, and it is also useful * in the standardization of potassium permanganate for quantitative oxidations.

To make a standard arsenite solution it is convenient to dissolve 4.95 grm. of arsenic trioxide in a concentrated solution of 4 grm. of potassium hydroxide, and to make up the solution to 1 liter for the $n/10$ or to 2 liters for the $n/20$ solution by adding 100 cm.^{3} or 200 cm.^{3} of a saturated solution of potassium hydrogen carbonate and water to complete the volume. The arsenic trioxide dissolves much more readily in the alkali hydroxide than in the acid carbonate alone, and the proportion given is not enough to entirely form dipotassium hydrogen arsenite, but more than enough to form potassium dihydrogen arsenite.† Should more alkali hydroxide be used in effecting solution, it should be neutralized by suitable amount of acid (sulphuric acid or hydrochloric acid) before the final addition of the acid carbonate in large excess; but it is important to obtain the neutralization without the addition of any indicator containing alcohol, and this is easily accomplished without the use of an indicator if attention is paid to the amount of alkali hydroxide employed.

With this standard arsenite solution the standardization of iodine in potassium iodide is effected by titration with or without starch as an indicator, according to circumstances.

The Starch Indicator for Free Iodine.

In the literature of iodometric titration frequent mention is made of the production of a red color, as well as a blue, when starch is used as an indicator, and numerous formulæ have been given for making a starch solution designed to give a blue color with iodine, and to keep without spoiling. An investigation by Hale ‡ into the occasion of the loss of iodine met with in titrimetric processes in connection with the formation of a red color, has led not only to the elimination of the loss, but also to a

* See page 41.

† Hale, Am. Jour. Sci., [4], xiii, 387.

‡ F. E. Hale, Am. Jour. Sci., [4], xiii, 379.

probable explanation of the cause of the red color, of the loss of iodine, and of their mutual relation.

Hale observed that, in titrating a decinormal solution of arsenite with a decinormal solution of iodine, the amount of iodine needed was greater when dependence was placed upon the occurrence of the starch blue following intermediate tints of red to determine the end-point, than when the titration was made without the indicator to the appearance of the yellow tinge due to the addition of a single drop of decinormal iodine in excess.

The loss of iodine in the interaction of decinormal solutions in presence of the starch indicator used in a certain series of experiments is shown in the statement below to be very considerable for the undiluted solutions and much less considerable when the reaction took place in dilute solution.

The Decinormal Solutions Undiluted.

$n/10$ As_2O_3 sol.	Reading with 1.25 cm.^{3} of starch solution to a deep blue. Nearly $n/10$ I.	Iodine reading calculated. Nearly $n/10$ I.	Error.
cm.^{3}	cm.^{3}	cm.^{3}	cm.^{3}
5	5 23	5.14	0.09
10	10 37	10.28	0 09
15	15.50	15.42	0.08
20	20.70	20.56	0 14
25	25.85	25.70	0.15
30	31.00	30.84	0.16
35	36.12	35.98	0.14
40	41.20	41.08	0 12
45	46.39	46 26	0.13
50	51.54	51.40	0.14

Final Volume 110 cm.^{3}

$n/10$ As_2O_3 sol.	Reading with 1.25 cm.^{3} of starch solution to a deep blue. Nearly $n/10$ I.	Iodine reading calculated. Nearly $n/10$ I.	Error.
cm.^{3}	cm.^{3}	cm.^{3}	cm.^{3}
1	1.05	1.03	0 02
2	2.10	2.05	0 05
3	3.15	3.08	0.07
5	5.20	5.14	0 06
7	7.25	7.20	0 05
10	10.32	10.28	0.04
15	15.47	15.42	0 05
20	20.60	20.56	0 04
35	36.02	35.98	0.04

If the starch solution be added after the yellow color of free iodine is visible a fine blue is produced by a preparation which would give the intermediate colors if it were present during the entire titration. This fact shows that some cause for the production of the red lies in the titration. It excludes any explanation of the loss of iodine by the formation of iodate. It excludes any explanation of the red color based upon the assumed formation of some arsenic acid compound with starch.

It is well known that malt extract, and many chemical reagents such as hydrochloric acid, potassium hydroxide, nitric acid, etc., readily hydrolyze starch more or less completely through a series of bodies, called dextrins, to a final product, one of the sugars. One of the first of these dextrins is erythrodextrin, which is colored red with iodine. It seems plausible that water or the alkali, acid potassium carbonate, might cause hydrolysis under certain conditions. As there is a loss of iodine, indications would point to an oxidizing action, and a hydrolysis may be possible because of such oxidizing action. It is also possible that the arsenious acid (or antimonious acid) becomes joined to the starch, in some such way as antimonious acid attaches to acid potassium tartrate, and that this compound is then easily hydrolyzed. A possible indication of this lies in the anomalous fact that the blue color fades first and the red last in titration in alkaline solution, though a few drops of a starch solution added to a solution of erythrodextrin, colored red with iodine, will develop the blue starch iodide.

Experimental evidence seems to substantiate the following points:

1st. The loss of iodine and the production of a red color does not take place if an absolutely pure and freshly made starch solution is employed.

2d. Ordinary starch contains, usually, at least two impurities, one coloring red with iodine, the other coloring blue, the latter being readily changed under the influence of oxygen and acid potassium carbonate to the former. These impurities tend likewise to form in pure starch, whether solid or in solution.

3d. The impurity coloring blue with iodine is identical with, or analogous to, amidulin, made by saliva digestion of pure starch, and the impurity coloring red with iodine is erythrodextrin, the second product of saliva digestion of pure starch.

4th. The loss of iodine is due to the formation of erythrodextrin from this amidulin-like body, and erythrodextrin does not use up iodine by any transformation to achroodextrins.

So it appears that the colors found in iodometric titrations in which ordinary starch is used as an indicator are probably due to the admixtures of the starch blue or possibly of the amidulin blue with the red of erythrodextrin derived from amidulin by hydrolysis initiated by the oxidizing effects of the iodine. Pure starch, containing neither amidulin nor erythrodextrin, gives only blue in the iodometric titration. Starch, on the other hand, which has undergone partial hydrolysis, is likely to contain both amidulin and erythrodextrin. It is not strange that both amidulin and erythrodextrin should be present as impurity in starch, since they both stand in the order named as the first two dextrins produced from starch, as shown by saliva digestion of starch, as also by malt-extract digestion of starch. Starch both in solid state and in solution tends to pass through these stages of hydrolysis. Germ growth rapidly appears in solutions of pure amidulin and pure erythrodextrin, with the destruction of these bodies to form dextrins lower in the series.

Pure starch causes no red color, nor loss of iodine, in alkaline titration of arsenite solution or of tartar emetic. If any purplish tinge occasionally occurs, it is no hindrance to the reading and causes no appreciable loss of iodine.

With an impure starch, the reading from the first permanent color, whether red or blue, is nearest to the correct value. The readings may be compared with plain iodine readings and a correction applied, since the loss for a constant quantity of starch proves to be constant in the titration of 20 $cm.^3$ to 50 $cm.^3$ of arsenite solution. Titration should be made at considerable dilution — e.g., 150 $cm.^3$ to 200 $cm.^3$ (since the production of red is at a minimum and the loss of iodine small at high dilutions) — and in presence of a suitable amount of potassium iodide to render the reaction delicate. Whenever it is practicable, however, the best method of using an impure starch is to make the titration without it up to the appearance of the yellow tinge of free iodine and then to add the starch. In this way the color comes out a clear blue and the exact adjustment may be easily made by alternate additions of a drop or two of $n/10$ arsenite and $n/10$ iodine.

Ordinary Preparation.

Hale * experimented with various preparations of starch. The starch solution was made by grinding 5 grm. of starch paste with a few cubic centimeters of cold water with the addition of 0.01 grm. of mercuric iodide, pouring into a liter of boiling water, and boiling five to ten minutes. Only the clear supernatant liquid was used.†

Preparation with Potassium Iodide.

In 25 $cm.^3$ of cold water 5 grm. of pure starch were ground with 2 grm. of potassium iodide and the mixture was poured into 75 $cm.^3$ of boiling water and boiled, the beaker being protected with asbestos. The mass became mucilaginous. After fifteen minutes the volume was increased to 500 $cm.^3$, and the boiling was continued for forty-five minutes. The solution was filtered, forty-eight hours being required, leaving a residue of jellylike consistency upon the filter. This method was suggested by the extreme delicacy which the presence of potassium iodide gives to the starch reaction, and by certain statements made in the literature, one that concentrated potassium iodide causes starch to swell and dissolve,‡ another that a solution of starch made by a somewhat similar method would keep a year without fermentation.§

Heating with Glycerin to Form Amylodextrin.

In 70 $cm.^3$ of pure glycerin 5 grm. of potato starch were heated at a temperature of 185°–190° C. for half an hour with constant stirring. The starch dissolved and the solution turned through yellow to a deep red. The solution was cooled to 120° C. and poured slowly and continuously into 200 $cm.^3$ of alcohol. The precipitate was thoroughly stirred, settled, and filtered while warm. It filtered readily and was washed with alcohol until the filtrate came through colorless. The colorless residue of amorphous amylodextrin was then dissolved in 500 $cm.^3$ of water heated to 60°–70° C.

Soluble Starch by Saliva Digestion (Amidulin).

In a little cold water 2 grm. of starch and 0.5 grm. of acid potassium carbonate were ground together and mixed with 200 $cm.^3$ of water kept boiling for a few minutes. The mixture was cooled to 40°–45° and treated with 10 $cm.^3$ of filtered saliva neutralized by 0.1 per cent hydrochloric acid with the use of a slip of litmus paper (first dipped in

* Loc. cit.

† Gastine's formula, Zeit. anal. Chem., xxviii, 339.

‡ Payen, Compt. rend., lxi, 512.

§ Zeit. anal. Chem., xxv, 37.

acetic acid and then washed) as indicator. When, in the course of three or four minutes, the solution had become entirely clear, it was boiled for ten minutes. In this process the addition of the alkali hinders the action from going beyond the first step of digestion: the boiling at the end destroys any further action of the saliva. Starch cellulose, which is said to produce a feeble red or brownish color with iodine,* is digested and destroyed by the saliva.

In the presence of a suitable amount of potassium iodide each of the first three preparations gives a sharp indication with a single drop of $n/10$ iodine; the amidulin is only a little less delicate. In the titration of 50 cm.³ of $n/10$ arsenite, enough potassium iodide is present in the $n/10$ iodine to give a sharp reaction at a volume of 125 cm.³ This is shown in the following statement. For comparison, a titration made with a large amount (25 cm.³) of ordinary impure starch is included.

Titrations with the Different Preparations of Starch.

Volume 125 cm.³.

$n/10$ As_2O_3 sol. cm.³	Nearly $n/10$ I sol. cm.³	Starch solution. cm.³					Color.
		KI starch.	Pure ordinary starch.	Amylo-dextrin.	Amid-ulin.	Impure ordinary starch.	
50	49.38	1	..	...	..	..	Permanent purplish.
50	49.40	1	..	...	..	..	Good blue.
50	49.40	1	..	...	..	..	Good blue.
50	49.40	..	2	...	..	..	Slow-fading purplish.
50	49.42	..	2	...	..	..	Good blue.
50	49.38	..	..	1.5	..	..	Permanent purplish.
50	49.40	..	..	1.5	..	..	Deep blue.
50	49.42	..	..	..	5	..	Permanent purplish.
50	49.44	..	..	...	5	..	Purple.
50	49.55	..	..	...	..	25	Abundant red.

In a series of parallel titrations, designated A and B in the next table, readings were made with the blue developed by the KI-starch indicator and with plain iodine, alternately, in order to eliminate accidental errors. Starch was added subsequently to corroborate the very pale iodine reading. The corrected readings were found by subtracting one drop from the actual

* Beilstein (First Edition), i, 1082, line 17.

reading. To render the plain iodine readings sharp a crystal of potassium iodide was added in the first two titrations.

Titrations with and without Starch.

Volume 125 cm.^{3}

$n/10$ As_2O_3.	Nearly $n/10$ I. Direct readings.		Nearly $n/10$ I. Corrected readings.		$n/10$ I sol. Absolute amount calculated from 50 cm.^{3}	$n/10$ I sol. Absolute errors in A, . . in B.	
	By iodine color, A.	By KI-starch blue, B.	A.	B.		A.	B.
cm.^{3}	cm.^{3}	cm.^{3}	cm.^{3}	cm.^{3}	cm.^{3}	cm.^{3}	cm.^{3}
..	1 drop*	1 drop					
5	4 94*	4.96	4.92	4.94	4.94	−0.02	0 00
10	9.88	9 90	9.86	9.88	9.88	−0.02	0 00
15	14.83	14.83	14.81	14.81	14.82	−0.01	−0.01
20	19.78	19.78	19.76	19.76	19.76	0.00	0.00
30	29.64	29.64	29.62	29.62	29.63	−0.01	−0.01
40	39.55	39 55	39.53	39.53	39.51	+0.02	+0 02
50	49 41	49.41	49.39	49.39	49.39	0.00	0.00

* A crystal of KI was added.

It is to be noted that the readings with plain iodine and with pure starch agree exactly except for the first two titrations, and here there is only a difference of a drop. The absolute errors are interesting, as they show how the absolute values fluctuate about a standard set by the 50-cm.^{3} readings. This fluctuation is limited to a drop plus or minus.

The statement has been made that starch from different sources has a varying power of absorbing iodine, e.g., that potato starch absorbs three times as much as rice starch.* To discover whether this fact has any bearing upon the use of the starch indicator, and at the same time to learn whether pure starch solutions made in the ordinary way (Gastine)† would give as delicate readings as that boiled with potassium iodide, solutions were made from pure potato starch, pure rice starch, pure arrowroot starch and a pure (so-called) soluble starch of unknown origin. The results of titration with these solutions as indicators are shown in the following statement.

It is at once seen that these values are coincident within a drop, and that all the starch solutions are within the limits set by a plain iodine reading on the one hand and a potassium iodide

* Girard, Ann. Chim., [6], xiii, 275.

† Loc. cit.

starch reading on the other, though there was no dilution of the standard solutions.

Starch from Various Sources.

No extra dilution. Volume about 110 cm.³

$n/10$ As_2O_3. cm.³	Nearly $n/10$ I sol. cm.³	$KHCO_3$ saturated cm.³	Starch solution. cm.³	Color
50	49 11 49 13	5	1 (Ordinary pure potato.)	Pale blue. Deep blue.
50	49 13	5	1 (Ordinary pure rice.)	Blue, slightly purplish.
50	49 14	5	1 (Ordinary pure soluble.)	Blue, slightly purplish.
50	49 15	5	1 (Ordinary pure arrow-root.)	Deep blue.
50	49.15	5	1 (KI pure starch.)	Good blue.
50	49 13	5	. . . .	Yellow.

If the starch is pure the amount of it used, within reasonable limits, is without effect upon the titration of $n/10$ arsenite by $n/10$ iodine, and the same may be said of the amount of acid carbonate, as shown in the following table.

Variation in the Amount of Starch.

$n/10$ As_2O_3. cm ³	Nearly $n/10$ I sol. cm.³	$KHCO_3$ saturated. cm.³	Starch solution * cm.³	Color.
10	9.82	5	1.5	Deep blue, purplish.
10	9 82	5	5	Deep blue.
10	9.82 9.84	5	10	Pale blue, purplish. Deep blue.
10	9.82	5	15	Deep blue.
10	9.84	5	20	Deep blue.
10	9.85	5	25	Deep blue.
10	9.82	5	1.5	Deep blue.
10	9.82	10	1.5	Deep blue.
10	9.83	15	1.5	Deep blue.
10	9.82	20	1.5	Deep blue, purplish.
10	9.81	25	1.5	Deep blue.

* Ordinary preparation; pure potato starch.

With pure starch, the end reaction of the titration of tartar emetic by iodine is likewise a pure blue. Following are Hale's results.

The average of the 10 cm.³ readings (absolute) multiplied by five equals 47.73. The absolute 50-cm.³ reading (47.75–0.02) equals 47.73. Evidently even tartar emetic causes no loss on pure starch, for the 50-cm.³ reading agrees with the plain iodine reading for the same amount and with the 10-cm.³ titrations.

End Reaction with Tartar Emetic.

Volume. cm.³	n/10 tartar emetic. cm.³	Nearly n/10 I sol. cm.³	Starch solution, pure potato. cm.³	$KHCO_3$. cm.³	Color.
100	10	9.58	1.5	10	Blue, no red.
100	10	9.58	1.5	10	Blue.
75	10	9 56	...	10	Yellow.
125	50	47 75	1.5	25	Blue (purplish tinge).
125	50	47.75	..	25	Yellow.

Hale* emphasizes the need of potassium iodide in suitable amount to bring out the delicacy of the blue end reaction, and the further necessity of not exceeding a suitable proportion, in order that the starch blue compound may not be modified by transformation to starch red. The proportions of potassium iodide and iodine entering into starch blue and starch red were carefully studied. According to experimental results, it would appear that the group $KI.I_4$ is characteristic of starch blue and the group $KI.I_2$ of starch red; that in the presence of a sufficiently concentrated solution of potassium iodide the group $KI.I_4$ changes by addition of KI to $2KI.I_2$, while dilution with water tends to bring about the reverse action.

Wholly apart from the consideration of theory, the influence of iodides upon the delicacy of the starch blue test for iodine is a matter of considerable practical importance analytically. Experience shows that 0.3 grm. of potassium iodide is sufficient in volumes not exceeding 300 cm.³ to make readings by the starch indicator sharp, and that under these conditions indications are quite as delicate at ordinary room temperatures as at the temperature of ice water. The addition of more potassium iodide renders the reading no sharper within that range of dilution, though the total volume has considerable proportionate influence upon the amount of iodine needed to bring out the indication. In the absence of potassium iodide in sufficient amount, the influence of temperature is very noticeable. Hydrochloric acid either in small amount or large amount does not render the reading sharp in absence of potassium iodide. These results are in agreement with the work of Lonnes.†

* Am. Chem. Jour., xxviii, 450. † Zeit. anal. Chem., 33, 409

Effect of Temperature when Potassium Iodide is Restricted.

Volume. cm.³	Potassium iodide not exceeding 0.3 grm.	Hydrochloride acid. (Sp.gr.1.12.) cm.³	Pure potato starch solution. cm.³	Temperature.	n/1000 iodine. cm.³	Color.
100	1 crystal	..	2	23°	0.45	Faint blue.
200	1 crystal	..	2	23°	0 65	Faint blue.
300	1 crystal	..	2	23°	0.80	Faint blue.
400	1 crystal	..	2	23°	1.30	Faint blue.
300	1 crystal	..	2	23°	0.80	Faint blue.
300	1 crystal	..	2	5°	0.70	Faint blue.
300	....	..	2	23°	9.20	Faint blue.
300	....	..	2	5°	1.50	Faint blue.
300	....	1	2	23°	4.15	Faint blue.
300	....	10	2	23°	4.00	Faint blue.

The influence of different amounts of potassium iodide in bringing out the starch iodide indication of iodine set free by reaction with gold chloride is of interest in this connection. At volumes lying between the limits of 25 cm.³ and 50 cm.³ 0.1 grm. of potassium iodide is an appropriate amount; at a volume of 15 cm.³, 0.01 grm. to 0.05 grm. of the iodide will do the work; and at lower dilutions even less of the iodide is effective.*

Standard Tartar Emetic.

Variation of Standard. Gruener† has shown that tartar emetic solutions containing about 16 grm. of tartar emetic, 20 grm. to 30 grm. of tartaric acid and 1 cm.³ of concentrated hydrochloric acid to the liter, will keep from five to twelve months without any change in strength. There is no deposit of antimonious oxide under these conditions, no oxidation and no signs of fungous growth. Gruener determined the strength of his tartar emetic solutions by titration with a decinormal iodine solution, standardized by decinormal arsenite. The mean of twenty-nine determinations showed 43.95 per cent of antimonious oxide in tartar emetic. Theory required 43.37 per cent ($Sb = 120$, $KSbOC_4H_4O_6.\frac{1}{2}H_2O = 332$). The cause of this discrepancy between arsenite and tartar emetic solutions made up as standards according to the accepted molecular formulas is a matter of considerable interest. One suggested explanation of

* See page 146.

† Hippolyte Gruener, Am. Jour. Sci., [3], xlvi, 206.

this difference is that the end reaction between starch and iodine is delayed until an excess of iodine is present. Hale* has shown, however, that a pure starch solution gives a sharp end reaction with both tartar emetic and arsenite solutions, and that while with impure starch there is a loss of iodine accompanied by the production of reddish hues in titrating tartar emetic, as shown by the difference between the readings made in the presence of potassium iodide by the yellow color of iodine and by the blue of starch iodide, yet it is no greater than in titrating arsenite solution in the presence of an impure starch. If only an impure starch is available, the reading should be made without starch, for the presence of potassium iodide renders very sharp the yellow color of the first excess of free iodine. This first reading may be afterwards corroborated by adding the starch solution, which will then give only a pure blue color. The above discrepancy must then be due to some other cause than delay of formation of the starch iodide. Hale† has given proof that this discrepancy is due to the ease with which tartar emetic loses its water of crystallization, and that in order to get a salt of the exact composition, $KSbOC_4H_4O_6.\frac{1}{2}H_2O$ (mol. wt. 332.15), certain conditions must be very closely observed.

Molecular weights of tartar emetic variously prepared, calculated from the results of iodometric titration, run in a series from that of crystalline tartar emetic almost to that the salt which has lost 1.5 molecules of water, passing through all intermediate stages, but never surely resting at any one spot. Two important stages are reached: when all the water of crystallization is gone, the anhydrous state; and when 0.5 molecule of water further has been lost, the first anhydride stage. The variation of molecular weight is shown in the following table.

The condition most easily and definitely reached is that of the hydrous crystalline salt. The greatest error met with in the recrystallized salt, if air-dried, is about +0.2 per cent, calculated on the ratio of antimony to tartar emetic, and that after standing in fine condition in a closed bottle for several weeks. Drying in the air bath does not yield a product, even with the most finely divided preparation,—that precipitated by alcohol,—which is sufficiently uniform to serve the purpose of a standard.

* F. E. Hale, Am. Jour. Sci., [4], xiii, 379.

† F. E. Hale, Jour. Am. Chem. Soc., xxiv, 828.

Hale recommends, therefore, the preparation of tartar emetic, in medium-sized crystals ($\frac{1}{32}$ to $\frac{1}{4}$ inch in diameter), rather than in the minutely crystallized condition.

Variations in Molecular Weight.

49.29 cm.^{3} (absolute amount) of *n*/10 iodine solution = 50 cm.^{3} of *n*/10 arsenic trioxide solution.

Preparation.	*n*/10 tartar emetic solution. cm.^{3}	*n*/10 iodine solution. cm.^{3}	Molecular weight.	Antimony. per cent.	Remarks.
Freshly crystallized............	50	49.29	332.15	36.13	
Crystals kept several weeks....	50	49.56	330.34	.....	
	50	49.58	330.21	36.35	
	Tartar emetic. grm.				
Medium-sized crystals, air-dried.....	0.5	29.82	330.58	.....	
Crystals, 4 days over H_2SO_4....	0.5	29.95	329.13	.....	
Crystals, 7 days over H_2SO_4....	0.5	29.98	328.77	.....	
Crystals, 16 days over H_2SO_4...	0.5	30.16	327.15	.....	
Crystals, 4 hours at 95°–130°...	0.5	30.42	324.50	.....	
Crystals, $7\frac{1}{2}$ hours at 104°–130°..	0.5	30.51	323.15	37.14	(Anhydrous.)
Crystals.. ......at 104°–130°..	0.5	31.27	315.29	.....	
	...	.....	314.15	.....	($\frac{1}{2}H_2O$ anhydride.)
Crystals.. ......at 160°–165°..	0.5	31.88	309.28	.....	
Crystals, 2 hours at 160°–165°..	0.5	31.93	308.80	.....	
	...	.....	305 15	. . .	(H_2O anhydride — all hydroxyls gone.)

Preparation. Enough tartar emetic is dissolved in about 300 cm.^{3} of boiling water to make a concentrated but not a saturated solution. This hot solution is filtered into flat crystallizing dishes and allowed to crystallize over night. The crystallization should not be too rapid. The crystals are filtered off by suction, washed twice with distilled water, kept under suction for about five or ten minutes more, and then air-dried from one and a half to four hours at room temperature, not above 25° and preferably lower, in a clear, dry air. The crystals may be considered dry an hour or two after they cease to show the slightest tendency to cling to a glass rod used as a stirrer.

Tartar emetic may be prepared in this manner, which, with good starch, shows practically the theoretical value when tested against a standard arsenite solution by titration of both with iodine. As has been pointed out, if pure starch is not available the first reading should be made without starch in presence of potassium iodide, and this corroborated after adding the starch solution, which will then give only a pure blue.

A comparison of several preparations of tartar emetic with standard arsenite is given in the table.

Comparison of Tartar Emetic with Standard Arsenite.

n/10 solution.	Crystals, air-dried.		n/10 tartar emetic solution.	n/10 arsenic trioxide solution	Approximately n/10 iodine solution.	Potassium acid carbonate solution.	Starch solution.	Color.
	hours.	temp.	cm.³	cm.³	cm.³	cm.³	cm.³	
Tartar emetic I....	1½	19°–24°	50	..	49.32	25	1	Medium blue.
Tartar emetic I....	4	19°–24°	50	..	49.42	25	1	Medium blue.
Tartar emetic II...	2½	19°–24°	50	..	49.43	25	1	Medium blue.
Arsenite...	..	....	.	50	49.43	5	1	Medium blue.
Tartar emetic III..	3	19°–24°	50	..	49.58	25	1	Deep blue.
Tartar emetic IV..	4	19°–24°	50	..	49.58	25	1	Medium blue.
Arsenite.........	..	....	.	50	49.61	5	1	Medium blue.

Processes of Oxidation.

Arsenic Trioxide as a Standard.

Carefully sublimed and anhydrous arsenic trioxide serves admirably for the standardization of potassium permanganate for use in direct oxidations, as well as in the processes of iodometric analysis. It is generally best to work with standard arsenite solution prepared as previously described,* and it is convenient, though not necessary, to have at hand a solution of iodine,† also standardized against the arsenite solution.

Standardization without Iodine. The standardization of the permanganate solution, made up to an approximate value, is easily accomplished by running a suitable amount of it into a solution of potassium iodide contained in a reaction bottle ‡ and acidified with dilute sulphuric acid, neutralizing with acid potassium

* See page 29.
† Ibid.
‡ See page 6.

carbonate, and titrating by the standard arsenite the iodine set free by the permanganate. In this operation it is best to dispense with the starch indicator usually employed to fix the end reaction. The vanishing point of the color of free iodine is itself sufficiently definite, even at a dilution of 300 $cm.^3$, and the disappearance of color is much sharper than that of the blue starch iodide.*

Standardization with the Aid of Iodine. If a solution of iodine standardized against the arsenite solution is at hand, the process just described may be modified so that there is no danger of overrunning the end-point in a single titration. In this procedure, an excess of the standard arsenite, taken in known amount, is added at once after the reaction of the permanganate upon the iodide in presence of acid, and is followed by the acid carbonate. The excess of arsenite is determined by titration with iodine in presence of starch. Should too much iodine be added in the titration, it is, of course, only necessary to add another measured amount of arsenite and then to repeat the titration by iodine.†

The Gravimetric Standardization of Permanganate.

The standardization of a permanganate solution may be made gravimetrically by adding a suitable amount of the solution to an excess of potassium iodide acidified with hydrochloric acid, shaking the mixture with a weighed amount of specially prepared silver in an atmosphere of hydrogen, collecting upon asbestos in the perforated crucible the residue of silver and silver iodide, drying, and weighing, according to the procedure previously described.‡

The Loss of Oxygen in Oxidations by Potassium Permanganate.

Concentration of Acid. A statement made many years ago, that in the interaction of oxalic acid and potassium permanganate free oxygen is always a product,§ met with adverse criticism; but subsequently a similar effect was noted by Brauner ‖

* Gooch and Peters, Am. Jour. Sci., [4], viii, 125.

† Gooch and Gilbert, Am. Jour. Sci., [4], xv, 390.

‡ See page 27.

§ Francis Jones, Jour. Chem. Soc., 1878, 95.

‖ Jour. Chem. Soc. (1891), 238.

in the action of permanganate upon tellurous acid dissolved in sulphuric acid, with the additional observation that the evolution of oxygen is proportional to the amount of sulphuric acid employed, and that in alkaline solution little evidence of such an effect appears. Recognizing that the production of permanganic acid, free oxygen and ozone, by the action of strong sulphuric acid upon permanganate in absence of oxidizable material, is a common phenomenon, and that the formation of a precipitate consisting largely of hydrated manganese dioxide by the action of hot dilute sulphuric acid upon the permanganate in aqueous solution is likewise well known, Gooch and Danner* have investigated the action of sulphuric acid in different concentrations upon permanganate in solution with a view to determining how far such action may be directly or indirectly responsible for the liberation of free oxygen in processes of oxidation.

In certain experiments tubes of suitable size and length, holding from 100 cm.^{3} to 200 cm.^{3}, were sealed at one end, filled completely with the mixtures of acid and permanganate, inverted, and allowed to stand with the lower and open end submerged in liquid of the exact composition of that which filled them. The details and results of these experiments are recorded in the statement below.

Time elapsed.	Gas from 100 cm.^{3}	Appearance.	Time elapsed.	Gas from 100 cm.^{3}	Appearance.
A. H_2SO_4 [1 : 1]=50 per cent.			B. H_2SO_4 [1 : 1]=25 per cent.		
5 min.	0.1 cm.^{3}	No change.	5 min.	Small bubble.	No change.
1 hour.	1 1 cm.^{3}	No change.			
1 day.	14 cm.^{3}	Red brown.			
3 days.	15 3 cm.^{3}	Light brown.			
4 days.	15.6 cm.^{3}	Light brown.	3 days.	9.6 cm.^{3}	Reddish purple. Turbid.
7 days.		Brown, turbid.			
8 days.	16 cm.^{3}	Clearing by precipitation.	7 days.	15.1 cm.^{3}	Reddish pink. Clearing by precipitation.
15 days.	17 3 cm.^{3}	Clear, straw-colored.	15 days.	18 cm.^{3}	Nearly clear.
17 days.	17 4 cm.^{3}	Clear, straw-colored.	35 days.	18 4 cm.^{3}	Clear and colorless.
35 days.	17.5 cm.^{3}	Clear, straw-colored.			
C. H_2SO_4 [1 : 1]=12.5 per cent.			D. H_2SO_4 [1 : 1]=6.25 per cent.		
1 hour.	Small bubble.	No change.	1 hour.	Small bubble.	No change.
1 day.	Bubble.	No change.	1 day.	Bubble.	No change.
3 days.	Bubble larger.	No change.	3 days.	Bubble larger.	No change.
14 days.	7 1 cm.^{3}	Color lighter.	14 days.	1 cm.^{3}	No change.
37 days.	11 cm.^{3}	Color lighter.	37 days.	3 cm.^{3}	No change.
44 days.	12 cm.^{3}	Color lighter.	44 days.	5 cm.^{3}	Little change.

* F. A. Gooch and E. W. Danner, Am. Jour. Sci., [3], xliv, 301.

In other experiments note was made of changes in color and formation of precipitates in 100-cm.[3] portions of liquid containing 10 cm.[3] of decinormal permanganate and varying proportions of acid during five days' standing, and the degree of decomposition of the permanganate was finally determined by adding a small excess of oxalic acid to the mixtures contained in Erlenmeyer flasks, heating to about 80° C., and titrating with permanganate the residual oxalic acid.

Percentage of H_2SO_4 [1 : 1]	Time elapsed.					Percentage of $KMnO_4$ decomposed.
	1 day	2 days.	3 days.	4 days.	5 days.	
10	Color unchanged.	Color unchanged.	Color unchanged.	Color unchanged. Slight sediment.	Color unchanged. Slight sediment. Slight scum.	3 6
20	Color unchanged.	Color unchanged.	Color unchanged.	Color unchanged. Slight sediment.	Color unchanged. Slight sediment.	7 4
30	Color unchanged.	Color unchanged.	Color unchanged.	Reddish tinge. Slight sediment.	Reddish tinge.	6.9
40	Color unchanged.	Color unchanged.	Tinged with reddish brown	Reddish brown.	Reddish brown.	39 2
50	Color unchanged.	Color unchanged.	Reddish brown.	Reddish brown.	Red brown.	57 4
60	Color redder.	Color redder.	Reddish brown.	Sherry brown.	Reddish olive.	58.9
70	Color redder.	Color redder.	Sherry brown.	Reddish olive.	Reddish olive.	61 1

In the first five experiments little change of tint was noted upon the addition of the oxalic acid to the cold solution, but in the last two experiments the reddish-olive color became at once distinctly red — presumably because the higher sulphate of manganese was attacked in the cold by the oxalic acid (as Brauner has shown), and so the natural color of the permanganate was permitted to assert itself. The extreme decomposition — that which took place in the last experiment, in which 70 per cent of the [1 : 1] acid was present — corresponds nearly to the reduction of the entire amount of permanganate present to the condition of oxidation of MnO_2 which is known to exist in combination with sulphuric acid in the form of a higher manganic sulphate. It is to be noted that the separation of the insoluble higher oxide took place only when the percentage of acid was low.

In still another series of experiments the solution of permanganate was mixed with sulphuric acid previously diluted with an equal volume of water, and cooled; and after the lapse of time indicated oxalic acid was added in quantity a little more than sufficient to bleach the permanganate. The solution was warmed to about 80° C., and the residual oxalic acid titrated by gradual addition of more permanganate. The difference between the amount of permanganate needed under the conditions to destroy the known amount of oxalic acid, and that used in the determination of the standard, should measure the oxygen lost and the permanganate decomposed under the action of the sulphuric acid. The results and details of these experiments are given below.

Residual Permanganate Reduced by Oxalic Acid at 80°

H_2SO_4 (1 : 1). cm.³	Water cm.³	$KMnO_4$ in decinormal solution cm.³	Percentage of H_2SO_4 (1 : 1) in solution during action	Percentage of $KMnO_4$ decomposed.
A. Treated immediately.				
2	8	10	10	0
4	6	10	20	0
6	4	10	30	0.5
8	2	10	40	1.6
10		10	50	1.9
B. Treated after standing eight hours at ordinary temperature.				
2	8	10	10	−0.3
4	6	10	20	−0.3
6	4	10	30	1.3
8	2	10	40	5.3
10	..	10	50	15.7
C. Treated after standing five days at ordinary temperature.				
2	8	10	10	4.0
4	6	10	20	21.6
6	4	10	30	49.7
8	2	10	40	55.9
10	.	10	50	56.4
D. Treated after standing one and one-half hours at 80°–90° C.				
2	8	10	10	1.3
4	6	10	20	43.8
6	4	10	30	35.9
8	2	10	40	49.1
10	..	10	50	55.3

It is obvious that the decomposition of the permanganate increases directly in each series of experiments with the increase in the proportion of sulphuric acid, that the amount of decomposition is greater as the time of action is extended, and that increase of temperature heightens the change. It is noted in particular that the presence of ten per cent of [1 : 1] sulphuric acid induces at the ordinary temperature no immediate decomposition of the permanganate, none in eight hours, and a breaking down amounting to four per cent in five days; and that the presence of fifty per cent of acid of the same strength occasions the decomposition of about two per cent at once, fifteen per cent in eight hours, and more than half the entire amount of permanganate in the course of five days. It is evident, also, that twenty per cent of the [1 : 1] acid produces no appreciable effect at ordinary temperatures and under exposures of a few hours only. The effect of heating the mixture of acid and permanganate to 80° C. for an hour and a half is closely comparable with that brought about by the five days' action at the ordinary temperature. It is, of course, probable that some decomposition of the permanganate by the sulphuric acid is brought about after the addition of the oxalic acid during the warming of the mixture up to the temperature at which the oxalic acid and permanganate interact.

Residual Permanganate Reduced by Ferrous Sulphate.

H_2SO_4 [1 : 1]. cm.³	Water. cm.³	$KMnO_4$ in decinormal solution. cm.³	Percentage of H_2SO_4 [1 : 1] in solution during action.	Percentage of $KMnO_4$ decomposed.
A. Treated at once: Volume, 20 cm.³				
2	8	10	10	0.0
4	6	10	20	0.2
6	4	10	30	0.1
8	2	10	40	0.1
10	..	10	50	0.3
B. Treated at once: Volume, 100 cm.³				
10	80	10	10	0 1
20	70	10	20	0.1
30	60	10	30	0 0
40	50	10	40	0.5
50	40	10	50	1.3
60	30	10	60	3.0
70	20	10	70	5.0
80	10	10	80	3.3
90	..	10	90	8.1

Experiments in which ferrous sulphate is used at the ordinary temperature to effect the reduction of the residual permanganate, instead of oxalic acid at the higher temperature, show a lower degree of decomposition, as is natural. The increase in the amount of decomposition as the proportions of sulphuric acid [1 : 1] are advanced beyond 50 per cent by volume is striking. The results of these experiments appear in the accompanying table.

It appears, therefore, that when potassium permanganate and sulphuric acid are brought into solution together under these conditions, there is likelihood of a reduction of the former, which is greater as the strength of the acid is increased, as the temperature is raised, and as the duration of action is extended. It appears further, at least when the acid is not present in proportion greater than 50 per cent of the [1 : 1] mixture, that in the early stages of the action the oxygen lost to the permanganate is liberated, and that later on the decomposition of the permanganate results in the precipitation of manganese as a higher oxide or in the formation of a higher sulphate. The first effect of the mutual action of the acid and the permanganate is to set free permanganic acid, which, being unstable, breaks up with the results described.

The bearing of these observations and inferences upon the question of the action of potassium permanganate during oxidations carried on in the presence of sulphuric acid is obvious; for, if the aqueous acid is able to liberate permanganic acid in such proportions as to be spontaneously unstable, it is reasonable to presume that any reducing substance present at the time of such action may, by virtue of its attractive action upon the oxygen of many more molecules of the permanganic acid than would be necessary to supply the exact amount needed for perfect oxidation, tend to increase the general instability of the already unstable molecules and so set up a far-reaching decomposition. These considerations throw light upon the phenomena observed by Brauner* in the oxidation of tellurous oxide in presence of sulphuric acid; and the fact that the liberation of free oxygen in this special case is more noticeable than in the oxidation of ferrous salts or oxalic acid, for example, is explicable in the light of Brauner's observation that the attraction of tellurous oxide for oxygen is greatly inferior to that of these substances for

* Jour. Chem. Soc., 1891, 238.

oxygen — not sufficient, in fact, to break up so unstable a substance as manganic sulphate, which is at once reduced by ferrous salts or oxalic acid.

The practical lesson to be drawn from the investigation is the desirability of keeping the acid present in oxidations effected by the agency of permanganate at the lowest limit consistent with perfect oxidation; the time of digestion with an excess of permanganate as small as may be; and the temperature, if possible, not above the ordinary room temperature.

As to the correlative question of the liberation of oxygen during oxidations by potassium permanganate in alkaline solution, experience in the collection of the gas liberated in oxidations effected in presence of acid leads to distrust of the evidence of such experiments unless the amount of gas liberated is considerable. While, on the one hand, small quantities of liberated gas may be so completely absorbed as not to appear free at all, it often happens, on the other hand, that the simple admixture of unlike liquids—such, for example, as a solution of potassium permanganate with sulphuric acid of strength insufficient to liberate oxygen — may bring about a very appreciable liberation of dissolved gases. So far as appears, however, the affirmation that oxygen is liberated in oxidations by potassium permanganate in alkaline solutions rests upon evidence of that nature only.

Hydrochloric Acid with Ferrous Salts. Löwenthal and Lenssen* were the first to show that the titration of a ferrous salt by potassium permanganate in the presence of hydrochloric acid, according to the process of Margueritte,† is vitiated by the evolution of chlorine outside the main reaction, and to point out that a remedy for the difficulty is to be found in the titration of the ferrous salt in divided portions, other equal volumes of the ferrous solution being added to the liquid in which the first titration is accomplished until the amount of iron indicated by successive titrations becomes constant. Kessler‡ showed the restraining influence of certain sulphates, of manganous sulphate in particular, upon the irregular and undesirable interaction of the permanganate and hydrochloric acid, and Zimmermann,§ in

* Zeit. anal. Chem., i, 329.
† Ann. Chim. Phys., [3], xviii, 244.
‡ Ann. Phys. Chem., cxviii, 48; cxix, 225–226.
§ Ann. Chem., ccxiii, 302.

apparent ignorance of Kessler's forgotten proposal, advocated the introduction of a manganous salt, best the sulphate, into the ferrous salt to be determined, thus accomplishing the purpose of the empirical procedure of Löwenthal and Lenssen. The protective influence of the manganous salt turns apparently, as Zimmermann suggested, upon the initiation of Guyard's reaction, according to which the permanganate and manganous salt interact to form a higher oxide of manganese capable of oxidizing the ferrous salt, but slow to act upon the hydrochloric acid.* According to Volhard,† the reaction of Guyard is favored and hastened by heat and concentration of the solution, while it is delayed by acidity and dilution; but even in solutions containing very little manganous salt and a considerable quantity of free acid, the faint rose color developed by the careful addition of permanganate ultimately vanishes until every trace of the manganous salt is precipitated. When a considerable amount of the salt is present interaction follows immediately the introduction of the permanganate.

In titrations of a ferrous salt by permanganate, Zimmermann advocates the use of 4 grm. of manganous sulphate uniformly. In putting this matter to the test, Gooch and Peters‡ have found that as much as 5 grm. of manganous sulphate may be present in 135 $cm.^3$ of the liquid, containing about 5 $cm.^3$ of hydrochloric acid of full strength, without interfering with the regularity of the titration; and the effect is trivial even when the amount of manganous sulphate reaches 10 grm. In all cases, however, in which the larger amounts of manganous salt are present, the end reaction is marked by the advent of a brownish-red precipitate rather than the clear pink of the soluble permanganate; and it is obvious that in case a substance to be oxidized were not active enough to act with rapidity upon the product of the Guyard reaction, difficulty might follow the failure to adjust the conditions more particularly.

Regularity of action is also noted when manganous chloride is substituted for the sulphate, and in this respect the results accord with those of Zimmermann and differ from those of Wagner.§

* See also Manchot, Ann. Chem. (1902), cccxxv, 105.

† Ann. Chem. (1879), cxcviii, 318.

‡ F. A. Gooch and C. A. Peters, Am. Jour. Sci., [4], vii, 461.

§ Zeit. physikal. Chem., 28, 33.

Total volume at beginning of titration. cm.³	HCl. (Sp. gr. 1.09). cm.³	$FeCl_2$. cm.³	$KMnO_4$ n/10. cm.³	$MnSO_4.5H_2O$. grm.	$MnCl_2.4H_2O$. grm.
135	10	25	21.70	1	.
135	10	25	21.70	3	..
135	10	25	21.70	5	..
135	10	25	21.75	7	..
135	10	25	21.75	10	..
145	20	25	21.75	10	..
175	50	25	21.75	10	..
135	10	25	21.70	...	1
135	10	25	21.70	..	2
145	20	25	21.70	...	2
155	30	25	21.75	...	3
165	40	25	21.70	...	4

Hydrochloric Acid with Oxalic Acid. It has been stated* that hydrochloric acid interferes in no way with the titration of oxalic acid by permanganate. Gooch and Peters find, however, that in such titrations there is a small though real waste of permanganate proportionate to the amount of hydrochloric acid present. This fact is brought out clearly in the comparison of experimental results in the following table.

Temperature at Beginning about 80° C.

Approximate volume at beginning of titration. cm.³	H_2SO_4 [1 : 1]. cm.³	HCl. (Sp. gr. 1.09.) cm.³	Ammonium oxalate n/10. cm.³	$KMnO_4$. n/10. cm.³	Variation from mean of A taken as standard. cm.³
			A.		
200	5	....	50	47.50	0.00
200	5	....	50	47.50	0.00
200	10	....	50	47.50	0.00
200	10	....	50	47.50	0.00
200	25	...	50	47.50	0.00
200	25	....	50	47.50	0.00
			B.		
150	10	2.5	25	23.80	+0.05
150	10	2.5	25	23.90	+0.15
150	10	5.0	25	23.90	+0.15
150	10	10.0	25	24.00	+0.25
500	5	....	25	23.80	+0.05
500	10	10.0	25	24.00	+0.25
500	10	10.0	25	24.10	+0.35

* Fleischer, Volumetric Analysis, Trans. Muir., p. 71. Zimmermann, loc. cit.

Temperature 20°–26° C.

Volume at beginning of titration.	H_2SO_4 [1 : 1]. cm.³	HCl. (Sp. gr. 1.09) cm.³	Ammonium oxalate, n/10. cm.³	$KMnO_4$. n/10. cm.³	$MnSO_4$.-$5H_2O$. grm.	$MnCl_2$.-$4H_2O$, grm.	Variation from standard. cm.³
130	..	10	25	23.90	.0040	......	+0.15
130	..	10	25	23 90	0120	......	+0.15
130	..	10	25	23.80	0250	.. ...	+0 05
130	..	10	25	23.75	0400	.... .	+0.00
130	..	10	25	23.76	0500	......	+0.01
130	..	10	25	23 70	1000	......	−0.05
130	..	10	25	23 75	2000	. ....	0.00
130	..	10	25	24.20	. ..	.0200	+0.45
130	.	10	25	23.95	. ....	.0200	+0.20
130	..	10	25	23.80	. ....	.0400	+0.05
130	..	20	25	23.75	......	.0400	0.00
130	..	30	25	23.75	....	.0400	0 00
130	..	10	25	23.75	1 0000	.. ...	0.00
130	..	10	25	23.75	2.0000	.. ...	0.00
130	..	10	25	23.75	3.0000	. ... .	0.00
130	1	..	25	23 72	.....	1.0000	−0.03
130	1	..	25	23.74	......	2.0000	−0.01
130	1	..	25	23.72	......	3.0000	−0.03
130	2	..	25	23.70	......	0.5000	−0.05
130	3	..	25	23.75	......	0.5000	0.00
Temperature about 80°.							
145	10	10	25	23.90	0.5000	. . .	+0.15
145	10	10	25	23.70	1.0000	.. ...	−0.05
500	10	10	25	23.75	1.0000	. . .	0.00
500	..	10	25	23.70	1.0000	....	−0.05
500	..	10	25	24.10	0.5000	......	+0.35

The error introduced by the presence of hydrochloric acid during the action of the permanganate upon oxalic acid may be obviated by the introduction of a manganous salt, but the persistence of the Guyard reaction is liable to interfere with the end reaction of oxidation of oxalic acid unless an adjustment is made between the quantity of the manganous salt, the amount of acid and the dilution. It appears, also, that for a given dilution and strength of acid less manganous salt is needed in a cold solution than in a hot solution. Thus, in the hot solution, at a dilution of 145 cm.³ to 500 cm.³ with 5 cm.³ of strong hydrochloric acid, with or without sulphuric acid, 1 grm. of manganous sulphate must be present; while in the cold solution, 0.04 grm. of either the sulphate or chloride is enough to secure adequate protective effect. Experience showed, however, that 1.0 grm. of the manganous salt should be present in order to push the reaction with reasonable speed, and that this amount is enough to so affect the conditions of equilibrium that titrations in moderate volumes (100 cm.^{3} to 500 cm.^{3}) and in presence of hydrochloric acid (5 cm.^{3} to 15 cm.^{3} of the strong acid) may be conducted with safety and reasonable rapidity, either with or without sulphuric acid, at the ordinary atmospheric temperature.

Experimental results are given in the preceding table.

Effect of Other Chlorides. It has been shown in the preceding account that when potassium permanganate and hydrochloric acid react at high concentrations chlorine is evolved, and when oxidations are brought about by permanganate in presence of sulphuric acid and small amounts of chlorides, the tendency toward evolution of chlorine is in evidence. This fact must be taken into consideration in analytical operations involving such conditions. It has also been stated* that certain chlorides act catalytically to induce further decomposition of the permanganate than would take place under the conditions were hydrochloric acid the only chloride present; but Brown† has shown that this assertion is a mistake due to faulty analytical procedure. When a definite quantity of potassium permanganate is digested with a definite quantity of normal hydrochloric acid under defined conditions of time and temperature, and the resulting mixture is treated with a definite amount of oxalic acid, the excess of which is determined by titration with permanganate, the data are at hand for calculating the amount of permanganate apparently reduced during the digestion.

Wagner has found that when a small amount of $n/10$ ferric chloride is added to one such mixture before the digestion and an equivalent amount of $n/10$ hydrochloric acid to another such mixture, more permanganate is required in the final titration of the excess of oxalic acid in the mixture which contains the ferric salt than in that which contains no ferric salt, a fact which seems at the outset to imply more decomposition of permanganate when the digestion takes place in presence of the ferric salt. Brown shows, however, that when the chlorine produced in the reaction is removed by a current of carbon dioxide or air before

* Wagner, Zeit. physikal. Chem., 28, 33.

† James Brown, Am. Jour. Sci., [4], xix, 31.

the addition of the oxalic acid, the same quantity of permanganate is required to bring about the final coloration whether ferric chloride is present or not.

The conclusion must be drawn, then, that Wagner's experiments in no way show the catalytic effect of ferric chloride in the interaction between hydrochloric acid and potassium permanganate, nor do they furnish evidence in support of the assumed formation of chlor-ferrous acid. They afford simply an indication of the greater or less retention of chlorine in solution, and the greater or less oxidizing action of this chlorine on the oxalic acid in the presence or absence of ferric chloride.

The following tables give experimental results.

Digestion without the Removal of Chlorine.
(9 91 cm.³ $H_2C_2O_4$ = 20.25 cm.³ $KMnO_4$.)

n/1 HCl.	*n*/10HCl.	*n*/10 $FeCl_3$.	$KMnO_4$ before digestion.	Temperature.	Time of digestion	*n*/10 $H_2C_2O_4$.	$KMnO_4$ to color.	KMnO apparently reduced during digestion.
cm.³	cm.³	cm.³	cm.³	C°	min.	cm.³	cm.³	cm.³
100	9 91	....	9 91	50	60	9.91	15.89	5.55
100	9.91	....	9 91	50	60	9.91	15.11	4.77
100	9.91	....	9.91	50	60	9.91	15.15	4.81
100	9.91	....	9.91	50	60	9.91	15.07	4 73
100	9.91	...	9 91	50	60	9.91	15.13	4.79
100	9.91	....	9.91	50	60	9 91	15.07	4.73
100	9.91	...	9 91	50	60	9.91	15.08	4.74
100	9.91	....	9.91	50	60	9.91	15.02	4.68
100	9.91	....	9 91	50	60	9 91	14.85	4.51
100	9.91	....	9 91	50	60	9.91	14.40	4.06
100	9.91	....	9 91	50	60	9 91	15.05	4.71
100	....	9.91	9.91	50	60	9.91	15.60	5.26
100	.. .	9.91	9.91	50	60	9.91	15.35	5 01
100	....	9 91	9.91	50	60	9.91	15.32	4 98
100	....	9 91	9.91	50	60	9.91	15.88	5 54
100	....	9 91	9.91	50	60	9 91	15.42	5.08
100	....	9.91	9.91	50	60	9 91	15.45	5 11
100	....	9.91	9.91	50	60	9.91	15.95	5.61
100	....	9 91	9 91	50	60	9.91	15.41	5.07
100	....	9 91	9 91	50	60	9.91	15.95	5.61
100	....	9.91	9 91	50	60	9.91	16.65	6.31
100	....	9.91	9.91	50	60	9.91	15.75	5 41
100	....	9.91	9.91	50	60	9.91	15.79	5.45

In all these experiments the potassium permanganate was entirely destroyed, the hydrochloric acid present being capable of breaking down many times as much permanganate under identical conditions of time and temperature.

Chlorine Removed by a Current of Air during Digestion.

(9.90 cm.³ approximately $n/10$ $H_2C_2O_4$ = 20.09 cm.³ $KMnO_4$.)

$n/1$ HCl. cm.³	$n/10$ HCl. cm.³	$n/10$ $FeCl_3$. cm.³	$KMnO_4$ before digestion. cm.³	Temperature. C°.	Time of digestion. min.	Residual $KMnO_4$ color after digestion.	Cl test after aëration.	$H_2C_2O_4$. cm.³	$KMnO_4$ to color. cm.³	$KMnO_4$ apparently reduced during digestion. cm.³
100	9.90	....	9.90	50	60	none	none	9 90	18 88	8 69
100	9.90	....	9.90	50	60	none	none	9 90	18 87	8 68
100	9.90	....	9.90	50	60	none	none	9 90	18.80	8 61
100	9.90	.. .	9.90	50	60	none	none	9.90	18.81	8 62
100	9.90	....	9.90	50	60	none	none	9 90	18.80	8 61
100	9.90	....	9.90	50	60	none	none	9.90	18.82	8.63
100	9.90	..	9.90	50	30	none	none	9.90	18.77	8.58
100	9 90	....	9.90	50	30	none	doubtful	9.90	18.70	8 51
100	9.90	..	9.90	50	15	none	very faint	9 90	18.70	8.51
100	9.90	. .	9.90	50	15	none	very faint	9 90	18.68	8.49
100	..	9.90	9.90	50	60	none	none	9 90	18.87	8.68
100	....	9.90	9.90	50	60	none	none	9.90	18.85	8.66
100	....	9.90	9.90	50	60	none	none	9.90	18.81	8.62
100	....	9.90	9.90	50	60	none	none	9 90	18.81	8 62
100	....	9.90	9.90	50	60	none	none	9 90	18.87	8.68
100	....	9.90	9 90	50	60	none	none	9.90	18.85	8.66
100	....	9.90	9.90	50	30	none	none	9.90	18.80	8.61
100	.. .	9.90	9.90	50	30	none	doubtful	9.90	18.72	8.53
100	....	9.90	9.90	50	15	none	very faint	9.90	18.65	8.46
100	.	9.90	9 90	50	15	none	very faint	9.90	18.67	8.48

With cadmium chloride and gold chloride, the apparently catalytic effect is due entirely to chlorine retained in solution, while with chromic chloride and platinic chloride increased effects are due partly to chlorine retained in solution and partly to the total reduction of residual oxides of manganese.

Acidimetry and Alkalimetry.

The Use of Succinic Acid as the Standard in Neutralization Processes.

Methods for preparation and the use of succinic acid as a standard in the titration of an alkali hydroxide have been studied by Phelps and Hubbard.* Succinic acid was prepared by hydrolysis of the pure ethyl ester, by hydration of the pure anhydride, by recrystallization of the commercial acid from hot water, and by recrystallization of the commercial acid from hot water

* I. K. Phelps and J. L. Hubbard, Am. Jour. Sci., [4], xxiii, 211.

containing nitric acid. The product obtained by each of these methods dries to a constant weight in air, and loses nothing on further standing over sulphuric acid.

From ethyl succinic ester, boiling at 213.3°–213.5° C. under a barometric pressure of 749 mm., pure succinic acid was obtained by boiling it for four hours on a return condenser with nitric acid and water in these proportions: 20 cm.^{3} of succinic ester, 200 cm.^{3} of water, three drops of nitric acid. The solution was evaporated to crystallization, and, after filtering off from the mother liquor, the solid product was recrystallized from distilled water. The crystals dried carefully in the open air to constant weight melted in an open capillary tube at 182.3°.

Ammonium Hydroxide against Succinic Acid.

Succinic acid.	HCl value of succinic acid.		
	Found.	Theory.	Error.
grm.	grm.	grm.	grm.
Acid from ester.			
0 2000	0 1235	0.1235	0 0000
0 2000	0.1235	0 1235	0.0000
0.2000	0.1235	0 1235	0.0000
0.2000	0.1235	0.1235	0 0000
Acid from anhydride.			
0.2000	0.1231	0.1235	−0 0004
0.2000	0.1233	0.1235	−0 0002
0.2000	0.1234	0.1235	−0 0001
0.2000	0.1234	0.1235	−0.0001
Acid recrystallized from water.			
0.2000	0.1231	0.1235	−0.0004
0.2000	0.1231	0.1235	−0.0004
0.2000	0.1231	0.1235	−0.0004
0.2000	0.1231	0.1235	−0.0004
0.2000	0.1231	0.1235	−0.0004
0.2000	0.1231	0.1235	−0.0004
Acid recrystallized from water containing nitric acid.			
0.2000	0.1233	0 1235	−0.0002
0.2000	0.1233	0.1235	−0.0002

For the preparation of the acid from succinic anhydride the anhydride was purified by recrystallizing from absolute alcohol until the crystals obtained, after carefully drying, melted sharply at 119° C. The pure anhydride obtained in this manner was converted to the acid by dissolving it in distilled water heated to the boiling point. The crystals formed on cooling the solution were filtered off and dried in air and over sulphuric acid. The melting point of the product was 182.8°.

Preparations made by dissolving the acid of commerce in distilled water at the boiling point, crystallizing by cooling, and drying in the air, melted at 181.7°. The acid made by dissolving the succinic acid of commerce in boiling water, adding nitric acid, crystallizing, and drying, melted at 182.3°.

Tests of these preparations of succinic acid were made by titrating solutions of weighed amounts by approximately $n/10$ ammonium hydroxide, with cochineal as an indicator, the ammonium hydroxide having been previously standardized against approximately $n/10$ hydrochloric acid, the exact strength of which had been determined gravimetrically by precipitation with special precautions by silver nitrate. The results are given on page 55.

Organic Acids and Acid Anhydrides as Standards in Neutralization Processes.

Phelps and Weed* have shown that, with phenolphthalein as an indicator, pure sodium hydroxide in solution and pure barium hydroxide in solution may be determined very exactly by titration against weighed amounts of succinic acid, succinic anhydride, malonic acid, benzoic acid, phthalic acid and phthalic anhydride used as standards. Following are experimental results (pp. 57, 58).

Phelps and Weed point out that these organic acids and acid anhydrides, in pure state, make standards in alkalimetry and acidimetry, as accurate as the best previous standard,—hydrochloric acid determined gravimetrically as silver chloride. The most serviceable are those most readily soluble in water,—succinic and malonic acids.

Phelps and Weed† point out, also, that these organic acids and anhydrides — since they may be used to fix the standards of alkali hydroxides, and these to fix the standard of hydrochloric

* I. K. Phelps and L. H. Weed, Am. Jour. Sci., [4], xxvi, 138.

† Am. Jour. Sci., [4], xxvi, 143.

Sodium Hydroxide and Barium Hydroxide against Succinic Acid and Succinic Anhydride.

	Succinic acid. grm.	Succinic anhydride. grm.	HCl value of NaOH used. grm.	HCl value of BaO_2H_2 used. grm.	Theory in terms of HCl. grm.	Error in terms of HCl. grm.
1	0.2000	......	0 1236	......	0.1235	+0 0001
1	0 2000	......	0 1238	......	0.1235	+0 0003
1	0.2000	......	0 1237	......	0 1235	+0 0002
1	0.2000	......	0 1236	......	0 1235	+0 0001
1	0.2000	......	0.1236	......	0.1235	+0 0001
2	0.2000	......	0 1237	......	0.1235	+0.0002
2	0.2000	......	0 1237	......	0.1235	+0 0002
2	0.2000	......	0 1237	......	0.1235	+0.0002
3	0 2000	......	0.1237	......	0.1235	+0 0002
3	0 2000	......	0.1237	......	0.1235	+0.0002
1	0 2000	......	......	0.1238	0 1235	+0.0003
1	0.2000	......	......	0.1237	0.1235	+0.0002
1	0.2000	......	......	0.1235	0.1235	0 0000
1	0.2000	......	......	0.1236	0 1235	+0.0001
	......	0 2000	0.1458	......	0.1458	0.0000
	......	0 2000	0 1458	......	0.1458	0.0000
	......	0 2000	0.1459	......	0.1458	+0.0001
	......	0 2000	0.1458	......	0.1458	0.0000
	......	0 2000	......	0.1457	0.1458	−0.0001
	......	0.2000	......	0.1456	0 1458	−0.0002
	......	0 2000	......	0.1459	0.1458	+0.0001
	......	0.2000	......	0.1458	0.1458	0.0000

1. Freshly made the ester and dried over sulphuric acid.
2. Dried for a year over sulphuric acid.
3. Dried for a year over calcium chloride.

*Sodium Hydroxide and Barium Hydroxide against Malonic Acid.**

Malonic acid. grm.	HCl value of NaOH used. grm.	HCl value of BaO_2H_2 used. grm.	Theory in terms of HCl. grm.	Error in terms of HCl. grm.
0.2000	0 1404	......	0.1402	+0.0002
0.2000	0.1403	......	0.1402	+0.0001
0.2000	0.1402	......	0.1402	0.0000
0.2000	0.1401	......	0.1402	−0.0001
0.2000	......	0.1401	0 1402	−0.0001
0.2000	......	0.1400	0.1402	−0.0002
0.2000	......	0.1402	0.1402	0.0000
0.2000	......	0.1400	0.1402	−0.0002

* Prepared from malonic ester by hydrolysis, recrystallized from water, and dried over sulphuric acid.

*Sodium Hydroxide and Barium Hydroxide against Benzoic Acid.**

Benzoic acid. grm.	HCl value of NaOH used. grm.	HCl value of BaO_2H_2 used. grm.	Theory in terms of HCl. grm.	Error in terms of HCl. grm.
0 2000	0 0598	......	0 0597	+0 0001
0.2000	0.0599	......	0 0597	+0.0002
0.2000	0.0597	......	0 0597	0 0000
0.2000	0.0598	......	0.0597	+0 0001
0.2000	......	0.0598	0.0597	+0.0001
0.2000	......	0 0597	0.0597	0.0000
0.2000	......	0 0597	0 0597	0.0000
0.2000	......	0.0597	0.0597	0.0000

* Prepared by treating benzoic ester with sodium hydroxide, acidifying with hydrochloric acid, and twice crystallizing from water the precipitate, and drying over sulphuric acid. In these experiments, alkali in amount nearly sufficient for neutralization was run into the flask upon the acid, which was then heated to bring about solution of the acid.

Sodium Hydroxide and Barium Hydroxide against Phthalic Acid and Phthalic Anhydride.†*

Phthalic acid. grm.	Phthalic anhydride grm.	HCl value of NaOH used. grm.	HCl value of BaO_2H_2 used. grm.	Theory in terms of HCl. grm.	Error in terms of HCl. grm.
0.2000	......	0.0880	......	0.0878	+0.0002
0.2000‡	......	0.0880	......	0 0878	+0.0002
0.2000	......	0.0879	......	0 0878	+0.0001
0.2000	......	0.0878	......	0.0878	0.0000
0.2000	......	......	0 0876	0.0878	−0 0002
0.2000	......	......	0 0877	0.0878	−0.0001
0.2000	......	......	0.0878	0.0878	0.0000
0.2000‡	......	......	0.0879	0.0878	+0 0001
......	0.2000	0.0986	......	0.0985	+0.0001
......	0.2000	0.0985	......	0.0985	0.0000
......	0.2000	0.0986	......	0 0985	+0.0001
......	0.2000	0.0987	......	0.0985	+0 0002
......	0.2000	......	0 0986	0.0985	+0.0001
......	0.2000	......	0.0985	0.0985	0.0000
......	0.2000	......	0 0986	0.0985	+0.0001
......	0.2000	......	0.0987	0.0985	+0.0002

* Prepared from the commercial anhydride by dissolving in boiling water, crystallizing from the filtered solution by cooling, and drying finally over sulphuric acid.

† Prepared by distilling *in vacuo* the phthalic anhydride of commerce.

‡ Titrated at the room temperature; all others titrated after heating to secure solubility.

acid or sulphuric acid, which in turn (under definite conditions of concentration, as will be shown*) may set free from an iodide-iodate mixture an exactly equivalent amount of iodine — may be used indirectly to set the standards used in iodometric analysis.

The Use of the Iodide-Iodate Mixture and the Estimation of the Iodine Evolved.

Determination of Free Acids. It is well known that when a free mineral acid is added to a neutral mixture of metallic iodate and iodide, the iodate is reduced and iodine is liberated according to the equation:

$$RIO_3 + 5\,RI + 3\,H_2SO_4 = 3\,I_2 + 3\,R_2SO_4 + 3\,H_2O.$$

This reaction is complete and nonreversible under suitable conditions and may therefore be applied to the estimation of amounts of iodate, iodide or mineral acid present in an unknown solution. A solution of iodate to be analyzed is mixed with an excess of iodide and mineral acid, the resulting free iodine estimated by directly titrating with sodium thiosulphate or arsenious acid after neutralization and one-sixth of the amount found taken as equivalent to the iodate originally present.† Similarly, a solution of iodide to be analyzed is mixed with an excess of iodate and mineral acid, the resulting free iodine estimated by directly titrating in alkaline solution with arsenious acid, and five-sixths of its amount taken as equivalent to the iodide originally present.‡ The solution of mineral acid to be analyzed is mixed with an excess of iodate and iodide, the resulting free iodine estimated by directly titrating with sodium thiosulphate, and its entire amount taken as equivalent to the amount of mineral acid originally present.§ Gröger has applied the last-mentioned method to the direct analysis of various mineral acids and has also indirectly analyzed solutions of alkali hydroxides and carbonates by adding the solution to be analyzed to a measured volume of mineral acid, previously standardized by the above method, and estimating the small excess of free mineral acid that finally remains

* See page 60.

† Rammelsberg, Ann. Phys., cxxxv, 493; Walker, Am. Jour. Sci., [4], iv, 235.

‡ Gooch and Walker, cf. page 454.

§ Kjeldahl, Zeit. anal. Chem., xxii, 366; Furry, Am. Chem. Jour., vi, 341; Gröger, Zeit. angew. Chem., 1894, 52.

by the same method. The only difficulty with the Gröger process lies in the fact that in dilute solutions, as shown by Furry,* the end-point of the final reaction between iodine and sodium thiosulphate is somewhat obscured by a peculiar back-play of color due to a continuous slow liberation of iodine in the system.

When the suitably concentrated iodide-iodate mixture (1 grm. of KI and 0.166 grm. of KIO_3 in 50 cm.³) is acted upon by a definite quantity of approximately decinormal hydrochloric acid in a total volume not exceeding 100 cm.³, the iodine set free, as determined by titration with *n*/10 sodium thiosulphate (standardized against iodine of strength fixed by titration against weighed arsenious oxide), measures very exactly, according to Phelps and Weed,† the acid taken. The recorded results of experiments, in which the end reaction was brought about by iodine added directly or set free by addition of more standard acid after the bleaching by thiosulphate, are exceedingly good.

Volume about 100 cm.³

HCl values.				
HCl solution used. grm.	$Na_2S_2O_3$ solution used. grm.	Iodine solution used. grm.	HCl found. grm.	Error in HCl. grm.
0.1074	0.1075	.....	0.1075	+0.0001
0.1074	0.1074	.....	0.1074	0.0000
0.1074	0.1075	.....	0.1075	+0.0001
0.0520	0.0520	......	0.0520	0.0000
0.1560	0.1562	......	0.1562	+0.0002
0.0645	0.0712	0.0068	0.0644	−0.0001
0.0968	0.1068	0.0101	0.0967	−0.0001
0.0645	0.0712	0.0067	0.0645	0.0000
0.0968	0.1068	0.0101	0.0967	−0.0001
0.0484	0.0534	0.0050	0.0484	0.0000
0.0645	0.0748	0.0104	0.0644	−0.0001
0.0484	0.0534	0.0050	0.0484	0.0000
0.0645	0.0712	0.0066	0.0646	+0.0001

Determination of Alkali Hydroxides and Carbonates. Alkali hydroxides may be determined by the action of an excess of standard hydrochloric acid or sulphuric acid, the excess being determined by estimation of the iodine set free by the iodide-iodate mixture at suitable concentrations. Alkali carbonates may also be similarly

* Am. Chem. Jour., vi, 341.

† Am. Jour. Sci., [4], xxvi, 143.

determined by treatment with an excess of sulphuric acid, boiling to remove carbon dioxide from the solution containing the nonvolatile acid, and determination of iodine liberated by the action of the iodide-iodate mixture in a total volume of about 100 cm.[3] Results obtained by this treatment of sodium hydroxide first treated with carbon dioxide are given in the following statement, and, for comparison, results obtained by titration of the excess of standard acid by standard sodium hydroxide.

Volume not Exceeding 100 cm.[3]

Treatment with CO_2.	HCl values.					Difference in terms of HCl.
	NaOH solution used.	H_2SO_4 solution used.	NaOH solution to coloration.	$Na_2S_2O_3$ solution used.	Iodine solution to coloration.	
min.	grm.	grm.	grm.	grm.	grm.	grm.
15	0.1012	0.1306	0.0293	. ...	. ...	+0.0001
30	0.1012	0.1219	0.0205	.. ...	. ...	+0 0002
15	0.1349	0 1524	0 0172	. . .	......	+0 0003
30	0.1349	0.1568	0.0216	. ...	. ....	+0 0003
15	0 1012	0.1306	. .	0.0302	0.0013	+0.0005
30	0 1012	0.1306	. ..	0.0302	0 0013	+0 0005
15	0.1349	0.1742	. ..	0.0392	0 0004	+0 0005
35	0.1349	0.1742	......	0 0552	0.0162	+0.0003

Determination of Acids Liberated in Hydrolysis. When certain salts are brought into association with water, a tendency to the formation of more basic products and free acid in consequence of the hydrolytic action of water becomes evident. Such action goes on until an equilibrium is reached between the factors and products of reaction.

Action between Iodide-Iodate Mixture and Certain Salts.—In the presence of the iodide-iodate mixture, free acid, a product of the hydrolytic action, may be constantly removed, and so the hydrolysis may proceed further, the iodine set free in reaction of the iodide-iodate mixture upon the free acid being obviously a measure of the degree of hydrolytic action. Such action may proceed to the complete hydrolysis of the salt or may cease at an earlier stage, depending upon the nature of the salt. The behavior of certain salts in presence of the iodide-iodate mixture has been studied experimentally by Moody.* In Moody's experiments, a suitable amount of the salt to be tested is put

* Seth E. Moody, Am. Jour. Sci., [4], xx, 181; xxii, 176, 379, 483.

into the Voit flask (B) of the apparatus shown in Fig. 3,* and 10 cm.³ of an iodide-iodate mixture (1 grm. KI. : 0.3 grm. KIO_3) are added, the Drexel flask (C) and trap are charged with a solution containing 3 grm. of potassium iodide, hydrogen is passed from the generator through the apparatus, and the contents of the Voit flask are heated until (according to circumstances) all or nearly all the liberated iodine is collected in the Drexel flask. The free iodine, whether in the receiver or remaining in small amount in the distilling flask, is titrated with sodium thiosulphate.

Aluminium Sulphate. — Proceeding in this manner, Moody found that aluminium sulphate, though only partially hydrolyzed at ordinary temperatures, is completely broken up by heating, according to the reaction

$$Al_2(SO_4)_3 + 5\,KI + KIO_3 + 3\,H_2O = 2\,Al(OH)_3 + 3\,K_2SO_4 + 3\,I_2,$$

and that aluminium chloride behaves similarly.

The following results were obtained with potassium alum, the potassium sulphate not being susceptible to hydrolytic action.

Approx. n/10 aluminium potassium alum. cm.³	KIO_3. grm.	KI. grm.	Time in minutes.	Approx. n/10 $Na_2S_2O_3$. cm.³	Al_2O_3 calculated from iodine found. grm.	Al_2O_3 precipitated and weighed. grm.	Difference. grm.
25	0.3	1.0	30	24.55	0.0410	0.0414	−0.0004
25	0.3	1.0	30	24.60	0.0411	0.0416	−0.0005
25	0.3	1.0	25	24.50	0.0409	0.0414	−0.0005
25	0.3	1.0	30	24.70	0.0413	0.0416	−0.0003
25	0.3	1.0	35	24.50	0.0409	0.0415	−0.0006
25	0.3	1.0	30	24.55	0.0410	0.0415	−0.0005
25	0.3	1.0	25	24.50	0.0409	0.0415	−0.0006

Upon the presumption that the aluminium salt taken is perfectly neutral, the reaction affords means for determining aluminium iodometrically.

Upon repeating the experiment with similar amounts of an ammonium alum it was found that the amounts of iodine liberated were in the average somewhat excessive, corresponding to 0.0006 grm. of Al_2O_3 more than the theory called for. The process is, therefore, less exact in the presence of ammonium salts. In fact,

* See page 4.

as will be shown, ammonium sulphate in the amounts taken may be completely hydrolyzed in the course of three hours.

Chromic Sulphate, Tin Chloride. — Chromic sulphate, taken in the form of chrome alum, undergoes complete hydrolysis in presence of the iodide-iodate mixture with precipitation of chromic hydroxide, as does the double potassium tin chloride, $SnCl_4.2KCl$.

Iron Sulphate. — Ferric sulphate reacts like aluminium sulphate according to the equation

$$Fe_2(SO_4)_3 + 5\,KI + KIO_3 + 3\,H_2O = 2\,Fe(OH)_3 + 3\,K_2SO_4 + 3\,I_2.$$

The hydrolysis of ferrous sulphate is accompanied by oxidation of the ferrous hydroxide at the expense of the iodate, as follows:

$$3\,FeSO_4 + 5\,KI + KIO_3 + 3\,H_2O = 3\,Fe(OH)_2 + 3\,K_2SO_4 + 3\,I_2.$$
$$6\,Fe(OH)_2 + KIO_3 + 3\,H_2O = 6\,Fe(OH)_3 + KI.$$

From the experimental tests it appears, however, that the hydrolysis of ferrous sulphate is complete in the presence of the iodide-iodate mixture, and that the iodine set free is an exact measure of the SO_4 ion present and of the iron in the ferrous sulphate of ideal composition.

Volume.	KI.	KIO_3.	Time in minutes.	Approx. $n/10$ $Na_2S_2O_3$.	Iodine found.	Iodine value of $FeSO_4$ taken.	Difference.
cm.³	grm.	grm.		cm.³	grm.	grm.	grm.
40	1.0	0.45	30	26.67	0.3324	0 3322	+0.0002
40	1 0	0.45	30	26.68	0.3325	0 3322	+0 0003
40	1.0	0.45	30	26.65	0.3321	0.3322	−0 0001
40	1.0	0.45	30	26.67	0 3324	0.3322	+0 0002
40	1.0	0 45	30	26.66	0.3323	0 3322	+0.0001

Cobalt Sulphate. — Cobalt sulphate when similarly hydrolyzed is likewise oxidized at the expense of the iodate, the reactions following similar lines.

$$3\,CoSO_4 + 5\,KI + KIO_3 + 3\,H_2O = 3\,Co(OH)_2 + 3\,K_2SO_4 + 3\,I_2.$$
$$6\,Co(OH)_2 + KIO_3 + 3\,H_2O = 6\,Co(OH)_3 + KI.$$

The iodine value obtained in test experiments, upon the hypothesis that cobaltous sulphate is completely hydrolyzed, is closely comparable with the iodine equivalent of the cobalt found by the electrolytic deposition of the metal.

Volume.	KI.	KIO_3.	Time in hours.	Approx. $n/10$ $Na_2S_2O_3$.	Iodine value found.	Iodine value of $CoSO_4$ taken.	Difference.
cm.[3]	grm.	grm.		cm.[3]	grm.	grm.	grm.
40	1.0	0.45	4	17.80	0.2244	0.2242	+0.0002
40	1.0	0.45	3½	17 78	0.2242	0.2242	0.0000
40	1.0	0.45	3½	17 75	0.2238	0.2242	−0.0004
40	1.0	0.45	4	17.79	0 2243	0.2242	+0 0001
40	1.0	0.45	4	17.79	0.2243	0.2242	+0 0001
40	1.0	0.45	4	17.78	0.2242	0.2242	0.0000

Nickel Sulphate. — Nickel sulphate, like cobalt sulphate, is hydrolyzed completely, after a considerable time, in the presence of the iodide-iodate mixture, likewise yielding iodine, which may be collected similarly and estimated as a measure of the nickel present. Nickelous hydroxide formed in the reaction remains, however, unoxidized by potassium iodate in neutral solution, and therefore the following equation shows the final products:

$$3\,NiSO_4 + 5\,KI + KIO_3 + 3\,H_2O = 3\,Ni(OH)_2 + 3\,K_2SO_4 + 3\,I_2.$$

Volume.	KI.	KIO_3.	Time in hours.	Approx. $n/10$ $Na_2S_2O_3$.	Iodine found.	Iodine value of $NiSO_4$ taken.	Difference.
cm.[3]	grm.	grm.		cm.[3]	grm.	grm.	grm.
40	1.0	0.45	3	17.87	0.2254	0.2255	−0 0001
40	1.0	0.45	3	17.88	0.2256	0.2255	+0.0001
40	1.0	0.45	3	17.84	0.2250	0.2255	−0 0005
40	1.0	0.45	3	17.87	0.2254	0.2255	−0.0001
40	1.0	0.45	3	17.83	0.2249	0.2255	−0.0006

Thus it appears that nickel sulphate may be completely hydrolyzed in the presence of the iodide-iodate mixture, and that the nickel of nickel sulphate of ideal composition can be estimated from the amount of iodine liberated in the action of that salt upon the iodide-iodate mixture.

Zinc Sulphate. — Like the sulphates of nickel, cobalt, iron, aluminium and chromium, zinc sulphate is hydrolyzed in presence of the iodide-iodate mixture, but unlike the sulphates of the other elements mentioned the reaction stops short of complete hydrolysis, as shown in the results of the table. The reaction may be expressed by the equation

$$15\,ZnSO_4 + 20\,KI + 4\,KIO_3 + 12\,H_2O$$
$$= 3\,Zn_5(OH)_8SO_4 + 12\,K_2SO_4 + 12\,I_2.$$

Volume.	KI.	KIO_3.	Time in hours.	Approx $n/10$ $Na_2S_2O_3$.	Iodine found.	Equivalent of SO_3 found.	SO_3 present.
cm.³	grm.	grm.		cm.³	grm	grm.	grm.
40	1.0	0.45	½	27 8	0 3557	0 1123	0 1408
40	1 0	0.45	½	28 0	0.3582	0 1130	0.1408
40	1.0	0 45	3	27 8	0.3557	0.1123	0.1408
40	1.0	0.45	3	27.8	0 3557	0.1123	0.1408

The mean percentage of hydrolysis here found is 79.90. It appears that a basic sulphate containing 5 of Zn to one of SO_4 is formed, and so definitely that from the iodine liberated the zinc content may be calculated with accuracy.

Ammonium Sulphate. — When solutions of ammonium sulphate are subjected to heat the salt is hydrolyzed, and as the acid is increased, either as a direct product of this hydrolysis or by addition, further dissociation is inhibited. The decrease in dissociation is dependent upon the increase of the acid, and when sufficient acid is present further hydrolysis is entirely prevented. The amount of hydrolysis is, however, small under the most favorable conditions.*

Moody has studied the effect of the iodide-iodate mixture in eliminating the acid as it is produced. In experiments made in the manner described above it was found to be impossible to collect the iodine in the Drexel flask used as a receiver when charged with potassium iodide only, although it was evident that much iodine came over. It appeared, upon investigation, that ammonium iodide and ammonium iodate were formed by reaction in the receiver between the liberated iodine and the ammonia also volatilized, and to obviate the difficulty sulphuric acid was added to the contents of the receiver into which the distillate was passed. Under these conditions iodine is obtained in amount corresponding to that which should be eliminated when the ammonium sulphate is entirely hydrolyzed.

* Bruck, Dissertation, Giessen, 1903.

This is shown in the subjoined table:

Volume.	KI.	KIO_3.	Time in hours.	H_2SO_4 (1 : 1) in the receiver.	Approx. $n/10$ $Na_2S_2O_3$.	Iodine found.	Iodine value of $(NH_4)_2SO_4$ taken.	Difference.
cm.³	grm.	grm.		cm.³	cm.³	grm.	grm.	grm.
35	1 0	0.30	3	40	38.25	0.4769	0 4773	−0 0004
35	1 0	0.30	3	40	38.25	0.4769	0.4773	−0.0004
35	1.0	0.30	3	40	38.30	0.4775	0.4773	+0 0002
35	1.0	0.30	3	40	38.25	0 4769	0.4773	−0 0004
35	1 0	0.30	3	40	38.23	0.4766	0 4773	−0.0007

In another series of experiments, the apparatus of Fig. 4* was used. In these experiments the mixture was boiled in the first Voit flask and the distillate passed from the first flask, V^1, through the second flask, V^2, containing 50 cm.³ of $n/10$ H_2SO_4 to take up the ammonia, and then into the receiver containing potassium iodide without acid. The sulphuric acid remaining free in the second Voit flask at the end of the operation was determined by titration of the iodine liberated upon the addition of the iodide-iodate mixture, the difference between this amount of iodine and the iodine equivalent of the sulphuric acid used being the measure of the ammonia absorbed. The iodine passing to the receiver was determined as usual by titration with sodium thiosulphate.

The results of these experiments, given below, show that the sulphuric acid neutralized in the Voit flask is a measure of the ammonia, while the iodine in the Drexel flask corresponds to the sulphuric acid of the ammonium sulphate.

Similar results were obtained with ammonium chloride.

Volume, 45 cm.³; KI, 1 grm.; KIO_3, 0.6 grm.; Time, 3 to 3½ hours.

Iodine value of $(NH_4)_2SO_4$ taken.	Iodine equivalent of ammonia absorbed in Voit flask.			Iodine estimated in Drexel flask.		
	Approx. $n/10$ $Na_2S_2O_3$.	I.	Difference.	Approx. $n/10$ $Na_2S_2O_3$.	I.	Difference.
grm.	cm.³	grm.	grm.	cm.³	grm.	grm.
0.4773	38.15	0.4757	−0.0016	38 23	0.4767	−0.0006
0.4773	38.20	0.4763	−0.0010	38.25	0.4769	−0.0004
0.4773	38.15	0.4757	−0.0016	38.20	0.4763	−0.0010
0.4773	38.20	0.4763	−0.0010	38.27	0.4771	−0.0002
0.4773	38.17	0.4759	−0.0014	38.20	0 4763	−0.0010
0.4773	38.15	0.4757	−0 0016	38.20	0.4763	−0.0010
0.4773	38.20	0.4763	−0 0010	38.25	0 4769	−0.0004

* See page 5.

This procedure is not presented as an analytical method for determining ammonia or the acid-ion, but to show that the effects of hydrolysis must not be ignored when ammonium salts are heated in solution with the iodide-iodate mixture.

Alums: Basic Alumina and Free Acid. Moody has shown * how the phenomena of hydrolysis may be applied to the determination of basic alumina and free acid in the analysis of alums and commercial aluminium sulphates, which, beside aluminium sulphate, may contain ferrous sulphate, ferric sulphate, and zinc sulphate as impurities. Potassium sulphate and sodium sulphate, if present, do not set free iodine from the boiling solution containing the iodide-iodate mixture. The determinations of the ferrous iron, the ferric iron, the zinc, and the ammonia furnish data from which the equivalent amounts of sulphuric acid, to be taken into account in the reckoning of the free acid or basic alumina, may be calculated. The behavior of these commercial products toward the iodide-iodate mixture affords, therefore, an easy method of determining basic alumina or free acid, as the case may be.

Following are the details of treatment:

I. A sample of 15 grm. is weighed and treated with water. The solution is filtered and made up to 1 liter. The material which does not dissolve is dried at 100° and weighed as *insoluble material.*

II. Of the solution, a portion of 25 cm.^{3} is titrated directly with standard potassium permanganate to find the amount of iron in the ferrous salt, and from this is calculated the *ferrous oxide.*

III. Of the solution, another portion of 25 cm.^{3} is treated with zinc to reduce the ferric salt and then titrated with permanganate to give the total iron. From the difference between the total iron and the ferrous iron is calculated the *ferric oxide.*

IV. Of the solution, a portion of 25 cm.^{3} is diluted to 50 cm.^{3}, treated with 3 grm. of sodium acetate and 1 cm.^{3} of acetic acid, and electrolyzed with the use of the rotating cathode † by a current of about 2 amperes for 30 minutes. The deposit of zinc, including some iron, is washed with alcohol, dried and weighed.

* Am. Jour. Sci., [4], xxii, 483.

† See page 11.

The solution of the deposit in sulphuric acid is titrated with permanganate, and the amount of iron thus found is deducted from the total weight of the deposit to give the amount of zinc. From the zinc is calculated the *zinc oxide.*

V. Of the solution, a portion of 25 cm.^{3} is drawn from a burette into the Voit flask of the distillation apparatus, a solution (10 cm.^{3}) containing 0.3 grm. of potassium iodate and 1 grm. of potassium iodide is added, the mixture boiled, and the iodine, collected in the receiver charged with water containing 3 grm. of potassium iodide (and acidified with sulphuric acid in case the substance contains ammonia), is titrated with sodium thiosulphate.

The iodine set free corresponds to the various oxides, to ammonia and to sulphuric acid in the following proportions:

Al_2O_3	:	6 I;
Fe_2O_3	:	6 I;
FeO	:	2 I;
5 ZnO	:	8 I;
NH_3	:	I;
H_2SO_4	:	2 I;

it is the *total iodine.*

VI. Of the solution, a portion of 25 cm.^{3} is treated in an open beaker with 1 grm. of potassium iodide and 0.3 grm. of potassium iodate, the mixture is boiled until nearly all free iodine is expelled, the precipitate is filtered on paper, ignited carefully, and weighed as Al_2O_3, Fe_2O_3 and ZnO, the *total oxides.*

Given the *total iodine* liberated in the distillation process, the weight of the *total oxides* obtained in the parallel boiling process, the *ferrous oxide* and *ferric oxide* by the permanganate titrations and the *zinc oxide* deduced from the corrected electrolytic determination, the *total alumina* and the *basic alumina* or *free acid* (as the case may be) are easily calculated.

Total oxides − (ferric oxide + ferrous oxide + zinc oxide) = *total alumina.*

$$\left(\frac{6 \times 126.97}{102.2}\right) \text{ or } (7.454) \times \text{total alumina} =$$

iodine equivalent to total alumina.

$\left(\frac{6 \times 126.97}{159.8}\right)$ or $(4.767) \times$ ferric oxide =

iodine liberated by ferric sulphate.

$\left(\frac{2 \times 126.97}{71.9}\right)$ or $(3.532) \times$ ferrous oxide =

iodine liberated by ferrous sulphate.

$\left(\frac{8 \times 126.97}{5 \times 81.4}\right)$ or $(2.496) \times$ zinc oxide =

iodine liberated by zinc sulphate.

$\left(\frac{126.97}{17}\right)$ or $(7.469) \times$ ammonia =

iodine liberated by ammonium sulphate.

Total iodine − (iodine corresponding to total alumina, ferric sulphate, ferrous sulphate, zinc sulphate, ammonium sulphate) = differential iodine.

Differential iodine (if positive)

$$\times \left(\frac{98.08}{2 \times 126.97}\right) \text{ or } (0.386) = \textit{free acid.}$$

Differential iodine (if negative)

$$\times \left(\frac{102.2}{6 \times 126.97}\right) \text{ or } (0.134) = \textit{basic alumina.}$$

The results of analyses of four specimens of alums are given in the following table:

Percentage Composition.

	Al_2O_3. Total.	Al_2O_3. Combined.	Al_2O_3. Basic	FeO.*	ZnO.	NH_3.	Insoluble material.
No. I.							
(1)	14 48	13.49	0.99	0.43	3 70	None.	0.61
(2)	14 28	13.46	0 82	0 44	3 70	None.	0.61
No. II:							
(1)	15.94	14 21	1.73	0 43	1.47	None.	0.21
(2)	15 90	14.34	1.56	0.45	1.34	None.	0.21
No. III.							
(1)	15 59	14.81	0.78	0 34	0.73	None.	0.71
(2)	15 97	14.80	1.17	0.36	0.82	None.	0.71
D:							
(1)	16.59	15.24	1.35	0.24	0.11	None.	0.61
(2)	16.37	15.18	1.19	0.24	0.19	None.	0.61

* With a trace of Fe_2O_3.

The Use of the Bromide-bromate Mixture and the Estimation of the Bromine Evolved.

The reaction of a mixture of potassium bromide and potassium bromate upon aluminium sulphate has been studied by Gooch and Osborne* with the aid of the apparatus previously described.† Assuming that the acidic ion is entirely liberated from the aluminium salt, the reaction should follow the equation

$$2\ KAl(SO_4)_2 + 5\ KBr + KBrO_3 + 3\ H_2O = 2\ Al(OH)_3 + 4\ K_2SO_4 + 3\ Br_2.$$

The precipitation proves to be complete, or nearly so; but the process of hydrolysis does not go easily to the point of forming aluminium hydroxide. A reasonable excess of the bromide-bromate mixture is able in a moderate time to carry the hydrolysis of aluminium sulphate to a fairly definite point corresponding nearly to the removal of five-sixths of the acidic ion, while the iodide-iodate mixture under similar conditions of action removes practically all the acidic ion.

With a very large increase in the concentration of the bromide and bromate and prolonged boiling, the completion of the hydrolysis to the point of liberating bromine equivalent to the entire amount of the acidic ion is very nearly realized.

Like the iodide-iodate mixture, the bromide-bromate mixture is a very delicate indicator of free acid; 0.00018 grm. of sulphuric acid proving to be sufficient to liberate bromine from the bromide-bromate mixture when boiled in the Voit flask under the experimental conditions.

Some experiments to test the effect of a mixture of potassium chloride and potassium chlorate upon aluminium sulphate indicated that the hydrolysis under otherwise similar conditions is very slight compared with that produced by the bromide-bromate mixture or by the iodide-iodate mixture.

The Reaction of Iodine with Alkali Hydroxides.

When the solution of a metallic hydroxide is acted upon by iodine at a temperature high enough to decompose the small amounts of hypoiodites that might otherwise be present, the final

* F. A. Gooch and R. W. Osborne, Am. Jour. Sci., [4], xxiv, 167.

† See pages 4, 61.

action results in the formation of an exactly neutral mixture of iodate and iodide according to the equation

$$6\,KOH + 3\,I_2 = RIO_3 + 5\,RI + 3\,H_2O.$$

In a process described elsewhere* for the determination of carbon dioxide, Phelps† has applied this reaction to the determination of barium hydroxide. In this process, standard iodine is made to act in a suitable apparatus upon the barium hydroxide and the excess of it is determined.

Instead of determining, according to the procedure of Phelps, the amount of iodine left over in the action of an excess of that element upon the alkali hydroxide, Walker and Gillespie‡ prefer to expel, by boiling, all free iodine remaining after action and to determine the amount of iodine liberated by acting with an acid from the residual mixture of iodate and iodide. According to the procedure recommended, an excess of decinormal iodine is drawn into an Erlenmeyer beaker and the desired amount of alkali hydroxide is run in rapidly. The beaker, closed by a little trap,§ made of a calcium chloride drying tube, to prevent appreciable loss by spattering, is placed over a low flame, and the contents boiled until the last trace of the excess of iodine has been volatilized from the solution and the trap. The volume is carefully regulated before and during the boiling, being kept as small as possible, usually amounting to about 100 $cm.^3$ at the start and 35 $cm.^3$ at the close. In the case of barium hydroxide, care has to be taken to keep the dilution sufficient to prevent the separation of the crystalline barium iodate, which is soluble with difficulty. To steady the ebullition a little spiral of platinum is introduced into the beaker. After the boiling, the colorless solution, containing a neutral mixture of iodate and iodide, is cooled in running water, and treated with 10 $cm.^3$ of dilute acid — in the case of barium hydroxide, dilute [1 : 3] hydrochloric acid, to save the inconvenience of working in the presence of precipitated barium sulphate; with potassium hydroxide, dilute [1 : 3] sulphuric acid. The liberated iodine is titrated

* See page 231.

† Am. Jour. Sci., [4], ii, 70.

‡ Claude F. Walker and David H. M. Gillespie, Am. Jour. Sci., [4], vi, 455.

§ See Fig. 6.

directly with sodium thiosulphate, in the presence of starch.

Tests of this method in comparison with that of Phelps showed a fair agreement, though the results of both methods were invariably lower by a small, nearly constant amount than those obtained by gravimetric estimations and by the Gröger process.* This error of the Phelps process and its modification is probably due to the action of atmospheric carbon dioxide on the hydroxide solution during the short time it is exposed. While it affects considerably the value of the method as a means of accurately determining the absolute amount of hydroxide present in a given volume of solution, it apparently does not so seriously affect the accuracy of the differential method founded on the original Phelps process or its modification.

Analyses of n/10 Hydrochloric Acid Solution.

(By adding to excess of *n*/10 $Ba(OH)_2$, boiling with excess of iodine to colorlessness, and acidifying the residue.)

HCl taken. cm.^{3}	$Ba(OH)_2$ taken. grm.	$Ba(OH)_2$ neutralized by HCl. grm.	HCl found. grm.	HCl by Groger method. grm.	Variation. grm.
15	0 17	0.1128	0.0480	0.0481	+0 0001
15	0.17	0 1118	0.0475	0.0481	−0 0006
15	0.17	0.1112	0 0473	0.0481	−0 0008
25	0.26	0.1860	0.0791	0.0801	−0.0010
25	0 26	0.1866	0.0794	0 0801	−0.0007
35	0 34	0.2634	0.1120	0 1121	−0 0001
35	0.34	0.2603	0.1107	0.1121	−0.0014

Analyses of n/10 Hydrochloric Acid Solution.

(By adding to excess of *n*/10 KOH, boiling with excess of iodine to colorlessness, and acidifying the residue.)

HCl taken. cm.^{3}	KOH taken. grm.	KOH neutralized by HCl. grm.	HCl found. grm.	HCl by Gröger method. grm.	Variation. grm.
20	0.14	0.0972	0 0633	0.0641	−0.0008
20	0.14	0.0975	0.0634	0.0641	−0 0007
25	0.14	0.1222	0 0795	0.0801	−0.0006
25	0.14	0.1207	0 0785	0.0801	−0.0016

In applying the Walker-Gillespie modification to the indirect determination of hydrochloric acid and sulphuric acid, the acid

* See page 59.

solution to be analyzed is drawn into an Erlenmeyer beaker, an excess of decinormal iodine is added, and a measured excess of standardized alkali is introduced. The beaker is trapped, the solution boiled, and the residue cooled, acidulated, and titrated with thiosulphate in presence of starch. Test results of the process are given in the accompanying tables.

Analyses of n/10 Sulphuric Acid Solution.

(By adding to excess of $n/10$ $Ba(OH)_2$, boiling with excess of iodine to decoloration, and acidifying the residue.)

	H_2SO_4 taken.	$Ba(OH)_2$ taken.	$Ba(OH)_2$ neutralized by H_2SO_4.	H_2SO_4 found.	H_2SO_4 by Groger method.	Variation.
	cm.^{3}	grm.	grm.	grm.	grm.	grm.
(1)	10	0.21	0 0884	0.0506	0.0496	+0.0010
(2)	10	0.21	0 0880	0.0503	0.0496	+0 0007
(3)	15	0.30	0 1328	0 0754	0.0745	+0.0009
(4)	15	0.30	0 1313	0 0751	0.0745	+0.0006
(5)	25	0 43	0 2168	0 1239	0 1240	−0 0001
(6)	30	0.43	0.2600	0.1481	0.1489	−0.0008

The reaction between iodine and the hydroxides of potassium and barium in hot solution is, therefore, regular and complete under analytical conditions, and not appreciably affected by the mass action of considerable excesses of iodine. It is best applied in analysis by treating the alkali with an excess of iodine, removing this excess by boiling, and estimating the iodine in the residue. While mechanical difficulties and the interfering action of carbon dioxide may affect the extreme accuracy of the process as a direct means for analyzing alkalies, it may be indirectly applied with fair accuracy to the analysis of various acids and possibly to other compounds. The reaction between iodine and alkali carbonates, on the contrary, is irregular and cannot be made the basis of an analytical process.

CHAPTER II.

THE ALKALI METALS.

SODIUM.

The Detection of Sodium.

THE application of the spectroscopic method to the detection of potassium and sodium in ordinary analysis is unsatisfactory on account of its failure, except in delicate quantitative comparisons, to give any idea as to the quantity of either element indicated; and, since the most minute quantity of either element is sufficient to produce its characteristic line in the spectroscope, and many of the reagents employed in analysis contain a trace of alkali, the spectroscopic indication is often misleading. While to the careful observer the presence or absence of potassium in appreciable amount is revealed, the evidence as to the quantity of the ubiquitous element sodium is practically worthless.

Kreider and Breckenridge* have developed a method for the detection of sodium based upon the utilization of the perchlorate method for the preliminary separation of potassium. The insolubility of potassium perchlorate and the easy solubility of sodium perchlorate in 97 per cent alcohol afford means for the separation of these elements as well as for the identification of the former.† By converting to the chloride the sodium perchlorate in the alcoholic filtrate from the precipitated potassium salt, exceedingly small amounts of sodium may be detected. For this purpose the passing of gaseous hydrochloric acid into the alcoholic solution of the sodium perchlorate, kept cool, has proved most effectual, the dehydrating effect of the acid upon the alcohol greatly increasing the insolubility of the sodium chloride. The delicacy of this test for sodium is shown in the results of the table.

By the use of 10 $cm.^3$ of 97 per cent alcohol with gaseous hydrochloric acid 0.0003 grm. of sodium oxide can be found with

* D. Albert Kreider and J. E. Breckenridge, Am. Jour. Sci., [4], ii, 263.

† See page 88.

certainty, and when the alcohol is allowed to become saturated with the gas even 0.00006 grm. The quantity of alcohol, 10 cm.^{3}, is sufficient for all purposes, since this amount will dissolve about 2 grm. of sodium perchlorate; but even in 40 cm.^{3} 0.0002 grm. of sodium oxide may be seen distinctly. It is evident that this method may also be applied to the quantitative determination of sodium.

Test for Sodium.

$NaClO_4$ taken. grm.	Na_2O equivalent. grm.	97 per cent alcohol. cm.^{3}	Indication.
0.0100	0.00250	10	Very strong.
0.0050	0.00125	10	Strong.
0.0040	0 00100	10	Strong.
0.0030	0 00075	10	Strong.
0.0030	0 00075	10	Good.
0.0020	0.00050	10	Good.
0.0020	0.00050	10	Good.
0 0010	0.00025	10	Good.
0.0005	0.00012	10	Trace.
0.0003	0.00006	10	Trace.
0.0001	0.00003	10	None.
0.0000	0.00000	10	None.
0.0010	0.00025	40	Distinct.

In the following table are recorded the results of experiments in which the perchlorates of sodium and potassium were separated by 97 per cent alcohol and the sodium test made upon the alcoholic solution. The sodium and potassium salts were drawn from separate standard solutions of the purified perchlorates.

Separation of Potassium with Test for Sodium.

$KClO_4$ taken. grm.	K_2O equivalent. grm.	$NaClO_4$ taken. grm.	Na_2O equivalent. grm.	Indication for potassium.	Indication for sodium.
0.0500	0.01699	0.0500	0.01250	Strong.	Strong.
0.0200	0.00680	0.0200	0.00500	Strong.	Strong.
0.0100	0.00340	0.0100	0.00250	Strong.	Strong.
0.0050	0.00170	0.0050	0.00125	Strong.	Strong.
0.0040	0.00136	0.0040	0.00100	Good.	Good.
0.0030	0.00102	0.0030	0.00075	Good.	Good.
0.0020	0.00068	0.0020	0.0005	Good.	Good.
0.0010	0.00034	0.0010	0.00025	Good.	Good.
0.0005	0.00017	0.0005	0.00012	Trace.	Trace.
0.0003	0.00010	0 0003	0.00007	Trace.	Trace.
0.0001	0.00003	0.0001	0.00003	Faintest trace.	None.
0.0000	0.00000	0.0100	0.00250	None.	Strong.
0.0100	0.00340	0 0000	0.00000	Strong.	None.

After evaporating to dryness on the steam bath, the residue was treated with 10 cm.[3] of 97 per cent alcohol, the insoluble potassium perchlorate was removed by filtering through a dry paper filter and dry funnel into a dry test tube, and the filtrate saturated with gaseous hydrochloric acid.

The results show that sodium and potassium perchlorates when associated in any proportion may be separated by 97 per cent alcohol with exactness, and that the sodium may be indicated in the filtrate with great delicacy and certainty by the action of gaseous hydrochloric acid.

For the conversion of other salts of sodium and potassium to perchlorates for the purpose of making the test for sodium, it is obvious that perchloric acid free from sodium must be employed. The perchloric acid prepared according to the method described later,* — by heating sodium chlorate to form the perchlorate, destroying any residual chlorate by treatment with the strongest hydrochloric acid, separating sodium chloride by filtration on asbestos in the filtering crucible, and removing the excess of hydrochloric acid by evaporation, — while answering perfectly well for the detection of potassium, is inapplicable to the test for sodium, because of the small amount of this element which the acid always contains on account of the partial solubility of sodium chloride in hydrochloric acid. This perchloric acid must, therefore, be purified by distillation, and to prevent loss by decomposition the process must be carried on under diminished pressure. To obviate violent ebullition, only acid previously concentrated to the fuming point should be subjected to the distillation process, and this only in small amounts. Rubber stoppers or connectors are not to be used where the acid may condense upon them and flow back into the flask, since oxidizable matter carried back causes explosions which vary in force and seriousness according to circumstances.

To construct a suitable apparatus a strong side-neck flask is selected, the bottom covered to the depth of 1 cm.[3] with fine chips of porcelain, and into the neck is sealed a stoppered funnel reaching well into the bulb. The stopcock of this funnel is carefully cleansed and lubricated with metaphosphoric acid obtained by boiling sirupy orthophosphoric acid until the temperature of 350° C. has been attained. The side neck of the

* Page 89.

flask is inclined upward for a short distance before being bent into the receiver, with which it is connected by a rubber stopper through which the tube extends for a safe distance. An ordinary bottle of 250 $cm.^3$ capacity serves for a receiver. This is closed by a doubly perforated stopper, and through one of the perforations the adapter from the condenser is inserted, while through the other connection is made with a small glass bulb and absorption tube, filled with stick potash to take up any chlorine evolved in the inevitable slight decomposition of the acid, which in turn is connected with an automatic mercury pump. The whole flask is surrounded by a cylinder of thin sheet iron closed below, while the upper opening is protected by an asbestos cover in order that the heat may be uniformly applied up to the point at which condensed acid flows into the receiver.

In making a distillation, 3 $cm.^3$ or 4 $cm.^3$ of the concentrated acid are admitted to the flask through the stoppered funnel, the pump is started, and when the pressure has been reduced to about 8 mm. the heat is raised to about 130° and the distillation begins. During the process the pressure is kept at about 3 mm. to 5 mm., and the acid is admitted at about the same rate at which the distilled product drops from the condenser. Under the conditions described, the process will yield per hour from 25 $cm.^3$ to 40 $cm.^3$ of the dihydrate of perchloric acid, the most concentrated form in which the acid is stable. Of this product 0.1 grm. of potassium oxide requires for precipitation 0.16 $cm.^3$.

With pure perchloric acid at hand the separation of the sodium salt from a mixture of the pure chlorides of sodium and potassium, preparatory to making the sodium test, requires only treatment of the mixture with perchloric acid, evaporation on the water bath until the white fumes of perchloric acid appear, treatment of the residue with 97 per cent alcohol, and filtration. In general, however, it is necessary to remove certain interfering substances before applying the method. While potassium may be safely tested for in the presence of other bases and acids, except ammonium, cæsium and rubidium, and sulphuric acid, there are many elements the insolubility of whose chlorides in alcohol necessitates their removal before testing for sodium. But among the common alkalies ammonium is the only one whose presence is objectionable. Lithium does not affect the test for either potassium or sodium.

In the experiments made with potassium and sodium salts associated with salts of other common elements, the results of which are recorded below, the following treatment was adopted. The several groups of bases were successively removed in the ordinary way: Hydrogen sulphide in ammoniacal solution removed the lead, mercury, copper and zinc. Barium and calcium were removed by ammonium carbonate, the final filtrate being evaporated and the residue ignited to volatilize ammonium salts. The residue was dissolved and treated with barium hydroxide for the removal of magnesium, and, after filtering, the barium was again removed by ammonium carbonate, and the filtrate evaporated. The residue was then ignited as before, treated with 10 cm.^{3} of boiling water and the solution filtered in order to remove the organic matter usually found at this stage of the treatment. To the filtrate was added 0.1 to 0.5 cm.^{3} of pure perchloric acid, about 1.7 sp. gr., according to the amount of residue, and the mixture was evaporated over the steam bath until the white fumes of perchloric acid appeared. When the quantity of sodium is large it is safer to evaporate several times in order to secure a complete conversion to the perchlorate.

Potassium and Sodium in Mixtures of Salts of Other Elements.

Pb, Cu, Al, Fe, Zn, Ba, Ca and Mg as nitrates. grm.	K_2O taken. grm.	Na_2O taken. grm.	Indication for potassium.	Indication for sodium.
0.0500 of each.	0.0000	0.0000	Faintest trace.	Trace.
0.0500 of each.	0.0017	0.0012	Good.	Good.
0.1000 of each.	0.0000	0.0000	Faintest trace.	Trace.
0.1000 of each.	0.0000	0.0005	Faintest trace.	Good.

The fact that minute traces of sodium and potassium are found in the blank tests is to be expected from the delicacy of the method when it is remembered that but very few of the so-called chemically pure reagents are absolutely free from sodium, and that even distilled water kept in glass vessels contains a trace of the alkali elements. However, the indication for sodium in the blank tests appeared only as a cloudiness and after complete saturation. When the quantity of sodium oxide present is not less than 0.0005 grm. the precipitate appears in granular form and before the alcohol is completely saturated. The method is all that could be desired for the qualitative determination of sodium.

The Estimation of Sodium as the Pyrosulphate.

Sodium sulphate in water solution may be recovered by simple evaporation and ignition. When free sulphuric acid is also present the acid sulphate is first formed, then, as the temperature rises, the pyrosulphate, and gradually, at red heat, the neutral sulphate. In quantitative estimations it is usual to weigh in the form of neutral sulphate and to hasten conversion to this condition by making the final ignition in the atmosphere produced by ammonium carbonate either projected into the hot crucible,* or placed in the crucible before the application of heat.†

In a study of the behavior of the sulphates of the alkali elements, Browning ‡ has shown that by holding the temperature between 250° and 270° during the heating of sodium sulphate with the excess of sulphuric acid, sodium pyrosulphate, $Na_2S_2O_7$, may be formed with a degree of exactness which makes it possible to estimate sodium as that salt.

Results of this procedure are given in the table, and, for comparison, the results obtained in the usual process of weighing as the neutral sulphate after moistening with ammonium hydroxide and then igniting strongly.

Sodium as the Pyrosulphate.

NaCl taken. grm.	$Na_2S_2O_7$ calculated. grm.	$Na_2S_2O_7$ found. grm.	Error. grm.	Na_2SO_4 calculated. grm.	Na_2SO_4 found. grm.	Error. grm.
0.1042	0.1978	0.1972	−0.0006	0.1266	0.1254	−0.0012
0.1028	0.1952	0.1952	0.0000	......	......	....
0.1093	0.2075	0.2065	−0.0010	0 1328	0.1320	−0.0008
0.1402	0.2662	0.2651	−0 0011	0.1703	0.1696	−0.0007

Salts of potassium also yield when similarly treated the pyrosulphate,§ while salts of cæsium and rubidium remain

* Fresenius, Quant. Anal., trans. Cohn, pages 161, 165.

† Treadwell, Anal. Chem., trans. Hall (1911), page 42.

‡ Philip E. Browning, Am. Jour. Sci., [4], xii, 301.

§ See page 92.

under the conditions of temperature in the condition of acid sulphates.* Lithium salts apparently do not yield the acid sulphate or the pyrosulphate in stable form.

POTASSIUM.

The Spectroscopic Detection and Determination of Potassium.

Bunsen and Kirchoff originally determined the delicacy of the spectroscopic test for potassium by exploding in a darkened room a mixture of potassium chlorate with milk sugar, and observing the amount of finely divided chloride which it was necessary to diffuse through the given space in order to bring out unmistakably the spectrum of the metal. These investigators were able to state that the presence of no more than $\frac{1}{1000}$ of a milligram of the potassium salt is sufficient to give to the flame the characteristic spectrum of the element. By similar methods, the delicacy of the tests for lithium carbonate and sodium chlorate were shown to be a thousand times and three thousand times as delicate respectively. The practical detection of lithium and sodium spectroscopically is extremely easy and satisfactory, the only difficulty being that the exceeding delicacy of the sodium test and the ubiquitousness of sodium salts often make a decision doubtful as to whether that element is present appreciably in the substance under examination, or by accident. With potassium the case is different.

Detection of Potassium. Gooch and Hart † have succeeded in showing that while the simple method in vogue for developing the luminosity of lithium and sodium — the dipping of a single loop of platinum wire in the liquid or solid substance, and the placing of the loop in the Bunsen flame — is unsatisfactory because so great a proportion of the material is dispersed before the heat of the flame effects the dissociation of the salt, much better results may be obtained by making use of more powerful flames and substituting for the single loop the hollow coils of platinum wire first recommended by Truchot ‡ for the quantitative determination of lithium. Such coils are easily made by winding the wire somewhat obliquely about a rod of suitable size, pressing the

* See page 106.

† F. A. Gooch and T. S. Hart, Am. Jour. Sci., [3], xlii, 448.

‡ Compt. rend., lxxviii, 1022.

coils close together, and gathering the free ends into a twisted handle. The size of the coils is adjustable without difficulty, so that each coil may be made to hold almost exactly any appropriate amount, and to take up this amount with very little variation in successive fillings, provided only that the precaution be taken in the process of filling to plunge the coil while hot into the liquid, and to keep its axis inclined obliquely to the surface of the liquid while withdrawing it.

The coils, after use, may be conveniently cleaned by heating them in the flame of an annular burner beneath which is burned in a small lamp a 5 per cent solution of chloroform in alcohol, the products of combustion of the alcohol and chloroform being conveyed to the interior of the flame from below by a glass funnel fitted by a cork to the tube of the burner. This arrangement of apparatus gives a hot, colorless flame through which hydrochloric acid is constantly diffused in condition to clean the wires completely and without attention. How closely the capacity of such coils may be adjusted, and how uniformly they may be filled, is shown in the figures of the accompanying record.

Capacity of Coils.

	I. grm.	II. grm.	III. grm.	IV. grm.	V. grm.	VI. grm.
Weight of filled coil........	0.1996	0.2780	0 2794	0.2844	0.3572	0.3296
Weight of filled coil........	0 1996	0.2780	0.2794	0.2845	0.3571	0.3296
Weight of filled coil.......	0 1996	0.2780	0 2794	0.2844	0.3572	0.3298
Weight of filled coil........	0 1996	0.2780	0.2794	0.2845	0.3571	0.3298
Weight of filled coil	0.1996	0.2781	0 2794	0.2844	0 3571	0.3296
Weight of empty coil.......	0.1986	0.2760	0 2764	0.2804	0 3521	0.3100
Weight of contents (mean)..	0.0010	0.00202	0.0030	0.00404	0 00504	0 01968

It is plain that these coils afford simple means of taking up known amounts of material in solution. By gentle heating the liquid may be evaporated and the solid material left thinly spread and in condition to be acted upon with effect when brought to the flame. The evaporation may be conducted with little danger of loss of material by holding the handle of the coil across the flame with the coil proper at a safe distance outside; or, preferably, by exposing the coils over a hot radiator, the handles resting upon a flat asbestos ring.

In test experiments with such coils, use was made of a Müncke burner giving a flame 3 cm. wide at the base and 20 cm. in height.

The coil was introduced, after thorough drying, just within the outer mantle, on the side next the spectroscope, with the axis transverse to the slit of the spectroscope and the handle across the body of the flame. The spectroscope used was a well-made single-prism instrument with adjustable slit and scale. Observations were made in the ordinary diffused light of the laboratory, with the eye in use shielded, the eye not in use covered, and the scale illuminated to the lowest degree of visibility.

Upon experimenting with the apparatus described, it was found that coils holding 0.02 grm. of water, measuring 2 mm. in diameter by 1 cm. in length, made of No. 28 wire (0.32 mm. in diameter), and wound in about thirty turns, were well adapted to the purpose. With these coils and the flame adjusted to a height of 20 cm., $\frac{1}{750}$ mg. of potassium to the coilful produces a line distinctly visible with a slit of 0.18 mm., and $\frac{1}{1000}$ mg. with a slit of 0.23 mm.; and it is evident that this practical method of producing the spectrum of potassium gives results of a delicacy approaching that indicated in the experiments of Bunsen and Kirchhoff.

These determinations were made with pure potassium chloride carefully prepared from the chlorate, but in practical analysis it almost always happens that sodium is also present. Experiments were therefore made to determine the influence of varying amounts of the latter upon the visibility of the potassium line. The dilution of the potassium chloride was adjusted nearly to the last limit of visibility, so that a coilful of the liquid should contain $\frac{1}{700}$ mg., or $\frac{1}{1000}$ mg. of the element, according as the slit was 0.18 mm. or 0.23 mm. wide; to this solution were added weighed amounts of pure sodium chloride twice reprecipitated and washed by hydrochloric acid; and the spectroscopic tests were carried out as before, the sodium line being kept within the field of view with the potassium line.

It is plain from the results (p. 83) that a considerable amount of sodium may be present in the flame, when the sodium line is in full view in the spectrum, and the slit adjusted to nearly the lowest limit of visibility of pure potassium, without interfering with the appearance of the potassium line, but that a quantity of sodium amounting to a hundred times that of the potassium may be sufficient to overpower entirely the spectrum of the potassium. The inference is plain that the proportion of sodium to potassium should not be permitted to reach 100:1 when it is desirable to bring out the full delicacy of the spectroscopic test with the sodium line in the field of view. When too great a proportion of sodium is present, its influence may be moderated by throwing the sodium line out of view, if the instrument in use possesses the necessary adjustment; otherwise, it is easy to effect a partial separation of the sodium chloride from the potassium chloride, before bringing the solution to the test, by precipitating with alcohol, and experience shows that the delicacy of the test for potassium is not impaired materially by such treatment of the mixed chlorides.

Effect of Sodium upon the Potassium Line.

Weight of K in a coil-ful. mg.	Weight of Na in a coilful. mg	Ratio of Na : K.	Width of slit. mm.	Number of trials.	Characteristic of line.
0.0010	0 0000	0 : 1	0.23	3	Visible.
0.0010	0 0020	2 : 1	0.23	3	Visible.
0 0010	0 0100	10 : 1	0.23	3	Visible.
0.0010	0.0200	20 : 1	0.23	3	Visible.
0 0010	0.0400	40 : 1	0 23	3	Visible.
0.0010	0.0500	50 : 1	0.23	4	Very faint or none.
0.0010	0.1000	100 : 1	0 23	3	None.
0.0010	0 2000	200 : 1	0.23	3	None.
0 0014	0.0000	0 : 1	0 18	3	Visible.
0.0014	0 0560	40 : 1	0 18	3	Visible.
0.0014	0 0700	50 : 1	0.18	3	Visible.
0 0014	0.1400	100 : 1	0 18	2	Visible.
0.0014	0.1400	100 : 1	0.18	2	None.

Certain experiments, in which the same method of manipulation was applied to the determination of potassium salts other than the chloride, indicate that the test is less delicate in the case of the sulphate, but rather more delicate in the case of the carbonate, the red line of potassium showing unmistakably in the latter case when only $\frac{1}{2000}$ of a milligram of potassium was present.

Determination of Potassium. It has also been shown by Gooch and Hart * that the quantitative determination of small amounts of potassium may be successfully accomplished by bringing an unknown solution of potassium chloride to the measured volume

* Loc. cit.

at which the residue left after evaporating the contents of a coil gives a line of the same strength as that produced by the residue of a coilful of a standard solution of potassium chloride and so determining the concentration of the unknown solution of finally measured volume. It is convenient to use several coils adjusted to the same capacity, and to clean, fill, dry and ignite them before the spectroscope in the manner previously described. From time to time the capacity of the coils should be readjusted, or else final comparison tests should be made with a single coil. It is essential that the eye of the observer should be kept as nearly as possible in the same condition of sensitiveness and in the same position in making the comparisons, and best to hold the eye at the observing telescope during the entire interval between the exposures, shading it carefully, and to light the comparison scale of the spectroscope to the faintest possible visibility sufficient to fix exactly the position in which the line is to be sought. It is important, too, that the trials of the test and standard should come as closely together as possible in point of time. The observations of a series should be made by the same individual, the preparation and exposure of the wires being made by another. It is not possible to attain the best results in such work single-handed. The dilution of the test solution is made conveniently, and with sufficient accuracy, in 100-cm.^{3} cylinders graduated to half cubic centimeters. It is advantageous to take a standard solution which corresponds to the presence of $\frac{1}{500}$ mg. of potassium to the coilful, and set the slit at a width sufficient to give lines for comparison bright enough to be visible without much effort. The mode of proceeding is to dilute the test solution until the line given by the potassium contained in a coilful is of the same brightness as that given by the same quantity of the standard solution. From the final volume of the test-solution the quantity of potassium present in it is directly calculable; for, since any given volume of the test solution at its final dilution contains exactly the same amount of potassium as the same volume of the standard solution, it is only necessary to multiply the number expressing the volume in cubic centimeters of the test solution by that of the weight in grams of the potassium contained in one cubic centimeter of the standard in order to obtain the weight in grams of potassium in the whole test solution.

The following is the record of the comparison of two unknown test solutions of pure potassium chloride with a standard solution containing 0.0001 grm. of the same salt to 1 cm.[3]

Determination of Potassium in Pure Potassium Chloride.

Experiment I.		Experiment II.	
Volume of test solution. cm.³	Characteristic of line compared with standard.	Volume of test solution. cm.³	Characteristic of line compared with standard.
20	Stronger.	30	Stronger.
50	Stronger.	60	Stronger.
100	Stronger.	82	Weaker.
110	Stronger.	70	Stronger.
120	Stronger.	76	Stronger.
150	Like.	78	Stronger.
200	Weaker.	80	Like.
160	Weaker.		
150	Like.		

Experiment I.		Experiment II.	
($150 \times 0.0001 = 0.0150$)		($80 \times 0.0001 = 0.0080$)	
Potassium found......	0 0150 grm.	Potassium found....	0 0080 grm.
Potassium taken......	0.0150 grm.	Potassium taken...	0.0080 grm.
Limits on either side.	0 0120 grm. 0.0160 grm.	Limits on either side	0.0078 grm. 0.0082 grm.
Error.....	0.0000 grm.	Error	0.0000 grm.

These results with the pure potassium salt show a degree of accuracy quite unexpected. In the former no attempt was made to approximate as closely as possible to the limits of dilution on both sides of the condition of equal brightness in test and standard, but in the latter great care was taken in this respect and the possible error does not exceed two and a half per cent of the entire amount of potassium involved. Similar experiments with potassium chloride in presence of varying amounts of sodium chloride, the sodium line being turned from the field of view, led to a recognition of the fact that sodium chloride tends to increase the brilliance of the potassium line, — the maximum strengthening effect of about 20 per cent occurring when the amount of sodium chloride stands to that of the potassium in the ratio of 10 : 1, — a phenomenon due to the chemical action of sodium dissociated in the flame. The effect of ammonium chloride, and of hydrochloric acid, in destroying the potassium light is well known, and is due, presumably, in very large degree, to prevention of the dissociation of the potassium chloride. The dissociated sodium should naturally reënforce the disintegrating action of heat upon the potassium chloride.

The complication introduced by the presence of any certain amount of the sodium salt in the test may be obviated by the addition of the same amount of the sodium salt to the standard, and experience shows that an unknown amount of the sodium salt in the test may be matched with a degree of accuracy sufficient for the end in view. The determination of potassium in the presence of sodium is performed, therefore, in three stages: first, the test solution is diluted until its potassium line matches approximately that of the standard made to contain in 1 cm.^{3} 0.0001 grm. of potassium and 0.0010 grm. of sodium chloride; secondly, sodium chloride is added to the solution thus diluted until the sodium lines of test and standard are brought to equality; and, finally, the potassium lines of test solution and standard solution are again brought into comparison. Following are the records of experiments made in this manner.

Determination of Potassium in Presence of Sodium.

Experiment I.

Part I.			Part II.			Part III.		
Volume of test solution. cm.^{3}	Width of slit. mm.	Characteristic of potassium line as compared with standard.	NaCl in 100 cm.^{3} of test solution. grm.	Width of slit. mm.	Characteristic of sodium line as compared with standard.	Volume of test solution. cm.^{3}	Width of slit. mm.	Characteristic of potassium line as compared with standard.
30	0.23	Stronger.	0.01*	0.18	Weaker.	108	0.23	Weaker.
70	0.23	Stronger.	0.03	0.18	Weaker.	108	0.23	Stronger.
100	0.23	Weaker.	0.05	0.18	Weaker.			Weaker.
			0.08	0.18	Weaker.	109	0.23	Stronger.
			0.09	0.18	Weaker.			Like.
			0.10	0.18	Like.			

* Originally present.

The test solution having been accidently over-diluted, its strength was increased by the addition of 0.0010 grm. of potassium, and this amount was added in the computation below to that originally in the test solution.

$$(109 \times 0.0001 = 0.0109)$$

Potassium found .... 0.0109 grm.
Potassium taken............ 0.0110 grm.
Error...................... 0.0001 grm. = 0.9 per cent.

Determination of Potassium in Presence of Sodium.

Experiment II.

Part I.			Part II.			Part III.		
Volume of test solution. cm.³	Width of slit. mm.	Characteristic of potassium line as compared with standard.	NaCl in 100 cm.³ of test solution. grm.	Width of slit. mm.	Characteristic of sodium line as compared with standard.	Volume of test solution. cm.³	Width of slit. mm.	Characteristic of potassium line as compared with standard.
40	0.23	Stronger.	0.025*	0.18	Weaker.	160	0.23	Stronger.
100	0.23	Stronger.	0.050	0.18	Weaker.	180	0.23	Stronger.
160	0.23	Weaker.	0.085	0.18	Weaker.	190	0.23	Stronger.
			0.100	0.18	Weaker.	200	0.23	Stronger.
			0.110	0.18	Like.	205	0.23	Weaker.
						210	0.23	Weaker.

* Originally present.

$$\left(\begin{matrix} 205 \times 0.0001 = 0.0205 \\ 200 \times 0.0001 = 0.0200 \end{matrix}\right\} \text{mean} = 0.02025 \Bigg)$$

Potassium found........ 0.02025 grm.
Potassium taken........ 0.02000 grm.
Error.................. 0.00025 grm. = 1.25 per cent.

Experiment III.

Part I.			Part II.			Part III.		
Volume of test solution. cm.³	Width of slit. mm.	Characteristic of potassium line as compared with standard.	NaCl in 100 cm.³ of test solution. grm.	Width of slit. mm.	Characteristic of sodium line as compared with standard.	Volume of test solution. cm.³	Width of slit. mm.	Characteristic of potassium line as compared with standard.
40	0.23	Stronger.	0.045*	0.18	Weaker.	110	0.23	Stronger.
80	0.23	Stronger.	0.082	0.18	Like.	120	0.23	Stronger.
100	0.23	Stronger.				130	0.23	Like.
110	0.23	Like.						

* Originally present.

$(130 \times 0.0001 = 0.0130)$

Potassium found........ 0.0130 grm.
Potassium taken........ 0.0140 grm.
Error.................. 0.0010 grm. = 7 per cent.

Determination of Potassium in Presence of Sodium.

Experiment IV.

Part I.			Part II.			Part III.		
Volume of test solution. cm.^{3}	Width of slit. mm.	Characteristic of potassium line as compared with standard.	NaCl in 100 cm.^{3} of test solution. grm.	Width of slit. mm.	Characteristic of sodium line as compared with standard.	Volume of test solution. cm.^{3}	Width of slit. mm.	Characteristic of potassium line as compared with standard.
						First.		
30	0.23	Stronger.	0 05*	0 18	Weaker.	100	0.23	Stronger.
90	0.23	Stronger.	0 07	0.80	Weaker.	120	0.23	Stronger.
100	0.23	Weaker. Like.	0.09	0 18	Weaker.	130	0.23	Stronger.
			0 10	0 18	Like. Stronger. Stronger.	140	0 23	Stronger. Weaker.
						Second.		
						120	0 23	Stronger.
						140	0 23	Stronger.
						150	0.23	Stronger. Weaker.
						160	0.23	Weaker.

* Originally present.

	First.	Second.
	(140 × 0 0001 = 0.0140)	(150 × 0.0001 = 0.0150)
Potassium found...	0.0180 grm.	0 0150
Potassium taken...	0.0150 grm.	0.0150
Error....	0.0010 grm. = 7 per cent	0.0000

Though not accurate in the highest degree when considerable amounts of potassium are to be estimated, the method is reasonably applicable to the determination of small quantities of that element.

The Separation and Determination of Potassium as the Perchlorate.

Kreider's method* for the preparation of perchloric acid has greatly facilitated the use of the perchlorate method for the estimation of potassium. This method — consisting essentially in the regulated heating of sodium chlorate (readily obtained in the market), treatment of the residue with strong hydrochloric acid to yield a precipitate of sodium chloride and a solution of perchloric acid containing a small amount of sodium chloride,

* D. Albert Kreider, Am. Jour. Sci., [3], xlix, 443.

and filtration of the mixture upon asbestos — may be detailed as follows:

A convenient quantity of sodium chlorate, from 100 to 300 grm., is melted in a glass retort or round-bottomed flask and gradually raised to a temperature at which oxygen is freely but not too rapidly evolved, and kept at this temperature for one and a half or two hours, until the thickening of the mass indicates the conversion of the chlorate to chloride and perchlorate; or, the retort may be connected with a gasometer and the end of the reaction determined by the volume of oxygen expelled, according to the equation

$$2\,NaClO_3 = NaCl + NaClO_4 + O_2.$$

The product thus obtained is washed from the retort to a capacious evaporating dish, where it is treated with sufficient hydrochloric acid to effect the complete reduction of the residual chlorate, which, if the ignition has been carefully conducted with well-distributed heat, will be present in but small amount. It is then evaporated to dryness on the steam bath, or more quickly over a direct flame, and with but little attention until a point near to dryness has been reached. Then stirring is found of great advantage in facilitating the volatilization of the remaining liquid and in breaking up the mass of salt; otherwise the perchlorate seems to solidify with a certain amount of water, and removal from the dish, without moistening and reheating, is impossible.

After trituration of the residue in a porcelain mortar, an excess of the strongest hydrochloric acid is added to the dry salt, preferably in a tall beaker, where there is less surface for the escape of hydrochloric acid and from which the acid may be decanted without disturbing the precipitated chloride. If the salt has been reduced to a very fine powder, by stirring energetically for a minute, the hydrochloric acid will set free the perchloric acid and precipitate the sodium as chloride, which in a few minutes settles, leaving a clear solution of the perchloric acid with the excess of hydrochloric acid. The clear supernatant liquid is then decanted upon asbestos in a perforated crucible, through which it may be rapidly drawn with the aid of suction, and the residue is again treated with the strongest hydrochloric acid. The liquid is again decanted, the salt is finally brought upon the filter, where it is washed with a little strong hydrochloric acid. A large platinum cone will be found more convenient than the crucible, because of its greater capacity and filtering surface.

The filtrate, containing the perchloric acid with the excess of hydrochloric acid and the small per cent of sodium chloride which is soluble in the latter, is then evaporated over the steam bath till all hydrochloric acid is expelled and the heavy white fumes of perchloric acid appear, when it is ready for use in potassium determinations. Evidently the acid is not chemically pure because the sodium chloride is not absolutely insoluble in hydrochloric acid; but a test with silver nitrate proves that the sodium, together with any other bases which may have gone through the filter, has been completely converted into perchlorate, and, unless the original chlorate contained potassium or the acid had been exposed to the fumes of ammonia, the residue of evaporation, which does not exceed 0.04 grm. in weight to 1 $cm.^3$, is easily and completely soluble in 97 per cent alcohol. Perchloric acid thus prepared was found to contain 0.9831 grm. of free anhydrous acid in 1 $cm.^3$.

Should the sodium chlorate used in the process contain potassium as an impurity, the mixture of sodium perchlorate and chloride, after being treated with hydrochloric acid for the reduction of the residual chlorate, is reduced to a fine powder, and well digested with 97 per cent alcohol, which dissolves the sodium perchlorate, but leaves the chloride as well as any potassium salt insoluble. The alcoholic solution of the perchlorate is then distilled from a large flask until the perchlorate begins to crystallize, when the heat is removed and the contents quickly emptied into an evaporating dish. The mixture is evaporated to dryness on the steam bath and the residue is treated with strong hydrochloric acid for the separation of the perchloric acid in the manner described above.

In applying perchloric acid, prepared by Kreider's method, to the determination of potassium according to the treatment suggested by Caspari,* very satisfactory results were obtained. Briefly, the method is as follows: The substance, free from sulphuric acid, is evaporated to the expulsion of free hydrochloric acid. The residue, stirred with 20 $cm.^3$ of hot water and then

* Zeit. angew. Chem., 1893, 68.

treated with perchloric acid in quantity not less than one and one-half times that required by the bases present, is evaporated with frequent stirring to a thick, sirup-like consistency, again dissolved in hot water and evaporated with continued stirring until all hydrochloric acid has been expelled and the fumes of perchloric acid appear. Further loss of perchloric acid is to be compensated for by addition of more. The cold mass is then well stirred with about 20 cm.^{3} of wash alcohol — 97 per cent alcohol containing 0.2 per cent by weight of pure perchloric acid — with precautions against reducing the potassium perchlorate crystals to too fine a powder. After settling, the alcohol is decanted on the asbestos filter and the residue is again washed with about the same amount of wash alcohol. The residual salt, freed from alcohol by gently heating, is dissolved in 10 cm.^{3} of hot water and a little perchloric acid. The solution is evaporated once more with stirring until fumes of perchloric acid rise. The precipitate is treated with 1 cm.^{3} of wash alcohol, transferred to the asbestos, preferably by a policeman, and washed with pure alcohol; the whole process requiring about 50 to 70 cm.^{3} of alcohol. It is then dried at about 130° C. and weighed.

The substitution of a perforated crucible for the truncated pipette employed by Caspari is advantageous; and asbestos capable of forming a close, compact felt should be selected, inasmuch as the perchlorate is in part unavoidably reduced, during the necessary stirring, to so fine a condition that it tends to run through the filter when under pressure.

A number of determinations made of potassium unmixed with other bases or nonvolatile acids are recorded in the following table:

Potassium in the Pure Salt.

KCl taken. grm.	Volume of filtrate. cm.^{3}	$KClO_4$ found. grm.	Error on $KClO_4$. grm.	Error on KCl. grm.	Error on K_2O. grm.
0.1000	54	0 1851	−0.0008	−0.0004	−0.0003
0.1000	58	0 1854	−0.0005	−0.0002	−0.0002
0.1000	51	0 1859	0.0000	0.0000	0.0000
0.1000	50	0 1854	−0 0005	−0 0002	−0.0002
0.1000	48	0.1859	0.0000	0.0000	0.0000
0.1000	52	0.1854	−0.0005	−0.0002	−0.0002

As Caspari has pointed out, sulphuric acid must be removed by precipitation as barium sulphate before the treatment with perchloric acid is attempted, and unless the precipitation is made in a strongly acid solution, some potassium is carried down with the barium. Phosphoric acid need not be previously removed; but to secure a nearly complete separation of this acid from the potassium, a considerable excess of perchloric acid should be left upon the potassium perchlorate before it is treated with the alcohol. When these conditions are carefully complied with, fairly good results may justly be expected. Below are given a number of the results obtained.

Potassium in Mixtures of Salts.

Compounds taken. grm.	Volume of filtrate. cm.^{3}	$KClO_4$ found. grm.	Error on $KClO_4$. grm.	Error on KCl. grm.	Error on K_2O. grm.
KCl=0.1000, $CaCO_3$=0.13, $MgSO_4$=0.13, Fe_2Cl_6=0.05, $Al_2(SO_4)_3$=0.05, MnO_2=0.05, $HNa_2PO_4.12H_2O$=0.40	50	0.1887	+0.0028	+0.0014	+0.0009*
	82	0.1875	+0.0016	+0.0008	+0.0005*
	80	0.1861	+0.0002	+0.0001	+0.0001†
	80	0.1843	−0.0016	−0.0008	−0.0005†
	92	0.1839	−0.0020	−0.0010	−0.0006†
	60	0.1854	−0.0005	−0.0002	−0.0002†

* The residue showed phosphoric acid plainly when tested.

† Only traces of phosphoric acid found in the residue.

In the last three experiments, in which the amount of perchloric acid employed was about three times that required to unite with the bases present, the phosphoric acid subsequently found with the potassium was hardly enough to appreciably affect the weight.

The Estimation of Potassium as the Pyrosulphate.

Browning* has shown that potassium sulphate, like sodium sulphate† when treated with sulphuric acid and submitted to a temperature ranging between 250° and 270°, takes very definitely the form of the pyrosulphate, $K_2S_2O_7$, and that potassium may be estimated as that salt. Under similar conditions cæsium and rubidium remain in the form of acid sulphates. Results of the procedure are given below in comparison with the results obtained upon moistening the pyrosulphate with ammonium hydroxide and igniting strongly to form the neutral sulphate.

* Philip E. Browning, Am. Jour. Sci., [4], xii, 301.

† See page 79.

Potassium Sulphates by Ignition.

KCl taken.	$K_2S_2O_7$ calculated.	$K_2S_2O_7$ found.	Error.	K_2SO_4 calculated.	K_2SO_4 found.	Error.
grm.	grm.	grm.	grm.	grm.	grm.	grm.
0.2172	0.3704	0.3698	−0 0006		. .	.. ..
0.1706	0.2909	0 2886	−0 0023	0 1993	0 1972	−0.0021
0 1192	0.2032	0.2022	−0 0010	0 1393	0 1381	−0.0012
0.1074	0.1830	0.1823	−0 0007		.	.. . .
0.1096	0.1868	0.1860	−0 0008		. ..	.. . .

The Volumetric Estimation of Potassium as the Cobalti-nitrite.

The use of sodium cobalti-nitrite to estimate potassium has been described by R. H. Adie and T. B. Wood,* who show results fairly accurate and favorably comparable with those obtained by the platinic chloride gravimetric method. In the process worked out by these investigators a solution of a potassium salt containing the equivalent of 0.5 per cent to 1 per cent of K_2O is acidified with acetic acid and precipitated by an excess of sodium cobalti-nitrite.† The mixture is allowed to stand at least a few hours, preferably over night, and is then filtered through a perforated crucible fitted with an asbestos felt. The precipitate is washed with 10 per cent acetic acid. According to Sutton, it is important that the precipitation should be made in a solution containing the equivalent of 0.5 per cent to 1 per cent of K_2O, since in solutions of lower concentration the precipitate comes down in a condition in which it is apt to run through the filter in washing. The precipitate is then decomposed by boiling in dilute sodium hydroxide, and the cobalt is removed as the hydroxide by filtration. The nitrites, which are a measure of the potassium in the precipitate, are estimated by titrating with standard potassium permanganate. Adie and Wood found by analysis that the composition of the potassium salt precipitated in presence of the excess of sodium cobalti-nitrite is represented by the formula $K_2NaCo(NO_2)_6.H_2O$, and that in their method a cubic centimeter of strictly $n/10$ potassium permanganate is equivalent to 0.000785 grm. of K_2O.

This process has been studied by Drushel‡ with a view to determining the best conditions for precipitating and filtering

* Jour. Chem. Soc., lxxvii, 1076; Sutton's Vol. Anal., 9th ed., page 62.

† Ibid.

‡ W. A Drushel, Am. Jour. Sci., [4], xxiv, 433.

the potassium cobalti-nitrite, and to shortening the work by oxidizing the precipitated cobalti-nitrite with potassium permanganate without the preliminary decomposition of the precipitate and removal of cobalt recommended by Adie and Wood, the excess of permanganate being reduced by standard oxalic acid, and the remaining oxalic acid titrated to color. In this treatment trivalent cobalt is reduced to the bivalent condition; the oxygen thus made available is equivalent to one-twelfth of that necessary to oxidize the nitrites. The factor used, therefore, in calculating the results from the direct titration should be twelve-elevenths of that given by Adie and Wood; that is, in titrating the precipitate without first separating the cobalt one cubic centimeter of strictly $n/10$ potassium permanganate is equivalent to 0.000857 grm. of K_2O.

By repeated experiments it was found that difficulty in filtration, as well as the necessity for allowing the precipitate to stand over night, may be avoided by evaporating the mixture nearly to dryness on the steam bath after adding the sodium cobalti-nitrite solution in considerable excess. Upon cooling the pasty residue becomes hard and dry. When treated with cold water the excess of sodium cobalti-nitrite dissolves, and the insoluble portion may be collected and freely washed without showing a tendency to pass through the filter.

Potassium in the Pure Salt. The application of the cobalti-nitrite method as worked out by Drushel is as follows: The solution of a potassium salt, containing not more than 0.2 grm. K_2O and free from ammonium salt, is treated with a rather large excess of sodium cobalti-nitrite solution acidified with acetic acid, and evaporated to a pasty condition over the steam bath. It is then cooled, treated with 50 cm.^{3} to 100 cm.^{3} of cold water and stirred until the excess of sodium cobalti-nitrite is dissolved, allowed to settle, and decanted through a perforated crucible fitted with an asbestos felt. The precipitate is washed two or three times by decantation, after which it is transferred to the crucible and thoroughly washed with cold water.* In the meantime a measured excess of standard potassium permanganate is diluted to ten times its volume and heated nearly to boiling.

* It was found later that a half-saturated sodium chloride solution is preferable to cold water for washing the precipitate, since it permits the use of a coarser asbestos felt in filtering without danger of loss.

Into this the precipitate and felt are transferred and stirred, after which the crucible is also put into the solution, since particles of the precipitate stick persistently to the sides of the crucible. After the oxidation has proceeded five or six minutes manganese hydroxide separates out and the color of the solution darkens. At this point 5 cm.^{3} to 25 cm.^{3} of sulphuric acid [1 : 7] are added, and the solution, after stirring, is allowed to stand a few minutes. Then a measured amount of standard oxalic acid, containing 50 cm.^{3} of strong sulphuric acid per liter, is run in from a burette, with care to add an excess. The temperature is maintained a little below the boiling point until the solution becomes colorless and the manganese hydroxide has completely dissolved. Titration is then effected by permanganate in the usual manner.

The results of the experimental tests with potassium chloride alone and in presence of salts of the calcium group are given below.

Potassium in Pure Potassium Chloride.

K_2O taken as KCl.	K_2O found.		Error in K_2O.	
	Gravimetrically.	Volumetrically.	Gravimetrically.	Volumetrically.
grm.	grm.	grm.	grm.	grm
0 0237	0.0240	0 0238	+0.0003	+0.0001
0.0237	0.0243	0 0242	+0.0006	+0 0005
0.0354	0.0359	0 0355	+0.0004	0.0000
0 0474	0.0478	0 0471	+0.0004	−0.0003
0.0048	0.0048	0.0050	0.0000	+0.0002
0.0024	0.0024	0 0023	0.0000	+0.0001
0.0005	......	0 0006	......	+0.0001
0.0015	......	0 0017	......	+0.0002
0 0355	......	0 0355	......	0.0000

In the first six experiments of this series the precipitate was dried at 115°, weighed, and then treated with permanganate.

Potassium in Mixtures of Salts.

$CaCl_2$.	$MgCl_2$ taken.	$BaCl_2$ taken.	$Sr(NO_3)_2$.	K_2O taken.	K_2O found.	Error.
grm.	grm.	grm.	grm.	grm.	grm.	grm.
0.2000	0.2000	......	......	0 0005	0 0007	+0.0002
0.3000	0 5000	......	......	0 0237	0 0234	−0 0003
0.5000	1.0000	......	......	0.0829	0 0824	−0 0005
0 5000	1.0000	......	0.5000	0.0711	0.0737	+0 0026
0.5000	1 0000	0.5000	0.5000	0.0474	0.0493	+0.0019
0.5000	1.0000	0 5000	......	0.0237	0.0251	+0.0014
0.5000	1.0000	.. .	......	0.0711	0.0713	+0.0002

The salts of calcium and magnesium do not influence the accuracy of the process, while the presence of salts of barium and strontium tends to high results.

Potassium in Fertilizers. The method* is applicable to the estimation of potassium in fertilizers. The process, as laid down, is as follows:

Ten grams of the fertilizer are placed in a 500-cm.³ flask, 300 cm.³ of water added, the contents boiled for 30 minutes, and ammonia water added to slight alkalinity. Enough ammonium oxalate is added to precipitate all the calcium, and, after cooling, the solution is made up to the mark on the neck of the flask and well shaken. The solution is then filtered through a dry filter into a dry flask, and 50-cm.³ portions of the filtrate are transferred with a pipette to platinum dishes, for estimation by the cobalti-nitrite method. After evaporating these portions to half their volume over the steam bath, 1 cm.³ of sulphuric acid [1:1] is added and the evaporation is continued as far as possible over the steam bath, and finally over a low flame. After the danger of spattering is over, the flame is increased and the charred organic matter is burned off, finally, over the blast lamp. The potassium sulphate is dissolved by adding a little water and heating over the steam bath, and the potassium is estimated as in the previously described treatment of the pure potassium salt.†

Potassium in Mixed Fertilizers.

Number.	K_2O by platinum chloride method.		K_2O by cobalti-nitrite method.	Water-soluble P_2O_5 in sample.
	Per cent.	Per cent.	Per cent.	Per cent.
1	5.22	5 18	5.18	4.16
2	6.53	6 56	6.56	3.10
3	2.23	2 24	2.24	7.82
4	8.68	8.64	8.78	0 94
5	6.37	6 42	6.38	6 62
6	6.08	6.13	6.13	5 61
7	4.08	4 02	4.02	3.15
8	4.62	4.66	4 67	2.43
9	1.68	1.67	1.77	6.03

* W. A. Drushel, Am. Jour. Sci., [4], xxiv, 437.

† See page 94.

In the preceding table are given results obtained by the method with nine fertilizers, and, for comparison, results (by two analysts) by the platinic chloride method.

Potassium in Soils.

In applying the cobalti-nitrite method to the estimation of potassium in soils, the general procedure may be outlined as follows:*

A weighed amount of dry soil is extracted with an excess of hydrochloric acid over the steam bath. The excess of acid is removed from the extract by evaporation. The bases which might interfere with the process are removed with sodium carbonate or ammonium hydroxide and ammonium oxalate. Ammonium salts and organic matter are removed by ignition.

Potassium in Soils.

Character of soil.	Soil taken. grm.	K_2O found. Platinum chloride method. grm.	K_2O found. Cobalti-nitrite method. grm.	K_2O found. Per cent.
Clay (1)	2 5	....	0.0028	0.11
Clay (2)	2 5	....	0.0035	0 14
Clay (3)	2 5	0.0035	...	0 14
Clay (1)	2.5	...	0.0100	0 39
Clay (2)	2 5	.....	0.0092	0 37
Clay (3)	2 5	0.0093	......	0.37
Loam (1)	2 5	....	0 0074	0 30
Loam (2)	2 5	. .	0.0068	0 27
Loam (3)	2 5	0.0075	......	0.30
Loam (1)	2 5	......	0 0060	0.24
Loam (2)	2 5	......	0.0058	0.23
Loam (3)	2 5	0 0058	......	0.23
Gravel (1)	2 5	. .	0 0042	0.17
Gravel (2)	2 5	0.0045	......	0.18
Gravel (1)	2.5	......	0.0047	0.19
Gravel (2)	2 5	......	0.0044	0.18
Gravel (3)	2 5	0.0050	......	0 20
Clay gravel (1)	2 5	.....	0.0048	0.19
Clay gravel (2)	2 5	0 0046	......	0.18
Clay gravel (3)	2 5	......	0.0045	0.18
Clay gravel (4)	2.5	......	0.0040	0.16
Clay gravel (5)	2 5	......	0.0044	0.18

Small amounts of phosphoric acid do not interfere. The residue is dissolved in a little water and a few drops of acetic acid,

* W. A. Drushel, Am. Jour. Sci., [4], xxvi, 329.

and the mixture evaporated with an excess of sodium cobalti-nitrite to a pasty condition, stirred up with cold water, and filtered upon asbestos in a perforated crucible. The precipitated potassium sodium cobalti-nitrite is washed with a half-saturated solution of sodium chloride, and treated with an excess of permanganate in hot dilute solution. The color of the permanganate is destroyed by an excess of standard acidulated oxalic acid, and the excess of oxalic acid titrated to color with permanganate.

The test analyses show an excellent agreement with one another and with the results of the platinum chloride method.

By using 10 grm. of soil for each estimation it should be possible to attain a higher degree of accuracy.

Potassium in Animal Fluids: Urine, Blood, Lymph, Milk. Drushel has also studied the application of the cobalti-nitrite method to the determination of potassium in animal fluids.* Of the constituents of urine, ammonia and the organic substances, especially urea, are the only ones which should interfere with the volumetric method as previously described. To remove these without the loss of potassium is apparently the only new problem in connection with the estimation of potassium in urine, and this is accomplished by the following procedure: Aliquot portions of urine of 10 to 50 cm.^{3} each are measured with pipettes or a burette into small platinum evaporating dishes, and evaporated to dryness over the steam bath in a good draft hood. The residues are best treated by acting upon them with 5 cm.^{3} to 10 cm.^{3} of a 9:1 nitric-sulphuric acid mixture in an evaporating dish kept covered until the first violent oxidation is over, evaporation to dryness, and ignition.

By this treatment the ignition of the residue from 50 cm.^{3} of urine may be readily made in 30 minutes without loss of material. The residue thus prepared is treated with a little water and a few drops of acetic acid to dissolve the alkalies, and from this point the process is carried out as in the application of the cobalti-nitrite method to the estimation of potassium in pure salts, as previously described.

The results obtained in the application of this method to a number of specimens of human urine are given in the following table.

* W. A. Drushel, Am. Jour. Sci., [4], xxvi, 555.

Potassium in Urine.

Urine taken.	Specific gravity.	Volume in 24 hours.	K found. Platinum chloride method.*	K found. Cobalti-nitrite method.	K in 24 hours.
cm.³		cm.³	grm.	grm.	grm.
10	1 025	950	.. .	0.0293	2 78
10	.....	...	..	0.0292	2 77
10	.....	....	0 0293	. ...	2 78
10	.....	...	0 0292	. ...	2 77
25	1.025	950	.. .	0.0740	2 81
25	...	. .	0.0747	.. ..	2 84
25	.....	....	......	0.0740	2 81
20	1.025	910	. ...	0.0757	3.44
20	.....	....	0.0752	. ..	3.42
20	.. .	....	.. ...	0.0764	3 47
20	1.024	1130	. ...	0 0663	3 74
20	....	. .	. ..	0 0662	3.74
20	.... .	. ..	0 0663	...	3 74
20	.....	.. .	0.0662	. ..	3.74
25	1.018	1500	.. ..	0.0425	2 55
50	.... .	.. .	0.0839	...	2.52
50	.... .	. ..	...	0 0843	2 53
25	.....	....	0 0424	. ...	2 54

* Modified Lindo-Gladding method, after removal of P_2O_5.

An additional difficulty presents itself in the presence of a large amount of protein material which cannot be removed by coagulation and filtration without a considerable loss of potassium. This is particularly true of the blood, where most of the potassium is intimately associated with the protein of the corpuscles. It is necessary therefore to decompose protein material by oxidation. For this purpose the nitric-sulphuric acid mixture works less satisfactorily than treatment with concentrated nitric acid, digestion on the steam bath, dilution, treatment of aliquot portions by evaporation, gentle ignition, addition of sulphuric acid and a final ignition; or liquid bromine may be substituted for nitric acid in the first oxidation.

Results obtained for potassium in circulating fluids are given in the following table.

Potassium in Blood and Lymph.

Nature of fluid.	Amount taken.	K_2O found.		
		Platinum chloride method.*	Cobalti-nitrite method.	Per cent.
	grm.	grm	grm.	
Defibrinated pig's blood†.	10 89	0 0227	0.0227	0 21
	11 21	.	0 0228	0 20
	20 33	. .	0 0391	0 19
	10 16	. . .	0 0203	0 21
	10.85	. . . . .	0 0211	0.20
	11 03	.	0 0236	0 21
Sheep's blood‡....	30.00	0 0174	0.0174	0.058
	30 00	.	0.0179	0 060
	30 00	0 0181	0 0181	0 060
	30.00	.	0.0181	0 060
	30 00	. . .	0 0180	0.060
Serum of dog's blood‡....	10 11	. .	0.0024	0 024
	10 04	0 0024	0 0024	0 024
	10 07	. .	0.0023	0 034
Dog's lymph‡........ ...	10 28	0 0018	0 0018	0.018
	10 01		0 0019	0.019
	10 00	. . .	0 0020	0.020
	10 03	.	0.0019	0.019
	10 12	.	0.0019	0.019
	10 32	0 0022	0.0022	0.021

* Modified Lindo-Gladding method, after removal of Ca, Fe, and P_2O_5.
† Oxidized by bromine.
‡ Oxidized by nitric acid.

In the estimation of potassium in milk suitable amounts are evaporated to dryness, oxidized with concentrated nitric acid, again evaporated to dryness, and ignited gently until nearly all organic matter has been burnt. The residue is moistened with concentrated sulphuric acid and again ignited. In the residue thus obtained the potassium may be estimated by the cobalti-nitrite method as applied to pure salts of potassium.

Results obtained by the method in the analysis of cow's milk are given on the following page.

The modifications introduced by Drushel into the cobalti-nitrite process — evaporation nearly to dryness and oxidation of the nitrite without previous removal of the cobalt — add greatly to its usefulness. The necessity of long standing is avoided, the precipitate may be filtered and washed without trouble and the manipulation previous to titration is much shortened.

Potassium in Milk.

Milk taken.	K_2O found.		
	Platinum chloride method *	Cobalti-nitrite method.	Per cent.
grm.	grm	grm.	
25 8	. . . .	0 0413	0 16
25 8	.	0.0432	0 17
25 8	0.0428	.	0 17
51 6		0 0833	0 16
25 7		0 0454	0 18
25 7		0.0457	0 18
25 7	0 0451	. . .	0 18

* Lindo-Gladding method, after removal of Ca and P_2O_5.

For small amounts of potassium fairly accurate results are obtained by using the permanganate factor calculated from Adie and Wood's formula for potassium sodium cobalti-nitrite. Sutton has suggested that more accurate results may be secured by obtaining a factor empirically from a pure potassium salt. The results recorded above were obtained, however, by using the theoretical factor calculated from the formula of Adie and Wood, $K_2NaCo(NO_2)_6.H_2O$, their analyses of potassium sodium cobalti-nitrite having been verified* by the analysis of a carefully prepared salt.

The chief sources of error in the method appear to be the slight solubility of the potassium sodium cobalti-nitrite, one part in 25,000 to 30,000 parts of water at room temperature, and the tendency of the precipitate to include traces of sodium cobalti-nitrite.

The method requires less time and labor than the chloroplatinate method, and is applicable in the presence of substances which form no insoluble cobalti-nitrites and which neither oxidize oxalic acid nor reduce potassium permanganate.

* Am. Jour. Sci., [4], xxvi, 562.

Rubidium and Caesium.

The Spectroscopic Determination of Rubidium.

The method of manipulation previously described for the spectroscopic determination of small amounts of potassium has been adapted by Gooch and Phinney* to the similar determination of rubidium.

In the work upon potassium the observations of the red line were made in the ordinary laboratory in diffused light, but preliminary experimentation upon the rubidium spectrum immediately developed the fact that the blue lines are better to work by in the case of this element, and that a dark room becomes a necessity. For the experiments described pure rubidium chloride was prepared by many fractional precipitations by alcohol out of aqueous solutions, and in settling the question as to the coils which should be used the choice fell upon the size holding 0.02 grm. of water and made of the No. 28 wire, the superior stiffness of these and consequent constancy in capacity giving them the advantage over smaller coils of finer wire, though the latter are capable of bringing out greater sensitiveness of the reaction. It was found, for example, that under the most favorable conditions as to height of flame and width of slit, 0.0002 mg. of rubidium chloride produced the blue lines at the last limit of visibility when the larger and heavier coil was in the flame; with a coil holding 0.006 grm. of water and made of very fine wire the more immediate volatilization of the chloride so increased the delicacy of the spectroscopic reaction that it was possible to see the lines from 0.00005 mg. of the salt. These figures serve as an indication of the possible delicacy of this method of producing spectra, but it should be remembered that all eyes do not see the rubidium lines with equal ease.

In comparative tests of brightness it was found best to employ as the standard the lines given by amounts of the chloride not exceeding 0.0005 mg. to 0.0007 mg., to set the slit at a width of 0.2 mm. and to bring the coils to the flame in sets of three — the first, usually a standard, serving to fix the position of the lines so that the comparative distinctness of the lines given by the other two might be the more readily determined. When that dilution had been found at which the test was barely brighter

* F. A. Gooch and J. I. Phinney, Am. Jour. Sci., [3], xliv, 392.

than the standard and that dilution at which the test was barely weaker than the standard, it was assumed that the mean of the numbers of cubic centimeters representing these two volumes might be taken as the volume at which the test and standard lines were equal. The amount of rubidium in the test solution was then calculated by multiplying the volume in cubic centimeters by the number of coilfuls in 1 cm.^{3} and the product by the amount of rubidium contained in a coilful of the standard solution.

The results of two experiments with pure rubidium chloride are given below.

Determination of Rubidium in Pure Rubidium Chloride.

Experiment I.

Standard. Rubidium in a coilful ($\frac{1}{50}$ cm.^{3}). mg.	Test (known to contain 10 mg. Rb). Volume in cm.^{3}	Line of test compared with standard.
0.0005	340	Brighter.
0.0005	370	Equally bright.
0.0005	370	Brighter.
0.0005	390	Weaker.
0.0005	390	Weaker.

Found, $\frac{370 + 390}{2} \times 50 \times 0.0005 =$ 9.5 mg.
Taken........................ 10.0 mg.
Error........................ 0.5 mg. = 5 per cent.

Experiment II.

Standard. Rubidium in a coilful ($\frac{1}{50}$ cm.^{3}). mg.	Test (known to contain 10 mg. Rb). Volume in cm.^{3}	Line of test compared with standard.
0.0005	300	Brighter.
0.0005	360	Equally bright.
0.0005	380	Brighter.
0.0005	380	Brighter.
0.0005	390	Brighter.
0.0005	400	Weaker.
0.0005	410	Weaker.

Found, $\frac{390 + 400}{2} \times 50 \times 0.0005 =$ 9.875 mg.
Taken........................ 10 0
Error........................ 0.125 mg. = 1.25 per cent.

These results make it plain that when the comparison is made between solutions of pure rubidium chloride the spectroscopic method is capable of yielding fair approximations to truth. In the practical determination of rubidium, however, the question of the effect of the presence of sodium and potassium which naturally accompany it is of importance.

It appears from practical tests that within limits the presence of sodium in the flame increases the brilliance of the rubidium spectrum. The brightness of the lines is raised under the conditions by a maximum of 50 per cent by the presence of sodium up to 40 per cent of the weight of the rubidium, and increase in the amount of sodium does not further influence the brightness of the lines until the proportion of sodium to rubidium is as ten to one; or, speaking broadly, the dissociating effect of sodium upon the rubidium chloride (to which the brightening noted is to be attributed) does not appear to be materially different whether one or a score of molecules of sodium chloride are present to one of the rubidium chloride. But when the proportion of sodium to rubidium much exceeds ten to one the glare of light diffused through the entire spectrum (though the sodium line itself may be cut off) begins to affect the vision, and as the increase advances ultimately extinguishes the rubidium lines.

It appears also that potassium chloride produces an effect similar to that of sodium chloride, the brightness of the rubidium line increasing by a maximum of 50 per cent when the potassium is present to between two-thirds and twice the amount of the rubidium; while the presence of potassium in the proportion five to one influences the visibility unfavorably, and in the proportion of thirty to one extinguishes the rubidium line in the glare of light. It is necessary therefore either to effect the separation of the rubidium from sodium and potassium, or else to bring test and standard to the same condition as regards the presence of these elements, before any reasonable degree of accuracy can be expected in the spectroscopic determination of rubidium as it ordinarily occurs in nature. The separation from sodium is easily accomplished by the conversion of the salts to the form of chloroplatinates; but for the quantitative separation of rubidium from potassium there is no good method known. The practical value of the spectroscopic reaction of rubidium for purposes of quantitative analysis depends, therefore, upon matching potassium lines as well as the rubidium lines (following the method outlined in the determination of potassium in presence of sodium), and so bringing the lines of test and standard equally under the influence of potassium. It has been shown that there is no difficulty in matching solutions of potassium by means of the red line, but the convenience of using the spectroscope without readjustment throughout an entire experiment makes a comparison by means of the blue line highly desirable and this has been found to be feasible. The details of a determination of rubidium in presence of a permissible amount of potassium are given in the following statement.

Determination of Rubidium in Presence of Potassium.

Standard solution containing { 0.0005 mg. Rubidium, 0.0015 mg. Potassium } to the coilful ($\frac{1}{50}$ cm.^{3}).

Test solution containing 8 mg. rubidium and no potassium.

Step 1.	Step 2.	Step 3.	Step 4.	Step 5.
Preliminary test for Rb.	Preliminary matching of K line.	Rematching of Rb line	Readjustment of K line.	Final matching of Rb line.
Test at 20 cm.^{3} gave Rb line like standard.	Test at 20 cm.^{3} gave K line like standard when 1 mg. of K had been added.	Test at 35 cm.^{3} gave Rb lines like standard.	Test at 35 cm.^{3} gave K line like standard when 2 mg. were present.	Test at 35 cm.^{3} gave Rb line like standard.

Found, $35 \times 50 \times 0.0005 = 0.875$ mg.
Taken.. .... = 0.8 mg.
Error...... . .... = 0.075 mg. = 9.4 per cent.

When the amount of potassium present is so great as to vitiate the test for rubidium a precipitation by alcohol may be utilized to remove the excessive amount of the potassium salt. The mixed chlorides are dissolved in the least possible quantity of water and treated with absolute alcohol; the precipitate is filtered off and washed with alcohol; the filtrate and washings are evaporated and the residue dissolved in a known volume of water is ready for the spectroscopic test. Results of experiments conducted in this manner follow.

Rubidium taken in the form of chloride. mg.	Potassium taken in the form of chloride. grm.	Rubidium found. mg.	Absolute error. mg.	Percentage error. per cent.
1	0.1	0 8	0.2	20
2	0 1	1.7	0.3	15
1	0.1	0 9	0 1	10

The error of the process is manifestly large, and only roughly approximate results can be hoped for when large amounts of rubidium are dealt with; but, in view of the fact that the only alternative is an indirect process, even this great error may not be prohibitive in the estimation of very small amounts of rubidium.*

The Estimation of Caesium and Rubidium as the Acid Sulphates.

Browning † has shown that by holding the temperature between 250° and 270° during treatment with sulphuric acid suitable salts of caesium and rubidium may be brought with a fair degree of certainty to the condition of the acid sulphates, which by treatment with ammonia and ignition at red heat yield the neutral sulphates. Salts of potassium and sodium, however, when heated at the same range of temperature, yield pyrosulphates reasonably stable under the conditions.‡ Lithium salts when treated similarly gave no evidence of the existence of a stable acid sulphate or pyrosulphate. Details of the experiments with the salts of rubidium and caesium are given below.

A weighed amount of caesium nitrate was placed in a previously weighed platinum crucible and treated with an excess of sulphuric acid. The crucible was then placed upon a steam bath until the water and nitric acid were largely expelled, and then removed to a radiator, consisting of a porcelain crucible fitted with a pipe-stem triangle so arranged that the bottom of the platinum crucible was about midway between the top and bottom of the porcelain crucible. This improvised radiator was set in an iron ring and a thermometer placed so that the mercury bulb was on a level with the bottom and close to the side of the platinum crucible. An ordinary Bunsen burner

* For an example of the practical use of this method, see The Excretion of Rubidium, Mendel and Slosson, Am. Jour. Physiol., xvi, 152.

† Philip E. Browning, Am. Jour. Sci., [4], xii, 301.

‡ See pages 79, 92.

served as the source of heat and the temperature was kept so far as possible between 250° C. and 270° C. After the fuming of sulphuric acid had ceased, the crucible and contents were removed to a desiccator, cooled and weighed. This process of heating was continued for half-hour periods until the weights were constant. The results given show that by regulating the heat at a temperature between 250° C. and 270° C. caesium may

Caesium Sulphates by Ignition.

$CsNO_3$ taken.	$CsHSO_4$ calculated.	First constant weight.	Second constant weight.	Error on $CsHSO_4$.	Cs_2SO_4 calculated.	Cs_2SO_4 found.	Error on Cs_2SO_4.
grm.	grm.	grm.	grm.	grm.	grm.	grm.	grm.
0.1706	0.2013	0.2054	0.2020	+0.0007	...	...	...
0.1706	0.2013	0.2010	...	−0.0003	...	...	...
0.1032	0.1217	0.1201	...	−0.0016	...	...	...
0.1032	0.1217	0.1252	0.1222	+0.0005	0.0961	0.0948	−0.0013
0.1218	0.1437	0.1458	...	+0.0021	0.1130	0.1118	−0.0012
0.1214	0.1435	0.1430	...	−0.0005	...	...	...
0.1214	0.1435	0.1422	...	−0.0013	...	...	...
0.1150	0.1356	0.1330	...	−0.0026	...	...	...
0.1056	0.1245	0.1272	0.1248	+0.0003	...	...	...
0.1056	0.1245	0.1252	...	+0.0007	...	...	...

be brought with a fair degree of certainty to the condition of the acid sulphate. As a check upon the results the acid sulphate was, in a few cases, treated with a little ammonium hydroxide, the excess of this was removed upon a steam bath and the neutral sulphate was obtained by ignition at a red heat to a constant weight. These determinations agree fairly well with the theory.

The same procedure was followed with rubidium, a pure rubidium chloride having been chosen as the starting point. The results are given below. No tendency was observed on the part of these elements to hold sulphuric acid in excess of the amount necessary for the formation of the acid sulphate.

Rubidium Sulphates by Ignition.

RbCl taken.	$RbHSO_4$ calculated.	$RbHSO_4$ found.	Error.	Rb_2SO_4 calculated.	Rb_2SO_4 found.	Error.
grm.	grm.	grm.	grm.	grm.	grm.	grm.
0.1252	0.1889	0.1878	−0.0011	...	...	...
0.1212	0.1829	0.1840	+0.0011	0.1460	0.1460	0.0000
0.1230	0.1856	0.1850	−0.0006	...	...	...
0.1230	0.1856	0.1858	+0.0002	0.1357	0.1350	−0.0007
0.1610	0.2430	0.2416	−0.0014	0.1777	0.1772	−0.0005
0.1360	0.2052	0.2032	−0.0020	0.1501	0.1490	−0.0011

CHAPTER III.

COPPER; SILVER; GOLD.

Copper.

The Gravimetric Determination of Copper as the Sulphocyanate.

As early as 1854 attention was drawn by Rivot* to the possibility of estimating copper gravimetrically by weighing as cuprous sulphocyanate, and to the advantages which the process afforded in separating copper from other metals. Rivot's procedure consisted in dissolving the substance to be analyzed in hydrochloric acid, reducing the copper with hypophosphorous or sulphurous acid, and precipitating with potassium sulphocyanate. The precipitate, dried at a moderate temperature, was weighed as cuprous sulphocyanate and then as a control converted by ignition with sulphur into cuprous sulphide and weighed in that condition. In spite of the evident advantages for certain purposes, Rivot's method, in its original form, has never come into general use, the chief reason for this being apparently the difficulty and inaccuracy attendant upon the weighing of the precipitate upon dried paper filters, a process which can hardly be depended upon unless managed with extreme care.

Van Name† has shown, however, that the process is accurate and easily managed if attention is given to the necessary conditions of concentration and acidity, and the precipitated cuprous sulphocyanate is filtered and weighed upon asbestos in the perforated crucible.

The table contains results obtained as follows: A suitable quantity of a standard copper sulphate solution was run from a burette, diluted to a convenient volume, a few cubic centimeters of a concentrated solution of ammonium bisulphite ‡ added, and the copper precipitated by an excess of ammonium sulpho-

* Compt. rend., xxxviii, 868.

† R. G. Van Name, Am. Jour. Sci., [4], x, 451; xiii, 20.

‡ Prepared by saturating strong aqueous ammonia with sulphur dioxide.

cyanate. After allowing the mixture to stand for a few minutes or hours, according to the amount of free acid present, the precipitate was collected upon asbestos in a weighed crucible, washed with cold water and dried at 110° until no further loss of weight took place.

Copper weighed as Cuprous Sulphocyanate.

Cu taken.	H_2SO_4 conc.	HNH_4SO_3 sat. sol	NH_4SCN approx. $n/10$.	Final volume.	Time of standing.	Cu found.	Error.
grm.	cm.³	cm.³	cm.³	cm.³	hours.	grm.	grm.
0.0795	None.	5	13	68	$\frac{1}{4}$	0.0795	0.0000
0.0795	None.	3	13	66	48	0.0793	−0.0002
0.0795	None.	3	25	78	$\frac{1}{2}$	0.0796	+0.0001
0.0795	None.	3	25	78	12	0.0796	+0.0001
0.0795	1.5	10	13	85	12	0.0792	−0.0003
0.0795	1.5	8	13	105	48	0.0785	−0.0010
0.0795	1.5	3	25	85	4	0.0783	−0.0012
0.0795	1.5	5	25	85	21	0.0795	0.0000
0.0795	5	5	25	85	3	0.0797	+0.0002
	HCl conc. cm.³						
0.0795	10	5	25	100	20	0.0795	0.0000
0.0795	25	10	25	100	28	0.0784	−0.0011

Larger amounts of copper may also be estimated in the same way, as the following table shows, but with a crucible of the ordinary size the process is more rapid and convenient when a smaller weight of copper is taken.

Copper weighed as Cuprous Sulphocyanate.

Cu taken.	H_2SO_4 conc.	NH_4SCN approx. $n/10$.	Final volume.	Cu found.	Error.
grm.	cm.³	cm.³	cm.³	grm.	grm.
0.3175	None.	60	500	0.3176	+0.0001
0.3175	None.	60	500	0.3177	+0.0002
0.3175	None.	60	500	0.3176	+0.0001
0.3175	10	100	500	0.3175	0.0000
	HCl conc. cm.³				
0.3175	20	100	500	0.3165	−0.0010

In solutions containing free acid the precipitation of the copper is greatly retarded, and the mixture should be allowed to stand for several hours, or, if the amount of acid is considerable, for at least twenty-four hours before filtering. Precipitation from a warm solution is permissible, but boiling the liquid causes the precipitate to turn brown with gradual loss in weight, and is therefore to be avoided.

The only difficulty which is likely to be encountered in the use of this method is a tendency, which sometimes appears, for traces of the precipitate to pass through the filter during the last stages of the washing. This tendency is most marked with precipitates from concentrated solutions containing little or no free acid. It may be reduced to an insignificant amount or entirely eliminated by employing one or more of the following expedients: (1) precipitating in dilute solution; (2) precipitating in the presence of free acid; (3) filtering and washing under light pressure, using a rather dense (not thick) asbestos mat; (4) washing with a decinormal solution of ammonium sulphocyanate. The last expedient is of little or no use when the precipitate is to be directly weighed, but is very satisfactory in separating copper from other substances. Precipitation in acid solution is the most effective method of obtaining precipitates which are easily filtered, but must be used with caution, for the errors from incomplete precipitation may easily exceed the mechanical losses which the acidity was employed to prevent.

The effect of hydrochloric acid of various concentrations upon the completeness of the precipitation was studied in a series of experiments,* in which the principal stress was laid upon determining the amounts of copper left in solution rather than the weights of the precipitates. The procedure was as follows: After filtering off the precipitate the copper in the filtrate was determined by evaporating the solution with nitric acid to a small bulk, heating in a platinum crucible over a radiator to expel sulphuric acid and decompose interfering substances, dissolving the residue in nitric acid, filtering, electrolyzing and weighing the copper. The electrolytic deposit was then redissolved and the copper estimated more accurately by a colorimetric method based on comparison of the intensity of the brown color produced upon the sample by potassium ferrocyanide with that of a variable standard of known copper content.

The results of this series of experiments may be summarized as follows: Allowance must be made for the amount of hydrochloric acid used up by interaction with the ammonium bisulphite

* Am. Jour. Sci., [4], xiii, 20.

solution forming ammonium chloride and sulphur dioxide. One cubic centimeter of a bisulphate solution, prepared by saturating strong aqueous ammonia with sulphur dioxide, may neutralize by this reaction about nine-tenths of a cubic centimeter of hydrochloric acid, sp. gr. 1.18. In order that the amount of copper left in solution may not exceed 0.1 mg. per 100 cm.³ of filtrate, the concentration of effective hydrochloric acid, i.e., that remaining after interaction with the bisulphite, stated in volume per cent of the concentrated acid (cubic centimeters of acid of sp. gr. 1.18 for 100 cm.³ final volume of solution), should not be above 0.8 when the excess of sulphocyanate employed is small (20 per cent above the theory), but may be as high as 3 per cent if ten times the theoretical amount of sulphocyanate be employed. A suitable degree of acidity for precipitating copper under ordinary conditions is given by 0.5 to 1.0 per cent of hydrochloric acid, expressed as above, using from five to ten times the theoretical amount of sulphocyanate.

As far as could be judged from a limited number of determinations made in the presence of sulphuric acid, the above holds true for the equivalent amount of sulphuric acid.

$Cu_2(SCN)_2$ taken. grm.	Volume of liquid. cm.³	HCl (sp. gr. 1.18). cm.³	NH_4SCN. grm.	Cu in filtrate. grm.
0.3	200	..	....	0 00035
0.3	200	..	....	0.00040
0.3	200	..	10.	0.0050
0.25	200	..	1.52*	0.00018
0.3	200	..	0.76†	0.00007
0.3	200	..	0.19‡	0.00004
0.3	200	2	....	0.0019
0.3	200	2	2.5	0.0013
0.3	200	2	1.77**	0 0009
0.3	200	2	0.19	0.0006
		NH_4Cl. grm.		
0 3	200	10	....	0 0031
0.3	200	10	0.19	0.00013
0.3	200	1††	....	0.00045
0.3	200	1	0.19	0.00005

* Solution $n/10$ in respect to NH_4SCN.
† Solution $n/20$ in respect to NH_4SCN.
‡ Solution $n/80$ in respect to NH_4SCN.
** HCl and NH_4SCN present in equivalent amounts.
†† Solution approximately $n/10$ in respect to NH_4Cl.

The effect of varying concentrations of different reagents involved in the process upon the solubility of cuprous sulphocyanate is shown in a rough way by the preceding table. As no stirring was employed the figures have no absolute value, but serve merely to give an idea of the relative magnitude of the solubilities in question.

Weighed amounts of cuprous sulphocyanate prepared by precipitation in the usual way, thoroughly washed, and dried at 105°, were allowed to stand in the solutions to be tested from 40 to 50 hours. After careful filtering through asbestos the copper in the clear filtrate was estimated by electrolysis, or, in cases where the amount was small, by the colorimetric method referred to above.

The solubility in presence of either hydrochloric acid, ammonium chloride or a large amount of ammonium sulphocyanate is considerable. It is lowest in dilute solutions of ammonium sulphocyanate, and the presence of a small amount of this salt lessens the solubility in hydrochloric acid, and in solutions of ammonium chloride.

Separation of Copper from Bismuth, Antimony, Tin and Arsenic.

From the nature of the process it is evident that it is much less likely to be interfered with by the presence of other metals than the other gravimetric methods for copper, and may, therefore, be directly applied with good results in many cases where the use of the electrolytic or the oxide method would involve a previous separation. Van Name* has tested the method for the separation of copper from bismuth, antimony, tin and arsenic.

Having copper present with these metals in a solution containing free hydrochloric acid, tartaric acid was added to aid in preventing the formation of insoluble products of hydrolytic action, and the copper then precipitated as cuprous sulphocyanate in the usual way.

All the determinations were allowed to stand fifteen hours or more before filtering to insure completeness of precipitation. The filtering was performed upon asbestos in a perforated crucible. The precipitate was thoroughly washed with cold water and dried at 105° to a constant weight.

In the following table are results obtained by this procedure. The acidity was kept within the limits shown above to be safe,

* Am. Jour. Sci., [4], xiii, 138.

and the amount of sulphocyanate used was in most cases about ten times the theory.

Copper in Presence of Bismuth, Tin, Antimony, and Arsenic.

Final Volume 200 cm.^{3}.

Cu. taken.	Bi.	HCl (sp. gr. about 1.17).	Tartaric acid.	HNH_4SO_3 sat. sol.	NH_4SCN approx. $n/10$.	$Cu_2(SCN)_2$ found.	Calculated as Cu.	Error.
grm.	grm.	cm.^{3}	grm.	cm.^{3}	cm.^{3}	grm.	grm.	grm.
0 0793	0.2	6	1	2	60	0.1504	0 0786	−0.0007
0.0793	0.1	6	1	2	100	0.1512	0 0790	−0 0003
0 0793	0.3	6	1	2	125	0.1515	0 0792	−0.0001
0.0793	0.2	6	1	2	125	0.1518	0.0793	0.0000
0.0793	0.2	6	1	2	125	0.1519	0.0794	+0.0001
0.0793*	0.2	6	1	2	230	0.1519	0 0794	+0.0001
	Sn taken as $SnCl_4$+HCl. grm.							
0.0793	0.2	5	1	2	40	0.1502	0 0785	−0.0008
0.0793	0.2	6	1	2	125	0.1514	0 0791	−0.0002
0.0793	0.2	5	1	2	130	0.1516	0 0792	−0.0001
	Taken as $SnCl_2$+HCl. grm.							
0.0793	0.2	6	1	2	125	0.1529	0.0799	+0.0006
	As. grm.							
0.0793	0.2	6	1	2	125	0.1523	0 0796	+0.0003
	Sb. grm.							
0.0793	0.2	6	2	2	125	0.1518	0.0793	0.0000
	As, Bi, Sb, Sn of each. grm.							
0.0795	0.1	6	2	2	130	0.1523	0.0796	+0.0001
0.0795	0.1	6	2	2	130	0.1525	0.0797	+0.0002

* Final volume 300 cm.^{3}

If bismuth is present in considerable amount, a good deal of hydrochloric acid is needed, and there is danger that interaction with the precipitants may reduce the acidity to the point where hydrolysis and precipitation of the bismuth begins. In such cases preliminary blank tests must be carried out to determine the minimum concentration of hydrochloric acid which may be employed under the conditions. With antimony the effectiveness of the tartaric acid is so great that this difficulty does not arise if enough tartaric acid is used. Tin in the stannous condition sometimes forms a slight precipitate of sulphur on standing in contact with the bisulphite, and it is, therefore, advisable to oxidize it at the outset to the stannic state.

It is evidently possible to estimate copper by this method in the presence of bismuth, antimony, tin and arsenic, either separately or in any combination. To separate copper from unknown quantities of bismuth, or from mixtures containing bismuth, the following procedure is recommended: Having the copper and bismuth in hydrochloric acid solution, add tartaric acid, and, after diluting if necessary, determine by blank tests with small aliquot portions of the solution how much ammonium bisulphite can be added to the whole without precipitating the bismuth. Then, keeping the bisulphite well within this limit, carry out the precipitation of the copper as already described, using a considerable excess of ammonium sulphocyanate. Where bismuth is absent, antimony and tin may be treated in the same way, but the latitude possible in the adjustment of the conditions is so much greater with these metals that preliminary tests will seldom be needed. For the separation from arsenic no special precautions are required.

The Determination of Copper as Cuprous Iodide and Separation from Cadmium.

The separation of copper from cadmium by the precipitation of the cuprous iodide by appropriate means has long been known. Pisani* mentions the fact that potassium iodide can be used to effect precipitations, and claims that a satisfactory separation can be made in this way. Flajolot†, stating that potassium iodide cannot be used as a precipitant on account of the solubility of cuprous iodide in that reagent, and that hydriodic acid cannot be employed if nitric acid is present, recommends that the solution containing copper be brought to acidity with sulphuric acid, that a considerable excess of sulphurous acid be added, and that the precipitation be effected by hydriodic acid. Kohner‡ reviews the various methods for the separation of copper from cadmium and states that the iodide method is impracticable on account of the solubility of cuprous iodide both in excess of hydriodic acid and in potassium iodide.

* Compt. rend., xlvii, 294.

† Jour. prakt. Chem., lxi, 105.

‡ Zeit. anal. Chem., xxvii, 203; Jour. Anal. Chem., iii, 339.

Browning* has fixed conditions under which accurate results may be obtained by this method.

According to the best procedure shown, the sulphates of copper and cadmium, in amount not exceeding 0.25 grm. of each metal, are dissolved in 25 cm.^{3} of water and treated with 1 grm. or 2 grm. of potassium iodide.† The mixture is evaporated to dryness to expel iodine and then treated with 100 cm.^{3} of water. Filtration is made under gentle pressure upon the asbestos felt in the perforated crucible. It is advisable, on account of the tendency of cuprous iodide to pass through the filter, to use a fairly thick felt and to keep it moist and under pressure during the filtration. The precipitate is washed thoroughly with either hot or cold water, dried at 120°–150° and weighed as cuprous iodide.

Copper Weighed as Cuprous Iodide.

Copper taken. grm.	Copper found (weighed as Cu_2I_2) grm.	Error on copper. grm.	KI used. grm.	Final volume of liquid. cm.^{3}
0.1194	0.1196	+0 0002	1	100
0.1191	0.1194	+0.0003	1	100
0.1193	0.1193	0.0000	2	100
0.0049	0.0045	−0 0004	2	100
0 0051	0.0047	−0.0004	2	100
0.1195	0.1195	0 0000	3	100
0.1192	0.1188	−0.0004	4	100

Copper Separated from Cadmium and Weighed as Cuprous Iodide.

Copper taken. grm.	Copper found (weighed as Cu_2I_2). grm.	Error on copper. grm.	Cadmium taken. grm.	Cadmium found (weighed as CdO). grm.	Error on cadmium. grm.	KI used. grm.	Found volume of liquid. cm.^{3}
0.2383	0 2386	+0.0003	0.0484	0 0490	+0.0006	2	100
0.1192	0 1185	−0 0007	0.2439	0 2430	+0.0001	2	100
0.1193	0.1194	+0.0001	0.1942	0.1942	0 0000	2	100
0 1201	0 1201	0 0000	0.2426	0.2428	+0.0002	2	100
0.1193	0.1193	0 0000	0.2436	0.2433	−0.0003	2	100
0.0239	0.0238	−0 0001	0.1934	0 1932	−0 0002	1	100
0.0236	0 0239	+0 0003	0.1942	0 1936	−0 0006	1	100
0.0239	0 0242	+0.0003	0.1444	0 1442	−0.0002	1	100
0 0238	0.0238	0 0000	0.1467	0 1461	−0 0006	1	100

* Philip E. Browning, Am. Jour. Sc., [3], xlvi, 280.

† For the results of study of the effect of excess of potassium iodide, free acid and concentration upon the solubility of cuprous iodide, see page 121.

The table contains results of experiments made according to this procedure with pure copper sulphate, and with copper sulphate and cadmium sulphate in mixture.

The Electrolytic Determination of Copper.

Gooch and Medway* have applied the rotating cathode to the rapid electrolytic determination of copper, making use of the apparatus shown in Fig. 13.†

The deposition of copper from a solution of the sulphate was first attempted, and the procedure was as follows: The solution, 50 cm.^{3} in volume, was placed in a 150-cm.^{3} beaker and acidulated to give better conductivity. The stand carrying the beaker was raised until the liquid covered about two-thirds of the crucible adjusted to the shaft, thus giving a cathode surface of about 30 cm.^{2} The anode was introduced and the motor started. The wires from the storage batteries were connected and the current was allowed to pass through the solution. The duration of the electrolysis was varied according to the strength of current used; but in each case, after the deposit was nearly complete, the current from the batteries was shut off, the motor stopped, and the sides of the beaker, the platinum anode and the crucible were carefully washed with a fine jet of water, the motor was again started and the current allowed to pass for the remaining time.

When the deposit was complete the crucible was removed and washed, first with water, then with alcohol, and finally was dried by passing it over a flame.

Sulphuric acid (6 or 7 drops of the dilute acid — 1 : 3) was generally used to acidulate the solution, since it was found that the copper was deposited in less time with sulphuric acid than with nitric acid present. Experiments in which small amounts of nitric acid (6 to 9 drops of the dilute acid — 1 : 3) were used show that the copper may also be deposited completely in presence of this acid.

The following tables show the results of a series of experiments made as described. The standard of the solution of copper sulphate was fixed by the usual slow method of electrolytic analysis.

* F. A. Gooch and H. E. Medway, Am. Jour. Sci., [4], xv, 320.

† See page 12.

Solution of $CuSO_4$ Acidulated with H_2SO_4.

Copper taken. grm.	Copper found grm.	Error. grm.	Current. amp.	N. D. $_{100}$	Time. min.
0.0651	0 0652	+0.0001	0.8	2.7	25
0.0651	0 0652	+0 0001	0 8	2.7	15
0 0651	0 0651	0 0000	1	3.3	10
0 0651	0 0649	−0.0002	1	3 3	10
0 0651	0 0648	−0 0003	1	3 3	10
0.1272	0 1272	0 0000	2.5	8 3	15
0 1272	0 1271	−0 0001	2.5	8.3	15
0 1272	0 1271	−0 0001	2.5	8 3	15
0.1272	0 1270	−0 0002	3	10	13
0 1272	0.1268	−0 0004	3	10	12
0 2548	0 2548	0 0000	3	10	20
0 2548	0 2548	0 0000	4	13 3	20
0 2548	0 2550	+0 0002	4	13 3	20
0 2548	0 2546	−0 0002	4	13.3	15
0.2548	0 2545	−0 0003	4	13 3	15

Solution of $CuSO_4$ Acidulated with HNO_3.

Copper taken. grm.	Copper found. grm	Error. grm.	Current. amp.	N. D. $_{100}$	Time. min.
0 0651	0 0648	−0 0003	1	3.3	35
0 0651	0.0652	+0 0001	0.8	2.7	30
0.0651	0 0650	−0 0001	0.8	2.7	25
0 0651	0 0649	−0 0002	1	3.3	25
0 0651	0 0650	−0.0001	0.8	2 7	25
0.0651	0 0652	+0.0001	1	3.3	35
0 0651	0 0648	−0 0003	1	3 3	30
0.0651	0 0650	−0.0001	1.5	5	25
0.0651	0 0650	−0.0001	1.5	5	25
0 0651	0.0647	−0 0004	1.8	6	20

It has been shown also by Medway* that a crucible of silver may be substituted for the platinum crucible as the rotating cathode, in the deposition of copper from the acidulated sulphate solution, with results that leave little to be desired on the score of accuracy.

Deposition upon the Silver Crucible.

Copper taken. grm.	Copper found. grm.	Error. grm.	Current. amp.	N. D. $_{100}$	Time. min.
0.1088	0 1086	−0.0002	2	6.6	15
0 1088	0.1090	+0.0002	2	6.6	15
0.1088	0 1084	−0 0004	1.5	5	15
0 1088	0.1085	−0.0003	2	6.6	15
0.1088	0 1080	−0.0008	2	6.6	15
0.1041	0.1041	0.0000	2	6.6	15
0.1041	0.1046	+0 0005	2	6.6	15
0.1041	0 1039	−0 0002	2	6.6	15

* Am. Jour. Sci., [4], xviii, 180.

To remove the copper from the crucible, the deposit is rubbed off as much as possible and the rest may be dissolved in a strong boiling solution of hydrochloric acid with but trifling loss of silver, as is shown in the statement of the results of two experiments given below:

	I.	II.
Weight of crucible before treatment............	36 0089	36.0062
Weight of crucible after treatment..............	36.0062	36.0041
Loss of silver............	0 0027	0.0021

It appears that the silver crucible may, with some economy and without sacrifice of accuracy, be substituted for the platinum crucible used as a rotating cathode in the electrolytic determination of copper.

The Iodometric Estimation of Copper.

When potassium iodide is added to a suitable solution of a cupric salt, cuprous iodide is precipitated, while iodine equivalent to the amount of iodine fixed in the cuprous iodide is liberated. This reaction has been made the basis of an iodometric method for the determination of copper, the first suggestion of such a method having apparently been made by De Haen in 1854. In this process cupric sulphate was treated in solution with potassium iodide and the free iodine determined by sulphurous acid according to Bunsen. From the amount of iodine thus found the copper was calculated, according to the equation

$$2\,CuSO_4 + 4\,KI = 2\,K_2SO_4 + Cu_2I_2 + I_2.$$

This method was mentioned in the following year by Mohr,* with the modification suggested by Schwarz that the free iodine be determined by sodium thiosulphate instead of by sulphurous acid. E. O. Brown,† apparently without knowledge of De Haen's previous work, proposed, in 1857, similar procedure, and in 1868 the method with slight modification was presented again by Rümpler.‡ Concerning the utility of the method opinions have

* Titrirmethode, page 387.

† Jour. Chem. Soc., x, 65.

‡ Jour. prakt. Chem., cv, 193.

varied. Mohr never favored it. As late as 1877, Mohr,* after quoting Meidinger to the effect that cuprous iodide freshly precipitated and washed is capable of taking up iodine, and Carl Mohr's criticism that potassium iodide acts upon cuprous iodide according to the concentration, states that the method is not exact and has nowhere found practical application. On the other hand, Fresenius† recommends the method for the determination of small amounts of copper, noting that ferric salts and other substances capable of setting free iodine from an acidified solution of potassium iodide must not be present, and indicates the most favorable procedure. The copper salt treated, he says, should be the sulphate, preferably in neutral solution, though a moderate amount of sulphuric acid is not objectionable. Much free sulphuric acid and all free nitric acid should be neutralized by sodium carbonate, and the precipitate dissolved in acetic acid, an excess of which does no harm in the iodometric process.

Of recent writers, some have favored the method, while others have commented upon it unfavorably. Low‡ has been outspoken in praise, to the extent of declaring a preference for this method in the most accurate technical work over all other methods, even the electrolytic method.

According to Low's earlier modification, metallic copper is dissolved in nitric acid, the solution is freed from nitrogen oxides by boiling, a considerable amount of zinc acetate is added, and in the solution having a volume of 50 $cm.^3$ an excess of solid potassium iodide is dissolved. Zinc acetate is preferred to sodium acetate to take up the free nitric acid. It is said that an excess of potassium iodide is necessary to insure rapidity of action and is harmless. According to the later modification of this method, Low prepares the cupric salt by dissolving the metal in nitric acid (sp. gr. about 1.20), boils the solution, adds bromine water to destroy the nitrogen oxides, boils to expel the bromine, treats with ammonium hydroxide in excess, adds acetic acid and boils again if necessary to get a clear solution. The advantage of using an excess of potassium iodide is emphasized, and the statement is made that unless an excess of this reagent is present the

* Titrirmethode, 5 Aufl., 288.

† Quant. Anal., 6te Aufl., 335, 1875.

‡ Jour. Am. Chem. Soc., 18, 468; 24, 1083.

reaction does not proceed to completion until the titration of the free iodine takes place. Low recommends the use of 1 grm. of potassium iodide, an excess of 0.6 grm., for every 0.075 grm. of copper.

As a result of elaborate study, Moser* has reached the conclusion that the reaction by which cuprous iodide is formed from potassium iodide and cupric sulphate in neutral solution is complete at very high concentration of the solution; that the completeness of the reaction is greatly affected by the volume of liquid; that the amount of potassium iodide employed is almost without influence either in neutral solution or in acid solution; and that the presence of free sulphuric acid even in large amounts or of hydrochloric acid present in amount equivalent to the cupric sulphate is advantageous. Moser recommends, therefore, the addition of sulphuric acid for the purpose of bringing the reaction to completion.

According to Fernekes and Koch,† an excess of acetic acid does not influence titrations, while a certain amount of potassium iodide — 1.5 grm. to 2 grm. for 0.0038 grm. of copper, and 2.5 grm. for 0.0939 grm. of copper — must be added to bring about complete action in a volume of 100 $cm.^3$

Quite recently Cantoni and Rosenstein‡ have tested the reaction between potassium iodide and a cupric salt under various conditions; but these investigators do not give the absolute values of the amounts of copper taken and found, merely recording the relative effects of varying conditions. From the record of their results it would appear that a fivefold increase of the minimum amount of potassium iodide added to portions of 100 $cm.^3$ of solution containing the same amount of copper salt is without influence upon the result; that increase of volume from 100 $cm.^3$ to 350 $cm.^3$, other conditions being the same, may affect the results by as much as 5 per cent of their value. The authors conclude that the method gives good results under properly controlled conditions.

So evidence and opinions as to the effect of various conditions in the process are contradictory, the chief matters of difference being the influence of an excess of potassium iodide used as the

* Zeit. anal. Chem., xliii, 597.

† Jour. Am. Chem. Soc., xxvii, 1229.

‡ Bull. Soc. Chim., [3], xxxv, 1067–73.

precipitant, the dilution at which the precipitation should take place, and the effect of acids upon the formation of the cuprous iodide.

These points have, therefore, been carefully investigated experimentally by Gooch and Heath,* with the results summarized below.

As to the use of potassium iodide in effecting the precipitation of cuprous iodide, it appears that the excess present has within limits an influence upon the result; that beyond the limits the addition of potassium iodide has no appreciable effect; and that the absolute amount of potassium iodide required increases with the dilution. An excess of potassium iodide ranging from 0.6 grm. to 1 grm. in a volume of 50 $cm.^3$, and from 3 grm. to 5 grm. in a volume of 100 $cm.^3$, will precipitate completely 0.0020 grm. of copper. In the practical application of these facts it must be borne in mind that it is the excess of potassium iodide and not the full amount added which is important. So it is reasonable to fix upon 2 grm. as the uniform amount of potassium iodide suitable for the precipitation of cuprous iodide equivalent to 0.2 grm. of copper in a volume of 50 $cm.^3$; and upon 5 grm. as the amount of potassium iodide suitable in neutral solutions having a volume of 100 $cm.^3$.

In a study of the effect of free acid upon potassium iodide alone, it appears that no more than 2 $cm.^3$ of concentrated sulphuric acid or hydrochloric acid may be present with 2 grm. of potassium iodide in 50 $cm.^3$ of solution without liberating an appreciable amount of iodine, and the presence of 1 $cm.^3$ of pure nitric acid causes error. The tendency to liberate iodine is manifestly less at the higher dilution, and it appears that in a volume of 100 $cm.^3$ of solution containing 5 grm. of potassium iodide 3 $cm.^3$ of concentrated sulphuric acid, hydrochloric acid or nitric acid free from nitrogen oxides may safely be present. Acetic acid of 50 per cent strength may apparently make up half the solution at either dilution. When either sulphuric acid, hydrochloric acid or nitric acid is present, obviously the higher dilution is preferable. In the determination of 0.0010 grm. of copper by titration of the iodine set free in a volume of 100 $cm.^3$ in presence of 5 grm. of potassium iodide, it appears that so much as 50 $cm.^3$ of 50 per cent acetic acid, 3 $cm.^3$ of sulphuric

* F. A. Gooch and F. H. Heath, Am. Jour. Sci., [4], xxiv, 67.

Iodometric Determination of Copper.

Copper taken as $Cu(NO_3)_2$.	KI.		Acid.	Volume.		Copper found.	Error.
	Present.	Approximate excess.		At beginning of titration.	At end of titration.		
grm.	grm.	grm.	cm.^{3}	cm.^{3}	cm.^{3}	grm.	grm.
Final volume between 110 cm.^{3} and 120 cm.^{3}.							
			H_2SO_4 conc.				
0.1200	5.0	4.5	2.5	100	119	0.1200	0.0000
0.0900	5.0	4.5	3.0	100	114	0.0903	+0.0003
0.0900	5.0	4.5	3.5	100	114	0 0905	+0.0005
			HCl conc.				
0.0900	5.0	4.5	2.0	100	117	0 0897	−0.0003
0.1200	5.0	4.5	2.0	100	119	0.1195	−0.0005
0.0900	5.0	4.5	3.0	100	114	0 0901	+0.0001
0.1200	5.0	4.5	3.0	100	119	0 1200	0.0000
0.1200	5.0	4.5	3.5	100	119	0 1197	−0 0003
0.0900	5.0	4.5	4.0	100	114	0.0903	+0.0003
			HNO_3 conc.				
0.0900	5.0	4.5	1.0	100	114	0 0900	0.0000
0.1050	5.0	4.5	1.5	100	117	0.1051	+0 0001
0.0900	5 0	4.5	2 5	100	114	0.0901	+0.0001
			50 per cent $HC_2H_3O_2$.				
0.1200	5.0	4.5	3 0	100	119	0.1195	−0.0005
0.0900	5.0	4.5	5.0	100	114	0 0898	−0.0002
0.1050	5.0	4 5	10.0	100	117	0.1048	−0.0002
Final volume increased by titration to 132 cm.^{3} and 150 cm.^{3}; KI correspondingly increased.							
			H_2SO_4 conc.				
0.2218	7.0	6.0	2	100	135	0.2214	−0.0004
0.3231	8.0	6.4	3	100	150	0 3226	−0 0005
			HCl conc.				
0.2023	7.0	6.0	2	100	132	0.2016	−0.0007
0.2581	7.8	6.7	3	100	141	0.2574	−0.0007
			HNO_3 conc. purified.				
0 2023	8.0	7.0	1	100	132	0 2017	−0 0006
0.2520	10.0	8.5	3	100	148	0.2512	−0.0008
			$HC_2H_3O_2$ 50 per cent.				
0.2125	7.5	...	5	100	133	0 2119	−0.0006
0.2064	8.0	...	8	100	132	0.2058	−0 0009

acid, 3 cm.³ of hydrochloric acid, or 3 cm.³ of nitric acid (free from nitrogen oxides) may be present without appreciable influence upon the indications of the process.

The best general procedure in determining by the iodometric method amounts of copper not exceeding about 0.3 grm. seems to be covered by the following directions: The solution of the cupric salt, containing no more than 3 cm.³ of concentrated sulphuric acid, hydrochloric acid or nitric acid (free from nitrogen oxides), or 25 cm.³ of 50 per cent acetic acid, is made up to a volume of 100 cm.³, 5 grm. of iodate-free potassium iodide added, and the titration of the free iodine made by sodium thiosulphate in the usual manner with the use of the starch indicator at the end. The $n/10$ sodium thiosulphate used in the titration adds appreciably to the initial 100 cm.³ of solution when much copper is estimated, and in case the end reaction has not appeared when so much as 25 cm.³ of the thiosulphate has been added, 2 grm. to 3 grm. more of potassium iodide should be added before continuing the titration.

The error of the process, properly conducted, should not exceed a few tenths of a milligram in terms of copper.

Results obtained by this procedure are given in the table.

The Determination of Copper by Titration of the Precipitated Oxalate with Potassium Permanganate.

Upon the well-known fact that copper oxalate is insoluble in water and scarcely attacked by moderate amounts of dilute nitric acid, Bournemann* has based an approximative method for separating copper from cadmium, by precipitation as the oxalate in the presence of nitric acid, and estimating the copper, after ignition, by any of the well-known gravimetric methods. Classen† describes a method for the separation of metals as oxalates by adding to the solution of the salt of the metals a dilute solution of potassium oxalate [1 : 6] and concentrated acetic acid to 80 per cent of the total volume. Regarding copper salts in particular, Classen states that precipitation takes place only in dilute solution, and then not completely.

According to the experience of Peters,‡ the precipitation of

* Chem. Ztg., xxiii, 565.

† Ber. Dtsch. Chem. Ges., x. b, 1316.

‡ Charles A. Peters, Am. Jour. Sci., [4], x, 359.

copper oxalate from solutions saturated with oxalic acid and containing at least 0.01 grm. of copper in 50 cm.^{3} of liquid may be made practically complete, the filtrate in such cases giving no blue color with ammonia, in a test tube viewed lengthwise, and only a faint brown color when the filtrate is neutralized, made acid with acetic acid, and tested with potassium ferrocyanide. Peters shows that copper may be determined quantitatively as the oxalate, by precipitation with oxalic acid and titration of the precipitate by potassium permanganate, and separated from certain other metals in the presence of nitric acid, by the addition of considerable amounts of oxalic acid, provided that the amount of copper present in solution exceeds a certain minimum value. It is shown that when the amount of copper present falls below the minimum either precipitation does not take place or it is incomplete. It is noted that the minimum is variable with the concentration of the precipitant, oxalic acid, and to some extent dependent upon the condition of the precipitant, the minimum being smaller when the oxalic acid is added in crystalline form rather than in solution to the liquid containing the copper salt. Peters' observations in respect to the effect of concentration and condition of the oxalic acid in solution may be summarized in the following statement:

Minimum amount of copper, taken as the sulphate, which must be present in order that nearly complete precipitation may take place.	Amount of oxalic acid used.		Volume of liquid.
	Crystalline.	In solution.	
grm.	grm.	grm.	cm.^{3}
0.010	5	5*	50
0.025	2	3.5	50
0 040	1	...	50
0 050	0.5	...	50

* Saturated solution added to the copper salt dissolved in the least amount of water.

It is noted that when a saturated solution of oxalic acid, containing 0.1 grm. of oxalic acid to 1 cm.^{3}, is slowly added to a drop of the copper solution containing 0.0003 grm. of copper, the precipitated oxalate first formed dissolves completely in a volume of 5 cm.^{3} of the precipitant.

When no added nitric acid is present precipitates formed in the hot solutions at a volume of 50 cm.^{3} may be filtered, either at once or after cooling, without loss; but the condition of the precipitate is better in the presence of nitric acid. When the nitric acid is added the mixture must stand before filtration—best over night. Results are unsatisfactory when ammonium nitrate is present.

Precipitation of Copper Oxalate. According to the procedure recommended by Peters, copper sulphate dissolved in approximately 50 cm.³ of hot water, to which nitric acid is added to the amount of 5 cm.³ when certain separations are to be effected, is treated in a small beaker with crystallized oxalic acid, 2 grm. to 3 grm., and allowed to stand over night. The precipitate is filtered off on asbestos and washed two or three times with small amounts of cold water and, still in the crucible, is returned to the beaker, 5 cm.³ or 10 cm.³ of dilute sulphuric acid [1 : 1] are added, together with a convenient amount of water, and, after heating the liquid to boiling, the oxalic acid is titrated with permanganate, the oxalate of copper dissolving readily as fast as the excess of oxalic acid is removed by the permanganate. The precipitate may also be dissolved in 10 cm.³ of strong hydrochloric acid and titrated at 30°–50° after the addition of 0.5 grm. of manganous chloride. Experimental results are given below.

Permanganate Titration of Copper Oxalate.

CuO taken as $CuSO_4$. grm.	Oxalic acid. grm.	HNO_3 (sp. gr. 1.40). cm.³	Volume at precipitation. cm.³	CuO found grm.	Error. grm.
0 1860	0.5	..	50	0 1864	+0 0004*
0.1860	0 5	...	50	0 1866	+0.0006
0 1860	0.5	..	50	0 1866	+0.0006
0 1860	1.0	..	50	0 1866	+0.0006
0.0398	1 0	..	50	0 0391	−0.0007
0.1860	0 5	5.0	55	0 1859	−0 0001
0.1860	0.5	5.0	55	0 1860	0.0000
0.1990	2.0	5.0	55	0 1989	−0.0001
0.1990	3.0	5.0	55	0 1990	0 0000

* The titration was made in the hydrochloric acid solution containing manganous chloride.

Solubility of Copper Oxalate. The fact noted by Peters, that small amounts of precipitated copper oxalate may be redissolved in a sufficient excess of the precipitant, points to an appreciable degree of solubility of the precipitate in the solution of oxalic acid. The observation that very considerable amounts of copper oxalate fail to come down at all until a certain minimum of the copper salt is present, although precipitation is nearly complete when that minimum is reached, indicates supersaturation of the solution by copper oxalate; while the capacity of the liquid for supersaturation is apparently limited to some extent by increase in concentration of the oxalic acid. The solubility coefficient of the copper oxalate under the conditions is made up, therefore, of at least two factors, of which one depends upon the normal solubility in the solution of oxalic acid which constitutes the medium of precipitation, while the other depends upon the solubility due to supersaturation. In order that small amounts of copper may be precipitated, it is necessary to eliminate, or at least to limit, the capacity of the medium for supersaturation; and in order that large amounts, as well as small amounts, of copper may be determined with the highest degree of accuracy, it is necessary to reduce to the lowest point the normal solubility of the oxalate under the conditions of precipitation.

Gooch and Ward * have studied the conditions under which small as well as large amounts of copper may be determined by the oxalate method. It is to be noted, in the first place, that the character of precipitated copper oxalate depends upon the conditions of precipitation. When oxalic acid is added to a cold concentrated solution of a salt of copper, the copper oxalate precipitated is of extreme fineness and tends to pass through the closest filters. The precipitate formed in hot solution is, on the other hand, crystalline and easily separated by filtration of this liquid. The solubility of the precipitate, as well as the ease with which it may be separated from the liquid, turns upon the conditions of precipitation and treatment. Throughout a series of experiments the error of the determination increases with the dilution. That the errors found in titration actually represent approximately the losses in copper for the smaller volumes, is shown by the electrolytic determination of copper in the filtrates from the precipitated oxalate. For a volume of 10 $cm.^3$, the average error in the titration of the oxalate precipitated, either from the solution of the sulphate or from a solution of the nitrate, is 0.0002 grm.; for 50 $cm.^3$ it is 0.0011 grm.; for 100 $cm.^3$, 0.0053 grm.; for 200 $cm.^3$, 0.0203 grm. For similar concentrations of the copper salt and of the oxalic acid, the deficiency in the copper

* F. A. Gooch and H. L. Ward, Am. Jour. Sci., [4], xxvii, 448.

indicated by titration of the precipitated oxalate increases more rapidly than the dilution, a fact which suggests some specific action of water, — perhaps hydration affecting the solubility, or hydrolysis affecting the composition, of the copper oxalate. That time and temperature are not essential factors in the precipitation of the oxalate at moderate dilution from solutions of the neutral salt, was shown by Peters in the following experiments which indicate also that the precipitates, whether thrown down in hot solution or in cold solution, possess after long standing the same degree of insolubility.

Effects of Temperature at Precipitation and Filtration; Filtration after Standing over Night.

Copper taken. grm	Volume. cm.^{3}	Oxalic acid. grm.	Copper found. grm.	Error. grm.	Precipitation.	Filtration.
0.0502	50	2.0	0.0491	0.0011	Hot.	Cold.
0.0502	50	2.0	0.0492	0.0010	Hot.	Hot.
0.0502	50	2.0	0.0490	0.0012	Cold.	Hot.
0.0502	50	2.0	0.0491	0.0011	Cold.	Cold.

If any part of the apparent loss of copper oxalate precipitated from solutions of oxalic acid is due to hydrolysis of the normal oxalate, and formation of a basic oxalate as the product of hydrolytic action, it should be possible to obviate such apparent loss by increasing the active acidity of the solution and thus inhibiting hydrolysis, provided the solubility of the normal oxalate is not made greater thereby. Experiment shows that beyond a reasonable degree of concentration the results are not affected by the use of oxalic acid up to the point of saturation of the solution, but that the apparent error is actually diminished by the presence of even very small amounts of sulphuric acid or nitric acid in the liquid, while, within reasonable limits, the addition of more acid produces no further effect. At the higher dilution the effect of the active acid is marked. At a volume of 100 cm.^{3} the average error of deficiency shown in absence of the stronger acids is cut in two by the addition of 0.1 cm.^{3} to 5 cm.^{3} of nitric acid, or of 0.5 cm.^{3} to 2 cm.^{3} of sulphuric acid. At the smaller volume of 50 cm.^{3} the effect is not so marked, but it is still obvious. These results favor strongly the hypothesis that copper oxalate is increasingly subject to hydrolysis as dilution increases, and that the tendency to form a basic salt may be checked by the presence of the stronger acids in suitable amounts. Even very large amounts of nitric acid produce a surprisingly small increase in the apparent solubility of the oxalate.

Losses due to solubility of copper oxalate may evidently be kept at low limits by restricting the volume of the solution of oxalic acid in which precipitation takes place; but too high concentration is likely to introduce error due to mechanical inclusion of oxalic acid in the precipitate. The natural alternative to a close restriction of the volume of the aqueous solution is the limitation of the solvent power of a larger volume of liquid by partially substituting for water some other miscible liquid less capable of dissolving the precipitated oxalate. In testing the effect of alcohol, suggested by Gibbs,* it was found that the presence of that medium, either with or without nitric acid, improves the results obtained at similar dilutions of the oxalic acid solution, and to about the same degree whether with or without nitric acid. So it would seem, if the effect of nitric acid is to prevent the formation of a basic salt, that alcohol not only diminishes the normal solubility, but checks hydrolytic action as well. In a volume of 100 $cm.^3$ containing 20 per cent of alcohol the error approximates −0.0010 grm.; and for a volume of 50 $cm.^3$ containing 50 per cent alcohol the error is reduced to −0.0003. The effect of nitric acid accompanying the alcohol is not marked.

In further experiments it was found that the addition of acetic acid, as proposed by Classen,† is even more effective than the use of alcohol, or of alcohol with nitric acid, but that when considerable amounts of copper are present the precipitates formed in solutions containing acetic acid are apt to be very finely divided and consequently difficult to filter. A better condition of the precipitate is obtained, however, if, with the acetic acid, there is also present a moderate amount of nitric acid. It appears that acetic acid, when present to the amount of 25 per cent of the liquid, produces in volumes of 100 $cm.^3$ about the same effect as alcohol, and, when present to the amount of 50 per cent, diminishes still further the solvent power of the medium for the oxalate.

* Am. Jour. Sci., [2], xliv, 214.

† Ber. Dtsch. chem. Ges., x, b, 1316.

The additional presence of nitric acid to 10 per cent of the entire volume does not materially affect the solubility. Sulphuric acid to 10 per cent of the volume of the liquid is without apparent effect upon the solubility of copper oxalate, provided the oxalic acid is also present in the proportion of 4 grm. to 100 cm.^{3} of the liquid. Treatment by oxalic acid in a medium consisting of acetic acid of half strength, with or without nitric to the extent of 10 per cent by volume, is plainly the best of the procedures studied for the complete precipitation of copper oxalate in ideal condition.

Prevention of Supersaturation. Gooch and Ward* have made use of various means in the effort to break up supersaturation of the precipitating medium when only small amounts of copper oxalate are present. The supersaturated solution was frozen, and the mass melted, following procedure which has been found to be successful in hastening the deposition of small amounts of ammonium magnesium arsenate;† the supersaturated solution was evaporated to dryness, and the residue extracted with water; alcohol was added to the solution of the copper salt before attempting precipitation by oxalic acid; acetic acid of 50 per cent strength was used as the medium in which precipitation was attempted by oxalic acid. From the experimental results it appears that by precipitating at a volume of 50 cm.^{3}, freezing, melting, and boiling, the condition of supersaturation may be broken up, the oxalate obtained being soluble in the proportion of about 0.0011 grm. to 50 cm.^{3} of liquid; that by precipitation at a volume of 50 cm.^{3}, evaporation to dryness, and extraction with the same volume of water, the copper may be recovered to an amount within about 0.0004 grm. of that taken; that treatment by oxalic acid in 50 per cent alcohol fails to precipitate about 0.0020 grm. of copper from amounts less than 0.0200 grm., while for amounts exceeding that limit the copper is nearly all recovered; and that in volumes of 50 cm.^{3} or 100 cm.^{3}, consisting of 50 per cent acetic acid, the copper oxalate is thrown down completely, the presence of nitric acid to the extent of 10 per cent making the filtration more effective without influencing the solubility, while even at a volume of 150 cm.^{3} the precipitation is complete provided the acetic acid makes up two-thirds of the volume. Acetic acid appears, therefore, to be most effective in

* Loc. cit.

† Gooch and Phelps, see page 290.

breaking up the condition of supersaturation as well as in diminishing the degree of normal solubility in the medium of precipitation.

Precipitation in Presence of Acetic Acid. According to the procedure recommended by Gooch and Ward, for small amounts as well as the larger amounts of copper, oxalic acid, 2 grm. or 4 grm., is added to the copper salt dissolved in 50 cm.^{3} or 100 cm.^{3}, respectively, of the 50 per cent solution of acetic acid containing 5 per cent to 10 per cent of nitric acid to induce a favorable condition for crystallization. After standing over night in contact with the solution, the precipitate is collected upon asbestos in a perforated crucible and washed carefully with small amounts of water. The crucible with its contents is placed in a beaker and covered with about 200 cm.^{3} of hot water containing 25 cm.^{3} of dilute sulphuric acid [1 : 3], and the solution is titrated with $n/10$ potassium permanganate. Results of this procedure are given in the table.

Filtration of Copper Oxalate Precipitated in Solutions Containing 50 per cent Acetic Acid and 5 to 10 per cent Nitric Acid.

Copper taken. grm.	Total volume. cm.^{3}	Oxalic acid. grm.	Acetic acid. cm.^{3}	Nitric acid. cm.^{3}	Copper found. grm.	Error. grm.
Volume 50 cm.^{3}.						
0 0010	50	2	25	5	0.0010	0.0000
0.0020	50	2	25	5	0 0021	+0 0001
0.0031	50	2	25	5	0 0033	+0 0002
0.0041	50	2	25	5	0 0042	+0 0001
0 0051	50	2	25	5	0 0049	−0 0002
0.0102	50	2	25	5	0 0103	+0 0001
0.0204	50	2	25	5	0 0204	0 0000
0.0511	50	2	25	5	0.0512	+0.0001
Volume 100 cm.^{3}.						
0 0031	100	4	50	5	0 0031	0 0000
0 0041	100	4	50	5	0.0041	0 0000
0 0051	100	4	50	5	0 0051	0.0000
0 0102	100	4	50	5	0 0103	+0 0001
0 0204	100	4	50	5	0 0196	−0 0008*
0.0511	100	4	50	5	0 0510	−0.0001
0 0511	110	4	50	10	0 0506	−0 0005
0 0511	100	4	50	10	0 0510	−0.0001
0.1530	100	4	50	10	0.1529	−0 0001
0.1530	100	4	50	10	0.1530	0.0000

* Filtration imperfect.

Separations by Peters' Procedure from Cadmium, Arsenic, Iron, Tin, Zinc.

Peters has shown * that copper exceeding the minimum amount of 0.01 grm. in 50 cm.^{3} (that is, in amount sufficient to be precipitated † according to his procedure), may be successfully separated from cadmium, arsenic, and iron taken as ferric nitrate. When iron is present as the sulphate the results are low. The separation from tin in the stannous form is fairly good for amounts of that element not exceeding 0.1 grm. In presence of the stannic salt the losses are considerable. Attempted separations from bismuth and antimony were unsuccessful, and the separation from zinc unsatisfactory on account of the tendency of zinc oxalate to fall with the copper oxalate. Experimental results are given in the table.

Permanganate Titration of Copper Oxalate in Separations.

CuO taken as $CuSO_4$. grm.	Element from which copper was separated. grm.	Oxalic acid. grm.	HNO_3 (sp. gr. 1.40). cm.^{3}	Volume at precipitation. cm.^{3}	CuO found. grm.	Error. grm.
			Cadmium.			
	CdO taken as $CdSO_4$.					
0.1990	0.10	2.0	5.0	60	0.1983	−0.0007
0.1990	0.20	2.0	5.0	65	0.1987	−0.0003
0.1990	0.30	2.0	5.0	70	0.1987	−0.0003
0.1990	0.40	2.0	5.0	75	0.1994	+0.0004
0.1990	0.50	2.0	5.0	80	0.1996	+0.0006
			Arsenic.			
	As_2O_3 taken as Na_3AsO_3.					
0.1990	0.10	2.0	...	55	0.1991	+0.0001
0.1990	0.20	2.0	...	60	0.1987	−0.0003
0.1990	0.50	2.0	...	75	0.1986	−0.0004
0.1990	0.20	2.0	5.0	60	0.1994	+0.0004
0.1990	0.20	2.0	5.0	75	0.1992	+0.0002
0.1990	0.60	2.0	5.0	85	0.1995	+0.0005
	As_2O_5 taken as H_2KAsO_4.					
0.1990	0.10	2.0	...	60	0.1985	−0.0005
0.1990	0.20	2.0	...	70	0.1990	0.0000
0.1990	0.10	2.0	5.0	65	0.1990	0.0000
0.1990	0.20	2.0	5.0	75	0.1992	+0.0002
0.1990	0.30	2.0	5.0	85	0.1985	−0.0005
0.2030	0.30	3.0	5.0	85	0.2026	−0.0004

* See page 125. † See page 124.

Permanganate Titration of Copper Oxalate in Separations.

CuO taken as $CuSO_4$. grm.	Element from which copper was separated. grm.	Oxalic acid. grm.	HNO_3 (sp. gr. 1.40). cm.^{3}	Volume at precipitation. cm.^{3}	CuO found. grm.	Error. grm.
			Iron.			
	Fe_2O_3 taken as $Fe(NO_3)_3$.					
0 1990	0 136	2 0	5 0	60	0.1987	−0 0003
0 1990	0 272	2 0	5 0	60	0.1983	−0 0007
0 1990	0 364	2 0	5 0	60	0.1988	−0 0002
0 1990	0.544	2.0	5 0	65	0.1971	−0.0019
			Tin.			
	Sn taken as $SnCl_2$+HCl.				Cu found.	
0 1990	0 0468	2.0	5 0	55	0 1979	−0 0011
0 1990	0 0936	2.0	5 0	60	0 2004	+0 0014
0 1990	0 0936	2.0	5 0	60	0.1992	+0 0002
0 1990	0 0936	2.0	5 0	60	0.1995	+0 0005
	Sn taken as $SnCl_4$.					
0 1990	0.10	2 0	5.0	55	0.1979	−0 0011
0.1990	0.10	2.0		55	0.1959	−0 0031
0.1990	0.20	2.0	5.0	55	0.1974	−0 0016
0.1990	0.50	2.0	5 0	60	0.1955	−0 0035
			Zinc.			
	ZnO taken as $ZnSO_4$.					
0 1990	0 028	2.0	5 0	60	0.2007	+0 0017
0.1990	0 057	2.0	5 0	65	0.2008	+0 0018
0.1990	0 057	2 0	5.0	65	0 2008	+0 0018
0.1990	0.085	2.0	5.0	70	0.2035	+0 0045

Separations by the Method of Desiccation.

Ward* has studied the effect of evaporation to dryness, to break up supersaturation, in separations of copper from cadmium, arsenic, and iron. In this process the boiling solution is treated with oxalic acid and nitric acid. The liquid and precipitate are evaporated to dryness, and the residue is extracted with cold water containing not too much nitric acid and, in separations from iron, more oxalic acid. The residual oxalate is filtered off on asbestos in the perforated crucible, washed with small amounts of water, treated

* H. L. Ward, Am. Jour. Sci., [4], xxxiii, 423.

with boiling water containing sulphuric acid, and titrated with permanganate. The details of the preferred treatment and the results are given in the accompanying tables.

Copper from Cadmium by Desiccation Process.

Copper present.	Cadmium present.	Nitric acid at precipitation.	Volume at precipitation.	Oxalic acid.	Liquid used in extraction.	Nitric acid in extraction.	Copper found.	Error.
grm.	grm.	cm.³	cm.³	grm.	cm.³	cm.³	grm.	grm.
0.0051	0.10	5	50	4	50	5	0.0048	−0.0003
0.0514	0.01	5	50	4	50	5	0.0507	−0.0007
0.0514	0.06	5	50	4	50	5	0.0506	−0.0008
0.0504	0.10	5	50	4	50	5	0.0502	−0.0002
0.0514	0.10	5	50	4	50	5	0.0508	−0.0006
0.0514	0.20	5	50	4	50	5	0.0508	−0.0006
0.0516	0.30	5	50	4	50	5	0.0507	−0.0009
0.1542	0.20	5	50	4	50	5	0.1537	−0.0005

Copper from Arsenic by Desiccation Process.

Copper present.	Arsenic as arsenate.	Volume at precipitation.	Oxalic acid.	Liquid used in extraction.	Nitric acid in extraction.	Copper found.	Error.
grm.	grm.	cm.³	grm.	cm.³	cm.³	grm.	grm.
0.0051	0.10	50	1	50	2	0.0047	−0.0004
0.0504	0.05	50	1	50	2	0.0499	−0.0005
0.0504	0.05	50	1	50	2	0.0501	−0.0003
0.0504	0.10	50	1	50	2	0.0503	−0.0001
0.0504	0.10	50	1	50	2	0.0497	−0.0007
0.0504	0.20	50	1	50	2	0.0498	−0.0006
0.1533	0.20	50	1	50	2	0.1528	−0.0005

Copper from Iron by Desiccation Process.

Copper present.	Iron present.	Volume at precipitation.	Oxalic acid in precipitation.	Liquid used in extraction.	Nitric acid in extraction.	Oxalic acid added in extraction	Copper found.	Error.
grm.	grm.	cm.³	grm.	cm.³	cm.³	grm.	grm.	grm.
0.0504	0.0393	50	1	50	2	2	0.0500	−0.0004
0.0504	0.0393	50	1	50	2	2	0.0501	−0.0003
0.0504	0.0393	50	1	50	2	2	0.0499	−0.0005
0.0504	0.0786	50	1	50	2	2	0.0499	−0.0005
0.0511	0.1000*	50	1	50	2	3	0.0506	−0.0005
0.1533	0.1000*	50	1	50	2	3	0.1527	−0.0006

* More than 0.1 grm. of iron apparently occasions greater solubility of the copper oxalate.

Separations in Presence of Acetic Acid.

WARD* investigated also the application of the process described above for precipitating copper oxalate in presence of 50 per cent acetic acid, and of nitric acid up to 10 per cent to favor crystallization, to separations of copper from cadmium, arsenic, iron and zinc. Details and results are given in the tables.

Copper from Cadmium in 50 per cent Acetic Acid.

Copper present. grm.	Cadmium present. grm.	Volume at precipitation. cm.[3]	Oxalic acid. grm.	Acetic acid. cm.[3]	Nitric acid. cm.[3]	Copper found. grm.	Error. grm.
0.0051	0.20	100	4	50	5	0.0049	−0 0002
0.0051	0.30	100	4	50	5	0.0053	+0 0002
0.0255	0 20	100	4	50	10	0.0257	+0.0002
0.0510	0 20	100	4	50	10	0.0512	+0.0002
0.0511	0 30	100	4	50	10	0.0520	+0.0009
0.1533	0 30	100	4	50	10	0.1556	+0.0023
0.1629	0 30	100	4	50*	10	0 1636	+0.0007

* The acetic acid was added after precipitation. This apparently makes a sharper separation of the cadmium from large amounts of copper.

Copper from Arsenic in 50 per cent Acetic Acid.

Copper present. grm.	Arsenic present as arsenate. grm.	Volume at precipitation. cm.[3]	Oxalic acid. grm.	Acetic acid. cm.[3]	Nitric acid. cm.[3]	Copper found. grm.	Error. grm.
0.0051	0 30	100	4	50	0.4	0.0054	+0 0003
0.0511	0.30	100	4	50	0 4	0 0505	−0 0006
0.1530	0.20	100	4	50	10	0 1530	0 0000
0.1535	0.30	100	5	50	10	0.1530	−0.0005

Copper from Iron in 50 per cent Acetic Acid.

Copper present. grm.	Iron present. grm.	Volume at precipitation. grm.	Oxalic acid. grm.	Acetic acid. grm.	Nitric acid. grm.	Copper found. grm.	Error. grm.
0 0510	0.188	100	4	50	..	0.0514	+0 0004
0 0510	0.188	100	4	50	2	0.0511	+0.0001
0.0510	0.188	100	4	50	5	0 0499	−0 0011
0 0511	0.100	100	4	50	10	0.0489	−0 0022
0 0510	0.188	100	4	50	10	0.0487	−0.0023

Much free nitric acid obviously occasions loss of copper oxalate. Ward advocates as the best procedure in separating

* Loc. cit.

copper from iron the nearly complete precipitation of the copper by oxalic acid added to the boiling solution devoid of other free acid, with the subsequent addition of acetic acid in amount equal to twice the volume of the solution, to complete the precipitation, after cooling. In this manner the results of the following table were obtained.

Copper from Iron by Precipitation in Water Solution with Subsequent Addition of Acetic Acid.

Copper present.	Iron present.	Volume at precipitation.	Oxalic acid	Acetic acid.	Copper found.	Error.
grm.	grm.	cm.^{3}	grm.	cm.^{3}	grm.	grm.
0.0051	0.31	50	6	100	0.0049	−0.0002
0.0051	0.45	50	6	100	0.0046	−0.0005
0.0543	0.15	50	6	100	0.0544	+0.0001
0.0543	0.21	50	6	100	0.0542	−0.0001
0.0543	0.31	50	6	100	0.0546	+0.0003
0.0543	0.45	100	12	200	0.0538	−0.0005
0.1629	0.45	50	6	100	0.1649	+0.0020
0.1629	0.45	100	12	200	0.1629	0.0000

When copper and iron are found together in acid solution, the free acid should either be removed by evaporation or neutralized by potassium hydroxide and the solution made faintly acid with acetic acid before precipitating the copper as oxalate. Ward has shown that the presence of moderate amounts of potassium sulphate, nitrate or chloride does not affect appreciably the analytical results. The presence of ammonium salts, must, however, be avoided, on account of the tendency of copper oxalate to form a soluble double ammonium oxalate.

Determination of Copper Associated with Lead. The oxalate of lead, though fairly soluble in nitric acid, falls with copper oxalate from the nitric acid solution, and so the separation of those elements by precipitation of copper oxalate from the acid solution is not feasible. Ward* has shown, however, that lead may be first precipitated completely as lead sulphate, and that then, either with or without previous removal of the precipitated sulphate, the copper may be determined by precipitation and titration of the oxalate. To the solution of lead and copper in the form of nitrates are added an equal volume of acetic acid and 3 cm.^{3}–5 cm.^{3} of sulphuric acid. The precipitated sulphate may be

* Loc. cit.

filtered off and weighed and the copper estimated in the filtrate by precipitation and titration as already described; or, if a determination of copper only is desired, the precipitation of the copper oxalate may be effected without removing the lead sulphate, sulphate and oxalate being filtered off together, treated with boiling dilute sulphuric acid and titrated to color with permanganate. Results of each procedure are given below.

Copper and Lead.

Copper present.	Lead present.	Sulphuric acid.	Acetic acid.	Volume	Oxalic acid.	Copper found.	Error.	Lead found.	Error.
grm.	grm	cm.³	cm.³	cm.³	grm.	grm.	grm.	grm.	grm.
Titration of copper oxalate precipitated after separation of lead sulphate.									
0 0511	0.0500	3	50	110	2	0.0513	+0 0002	0 0499	−0 0001
0 0511	0.1000	5	50	100	2	0.0508	−0 0003	0 0996	−0 0004
0 0511	0.1000	5	50	100	2	0.0508	−0 0003	0 0997	−0 0003
Titration of copper oxalate without separation of lead sulphate.									
0 0051	0.30	5	50	100	4	0.0052	+0 0001	.....	....
0.0511	0.10	5	50	100	2	0.0508	−0.0003	.....	....
0.0511	0 25	5	50	100	2	0.0511	0.0000	.....	....
0.0543	0.30	10	50	100	4	0.0537	−0.0006	.....	....
0.0511	0 30	3	50	100	2	0.0509	−0 0002	.....	....
0 0511	0 40	3	50	100	4	0.0508	−0.0003	.....	....
0 1022	0.30	10	50	100	2	0.1018	−0.0004	.....	....
0.1086	0 25	5	50	100	4	0.1081	−0.0005	.....	....
0.1533	0.20	5	50	100	2	0.1527	−0 0006	.....	....
0.1533	0 20	5	50	100	2	0.1530	−0 0003	.....	....

Silver.

The Gravimetric Determination of Silver as the Chromate.

The precipitation of silver chromate from the solution of a soluble chromate made faintly acid with acetic acid may be carried to completion by the addition of silver nitrate in considerable excess. The exact determination of the chromium of a soluble chromate or dichromate may therefore be effected by treating with silver nitrate the solution of either salt, adding ammonia to alkalinity and then acetic acid to faint acidity, transferring the precipitate and washing it in the filtering crucible with a dilute solution of silver nitrate until foreign material other than that reagent has been removed, finishing the washing with a small amount of water applied judiciously in portions, and weighing the dried or gently ignited residue of silver chromate.* The success of this process turns upon keeping the chromium at the moment of precipitation essentially in the form of chromate rather than dichromate, and in taking care that an excess of silver nitrate shall be present nearly to the end of the washing. Gooch and Bosworth† have investigated the conditions under which, in reversal of the process just described, silver may be precipitated completely as the chromate, and find that from solutions of silver nitrate alone or containing free nitric acid, potassium chromate precipitates silver chromate completely, provided enough potassium chromate is present to take up the nitric acid with formation of potassium nitrate or dichromate, as well as to form the silver salt. The precipitate, filtered at once or brought to better crystalline condition by dissolving it in ammonia and boiling the solution to small volume, may be transferred to the asbestos filter by dilute potassium chromate

Silver Weighed as Silver Chromate.

Silver taken as $AgNO_3$.		K_2CrO_4 used.		HNO_3.		Ag_2CrO_4 weighed.	Silver found.	Error in terms of silver.
Volume of solution.	Weight.	Volume of solution.	Weight.	Volume.	Weight.			
cm.³	grm.	cm.³	grm.	cm.³	grm.	grm.	grm.	grm.
Precipitation by K_2CrO_4.								
15	0.1652	50	0.3	..	....	0.2536	0.1649	−0.0003
10	0.1101	50	0.3	..	....	0.1693	0.1101	0.0000
25	0.1437	50	0.3	.	...	0.2200	0.1436	−0.0001
25	0.1437	50	0.3	..	...	0.2210	0.1437	0.0000
Precipitation by K_2CrO_4 in presence of HNO_3.								
25	0.1355	50	0.9	10	0.182	0.2091	0.1360	+0.0005
25	0.1355	50	0.9	10	0.182	0.2081	0.1353	−0.0002
25	0.1358	50	0.9	10	0.182	0.2090	0.1360	+0.0002
25	0.1355	50	0.9	10	0.182	0.2075	0.1349	−0.0006
25	0.1355	50	0.9	10	0.182	0.2090	0.1360	+0.0005
Precipitation by K_2CrO_4 in presence of HNO_3, treatment with NH_4OH, and boiling to a volume of 10 to 15 cm.³.								
25	0.1348	50	0.6	10	0.063	0.2076	0.1350	+0.0002
25	0.1348	50	0.6	10	0.063	0.2068	0.1344	−0.0004
25	0.1348	50	0.6	10	0.063	0.2072	0.1347	−0.0001
25	0.1348	50	0.6	10	0.063	0.2074	0.1348	0.0000
25	0.1348	50	0.6	10	0.063	0.2070	0.1346	−0.0002

* See page 406.

† F. A. Gooch and Rowland S. Bosworth, Am. Jour. Sci., [4], xxvii, 241.

and washed by small portions of water without appreciable loss. The weight of the residue of silver chromate, dried at gentle heat, may be taken as a measure of the silver originally present.

The Electrolytic Determination of Silver.

Silver may be determined by deposition of the metal upon the rotary cathode * from an ammoniacal cyanide solution in presence of a few grams of ammonium sulphate, the method of manipulation being precisely similar to that employed in the deposition of copper as previously described.†

The following table records determinations made in this manner‡ with a solution of silver nitrate standardized as the chloride.

Deposition of Silver from Ammoniacal Cyanide Solution.

Silver taken. grm.	Silver found. grm.	Error. grm.	Current. amp.	$N.D._{100}$.	Time. min.
0 0968	0.0966	−0.0002	1.8	6	15
0 0968	0 0967	−0 0001	1.9	6.3	15
0 0968	0.0965	−0.0003	1.8	6	15
0.0968	0.0969	+0 0001	2	6.7	10
0.0968	0 0965	−0 0003	3	10	8
0.1898	0.1901	+0 0003	2.5	8 3	10
0 1898	0 1898	0.0000	2.5	8.3	10
0.1898	0.1900	+0 0002	3	10	10
0.1898	0.1893	−0.0005	2.5	8.3	10

Gooch and Feiser § have determined silver by depositing it from an ammoniacal solution of the oxalate, as was done by Gooch and Read‖ in the preparation of silver-plated electrodes, in order to avoid all possible contamination of the silver deposit by nonvolatile material. In testing this process, measured amounts of the silver nitrate solution (25 cm.^{3} or 50 cm.^{3}) were drawn from a burette into a small beaker and treated with ammonium oxalate to complete precipitation. The silver oxalate was dissolved in a slight excess of ammonia, and this solution, diluted to 100 cm.^{3}, was electrolyzed with the rotary cathode and a current of 0.25 amp.–1.5 amp. and 4–7 volts. The cathode

* See Fig. 13.

† See page 116.

‡ Gooch and Medway, Am. Jour. Sci., [4], xv, 320.

§ F. A. Gooch and J. P. Feiser, Am. Jour. Sci., [4], xxxi, 109.

‖ F. A. Gooch and H. L. Read, Am. Jour. Sci., [4], xxviii, 544.

with the deposited silver was dried cautiously over a low flame and thereafter ignited to incipient redness. The details of individual experiments are given in the table.

Electrolysis of Silver Nitrate Dissolved in Ammonium Oxalate and Ammonia.

Ag in $AgNO_3$ taken. grm.	Ag found. grm.	Error. grm.	Current. Amp.	Current. Approx. N. D_{100}.	Current. Volt.	Time. min.	Revolutions per min.
			A crucible used as cathode.				
0.2687	0.2685	−0.0002*	1.5−1	5−3.3	6−7	25	500
0.2687	0.2687	0.0000*	1.5−1	5−3.3	6−7	30	500
0.2687	0.2684	−0.0003*	1−0.5	3.3−1.7	6−7	30	450
0.2687	0.2685	−0.0002*	1−0.5	3.3−1.7	4−6	30	450
0.3183	0.3181	−0.0002*	1.5−1	5−3.3	4−6	20	450
0.3183	0.3178	−0.0005†	1.5−1	5−3.3	4−6	10	450
			Gauze disks used as cathode.				
0.3183	0.3179	−0.0004*	1−0.5	0.5−0.25	4−6	20	400
0.3183	0.3182	−0.0001*	1−0.25	0.5−0.10	6−8	25	400
0.3183	0.3181	−0.0002*	1−0.25	0.5−0.10	5−8	25	400
0.3183	0.3180	−0.0003*	0.75−0.25	0.4−0.10	5−7	40	450
0.3183	0.3180	−0.0003*	1−0.25	0.5−0.10	4−6	40	450
0.3183	0.3176	−0.0007†	0.75−0.25	0.4−0.10	6	20	450
			Gauze cone used as cathode.				
0.2687	0.2686	−0.0001*	1.5−1	3−2	4−6	25	500
0.2687	0.2683	−0.0004*	1−25	2−0.5	6−7	30	450
0.2687	0.2684	−0.0003*	1−5	2−1	4−6	25	450
0.2687	0.2686	−0.0001*	1−0.25	2−0.5	4−6	25	450
0.5375	0.5373	−0.0002*	1.5−1	3−2	6−7	25	450
0.5375	0.5371	−0.0004*	1.5−1	3.2	6−7	25	500

* Deposition complete, as shown by H_2S test.
† Deposition incomplete, as shown by H_2S test.

The details of other experiments in which the silver was first precipitated as silver chloride and then deposited from the solution in ammonia and ammonium oxalate are given in the following table. In these experiments, in which the solutions were more strongly ammoniacal than those of the experiments of the preceding series, the deposits were dark and spongy, but they became lighter in color and more compact upon drying.

Electrolysis of Silver Chloride Dissolved in Ammonium Oxalate and Ammonia.

Ag in $AgNO_3$ taken. grm.	Ag found. grm.	Error. grm.	Current. Amp.	Current. Approx. $N.D._{100}$.	Current. Volt.	Time. min.	Revolutions per min
A crucible used as cathode.							
0 3191	0.3187	−0.0004	1.5–1	5–33	5–7	20	500
0.3191	0.3189	−0.0002	1 5–1	5–33	4–6	30	500
Gauze disks used as cathode.							
0.3191	0 3185	−0.0006*	1 5–1	0.75–0 5	5–7	15	500
0.3191	0.3189	−0.0002	1 5–1	0.75–0 5	5–7	25	500
0.3191	0.3190	−0 0001	1.5–1	0.75–0.5	4–6	35	500

* Time of electrolysis short.

It appears, therefore, that silver may be deposited from an ammoniacal solution of the oxalate, in presence of ammonium nitrate or ammonium chloride, in pure condition and in form suitable for quantitative estimation.

When the current density is high the deposit is apt to be voluminous, shrinking considerably upon drying, and this phenomenon was especially notable in deposits upon the comparatively small and smooth surface of the crucible. The best form of apparatus for this process appears to be the gauze cone set point downward, and so placed with relation to an annular platinum band used as the anode that the end of the axis, where the mechanical effect of rotation is least, shall not receive much of the deposit. Experiments made with stationary gauze electrodes were not successful, nor were those made with a dish cathode and stirring anode.

The Iodometric Estimation of Silver, based upon the Use of Potassium Chromate as a Precipitant.

With proper precautions, silver may be precipitated and accurately estimated as silver chromate.* The addition of a sufficient excess of potassium chromate to a solution of silver nitrate, even in the presence of small amounts of nitric acid, brings about a complete precipitation of the silver as silver chromate, and

* See page 136.

the precipitate thus obtained may be transferred to the asbestos filter by means of a dilute solution of potassium chromate and washed with small amounts of water without appreciable loss of silver chromate. Upon the basis of such exact precipitation of silver chromate by potassium chromate, Gooch and Bosworth * have accomplished the iodometric estimation of silver, both by the determination of the chromic acid ion of the potassium chromate which remains after the precipitation of the silver salt by a known amount of standard potassium chromate, and by the estimation of the chromic acid ion of the precipitated and washed silver chromate.

Precipitation of Silver Chromate and Determination of the Excess of the Precipitant.

Silver taken.		K_2CrO_4.		$NaNO_3$ present.	$Na_2S_2O_3$ used.	Silver found.	Error in terms of silver.
Volume of solution.	Weight.	Volume.	Weight.				
cm.^{3}	grm.	cm.^{3}.	grm.	grm.	cm.^{3}.	grm.	grm.
20	0.1261	..	0 3039		26.54	0 1262	+0.0001
25	0.1576		0.3039		22.61	0 1575	−0.0001
15	0.0946		0.3293		30.51	0 0946	0.0000
15	0.0946		0.3293		30 60	0 0940	−0 0006
15	0.0946		0.3293		30.57	0 0941	−0.0005
15	0 0946		0.3293	.	30 55	0 0945	−0 0001
19.98	0.1260	.	0.3293	.	32.20	0 1255	−0 0005
20	0.1261	..	0.3293	..	32.09	0 1263	+0.0002
20	0.1261	.	0.3293	..	32.10	0 1263	+0.0002
25	0.1576	..	0.3293	..	27 30	0 1576	0.0000
10	0.1101	37	0.2436	1	20 47	0 1107	+0.0006
10	0.1101	37	0.2436	1	20.53	0.1103	+0.0002
10	0.1101	25	0.1647	1	9 07	0 1097	−0 0004
10	0.1101	27	0.1778	1	11.02	0.1096	−0.0005
15	0.0862	30	0.1974	2	17.14	0.0859	−0.0003
15	0.0862	30	0.1974	1	17.05	0 0865	+0.0003
15	0.0862	30	0.1974	1	17.14	0 0859	−0 0003
25	0.1437	50	0.3294	10	28 53	0.1435	−0 0002

According to the first procedure a known amount of standard potassium chromate in excess of the amount needed to precipitate the silver is added to the solution of silver nitrate. The precipitate is dissolved in ammonia and reprecipitation brought about by boiling to a volume of 10 cm.^{3}–15 cm.^{3}. The second, crystalline precipitate is filtered upon asbestos and washed with the least possible amount of water applied in small portions successively. The filtrate is treated with potassium iodide and acidified with

* F. A. Gooch and Rowland S. Bosworth, Am. Jour. Sci., [4], xxvii, 302.

sulphuric acid. The iodine set free is titrated with sodium thiosulphate. The difference between the silver value of the iodine thus found and that of the potassium dichromate used is taken as the measure of the silver present. In the table are given the details of experiments performed in accordance with this procedure.

Inasmuch as relatively large amounts of potassium chromate are necessary to bring about complete precipitation of the silver chromate in the presence of nitric acid, the above procedure is less adapted to the determination of silver in a solution containing that acid than the second method whereby the precipitated and washed silver chromate is determined. In this second method the silver solution containing free nitric acid is treated with potassium chromate in excess of the amount necessary to take up the nitric acid with formation of potassium dichromate. The precipitate is dissolved in ammonia and reprecipitation is effected by boiling to a volume of 10 cm.^{3}–15 cm.^{3}. The second, crystalline precipitate is transferred to an asbestos filter by means of a dilute solution of potassium chromate, washed with the least possible amount of water applied in small portions successively, and dissolved in a few cm.^{3} of a strong solution of potassium iodide. The solution in potassium iodide is diluted and acidified with sulphuric acid. The iodine set free is titrated with sodium thiosulphate and taken as the measure of the silver present.

Results of this procedure are given in the following table.

Iodometric Determination of Precipitated Silver Chromate.

Silver taken.		HNO_3 present.	K_2CrO_4, weight.	$Na_2S_2O_3$ used.	Silver found.	Error in terms of silver.
Volume of solution.	Weight.					
cm.^{3}	grm.	grm.	grm.	cm.^{3}	grm.	grm.
25	0.1348	0.063	0 60	18 35	0.1342	−0.0006
20	0.1078	0 063	0.65	14.67	0.1073	−0 0005
15	0.0808	0 063	0.65	11 06	0.0809	+0.0001
15	0.0808	0.063	0.65	10.96	0 0802	−0 0006
20	0.1078	0.063	0.65	14.67	0.1073	−0 0005
30	0.1618	0.063	0.75	22.02	0.1610	−0 0008
20	0.1078	0.063	0.65	14.71	0.1075	−0 0003
25	0.1348	0.063	0.65	18.41	0 1347	−0 0001
25	0.1348	0.063	0.65	18.41	0 1347	−0 0001
20	0.1078	0.063	0.65	14.74	0.1078	0.0000

The Iodometric Determination of Silver Based upon the Reducing Action of Potassium Arsenite.

It is well known that an ammoniacal solution of silver arsenite deposits metallic silver when the ammonia is evaporated by boiling. During this reaction, by which the silver salt is reduced, the arsenious acid becomes oxidized to the higher condition of oxidation, where arsenic has a valence of five, according to the equation

$$2\,Ag_2O + As_2O_3 = As_2O_5 + 4\,Ag.$$

Bosworth * has shown that this reaction is quantitative and capable of serving as the basis of a reliable iodometric method for the determination of silver. To the solution of silver taken as the nitrate is added a known volume of a standard potassium arsenite solution in excess of the amount necessary to reduce the silver salt present. Ammonia is added in sufficient quantity to dissolve the precipitate formed, or 25 cm.^{3} of a saturated solution of sodium bicarbonate may be used instead of the ammonia. The resulting solution is diluted to 100 cm.^{3}, and boiled. The solution, out of which metallic silver separates, is cooled, slightly acidified, and then made alkaline with sodium bicarbonate. The excess of potassium arsenite is titrated with $n/10$ iodine. The silver value of the iodine used is subtracted from that of the potassium arsenite originally taken, and the result used as a measure of the silver present.

Results of this procedure are given in the table.

Reduction of Silver by Arsenite and Titration of the Excess.

Silver taken.	KH_2AsO_3 added.		I used.		Silver found.	Error in terms of silver.
		Silver value.		Silver value.		
grm.	cm.^{3}	grm.	cm.^{3}	grm.	grm.	grm.
Use of NH_4OH with filtering.						
0.1054	20	0.2000	8.53	0.0949	0.1051	−0.0003
0.1054	20	0.2000	8.52	0 0948	0.1052	−0.0002
0.1054	30	0 3000	17.42	0.1939	0.1061	+0.0007
0.1159	20	0.2000	7.60	0.0846	0.1154	−0.0005
0.1054	21	0.2100	9.37	0.1043	0.1057	+0.0003
0.1054	20	0.2000	8.48	0.0944	0.1056	+0.0002

* Rowland S. Bosworth, Am. Jour. Sci. [4], xxviii, 287.

Reduction of Silver by Arsenite and Titration of the Excess.

Silver taken.	KH_2AsO_3 added.		I used.		Silver found.	Error in terms of silver.
		Silver value.		Silver value		
grm.	cm.³	grm.	cm.³	grm.	grm.	grm.
Use of $NaHCO_3$ with filtering.						
0 1054	15	0.1618	5.65	0 0571	0.1047	−0.0007
0.1054	23	0.2481	14.14	0.1430	0.1051	−0 0003
0.1054	12	0.1295	2.40	0 0243	0.1052	−0 0002
0.1054	15	0.1618	5.60	0 0566	0.1052	−0 0002
0.1054	15	0.1618	5.55	0.0561	0.1057	+0.0003
0.1054	20	0.2158	10.91	0.1104	0.1054	0 0000
0.2635	35	0.3776	11 33	0 1146	0.2630	−0.0005
Use of NH_4OH. Titration carried on in presence of the precipitate.						
0.1054	20	0.2000	8 55	0.0952	0.1048	−0.0006
0.1054	20	0.2000	8.50	0 0946	0.1054	0.0000
0.1054	23	0.2300	11.28	0 1256	0.1044	−0.0010
0.1054	20	0 2000	8.45	0 0941	0.1059	+0 0005
0.1054	20	0.2000	8.48	0 0944	0.1056	+0.0002
Use of $NaHCO_3$. Titration carried on in presence of the precipitate.						
0.1054	18	0.1800	6.80	0 0757	0.1043	−0.0011
0.1054	17	0.1700	5.81	0.0647	0 1053	−0 0001
0.1054	15	0.1500	4.00	0 0445	0.1055	+0 0001
0.1054	21	0.2100	9.45	0.1052	0.1048	−0 0006
0.1054	25	0.2500	13 00	0.1447	0 1053	−0 0001
0.1054	31	0.3100	18 40	0.2048	0 1052	−0.0002
Use of $NaHCO_3$. 2 grm. of $NaNO_3$ present. Titration carried on in presence of precipitate.						
0.0949	21	0.2100	10.42	0.1160	0 0940	−0.0009
0.1054	21	0.2100	9.43	0.1050	0.1050	−0.0004
0.1265	20	0.2000	6.60	0.0735	0 1265	0.0000
0.1686	21	0.2100	3.80	0 0432	0.1678	−0.0008
0.1054	15	0.1500	4.08	0 0454	0.1046	−0.0008

This process proves to be applicable to the determination of the silver in freshly precipitated silver chloride, so that its range is thus extended to the determination of silver in many mixtures.

According to the procedure outlined, the freshly precipitated silver chloride is acted upon by ammonia until dissolved. The solution is diluted to 100 cm.³ and the reduction is effected, adding an excess of standard arsenite and boiling the mixture. The excess of arsenite is titrated according to the method described above.

Results of this procedure, including separations from copper and lead, are given in the following table.

Reduction of Silver Chloride after Separations.

Silver taken.	KH_2AsO_3 added.		I used.		Silver found.	Error in terms of silver.
		Silver value.		Silver value.		
grm.	cm.³	grm.	cm.³	grm.	grm.	grm.
Reduction of precipitated AgCl.						
0.1017	15	0.1619	5.40	0.0599	0.1020	+0.0003
0.1017	15	0.1619	5.44	0.0603	0.1016	−0.0001
0.1017	15	0.1619	5.40	0.0599	0.1020	+0.0003
0.1017	15	0.1619	5.42	0.0601	0.1018	+0.0001
0.1017	17	0.1834	7.44	0.0825	0.1009	−0.0008
Reduction of AgCl precipitated in the presence of 0.09 grm. of copper.						
0.1017	15	0.1619	5.41	0.0600	0.1019	+0.0002
0.1017	15	0.1619	5.44	0.0603	0.1016	−0.0001
0.1017	15	0.1619	5.39	0.0598	0.1021	+0.0004
Reduction of AgCl precipitated in the presence of 0.2 grm. of lead.						
0.1220	16	0.1726	4.57	0.0507	0.1219	−0.0001
0.1108	15	0.1619	4.60	0.0510	0.1109	+0.0001
Reduction of AgCl precipitated from a solution containing 0.09 grm. of copper and 0.2 grm. of lead.						
0.1017	15	0.1619	5.45	0.0604	0.1015	−0.0002

GOLD.

The Electrolytic Determination of Gold.

Medway* has shown that from an ammoniacal cyanide solution gold may be deposited rapidly and in good form upon the rotating crucible used as the cathode.

* H. E. Medway, Am. Jour. Sci., [4], xviii, 56.

Deposition from an Ammoniacal Cyanide Solution.

Gold taken. grm	Gold found. grm.	Error. grm.	Current. amp.	N. D.$_{100}$.	Time. min.
0.0695	0.0694	−0 0001	2	6.6	30
0.0695	0.0696	+0.0001	2	6.6	30
0 0598	0 0598	0.0000	0.5	1.8	30
0 0598	0 0598	0 0000	0 5	1 8	30
0 0598	0 05975	−0.00005	1	3.3	25

The Iodometric Estimation of Small Amounts of Gold.

Gooch and Morley* have shown that when potassium iodide reacts at suitable concentration upon small amounts of gold trichloride in solution, the reaction takes place regularly and in accordance with the theory that two molecules of the thiosulphate are the equivalent of two atoms of iodine and one atom of gold.

$$AuCl_3 + 2\,KI + 2\,Na_2S_2O_3 = AuCl + Na_2S_4O_6 + 2\,KCl + 2\,NaI.$$

The reduction of the auric salt, with the consequent liberation of iodine, is, however, conditioned by the volume of the solution, the mass of the iodine present, and the time of action. The following statement, in which each result is the average of several titrations in close agreement, shows the effect upon the immediate evolution of iodine brought about by adding varying amounts of water to the gold solution before introducing the iodide, and the effect of different amounts of iodide at different dilutions.

	Potassium iodide.					Gold chloride.	Volume before the addition of the thio-sulphate.
	0.01 grm.	0.02 grm.	0.05 grm.	0 1 grm.	0.2 grm.	grm.	cm.
Sodium thiosulphate, nearly *n*/100. cm.^{3}	0 81	0 81	0 81	0 82	0.84	0 00087	15
	0.77	0 78	0 80	0.81	0 81	"	25
	0 74	0 72	0.78	0 79	0.80	"	50
	0.61	0 61	0 68	0 76	0.79	"	100
	0 45	0 49	0.60	0.72	0.75	"	200

It is evident that for the smaller amounts of iodide the liberation of iodine decreases rapidly with the dilution. The larger amounts at the highest concentration show readings a trifle

* F. A. Gooch and Frederick H. Morley, Am. Jour. Sci., [4], viii, 261.

above the normal — perhaps because the well-known effect of concentrated solutions of a soluble iodide upon the delicacy of the starch end-color begins to appear. At volumes lying between the limit of 25 cm.³ and 50 cm.³, 0.1 grm. of potassium iodide is an appropriate amount to use; at a volume of 15 cm.³, 0.01 grm. to 0.05 grm. of the iodide will do the work; and at lower dilutions, as will appear in the tabular statements to follow, even less of the iodide is effective.

In carrying out the process of analysis, a convenient amount of the solution of gold chloride is drawn from a burette, potassium iodide is introduced in amounts always several times the theoretical equivalent of the gold, and more than enough to dissolve the aurous iodide precipitated at first, a sufficiency of clear starch indicator is added, the starch blue bleached by standardized thiosulphate, and standardized iodine added until the liquid assumes a faint rose color.

Experimental results are given in the following table:

Solutions Approximately n/100

Gold chloride =0.8710 grm. to 1 liter.
Sodium thiosulphate, nearly *n*/100, =1.7012 grm. to 1 liter.
Iodine, nearly *n*/100, =1 3697 grm. to 1 liter.
Volume at beginning of titration, approximately 50 cm.³.

$AuCl_3$ taken. cm.³	KI taken. grm.	$Na_2S_2O_3$ used. cm.³	Gold found. grm.	Theory for gold. grm.	Error. grm.
5	0 05	4.02	0 00426	0 00435	−0 00009
5	0 05	4.01	0 00425	0 00435	−0 00010
5	0 05	4.06	0 00431	0 00435	−0 00004
5	0 05	4.07	0 00432	0 00435	−0 00003
5	0 05	4.04	0 00428	0 00435	−0 00007
10	0 08	8.17	0 00867	0 00871	−0 00004
10	0 08	8.15	0 00864	0 00871	−0 00007
10	0 08	8.16	0 00865	0 00871	−0 00006
10	0.08	8.15	0 00864	0 00871	−0 00007
10	0 08	8 19	0 00869	0 00871	−0 00002
10	0 08	8 46	0 00897	0 00871	+0 00026
10	0.08	8.24	0 00874	0 00871	+0 00003

When approximately centinormal solutions of gold, iodine and thiosulphate are used, an error of 0.01 cm.³ in reading the volume corresponds to an error of 0.00001 grm. of gold. It is not to be expected that such readings can be trusted ordinarily to a higher degree of accuracy than 0.02 cm.³. In case all three solutions should be read to this limit of accuracy with the errors of all lying in the same direction, the summation of error would correspond to 0.00006 grm. of gold.

Errors in reading are, of course, reduced when *n*/1000 solutions are employed, but the use of *n*/1000 iodine necessitates a correction of 0.1 cm.³ for volumes not exceeding 30 cm.³, that being the amount necessary to bring out the rose color in blank tests containing no gold. After the introduction of 0.1 cm.³ of *n*/1000 iodine into a mixture of potassium iodide and starch indicator of volume not exceeding 30 cm.³, a single drop of the gold solution — equivalent to 0.000002 grm. of gold — gives a distinct rose color; before such adjustment of the solution five drops — equivalent to 0.000010 grm. of gold — must be added to develop the same color.

The following table gives the data of tests with such solutions.

Solutions Approximately n/1000

Gold chloride =0 0871 grm. to 1 liter.
Sodium thiosulphate, nearly *n*/1000 =0 17012 grm. to 1 liter.
Iodine, nearly *n*/1000 =0.13697 grm. to 1 liter.

$AuCl_3$ taken.	KI taken.	$Na_2S_2O_3$ used.	Gold taken.	Gold found.	Error.
cm.³	grm.	cm.³	grm.	grm.	grm.
10	0.01	8.39	0.000871	0 000890	+0 000019
9	0.01	7.45	0.000784	0 000790	+0 000006
8	0.01	6.30	0 000697	0 000668	−0 000029
7	0.008	5.50	0.000610	0.000583	−0 000027
6	0.008	5.12	0.000523	0 000543	+0 000020
5	0.005	4.23	0 000435	0 000449	+0.000014
4	0 005	3.38	0.000348	0 000358	+0 000010
3	0.003	2.55	0 000261	0 000270	+0.000009
2	0.003	1.71	0 000174	0 000181	+0 000007
1	0.003	0.90	0 000087	0.000095	+0.000008

In the practical application of the process to the determination of gold, the elementary form of that metal is the natural starting point. To get the metal into solution with chlorine water or mixed hydrochloric and nitric acids is an easy matter, but the removal of the excess of the oxidizer by evaporation without reducing some auric chloride to the aurous form is difficult. Free chlorine may, however, be removed from a solution of auric chloride, without reducing the auric salt, by treatment of the solution with ammonia in excess, boiling gently, acidifying with hydrochloric acid, and heating if necessary to redissolve the precipitate by ammonia, again treating with ammonia and heating, and once more acidifying. On the second addition of ammonia no precipitation usually takes place with these small amounts of gold.

The following table contains determinations made with a solution of pure gold leaf.

Iodometric Determination of Gold.

Gold chloride made by dissolving 0.0104 grm. of pure gold in the manner described and diluting to 200 cm.^{3}.

Sodium thiosulphate, nearly $n/1000$, =0.17012 grm. to 1 liter.
Iodine, nearly $n/1000$ =0.13697 grm. to 1 liter.
Potassium iodide =10 grm. to 1 liter.

Portions were treated with the potassium iodide without previous dilution.

$AuCl_3$ taken.	KI taken.	$Na_2S_2O_3$ used.	Gold taken.	Gold found.	Error.
cm.^{3}	grm.	cm.^{3}	grm.	grm.	grm.
1	0.005	0.55	0.000052	0.000058	+0.000006
1	0.005	0.55	0.000052	0.000058	+0.000006
2	0.005	1.06	0.000104	0.000112	+0.000008
2	0.005	1.08	0.000104	0.000114	+0.000010
5	0.01	2.45	0.000260	0.000260	0.000000
5	0.01	2.50	0.000260	0.000265	+0.000005
5	0.01	2.45	0.000260	0.000260	0.000000
5	0.01	2.50	0.000260	0.000265	+0.000005
5	0.01	2.50	0.000260	0.000265	+0.000005
10	0.02	4.86	0.000520	0.000515	−0.000005
10	0.02	4.85	0.000520	0.000517	−0.000003
10	0.02	4.90	0.000520	0.000520	0.000000
10	0.02	4.80	0.000520	0.000512	−0.000008
10	0.02	4.84	0.000520	0.000516	−0.000004

To show the range and error of the process, the results of these and other experiments recorded by Gooch and Morley may be summarized as follows:

Range of the Process.

Number of determinations.	Gold taken.	Strength of solutions.			Error, average.	Extremes.
		Iodine.	Thiosulphate.	Gold in 1 cm.^{3}		
	mg.			mg.	mg.	mg.
11	8.71–4.35	$n/100$	$n/100$	0.871	−0.05	+0.03 −0.1
20	0.871–0.087	$n/100$	$n/100$	0.0871	+0.02	+0.06 −0.02
10	0.871–0.087	$n/1000$	$n/1000$	0.0871	+0.004	+0.020 −0.029
14	0.520–0.052	$n/1000$	$n/1000$	0.052	+0.002	+0.01 −0.008

It is plain that the average experimental errors, due to all causes, do not very much exceed the errors which might naturally be expected to arise from errors of reading.

In repeating this work, Maxson * has obtained results of a similar order of accuracy. Maxson has also studied the possibility of reduction of the aurous iodide formed in the process and finds in periods much exceeding those required for the analytical operation no evidence of further action other than the formation of the aurous salt. Thus, aurous iodide, obtained by treating a solution of auric chloride, containing 0.0125 grm. of gold, with potassium iodide according to the directions of Gooch and Morley, adding starch and bleaching the starch iodide with sodium thiosulphate, shows no color of starch blue after the interval of an hour. Inasmuch as an interval of ten minutes is enough for the complete manipulation of a single determination, it is plain that the stability of the aurous iodide does not figure in the accuracy of the determination of the small amounts of gold for which the process was designed.

The Colorimetric Determination of Small Amounts of Gold.

A colorimetric method for the estimation of small amounts of gold has been based by Maxson † upon the coloration exhibited by suspensions of red colloidal gold.

Blake ‡ has shown that acetylene is the most suitable reagent for effecting the reduction of the auric salt with production of the red colloidal gold, the treatment consisting in drying the chloride at 170°, dissolving in ether, and pouring the etheral solution into water containing ether and saturated with acetylene gas. For the purposes of this method the simpler treatment with an aqueous solution of acetylene proved to be adequate. The procedure consists in preparing a red colloidal suspension containing a known amount of gold in a given volume and by means of measured amounts of this solution matching the color developed in a similar solution containing in measured volume the gold to be determined.

For making the comparisons of color a modified form of the apparatus proposed by Penfield for the colorimetric estimation

* Ralph N. Maxson, Am. Jour. Sci., [4], xvi, 155; xvii, 466.

† Ralph Nelson Maxson, Am. Jour. Sci., [4] xxi, 270.

‡ Am. Jour. Sci., [4], xvi, 381.

of titanium is used. This consists of comparison tubes set vertically in a dark box and illuminated from below. A mirror suitably situated beneath the box containing the tubes gives efficient illumination. Such an apparatus is cheaply and easily procured. With tubes having a diameter of 1 cm. and a length of 13 cm., and accurately graduated to hold 10 $cm.^3$, this simple colorimeter is capable of determining accurately very small amounts of gold.

The standard suspension is made by treating in a measuring flask a measured amount of a standardized solution of ordinary undried gold chloride with an aqueous solution of acetylene and, after the full development of color, making up to the mark.

A solution of pure gold to be examined is treated similarly with aqueous acetylene and made up to known volume. When small amounts are handled the volume of the solution should not exceed a few cubic centimeters, and only a small amount of the aqueous solution of acetylene should be added; otherwise the coloration may be partially or totally inhibited.

If traces of electrolyte are present, the coagulation of the red gold may sometimes be avoided by the addition of a few drops of ether to the cold solution.

It is a well-known fact that small amounts of electrolyte will rapidly change red gold to the blue modification. It is necessary, therefore, to conduct the comparisons in a room reasonably free from fumes, and to have all containing vessels free from soluble material. Flasks which have been treated with steam for a few minutes give the best results. Red suspensions contained in such flasks may show no trace of blue after an interval of several weeks.

The comparison of colors is carried out in the following manner: A measured amount of the suspension is drawn off into the left-hand tube and diluted to the mark with water; a suitable amount of water is then placed in the right-hand tube and the standard suspension drawn into the tube until the colors are seen to be identical. The amount of water to be used can be determined by preliminary experiment. The positions of the tubes are reversed before the final reading, and the mean taken.

In the table below is the record of experiments made to determine the range of amounts of gold capable of accurate estimation in such an apparatus with tubes of the dimensions described above. The comparisons were made with a red suspension, prepared by careful dilution of a more concentrated standard suspension, which contained 0.0000107 grm. of metal in 1 $cm.^3$.

Colorimetric Estimation of Gold.

Gold suspension taken. cm.³	Gold suspension used. cm.³	Gold used. grm.	Gold found. grm.	Error. grm.
9.50	9.05	0.000102	0.000097	−0.000005
8.00	7.59	0.000086	0.000082	−0.000004
7.00	6.89	0.000075	0.000074	−0.000001
6.00	5.83	0.000065	0.000063	−0.000002
5.00	4.84	0.000054	0.000052	−0.000002
4.00	3.88	0.000043	0.000042	−0.000001
3.00	2.47	0.000032	0.000027	−0.000005
2.00	1.82	0.000022	0.000020	−0.000002
1.00	0.93	0.000011	0.000010	−0.000001

The intensity of the color in the experiments ranged from a deep red to a faint pink. Further comparisons made with suspensions more dilute than those described above gave errors of magnitude increasing with the dilution. The amounts handled here are, then, the minimum quantities that can be accurately estimated with the apparatus described. It is obvious that if larger amounts of the metal are to be determined, tubes of greater dimensions should be used.

The application of such a method for the determination of gold naturally starts with that element. The weighed amount of metal, contained in a clean porcelain crucible, can be readily brought into solution with the aid of chlorine water or aqua regia and the excess of the solvent evaporated off upon the water bath.

CHAPTER IV.

BERYLLIUM; MAGNESIUM; CALCIUM; STRONTIUM; BARIUM.

BERYLLIUM.

Ammonium Beryllium Phosphate.

BERYLLIUM, like calcium, strontium and barium, cannot be estimated by precipitation as a double ammonium phosphate and subsequent ignition, as are magnesium, zinc and cadmium. The precipitated ammonium beryllium phosphate, as has been shown by Austin,* always contains triberyllium phosphate.

The Conversion of Beryllium Chloride to Beryllium Oxide.

Havens † has shown that small amounts of beryllium chloride may be easily converted to the oxide, without precipitation and filtration, by treatment with nitric acid and ignition.

The solution of the chloride is evaporated just to dryness on a radiator, care being taken not to heat to the volatilizing point of the salt, a few drops of concentrated nitric acid are added, the liquid is evaporated, and the residue heated gently at first and finally in the flame of the blast lamp. This conversion of beryllium chloride to beryllium nitrate may be carried on in platinum without attacking that metal appreciably, provided care be taken to remove all excess of hydrochloric acid and to add the nitric acid to the dry residue.

Conversion of Beryllium Chloride to the Oxide.

BeO taken in solution as $BeCl_2$. grm.	BeO found. grm.	Error. grm.
0.0483	0.0481	−0.0002
0.0483	0.0483	0.0000
0.1076	0.1085	+0.0009

* Martha Austin, Am. Jour. Sci., [4], viii, 207.

† F. S. Havens, Am. Jour. Sci., [4], iv, 112.

The Separation of Beryllium Oxide from Ferric Oxide.

Havens and Way* have separated beryllium oxide from ferric oxide by volatilization of the latter in a stream of gaseous hydrogen chloride charged with a little free chlorine, with care to avoid mechanical loss through too rapid volatilization of the iron.†

MAGNESIUM.

The Determination of Magnesium by Precipitation and Ignition of Ammonium Magnesium Carbonate.

According to Schaffgotsch,‡ the very concentrated solution of the sulphates, nitrates or chlorides of magnesium, sodium and potassium is treated with a concentrated solution of ammonium carbonate. The voluminous precipitate which first falls is acted upon by an excess of the precipitant, sometimes dissolving completely, and crystalline ammonium magnesium carbonate, $MgCO_3.(NH_4)_2CO_3.4H_2O$, is soon formed; after standing twenty-four hours the precipitate is filtered off, washed with the concentrated ammoniacal solution of ammonium carbonate, dried and strongly ignited. In the absence of salts of potassium, the residue is weighed at once as magnesium oxide, and from the filtrate sodium salts are recovered by evaporation. When a salt of potassium is originally present, with or without a salt of sodium, the ignited magnesium oxide is to be washed out and again ignited before weighing, and the washings are to be added to the filtrate containing the greater part of the alkalies.

Gooch and Eddy§ have shown that ammonium magnesium carbonate is noticeably soluble in Schaffgotsch's solution of full strength, and rather more so in the same reagent of half strength, and that an exact separation of magnesium from the alkalies, in solutions of reasonable volume, cannot be made without modification of the method. By suitable addition of alcohol, however, it has been found possible to make the precipitation complete and to effect the separation of magnesium from small amounts of alkali salt in one operation; when considerable quan-

* Franke Stuart Havens and Arthur Fitch Way, Am. Jour. Sci., [4], viii, 217.

† See page 507.

‡ Ann. Phys., civ, 482 (1858).

§ F. A. Gooch and Ernest A. Eddy, Am. Jour. Sci., [4], xxv, 444.

tities of alkali salt are present the separation may be made by two treatments.

The solution containing the salts of magnesium and the alkalies is brought to a volume of about 50 cm.[3] and an equal amount of absolute alcohol is added, precipitation is made by addition of 50 cm.[3] of the saturated ammoniacal ammonium carbonate solution containing 50 per cent alcohol, and the mixture is allowed to stand twenty minutes after stirring for five minutes. If the amount of alkali salt originally present is small, not exceeding 0.1 grm., the precipitate may be collected on asbestos in a perforated crucible, washed with the precipitant, dried, ignited, and weighed as magnesium oxide. When the amount of alkali salt originally present is larger, the precipitate may be freed from traces of the alkali salt by pouring off the supernatant liquid through the prepared asbestos filter, dissolving the precipitate, and precipitating ammonium magnesium carbonate as at first. This second precipitate, collected upon the filter originally used, leaves upon ignition practically pure magnesium oxide. The accompanying figures show excellent results.

Precipitation of Magnesium Ammonium Carbonate and Weighing of the Oxide.

MgO taken. grm.	NaCl taken. grm.	KCl taken. grm.	NH_4Cl taken. grm.	MgO weighed. grm.	Error MgO. grm	MgO found in filtrate as $Mg_2P_2O_7$. grm.	Volume of water solution. cm.[3]	Volume of alcohol added cm.[3]	Volume of precipitant. cm[3]	Volume of solution used in washing. cm.[3]
Single precipitation.										
0 1444			.	0.1443	−0 0001	0 0000	50	50	50	.
0 1444	.		.	0 1440	−0 0004	0 0002	50	50	50	.
0 1444		0 1		0.1445	+0 0001	0 0001	50	50	50	50
0 1444		0 1	.	0 1444	0 0000	0 0001	50	50	50	50
0 1444	0 1			0 1445	+0 0001	0 0002	50	50	50	50
0 1444	0 1			0 1449	+0 0005	0 0001	50	50	50	50
0 1444	0 2	.	.	0 1449	+0 0005	0 0002	50	50	50	50
0 1444	.	0 2		0 1461	+0 0017	0 0000	50	50	50	50
0 1444	.	.	3 0	0.1444	0 0000	0 0001	50	50	50	50
0 1444			3 0	0 1447	+0 0003	0 0002	50	50	50	50
Double precipitation.										
0 1444	0.2	..	...	0.1446	+0.0002	0.0002	50	50	50	50
0 1444	..	0 2	..	0 1442	−0.0002	0 0002	50	50	50	50

The Determination of Magnesium as the Pyrophosphate.

As Neubauer* had previously pointed out, the ideal ammonium magnesium phosphate, NH_4MgPO_4, which yields upon ignition the pyrophosphate, $Mg_2P_2O_7$, may, according to conditions of precipitation, be contaminated by the trimagnesic phosphate, $Mg_3P_2O_8$, or by a double phosphate, $(NH_4)_4Mg(PO_4)_2$, which upon ignition leaves magnesium metaphosphate. Gooch and Austin† have emphasized the effect of ammonium salts in bringing about the contamination of the precipitate which results in the formation of the metaphosphate on ignition, and have also pointed out that the use of strongly ammoniacal solutions and strongly ammoniacal wash-water is distinctly disadvantageous, as well as inconvenient. Having shown that as little as 0.0001 grm. of magnesium oxide may be precipitated in 500 $cm.^3$ of faintly ammoniacal solution (even when containing as much as 60 grm. of ammonium chloride or 100 $cm.^3$ of a saturated solution of ammonium oxalate), the authors recommend the use of faintly ammoniacal solutions and wash-water, and, to prevent as much as may be the formation of the phosphate containing excess of the ammonium ion, restriction of the amount of ammonium chloride present. When ammonium salts are present in quantity, as is the case in the ordinary course of analysis, the precipitate first thrown down by addition of ammonium sodium phosphate and ammonia in faint but distinct excess is settled, and the supernatant liquid is poured off through the filter used subsequently in collecting the precipitate. The precipitate is dissolved in the least possible amount of hydrochloric acid and thrown down again from the diluted solution by ammonia in slight excess. For safety, a little ammonium sodium phosphate may also be added. The precipitate is filtered off and washed with faintly but distinctly ammoniacal water, and, to avoid reduction, the ignition is made slowly and carefully so that all ammonia is expelled before the temperature is raised to redness.

When filtrations are made on asbestos in the perforated crucible, as was done in the experiments the results of which are recorded below, it is well to cap the crucible and moisten the

* Zeit. Angew. Chem., 1896, 439.

† F. A. Gooch and Martha Austin, Am. Jour. Sci., [4], vii, 187.

precipitate upon the felt with a drop of a saturated solution of ammonium nitrate before proceeding to dry and ignite. The accompanying figures show the accuracy which may be expected when precipitations are made in presence of varying amounts of ammonium salts.

Effect of Ammonium Salts in Respect to the Constitution of the Precipitate.

$Mg_2P_2O_7$ corresponding to $Mg(NO_3)_2$ taken	$Mg_2P_2O_7$ found.	Error in terms of $Mg_2P_2O_7$.	Error in terms of MgO.	NH_4Cl present.		$HNaNH_4PO_4 4H_2O$ used.	Volume	
				I.	II.		I.	II.
grm.	grm.	grm.	grm.	grm	grm.	grm.	cm.³	cm.³
0.5311	0.5312	+0.0001	0.0000	*	*	2.5	150	100
0.5311	0.5311	0.0000	0.0000	*	*	2.5	150	100
0.5311	0.5346	+0.0035	+0.0013	2	2	2.5	150	100
0.5311	0.5348	+0.0037	+0.0014	2	2	2.5	150	100
0.5311	0.5383	+0.0072	+0.0026	5	5	2.5	150	100
0.5311	0.5368	+0.0057	+0.0021	5	5	2.5	150	100
0.5311	0.5376	+0.0065	+0.0023	10	10	2.5	200	100
0.5311	0.5395	+0.0084	+0.0030	10	10	2.5	200	100
0.5311	0.5396	+0.0085	+0.0031	60	5	2.5	250	100
0.5311	0.5389	+0.0078	+0.0028	60	5	2.5	250	100

* Probably less than 1 grm.

Gooch and Austin point out that good results are obtained in one precipitation by the method of Wolcott Gibbs, proposed many years ago.* According to this method the boiling solution of the magnesium salt is treated with ammonium sodium phosphate, and ammonia is added after cooling. Even in presence of considerable amounts of ammonium chloride this process yields a phosphate of nearly ideal constitution if only the boiling be prolonged from three to five minutes. The greater part of the ammonium magnesium phosphate — about 90 per cent — forms in this process before free ammonia is added, and the ammonium which enters the phosphate thus formed is derived from the microcosmic salt, which must become correspondingly acidic. Under these conditions, the tendency to form an insoluble ammonium magnesium phosphate, richer in ammonium and poorer in magnesia than the normal salt, is slight. In the process of Gibbs, as well as in the modified precipitation process, the use of the faintly ammoniacal solution and wash-water is sufficient and advantageous.

* Am. Jour. Sci., [3], v, 114.

The Arsenate Process for the Separation of Magnesium and the Alkalies.

Browning and Drushel* have taken advantage of the insolubility of ammonium magnesium arsenate in ammoniacal solution, with the reducibility of arsenic acid by hydrobromic acid and volatility of arsenious bromide, to perfect a method for the separation of magnesia from the alkali metals, and the determination of these elements. From a solution of the chlorides of magnesium and potassium or sodium, the magnesium may be precipitated in a distinctly but not strongly ammoniacal solution by 40 per cent to 80 per cent excess of ammonium arsenate, with brisk stirring. When only a small amount of magnesium is present in a relatively large amount of solution, the precipitate forms slowly and becomes complete only on long standing. The precipitation of amounts of magnesium so small as not to be precipitated immediately by ammonium arsenate may be brought about by freezing the solution, a process previously shown to be applicable † in the precipitation of small amounts of arsenic acid by magnesia mixture. Precipitation may also be hastened by adding alcohol amounting to 15 per cent to 20 per cent of the mixture and filtering as soon as the precipitate settles completely. The precipitate is collected under moderate pressure in an ignited and weighed perforated crucible containing a close felt of fine asbestos. It is washed with 40 $cm.^3$ to 50 $cm.^3$ of ammoniacal water, after which it is dried at 125° to 140° and carefully ignited and weighed as magnesium pyroarsenate.

It is shown elsewhere‡ that arsenic acid may be reduced and volatilized by the action of hydrochloric acid and potassium bromide. The removal of the arsenic acid from the alkali salts is easily accomplished by the similar procedure of treating the mixture with hydrobromic acid or with ammonium bromide and hydrochloric acid and evaporating in an open dish under a good hood. The complete method as recommended for the estimation of magnesium and its removal from the alkalies, and the subsequent estimation of the alkalies, is as follows:

The magnesium is precipitated in a distinctly but not strongly ammoniacal solution by a 40 per cent to 80 per cent excess of

* Philip E. Browning and W. A. Drushel, Am. Jour. Sci., [4], xxiii, 293.

† See page 290.

‡ See page 316.

ammonium arsenate. The completeness of the precipitation may be hastened by freezing the solution in an ice-and-salt mixture or by adding alcohol to about 15 per cent to 20 per cent of the total volume of the solution, which may vary from 100 cm.³ to 250 cm.³ according to the amounts of salt present. The magnesium arsenate obtained is filtered on an asbestos felt contained in a perforated platinum crucible, the crucible and felt having been previously ignited and weighed, and is dried, ignited and weighed as the pyroarsenate.

Experimental results follow.

Magnesium and the Alkalies.

$(NH_4)_3AsO_4$ used calculated as As_2O_3.	Dilution.	NaCl or KCl converted to Na_2SO_4 or K_2SO_4 and calculated as Na_2O or K_2O.			$MgCl_2$ converted into $Mg_2As_2O_7$ and calculated as MgO.		
		Taken	Found.	Error	Taken.	Found	Error.
grm.	cm.³	grm.	grm.	grm	grm.	grm.	grm.
Precipitate stood 12 to 24 hours.							
0.1	100	0 1194	0 1191	−0 0003	0 0199	0 0197	−0 0002
0.2	150	0.1194	0 1196	+0 0002	0 0399	0 0397	−0.0002
0 45	250	0.1194	0 1195	+0 0001	0 0998	0 0998	0 0000
0 45	250	0 1194	0 1194	0 0000	0 0998	0 0997	−0 0001
0.45	250	0.2389	0 2385	−0 0004	0 0998	0 0999	+0.0001
0.4	250	0 0478	0.0481	+0 0003	0 1198	0 1193	−0 0005
0.35	250	0.0956	0 0957	+0 0001	0 0998	0 0996	−0.0002
0.35	250	0 0956	0 0957	+0 0001	0 0998	0 0994	−0 0004
0.45	250	0.0909	0 0915	+0 0006	0 0998	0 0993	−0.0005
0.1	100	0 0545	0 0549	+0 0004	0 0006	0 0004	−0.0002
Precipitation hastened by alcohol.							
0.1	100	0.1181	0 1184	+0.0003	0.0040	0 0038	−0 0002
0.1	100	0.1181	0 1184	+0 0003	0.0040	0 0038	−0 0002
0.1	100	. . . .	. . . . . .	. . . .	0.0040	0 0038	−0 0002
Precipitation hastened by freezing.							
0.1	100	0.1181	0.1184	+0.0003	0.0040	0 0040	0 0000
0.2	100	0.1181	0 1183	+0 0002	0.0040	0 0039	−0 0001
0.45	250	0.1181	0.1179	−0 0002	0.1002	0 1004	+0 0002

The filtrate is transferred from the filter flask to a platinum dish, and after the addition of 10 cm.³ of hydrochloric acid (sp. gr. 1.20) and about the same amount of hydrobromic acid (sp. gr. 1.3), or 1 to 3 grm. of ammonium bromide, is evaporated to dryness under a draft hood. The residue is gently ignited to remove the ammonium salts and transferred to a weighed platinum crucible with a small amount of water. A little sulphuric acid [1 : 1] is added, and the solution evaporated to remove the water and excess of sulphuric acid, by placing the crucible on a triangle in a porcelain crucible used as a radiator. After the sulphuric acid has ceased to fume, the crucible is removed from the radiator, and after ignition at the full heat of the Bunsen burner the alkali is weighed as the normal sulphate.

Calcium; Strontium; Barium.

The Detection of Barium and Strontium, Associated with Calcium and Lead.

In the ordinary procedure of qualitative analysis the alkali earth elements are usually separated by ammonium carbonate, after hydrogen sulphide and ammonium hydroxide have been used to remove the greater number of the bases. It has long been observed that a considerable part of the alkali earth, especially barium and strontium, fails to appear when the ammonium carbonate is added. The reasons given for this loss have been the oxidation of hydrogen sulphide or other sulphides to sulphates and the consequent precipitation of the alkali earth sulphates, the formation in alkaline solution of carbonates and the consequent precipitation of the carbonates, and the tendency of the large amounts of ammonium salts which collect during the analysis to interfere with the precipitation of the alkali earth carbonates by ammonium carbonate. Various precautions have been suggested to avoid these sources of error, such as the prompt removal of the excess of hydrogen sulphide by boiling, the use of freshly prepared hydroxides free from carbonate, and the removal of the ammonium salts by ignition before attempting to precipitate the alkali earth carbonates.

To obviate these difficulties, Browning and Blumenthal* have suggested the precipitation of the insoluble sulphates after the removal by hydrochloric acid of mercury in the mercurous condition, silver, and that amount of lead which may be precipitated; removal of the insoluble lead sulphate by treatment of the

* Philip E. Browning and Philip L. Blumenthal, Am. Jour. Sci., [4], xxxii, 246.

precipitate with ammonium acetate; reduction of the remaining insoluble sulphates by ignition with carbon; treatment of the residue with acetic acid; and testing of the solution for barium, strontium and calcium present as soluble acetates. The following method is suggested: The solution (about 10 $cm.^3$), which may contain mercury in the mercurous condition, silver, lead, barium, strontium and calcium, besides other elements, is treated with hydrochloric acid in faint excess and the precipitated chlorides are filtered off. To the filtrate are added about 5 grm. of ammonium acetate, and a 10 per cent solution of ammonium sulphate to complete precipitation. After gentle warming, the alkali earth sulphates are filtered off and washed with a saturated solution of ammonium acetate until the washings give no test for lead by hydrogen sulphide. The filtrate and washings are reserved for treatment by the ordinary course of analysis. To the precipitated sulphates on the paper a small amount of pure sugar carbon is added, the paper is rolled up, and the mass placed either in a porcelain crucible with a cover, or in a closed glass tube, and heated to full redness for a few minutes. The fused mass is treated with about 5 $cm.^3$ of 50 per cent acetic acid and warmed, when, if the alkali earth elements are present, an odor of hydrogen sulphide will generally be evident. The extract is thrown upon a filter and the residue washed with about 5 $cm.^3$ of water. The filtrate containing acid and water is treated with a few drops of a solution of potassium chromate to test for barium. The barium chromate is removed by filtration, and the filtrate boiled with sodium carbonate to precipitate strontium and calcium as the carbonates. If the precipitate of the carbonates is very small, it may be dissolved in hydrochloric acid and tested spectroscopically. If, however, it is not too minute in quantity, it should be dissolved in nitric acid after careful washing, and the strontium and calcium separated by dehydration with amyl alcohol.

The results follow in the table. All tests for strontium and calcium were confirmed by the spectroscope.

From these results it would appear that these tests for barium and strontium are effective to at least a milligram of each element and may with advantage precede the group precipitation by hydrogen sulphide in the ordinary course of qualitative analysis.

Pb present. grm.	Ba present. grm.	Sr present. grm.	Ca present. grm.	Indications.
0 0500	0.0500	0.0500	0 0500	Good tests for all.
0 0250	0.0250	0.0250	0 0250	Good tests for Pb, Ba and Sr. Ca faint.
0 0100	0.0100	0 0100	0.0100	Pb and Ba good Sr fair. Ca doubtful.
0 0050	0.0050	0.0050	0.0050	Pb and Ba good. Sr fair.
0 0010	0.0010	0.0010	0 0010	Pb good. Ba fair. Sr faint.
0 1000	0.0050	.. ...	..	Pb and Ba good.
0.1000	0.0010	.....	......	Pb good. Ba faint.

The Separation of Barium, Strontium and Calcium by the Action of Amyl Alcohol on the Nitrates.

Following in general known procedure for the separation of sodium and potassium from lithium by the use of amyl alcohol,* Browning† has developed methods for the detection of calcium and strontium in association, the separation and estimation of strontium associated with calcium, and the separation of barium associated with calcium or with strontium.

Experiments with pure strontium nitrate show that when the dry salt is treated by dissolving in a few drops of water, adding amyl alcohol, boiling until the water is expelled and the boiling point rises to the normal boiling temperature of the alcohol (128°–130°), filtering upon asbestos in the filtering crucible, washing with small amounts of previously boiled amyl alcohol, and heating to 150° in an air bath, nearly the entire original amount is recovered. The quantity which remains in solution in the dehydrated alcohol amounts very regularly to 0.0008 grm. of the nitrate or 0.0004 grm. of the oxide for every 10 cm.³ of liquid.

When calcium nitrate in water solution is similarly treated by boiling with amyl alcohol, the salt passes into solution with the exception of minute portions, not exceeding altogether 0.0003 grm. or 0.0004 grm., which separate on the surface of the container. This very slight residue (apparently the calcium salt of an acid formed by the action of nitric acid upon amyl alcohol), when dried, dissolved in dilute nitric acid, and again treated with amyl

* Gooch, Am. Chem. Jour., ix, 33.

† P. E. Browning, Am. Jour. Sci., [3], xliii, 50.

alcohol, separates out in the boiling, but, if ignited and then dissolved in a drop of dilute nitric acid, is not precipitated by subsequent boiling with amyl alcohol.

Detection of Strontium and Calcium. To detect strontium and calcium associated in the form of nitrates, the mixture, not exceeding 0.2 grm. (that being the limit of the solubility of calcium nitrate in 5 cm.^{3} of amyl alcohol) is put into a test tube and dissolved in a few drops of water, 5 cm.^{3} of amyl alcohol are added, and the boiling is carried on until the normal boiling point of the alcohol, 128°–130°, is reached. If strontium is present to the amount of 0.001 grm. of the oxide, a very decided separation takes place. If the amount is smaller, it cannot be readily distinguished from the residual spots deposited on the bottom of the tube by the calcium salt. The alcohol containing the calcium salt dissolved is decanted upon a dry filter paper in a dry funnel and the residue washed in the tube with about 5 cm.^{3} of absolute ethyl alcohol, this also being filtered into the tube containing the amyl alcohol. The filtrate is reserved to be tested for calcium. The residue, if so small that it may be a calcium deposit, is dried gently, ignited by agitating the tube over a flame, and dissolved in a drop of dilute nitric acid. To this solution 5 cm.^{3} of amyl alcohol are added and the boiling is repeated. Any amount of strontium above 0.0005 grm. of the oxide separates out distinctly, while the slight calcium residue does not reappear, as is shown in the accompanying record.

SrO taken. grm.	$Ca(NO_3)_2$ taken. grm.	Deposit after first boiling.	Deposit after second boiling.
......	0.1	Trace.	None.
......	0.2	Slight.	None.
......	0.2	Slight.	None.
......	0.2	Distinct.	Faintest trace.
0.0003	....	Faint trace.	Faint trace.
0.0003	....	Faint trace.	Faintest trace.
0.0005	....	Distinct.	Distinct.
0.0005	....	Distinct.	Distinct.
0.0010	....	Distinct.	Distinct.
0.0010	0.2	Distinct.	Distinct.
0.0005	0.1	Distinct.	Faintest trace.
0.0007	0.1	Distinct.	Faint trace.
0.0008	0.05	Distinct.	Distinct.

The test for calcium is made upon the filtrate and washings after the first boiling, by adding to the clear liquid about 2 cm.^{3}

of dilute sulphuric acid. In five minutes or less, calcium, if present to an amount exceeding 0.0001 grm., appears as a light, flocky precipitate, different in character and easily distinguishable from the faint cloudiness, gathering to a minute and granular precipitate, which results from the presence of a trace of strontium salt not precipitated in the boiling process. Following are the tests of this method.

$Sr(NO_3)_2$ taken. grm.	CaO taken. grm.	Result.
0.001	.....	Faint granular cloudiness.
0.1	....	Faint granular cloudiness.
0.2	......	Faint granular cloudiness.
.....	0 001	Decided flocky floating masses.
.....	0 0005	Decided flocky floating masses.
.....	0.0002	Decided flocky floating masses.
.....	0.0001	Plain flocky floating masses.
.....	0.00005	Faint flocky floating masses.
0.1	0.0005	Decided flocky floating masses.
0.1	0.0002	Decided flocky floating masses.
0.1	0.0001	Plain flocky floating masses.
0.2	0.0002	Decided flocky floating masses.

Separation and Estimation of Strontium and Calcium.

For the quantitative estimation of strontium and calcium,* the dry nitrates of these elements are dissolved in the least possible amount of water contained in a small beaker (50 cm.³ to 100 cm.³), a suitable amount of amyl alcohol (10 cm.³ to 30 cm.³) is added, the beaker is heated upon a wide piece of asbestos board so that inflammable fumes may not reach the flame below, and the mixture is boiled with a thermometer inserted until the normal boiling point of the alcohol (128° to 130°) is reached.

From the precipitated strontium nitrate the alcoholic solution is decanted through a weighed filtering crucible and asbestos felt, and the residue, dried at a gentle heat over a radiator, is dissolved in a drop or two of dilute nitric acid. Amyl alcohol is again added and the boiling repeated, experience having shown that the residue of the first boiling is apt to retain an appreciable amount of the calcium salt. The precipitate of the second boiling is filtered off upon the felt through which the solution had previously been decanted, washed with amyl alcohol, dried at 150°, and weighed as strontium nitrate $Sr(NO_3)_2$. Correction for the solubility of strontium nitrate in amyl alcohol is made according to the

* P. E. Browning, Am. Jour. Sci., [3], xliii, 50; and xliv, 462.

quantity of that reagent left after the boilings, — 0.0008 grm. of the nitrate or 0.0004 grm. of the oxide for every 10 cm.³ of amyl alcohol decanted or filtered off, exclusive of that used in washing, in which process no appreciable amount of the strontium salt is dissolved.

The calcium in the filtrate is determined as the sulphate by evaporation of the solution, ignition, treatment with sulphuric acid, and a final ignition. From the apparent amount of calcium sulphate found a correction of 0.0005 grm. for the included strontium sulphate is to be subtracted. Results of experimental tests are given in the tabular statement.

Separation and Estimation of Stronium and Calcium.

Final volumes 25 cm.³; two treatments.

SrO taken. grm.	SrO found.* grm	Error. grm.	CaO taken. grm.	CaO found.† grm.	Error. grm.
0.0148	0.0155	+0 0007	0 0256	0.0254	−0 0002
0 0183	0 0183	0 0000	0 1030	0.1015	−0 0015
0 0364	0 0366	+0 0002	0 0516	0 0511	−0 0005
0 0365	0 0365	0.0000	0.0515	0 0513	−0 0002
0.0493	0 0494	+0 0001	0 0515	0 0502	−0 0013
0.0497	0.0497	0.0000	0 0519	0 0511	−0.0008
0 0497	0 0503	+0.0006	0 0249	0 0245	−0 0004
0 0729	0 0732	+0 0003	0 0257	0 0251	−0 0006
0 0730	0.0732	+0 0002	0 0255	0 0255	0 0000
0 0744	0 0744	0.0000	0 0258	0 0260	+0.0002
0.0912	0.0910	−0 0002	0 1286	0.1276	−0 0010

* Corrected by addition of 0.0020 grm. for two treatments in final volumes of 25 cm.³.

† Corrected by subtraction of 0.0035 grm. for $SrSO_4$ included in $CaSO_4$ obtained by evaporation and ignition.

Final volumes 8 cm.³; two treatments.

SrO taken. grm.	SrO found.* grm.	Error. grm.	CaO taken. grm.	CaO found.† grm.	Error. grm.
0 0570	0 0571	+0.0001	0 0534	0.0536	+0.0002
0 0573	0.0573	0 0000	0 0534	0 0539	+0 0005
0 0285	0.0280	−0 0005	0 0272	0.0272	0 0000
0 0568	0 0566	−0.0002	0 0535	0 0533	−0 0002
0 0568	0 0567	−0 0001	0.0533	0.0531	−0 0002
0.0288	0 0286	+0.0002	0.0271	0.0268	−0 0003
0 1420	0 1422	+0 0002	0 0535	0 0540	+0.0005
0 1419	0.1422	+0.0003	0.0665	0.0665	0.0000
0 1135	0 1138	−0 0003	0.1066	0.1066	0 0000
0 1137	0 1132	−0 0005	0.1064	0.1066	+0.0002

* Corrected by addition of 0.0006 grm. for two treatments in final volumes of 8 cm.³.

† Corrected for 0.0010 grm. of $SrSO_4$ included in $CaSO_4$ obtained by evaporation and ignition.

Estimation of Barium and Calcium.

Procedure similar to that employed in the separation and estimation of strontium and calcium by the action of amyl alcohol on the nitrates may be applied to the separation of barium and calcium,* but, barium nitrate being almost entirely insoluble in the amyl alcohol, there is in this case no advantage in keeping the volume of alcohol at the lowest point. A convenient volume at the beginning of the dehydration is 30 cm.^{3}, and the results of one treatment are fully as satisfactory as those of the double treatment. In separating barium from calcium by this method, the dry nitrates are treated in 100-cm.^{3} beakers by dissolving in a few drops of water, adding 30 cm.^{3} of amyl alcohol, boiling until the normal boiling point of the alcohol is reached (128° to 130°), filtering by means of the perforated filtering crucible fitted with the asbestos felt, washing with previously boiled amyl alcohol, drying at 150° and weighing. The test results show the exactness of the method.

Estimation of Barium and Calcium.

BaO taken grm.	BaO found. grm.	Error. grm.	CaO taken. grm.	CaO found. grm.	Error. grm.
0.1410	0.1406	−0 0004	0.0112	0.0114	+0 0002
0.1300	0.1301	+0 0001	0.0926	0.0926	0.0000
0.1043	0 1049	+0.0006	0 0741	0.0736	−0 0005
0 0781	0 0781	0.0000	0.0556	0.0554	−0 0002
0 0525	0.0526	+0 0001	0 0373	0 0372	−0 0001

Estimation of Barium and Strontium together, and of Calcium.

The dry nitrates of the three elements, barium, strontium and calcium, are dissolved in the least possible amount of water, a suitable amount (15 cm.^{3} to 30 cm.^{3}) of amyl alcohol is added, and the mixture boiled until the normal boiling temperature of the alcohol is reached. The alcoholic solution is decanted from the precipitated barium nitrate and strontium nitrate through a weighed filtering crucible and asbestos felt. The residue is dried over a radiator, dissolved in a drop or two of dilute nitric acid, and again treated as before with amyl alcohol. The precipitate of the second boiling is filtered off upon the asbestos felt previously used, washed with amyl alcohol, dried at 150° and weighed. The calcium is determined as the sulphate in the combined

* P. E. Browning, Am. Jour. Sci., [3], xliii, 314.

filtrates and washings. Results, corrected for the solubility of the strontium nitrate (0.0008 grm. to 10 cm.^{3} of alcohol decanted or filtered, exclusive of washings) and for the contamination of the calcium sulphate by strontium sulphate (0.0005 grm. for 10 cm.^{3} of alcohol), are given in the table.

Estimation of Barium and Strontium, and Calcium.

$Ba(NO_3)_2$ and $Sr(NO_3)_2$ taken.	$Ba(NO_3)_2$ and $Sr(NO_3)_2$ found and corrected.	Error in nitrates.	Error averaged and calculated as oxide.	CaO taken.	CaO found.	Error
grm.	grm.	grm.	grm.	grm.	grm.	grm.
0.3941	0.3945	+0.0004*	+0.0002	0.0283	0.0277	−0.0006
0 1436	0.1442	+0.0006*	+0.0003	0.0568	0.0558	−0.0010
0.3163	0.3152	−0.0011*	−0.0006	0.0284	0.0274	−0.0010
0.1978	0.1987	+0.0009*	+0.0005	0.0285	0.0280	−0.0005
0.1948	0.1932	−0.0016†	−0.0008	0.0833	0.0835	+0.0002
0 1971	0.1971	0.0000*	0.0000	0.0830	0.0817	−0.0013
0.1973	0.1960	−0.0013*	−0.0007	0.0830	0.0824‡	−0.0006
0.1959	0.1970	+0.0011*	+0.0005	0.0830	0.0819	−0 0011
0 1971	0 1963	−0.0008†	−0 0004	0.0834	0.0831‡	−0 0003

* Final volume in each of two treatments, 30 cm ^{3}.
† Final volume in each of two treatments, 15 cm.^{3}.
‡ $CaSO_4$ precipitated and filtered: in other experiments obtained by evaporation.

*The Separation of Barium and Strontium by the Action of Amyl Alcohol on the Bromides.**

Methods upon which dependence can be placed for the separation of barium and strontium are few in number. The differences in solubility of the bromides of barium and strontium in amyl alcohol provide a method for a comparatively good separation and, when properly corrected, an exact determination of these elements. Anhydrous barium bromide dissolves in amyl alcohol to the extent of about 0.0013 grm. in 10 cm.^{3}, while the same quantity of amyl alcohol will take into solution approximately 0.2 grm. of strontium bromide. When a mixture of the dry salts is dissolved in water, and the water removed by boiling in a suitable amount of amyl alcohol, the barium bromide becomes nearly insoluble, while nearly all the strontium bromide goes into solution. The insoluble barium bromide cannot, however, be filtered off, washed, and dried to constant weight without decomposition, so it becomes necessary to determine the barium in

* Philip E. Browning, Am. Jour. Sci., [3], xliv, 459.

some other form. Moreover, a single treatment by boiling does not remove the strontium completely from the precipitated barium bromide. To separate barium and strontium taken as the bromides, therefore, the mixed salts are treated in a beaker by dissolving in the least possible amount of water, adding 10 $cm.^3$ of amyl alcohol, and boiling until the temperature rises to the normal boiling point of the alcohol (128° to 130°). The solution is decanted through a weighed filtering crucible and asbestos felt. The gently dried residue is dissolved in the minimum amount of dilute hydrobromic acid and again boiled with 10 $cm.^3$ of amyl alcohol. The precipitate is filtered off upon the asbestos felt previously used and dissolved from the felt with water. From the solution the barium is precipitated as the sulphate, which is dried, ignited and weighed. The strontium is precipitated from the united filtrates and washings by sulphuric acid after the addition of ethyl alcohol to insure thorough mixture. The results of tests of the method are given below.

Estimation of Barium and Strontium.

BaO taken. grm.	BaO found.* grm.	Error. grm.	SrO taken. grm.	SrO found.† grm.	Error. grm.
0.1212	0.1219	+0 0007	0.1068	0.1071	+0 0003
0.1215	0.1219	+0.0004	0.0358	0 0359	+0 0001
0 1220	0.1221	+0.0001	0.0353	0.0347	−0.0006
0.1212	0.1220	+0.0008	0 0363	0 0358	−0 0005
0.1219	0.1221	+0.0002	0 0361	0 0354	−0 0007
0.1211	0.1218	+0.0007	0 1126	0 1116	−0 0010
0.1319	0.1319	0 0000	0.0577	0.0586	+0 0009
0.0496	0.0492	−0 0004	0 0574	0 0579	+0 0005

* Corrected for 0 0025 BaO corresponding to the bromide dissolved in two treatments with 10 $cm.^3$ of amyl alcohol.

† Corrected by subtraction of 0 0040 grm. for the barium sulphate corresponding to dissolved barium bromide.

The method is rapid, and, while the correction to be applied, owing to the solubility of the barium salt, is large, it is definite.

The Estimation of Barium as the Sulphate.

In Presence of Hydrochloric Acid. In the ordinary mode of precipitating barium as barium sulphate, three conditions are carefully observed, — absence of excess of acid, slow mixing of the reagents, and standing twelve hours, or until the precipitate has completely subsided before filtration. Usually, in this process, the precipitate is thrown out in a finely divided, milky condition and settles very slowly. Mar* has observed, however, that the presence of hydrochloric acid influences the form in which the sulphate is deposited without affecting the completeness of precipitation, provided a sufficient excess of sulphuric acid is also present.

From a solution of 0.5 grm. of barium chloride in 400 cm.^{3} of water the precipitate appears immediately upon the addition of sulphuric acid, settling slowly, and this condition prevails also, even in hot solutions, when only one or two cubic centimeters of hydrochloric acid have been previously added. With 10 cm.^{3} to 15 cm.^{3} of strong hydrochloric acid in the solution heated to 85° or 90°, the precipitate settles clear in ten or twelve minutes, and is in excellent condition for filtration. When the solution contains 50 cm.^{3} of the acid, the precipitate settles clear in five minutes. Upon adding the sulphuric acid to such very acid solutions, no precipitate shows for a moment, but then it separates in beautiful crystalline condition and falls almost immediately. It can be safely filtered with or without pressure in ten minutes. In an instance cited, 2 grm. of barium chloride were precipitated in the presence of 30 cm.^{3} of hydrochloric acid, the precipitate was allowed to settle clear, and was then filtered and washed, the whole operation being completed in seven minutes. This rapid subsidence of the precipitate is seen in hot solutions only, 75° being the lowest temperature compatible with the attainment of good results, and 85° to 90° better.

Quantitative experiments, quoted below, show that precipitation is practically complete in 400 cm.^{3} of solution when sulphuric acid is added to the amount of 10 cm.^{3} of the 1 : 3 dilute acid (sp. gr. 1.28 — one of acid in a total volume of four) in presence of hydrochloric acid in amounts up to 150 cm.^{3} of the concentrated acid, sp. gr. 1.20. Considerable amounts of barium are precipitated at once, but when only a few milligrams are present complete formation of the precipitate requires more time. Two or three hours are in every case sufficient. In filtering on asbestos in the perforated crucible, as was done, care must be taken to use a very close felt, on account of the very minutely crystalline nature of the precipitate.

* F. W. Mar, Am. Jour. Sci., [3], xli, 288.

Precipitation in Presence of Hydrochloric Acid.

$BaCl_2.2H_2O$ taken.	Total volume.	HCl (sp. gr. 1.20).	Dilute H_2SO_4 (sp. gr. 1.28).	Time between precipitation and filtration.	$BaSO_4$ found.	Error.
grm.	cm.³	cm.³	cm.³	min.	grm.	grm.
0.0050	400	15	10	{ 5 60	{ 0.0023 0.0043	−0.0025 −0.0005
0.0050	400	15	10	5	0.0031	−0 0017
0.0050	400	15	10	10	0.0040	−0 0008
0.0100	400	15	10	10	0.0078	−0 0017
0.0100	400	15	10	15	0.0085	−0 0010
0.0100	400	15	10	30	0.0083	−0.0012
0.0100	400	15	10	60	0.0087	−0.0008
0.0030	400	15	10	120	0.0024	−0.0005
0 0050	400	15	10	150	0.0046	−0 0002
0.5014	400	15	10	10	0.4785	−0 0003
0.2227	400	15	10	10	0.2122	−0.0005
0.5003	400	15	10	10	0.4773	−0.0005
0.5046	400	15	10	10	0.4814	−0.0005
0.5016	400	15	10	10	0.4888	−0.0002
0.5004	400	150	10	10	0.4779	0.0000
0.5001	400	150	10	10	0.4776	0.0000

In Presence of Nitric Acid or Aqua Regia.

Browning* has investigated with similar results the effect of free nitric acid and aqua regia (3 : 1 mixture of hydrochloric acid and nitric acid) upon the precipitation of barium as the sulphate in presence of an excess of sulphuric acid. In a total volume of 100 cm.³ containing 10 cm.³ of dilute sulphuric acid [1 : 3] the barium sulphate falls with an average loss, after six hours' standing, of less than 0.0010 grm. in the presence of amounts of nitric acid up to 25 per cent of the entire volume. In aqua regia the solubility of barium sulphate is even less. Following are experimental results of these methods of treatment.

In this connection, the effect of the presence of a considerable amount of free nitric acid on the precipitation of barium as sulphate in cases where certain substances are present which under ordinary conditions tend to hold up the precipitate, is of interest. Fresenius† has demonstrated this property in the case of ammonium nitrate, Scheerer and Rube‡ have shown that meta-

* Philip E. Browning, Am. Jour. Sci., [3], xlv, 399.

† Zeit. anal. Chem., ix, 62.

‡ Jour. prakt. Chem., lxxv, 113–116.

Precipitation in Presence of Nitric Acid.

$BaSO_4$ equivalent to $Ba(NO_3)_2$ taken. grm.	$BaSO_4$ found. grm.	Error in terms of $BaSO_4$. grm.	Averages. grm.	Time between precipitation and filtration. hours.	Per cent by volume of strong HNO_3.	Total volume. cm.³
0.2540	0.2336	−0.0004		12	5	100
0.2489	0.2483	−0.0006	−0.0006	12	5	100
0.2495	0.2489	−0.0006		12	5	100
0.2492	0.2482	−0.0010		12	5	100
0.2486	0.2483	−0.0003	−0.0002	6	5	100
0.2490	0.2490	0.0000		6	5	100
0.2555	0.2546	−0.0009	−0.0006	1	5	100
0.2538	0.2534	−0.0004		1	5	100
0.2486	0.2477	−0.0009		12	25	100
0.2491	0.2490	−0.0001		12	25	100
0.2494	0.2484	−0.0010		12	25	100
0.2538	0.2535	−0.0003	−0.0008	12	25	100
0.2492	0.2484	−0.0008		12	25	100
0.2487	0.2471	−0.0016		12	25	100
0.3414	0.3407	−0.0007		12	25	100
0.2489	0.2481	−0.0008	−0.0007	6	25	100
0.2485	0.2478	−0.0007		6	25	100

Precipitation in Presence of Aqua Regia.

$BaSO_4$ equivalent to $Ba(NO_3)_2$ taken. grm.	$BaSO_4$ found. grm.	Error in terms of $BaSO_4$. grm.	Averages. grm.	Time between precipitation and filtration. hours.	Per cent by volume of strong aqua regia.	Total volume. cm.³
0.2539	0.2534	−0.0005		12	5	100
0.2540	0.2538	−0.0002	−0.0002	12	5	100
0.2490	0.2490	0.0000		12	5	100
0.2491	0.2492	+0.0001		12	5	100
0.2488	0.2484	−0.0004	−0.0001	6	5	100
0.3419	0.3421	+0.0002		6	5	100
0.2491	0.2485	−0.0006		12	25	100
0.1701	0.1697	−0.0004	−0.0003	12	25	100
0.1708	0.1705	−0.0003		12	25	100
0.1710	0.1710	0.0000		12	25	100
0.3415	0.3410	−0.0005	−0.0003	6	25	100
0.3418	0.3418	0.0000		6	25	100
0.3412	0.3405	−0.0007	−0.0007	1	25	100

phosphoric acid acts similarly, and Spiller* notes the same general effect where alkali citrates are present. Browning shows that these salts cause no apparent interference with the precipi-

* Chem. News, viii, 280, 281.

tation of barium in the presence of nitric acid amounting to one-tenth by volume of the entire liquid. The barium sulphate precipitated under such circumstances is, however, contaminated with foreign salts present and must be purified in order that the amount of barium actually present may be correctly indicated. The precipitate collected on paper and ignited, is, therefore, purified by dissolving it in sulphuric acid and recrystallizing according to the method of Mar, to be described.* Results of this treatment are given in the tabular statement.

Purification of the Precipitate.

Impurity present to the amount of 5 grm.	$BaSO_4$ equivalent to $Ba(NO_3)_2$ taken. grm.	Apparent amount of $BaSO_4$ found. grm.	$BaSO_4$ after purification. grm.	Error after purification. grm.	Percentage of strong HNO_3 by volume.
Ammonium nitrate. . . .	0.1710	0 1800	0 1702	−0 0008	10
Ammonium nitrate. . . .	0.3415	0.3440	0.3410	−0 0005	10
Ammonium citrate. . .	0.3412	0.3442	0.3407	−0.0005	10
Sodium citrate.	0.1360	0.1730	0.1366	+0.0006	10
Metaphosphoric acid. . . .	0.3461	0.3511	0.3470	+0.0009	10

Purification of Precipitated Barium Sulphate. When barium is precipitated as the sulphate the tendency of the precipitate to include foreign matter, if present, is very marked. It has been the custom to attempt the purification of barium sulphate contaminated by alkali salts by digesting in hydrochloric acid the washed precipitate. Phinney† has shown, however, that dilute hydrochloric acid alone dissolves barium sulphate itself, while mixtures of hydrochloric acid with enough sulphuric acid to prevent such solvent action do not completely remove the impurity; and Mar‡ has shown that the presence of hydrochloric acid, even in large excess, does not prevent contamination of the precipitate by alkali salts. After trying ineffectually the purification of the impure barium sulphate by solution in strong sulphuric acid and reprecipitation by water, Mar§ experimented with the crystallization of barium sulphate from its solution. The contaminated precipitate is dissolved in hot concentrated sulphuric acid and

* This page.

† J. I. Phinney, Am. Jour. Sci., [3], xlv, 468.

‡ Am. Jour. Sci., [3], xli, 293.

§ Loc. cit.

recovered from solution in crystalline form by evaporation of the acid. The crystallized sulphate is then washed upon a felt of asbestos in the filtering crucible, ignited and weighed. The evaporation may be effected over a radiator or by means of a ring burner; in either case, the process requires several hours. The operation may, however, be completed safely in a half-hour by the aid of the Hempel evaporating burner. Examples of the efficiency of the method are given in the tabular statements.

Degree of Contamination Found in Precipitated Barium Sulphate.

$BaCl_2.2H_2O$ taken. grm.	$BaSO_4$ found. grm.	Error. grm.	HCl in solution. cm.^{3}	Alkaline salts present
0 5092	0.5032	+0.0169	110	$KClO_3$ 3 grm.
0.5027	0.4907	+0 0107	10	$KClO_3$ 3 grm.
0.5026	0.4944	+0.0154	100	KCl 5 grm.
0.5045	0.4939	+0 0122	10	KCl 5 grm.
0.5020	0 4931	+0 0137	10	KCl 5 grm.
0.5013	0.4849	+0 0061	10	NaCl 5 grm.

Slow Evaporation over Radiator or by Ring Burner.

$BaCl_2.2H_2O$ taken. grm.	$BaSO_4$ found. grm.	Error. grm.
0.5029	0.4796	−0.0006
0.5008	0.4783	+0.0001
0.5038	0.4810	+0.0001
0.5087	0.4861	+0 0003
0.5025	0.4795	+0.0006

Rapid Evaporation by Hempel Burner.

$BaCl_2.2H_2O$ taken. grm.	$BaSO_4$ found. grm.	Error. grm.
0.5050	0 4824	+0 0002
0.5069	0.4838	0 0000
0.5041	0.4825	+0.0021
0.5021	0.4812	+0 0018
0.4033	0.4801	−0 0005

The results of applying this method to the purification of barium sulphate precipitated in presence of nitric acid from solutions containing citrates or a metaphosphate are given on page 172.

The Estimation of Barium as the Chloride.

Precipitation by Ether-Hydrochloric Acid Mixture. It has long been known that barium chloride is insoluble to a marked degree in hydrochloric acid, but the difficulty of filtering off the strong acid and washing the precipitate with strong acid prevented the early use of this characteristic of the chloride for the quantitative estimation of barium. The treatment of strong acid filtrates by means of the asbestos felt in the filtering crucible is now an easy matter, and the limits of insolubility of barium chloride in hydrochloric acid have been studied by Mar* with a view to developing a simple method for the separation of barium from calcium and magnesium. It has been shown that barium chloride is soluble to an extent not exceeding one part in 20,000 in pure, concentrated hydrochloric acid, the solubility increasing very rapidly with the diminution in the strength of the acid, while in concentrated hydrochloric acid containing ether the solubility falls to an amount not exceeding one part in about 120,000. To utilize this fact for the separation of barium from calcium and magnesium, Mar dissolves the chlorides of the earths in the least possible amount of boiling water and precipitates by 25 $cm.^3$ of concentrated hydrochloric acid with the addition of 5 $cm.^3$ of absolute ether after cooling. The acid is added drop by drop at first, as the precipitate is thus obtained in a coarse crystalline condition, filters very quickly, and is less liable to include foreign matter. After standing a few minutes the precipitate is filtered on an asbestos felt in a perforated crucible, washed with hydrochloric acid containing about 10 per cent of ether, and dried at 150°–200°. The method is accurate and rapid, and possesses the further advantage, when a number of determinations are to be made, that the precipitate may be dissolved off the felt by a little water, and, after ignition, the crucible and felt used again without reweighing. The felt upon which a half-dozen precipitates are thus treated may not change by so much as 0.0001 grm. in the process. The fumes of the strong acid cause no inconvenience if the filtration is performed in front of a good flue.

The figures of analysis, given below, indicate the accuracy of the process when applied to the pure barium salt.

* F. W. Mar, Am. Jour. Sci., [3], xliii, 521.

The Pure Barium Salt.

$BaCl_2.2H_2O$. grm.	HCl. cm.^{3}	Ether. cm.^{3}	$BaCl_2$. grm.	Error. grm.
0.5008	50	10	0.4267	−0.0002
0.5002	50	10	0.4257	−0.0007
0.4999	50	10	0.4252	−0.0009
0.4999	50	10	0.4258	−0.0003
0.5003	25	25	0.4259	−0.0005
0.5002	25	5	0.4262	−0.0002
0.5099	25	5	0.4344	−0.0003
0.5003	25	5	0.4261	−0.0003

Following are figures which show the results obtained in separating and determining barium when associated with magnesium and with calcium in mixtures of the chlorides.

Separation of Barium from Calcium and Magnesium.

$BaCl_2.2H_2O$. grm.	$CaCl_2$. grm.	HCl. cm.^{3}	Ether. cm.^{5}	$BaCl_2$. grm.	Error. grm.
0.5001	0.5	50	10	0.4250	−0.0013
0.4999	0.5	50	10	0.4250	−0.0011
0.5005	0.5	25	25	0.4260	−0.0006
0.5002	0.42	25	5	0.4258	−0.0004
0.5001	0.5	25	5	0.4255	−0.0008
0.5005	0.5	25	5	0.4251	−0.0015
0.5001	0.5	25	5	0.4254	−0.0009
0.5001	0.5	25	5	0.4258	−0.0005
0.5003	0.5	25	5	0.4261	−0.0004
0.1002	3.0	25	5	0.0842	−0.0012
0.0107	3.0	25	5	0.0080	−0.0005

$BaCl_2.2H_2O$. grm.	$MgCl_3.6HO_2$. grm.	HCl. cm.^{3}	Ether. cm.^{2}	$BaCl_2$. grm.	Error. grm.
0.4999	0.5	25	5	0.4253	−0.0007
0.5000	0.5	25	5	0.4257	−0.0005
0.1002	3.0	25	5	0.0844	−0.0010
0.0100	3.0	25	5	0.0077	−0.0008

Precipitation by Acetyl Chloride in Acetone. Gooch and Boynton* have given procedure for the precipitation of barium chloride from water solution and its separation from calcium and magnesium by the use of acetyl chloride to decompose the water of the solution according to the reaction

$$CH_3COCl + H_2O = CH_3COOH + HCl.$$

* F. A. Gooch and C. N. Boynton, Am. Jour. Sci., [4], xxxi, 212.

Inconvenient violence of the reaction is moderated by the addition of acetone which mixes in all proportions with both acetyl chloride and water, and by itself exerts no appreciable solvent action upon barium chloride.

When a mixture of acetone and acetyl chloride, preferably 4 : 1, is added slowly to a very concentrated solution of barium chloride in water, the water is attacked at once, hydrogen chloride is liberated, and precipitation begins immediately. If the temperature is kept down during the process by immersing in cool running water the vessel in which reaction takes place, no more than a mere trace of barium can be detected by sulphuric acid in the residue left after evaporating the liquid separated from the precipitate by filtration through asbestos. When, however, the temperature is allowed to rise, in consequence of the heat liberated in the reaction, an appreciable amount of barium may be found by sulphuric acid in the filtrate. It appears that when the acetone-acetyl chloride mixture [4 : 1] acts upon the cooled concentrated water solution of barium chloride the precipitate is the hydrous chloride, $BaCl_2.2H_2O$, only the water in excess of that needed to form the hydrous salt being immediately attacked; that acetyl chloride by itself produces only slight dehydration of this salt without marked solubility; and that prolonged action of an acetone-acetyl chloride mixture [2 : 1] results in appreciable dehydration and considerably increased solubility of the salt. When the acetone-acetyl chloride mixture is added without cooling to the water solution of barium chloride the heat of reaction favors dehydration of the hydrous salt, and the anhydrous salt may go into solution to the amount of several milligrams in 10 $cm.^3$ of the precipitating mixture. Upon filtering the mixture and treating the filtrate with acetone, with acetyl chloride, or with the acetone-acetyl chloride mixture, the dissolved anhydrous salt is not thrown out of solution, but the addition of a drop of water is sufficient to induce immediate precipitation in the form of the hydrous salt.

The best conditions for the quantitative precipitation of barium chloride by the acetone-acetyl chloride mixture are found in the use of minimum amounts of water, the preservation of ordinarily low temperature, a liberal proportion of acetone, and not too prolonged digestion of the precipitate in the excess of the precipitant. The salt to be analyzed is weighed out into a small beaker and dissolved in 1 cm.³ of water. The beaker is cooled by immersion in a water bath, preferably supplied with running water at a temperature of about 15°. To the cooled solution, constantly shaken, the acetone-acetyl chloride mixture is added from a dropping funnel at the rate of five drops to the second. The precipitate is filtered off upon asbestos in a perforated crucible, dried, or ignited, and weighed as the anhydrous chloride, $BaCl_2$. The best conditions studied for the handling of 0.1 grm. of hydrous barium chloride are the solution of the salt in 1 cm.³ of water, treatment with 30 cm.³ of the 4 : 1 mixture of acetone and acetyl chloride, washing with acetone, and drying in the air bath at 135° or at low redness.

Following are the results of experimental tests of the method applied to pure barium chloride.

The Pure Barium Salt.

$BaCl_2$ taken as $BaCl_2.2H_2O$.	$BaCl_2$ found.	Error.	Water to dissolve $BaCl_2.2H_2O$.	Amount of mixture and composition by volume.	
				To precipitate.	To wash.
grm.	grm.	grm.	cm.³		
0.0859	0.0859	0.0000†	1	5 cm.³ 2:1	10 cm.³ 2:1
0.0861	0.0854	−0.0007†	1	5 cm.³ 2:1	10 cm.³ 2:1
0.0861	0.0858	−0.0003†	1	5 cm.³ 2:1	10 cm.³ 2:1
0.0862	0.0854	−0.0008*	1	6 cm.³ 2:1	10 cm.³ 2:1
0.0857	0.0854	−0.0003*	1	6 cm.³ 2:1	10 cm.³ 2:1
0.0858	0.0860	+0.0002*	1	6 cm.³ 2:1	30 cm.³ 4:1
0.0860	0.0859	−0.0001*	1	6 cm.³ 2:1	30 cm.³ 4:1
0.0853	0.0850	−0.0003*	1	6 cm.³ 2:1	Acetone.
0.0854	0.0848	−0.0006*	1	6 cm.³ 2:1	Acetone.
0.0852	0.0851	−0.0001*	1	6 cm.³ 2:1	Acetone.
0.0857	0.0856	−0.0001†	1	6 cm.³ 2:1	Acetone.
0.0852	0.0845	−0.0007†	1	6 cm.³ 2:1	Acetone.
0.0855	0.0852	−0.0003†	1	6 cm.³ 2:1	Acetone.
0.0862	0.0862	0.0000†	1	30 cm.³ 4:1	Acetone.
0.0868	0.0868	0.0000†	1	30 cm.³ 4:1	Acetone.

* Ignited at low redness.
† Dried at 135° for 1½ hours.

Separation from Calcium and Magnesium. The application of these conditions to the separation of barium from moderate amounts of calcium and magnesium proves to be easily feasible. When acetone is added to the concentrated solution of calcium chloride or magnesium chloride in water two liquid layers are formed, the acetone above and the aqueous layer below; but the addition of a few drops of acetyl chloride renders the liquids miscible, while further addition causes no precipitation. When the 4:1 mixture of acetone and acetyl chloride is added at the rate of five drops in the second to the solution containing no more than 0.5 grm. of the calcium and magnesium salts, barium chloride is precipitated while calcium chloride and magnesium chloride are dissolved; but when the soluble chloride is present in the proportion of

Separation of Barium from Calcium.

$BaCl_2$ taken as $BaCl_2.2H_2O$. grm.	$CaCl_2.2H_2O$ taken. grm.	$BaCl_2$ found. grm.	Error. grm.	Water used to dissolve salts. cm.^{3}	Amount of mixture [4 : 1] used. cm.^{3}
0.0859	0.1000	0.0859	0.0000*	1	30
0.0867	0.1040	0.0867	0.0000*	1	30
0.0868	0.1022	0.0868	0.0000*	1	30
0.0865	0.1020	0.0865	0.0000*	1	30
0.0868	0.1017	0.0869	+0.0001*	1	30
0.0864	0.1016	0.0861	−0.0003*	1	30
0.0866	0.3025	0.0867	+0.0001*	1½	30
0.0859	0.5025	0.0859	0.0000*	2	30
0.0860	1.0020	0.0878	+0.0018*	3	30
0.0859	1.0020	0.0855	−0.0004†	2	30
0.0864	1.0035	0.0867	+0.0003†	2	30

* The precipitant was added at first at the rate of five drops in the second.
† The precipitant was added at the rate of two drops in the second at the outset and later of five drops in the second.

Separation of Barium from Magnesium.

$BaCl_2$ taken as $BaCl_2.2H_2O$. grm.	$MgCl_2.6H_2O$ taken. grm.	$BaCl_2$ found. grm.	Error. grm.	Water used to dissolve salts. cm.^{3}	Amount of mixture [4 : 1] used. cm.^{3}
0.0858	0.1000	0.0857	−0.0001*	1	30
0.0869	0.1025	0.0870	+0.0001*	1	30
0.0858	0.1025	0.0858	0.0000*	1	30
0.0862	0.1010	0.0863	+0.0001*	1	30
0.0858	0.1006	0.0860	+0.0002*	1	30
0.0860	0.1020	0.0859	−0.0001*	1	30
0.0860	0.1010	0.0862	+0.0002*	1	30
0.0865	0.3010	0.0867	+0.0002*	1½	30
0.0864	0.5000	0.0867	+0.0003*	2	30
0.0868	1.0015	0.0878	+0.0010*	3	30
0.0853	1.0010	0.0854	+0.0001†	3	30

* The precipitant was added at the rate of five drops in the second.
† The precipitant was added at first at the rate of two drops in the second and later of five drops in the second.

1.0 grm. to 0.1 grm. of the barium chloride, the rate of addition of the precipitating mixture should not be greater than two drops in the second at the start in order to avoid inclusion of the soluble salt in the insoluble barium salt. Even in such cases the mixture may be added at the rate of five drops in the second, after the greater part of the barium is down. The experimental results obtained in the separation of 0.1 grm. of the barium salt from 0.5 grm. of calcium and magnesium salts are excellent.

When the 4:1 mixture of acetone and acetyl chloride is added to the concentrated water solution of 0.1 grm. of strontium chloride a partial precipitation of the hydrous chloride, $SrCl_2.2H_2O$, takes place.

The action of a 4:1 mixture of acetone and acetyl chloride upon the concentrated solution of the chlorides affords easy and exact means for the separation and estimation of barium associated with calcium and magnesium. It is not recommended for the separation of barium from strontium.

The Precipitation of Barium Bromide by Ether-Hydrobromic Acid Mixture.

Thorne * has shown that barium bromide dissolved in the least possible amount of water is completely precipitated by a mixture of concentrated hydrobromic acid and ether in equal parts, and that the precipitate may be obtained of normal constitution, $BaBr_2$, and weighed as such if, after filtering upon asbestos in the perforated crucible, it is treated with ammonium bromide, and then gradually heated to 250°. Following are the results of test experiments made in this manner.

The Pure Barium Salt.

$BaBr_2.2H_2O$ taken. grm.	HBr and ether [1 : 1]. $cm.^3$	$BaBr_2$ found. grm.	$BaBr_2$ calculated. grm.	Error. grm.
0.2008	30	0.1793	0.1790	+0.0003
0.2041	30	0.1822	0.1820	+0.0002
0 2047	30	0.1821	0.1825	−0.0004
0.2171	30	0.1937	0.1936	+0.0001
0.3101	30	0.2768	0.2765	+0.0003
0.5035	30	0.4496	0.4490	+0.0006
0.5015	30	0.4476	0.4473	+0.0003

* Norman C. Thorne, Am. Jour. Sci., [4], xviii, 441.

Barium is precipitated completely either by hydrobromic acid or by hydrochloric acid* in mixture with ether, the precipitate falling as bromide or chloride in proportions according with the relative amounts of these acids present. In the presence of a great excess of hydrobromic acid the salt precipitated will be essentially bromide even if the original salt is taken in the form of the chloride. Salts of calcium and of magnesium remain in solution.

Separation of Barium from Calcium and Magnesium.

$BaCl_2.2H_2O$ taken. grm.	$CaCO_3$. grm.	$MgCO_3$. grm.	HBr and ether [1:1]. cm.[3]	$BaBr_2$ found. cm.[3]	Theory as $BaBr_2$. grm.	Error in $BaBr_2$. grm.
0.2253	.....	.....	30	0.2744	0.2741	+0.0003
0.2088	.....	....	30	0.2538	0.2540	−0.0002
0.3273	...	....	30	0.3975	0.3982	−0.0007
0.3177	.....	......	30	0.3864	0.3865	−0.0001
0.5041	0.5000	......	30	0.6134	0.6143	−0.0009
0.5083	0.5000	......	30	0.6185	0.6191	−0.0006
0.5046	0.5000	......	30	0.6139	0.6136	+0.0003
0.5022	0.5000	......	30	0.6110	0.6104	+0.0006
0.5018	0.5000	......	30	0.6106	0.6108	−0.0002
0.5007	.....	0.3000	30	0.6087	0.6092	−0.0005
0.5048	.....	0.3000	30	0.6144	0.6142	+0.0002

The Estimation of Calcium, Strontium and Barium, Precipitated as Oxalates.

The very high degree of insolubility which makes possible the well-known and exact process for the determination of calcium by precipitation as oxalate from ammoniacal water solutions, and weighing, is not directly applicable to strontium and barium on account of the greater solubility of the oxalates of those elements. Strontium oxalate is soluble in 12,000 parts of water,† while one part of barium oxalate dissolves in less than 3000 parts of cold water.‡

Gravimetric Determination of Strontium and Barium.

Peters § has shown, however, that strontium salts may be precipitated by ammonium oxalate with practical completeness in a solution containing one-fifth of its volume of 85 per cent alcohol, and with approximate

* Mar, page 174.

† Souchay and Lenssen, Ann. Chem., cii, 35.

‡ Souchay and Lenssen, Ann. Chem., xc, 102.

§ Am. Jour. Sci., [4], xii, 223.

completeness from water solutions at a dilution not exceeding 250 cm.[3] in presence of an amount of ammonium oxalate several times larger than that required theoretically; and that barium oxalate is precipitated almost completely from a solution containing one-third of its volume of 85 per cent alcohol. The precipitates thrown down hot and allowed to stand over night may be filtered off on asbestos in the perforated crucible, ignited a few minutes in the flame of a Bunsen burner and weighed as carbonate, or, after treatment with sulphuric acid, as sulphate. The results thus obtained are fairly accurate, as shown.

Precipitation in Approximately 17 per cent Alcohol.

SrO taken as $Sr(NO_3)_2$ grm.	SrO calculated from $SrCO_3$ found. grm.	Difference. grm.
0 1120	0.1113	−0.0007
0 1120	0 1116	−0.0004
0.2435	0.2425	−0.0010

Precipitation in Approximately 25 per cent Alcohol.

BaO taken as $Ba(NO_3)_2$. grm.	BaO found as $BaCO_3$. grm.	Difference. grm.
0.2912	0.2909	−0.0003
0.2912	0.2901	−0.0011
0.2912	0.2901	−0.0011

Titration of Oxalates by Potassium Permanganate.

Peters* has also shown that calcium, strontium and barium may be accurately estimated by titration of the oxalates in hydrochloric acid solution by potassium permanganate in presence of a manganous salt.†

Calcium oxalate is precipitated from the boiling hot solution with ammonium oxalate, and allowed to stand twelve hours. The supernatant liquid is decanted upon asbestos in the filtering crucible. The precipitate is washed two or three times, by decantation, with 50 cm.³–100 cm.³ of cold water and brought on the felt with care to avoid extended washing with hot water after all the precipitant, ammonium oxalate, has been removed. The crucible containing the precipitate is returned to the beaker,

* Am. Jour. Sci., [4], xii, 216.

† See page 50.

100 cm.³–200 cm.³ of water added, together with 5 cm.³–10 cm.³ of strong hydrochloric acid and 0.5 grm.–1.0 grm of manganous chloride, and the oxalic acid titrated at a temperature of 35°–45°. The results given are obviously excellent, and show that calcium, taken as the oxalate, may be estimated by potassium permanganate in the presence of hydrochloric acid and a manganous salt.

Precipitation from Water Solution.

CaO taken as $CaCl_2$. grm.	Ammonium oxalate. grm.	Volume at precipitation. cm.³	CaO found. grm.	Error. grm.
0 0656	0.3	100	0 0657	+0.0001
0 0656	0 3	100	0 0656	0.0000
0 0656	0.3	150	0 0658	+0.0002
0 0656	0.3	100	0 0655	−0 0001
0 0985	0.5	175	0.0981	−0.0004
0 1313	0 6	150	0 1315	+0.0002
0 1313	0.6	200	0 1315	+0 0002

Peters showed also that sulphuric acid may be employed in place of hydrochloric acid and manganous chloride when the dilution at titration is sufficient.

In precipitating strontium as the oxalate in alcoholic solution, ammonium oxalate is added to the hot solution with 85 per cent alcohol amounting to one-fifth to one-third of the total volume, the mixture is allowed to stand over night, and the clear liquid decanted upon an asbestos filter. The precipitate is washed with a mixture of equal parts of 85 per cent alcohol and water, transferred to the filter, dried in the filtering crucible over a flame to free it from alcohol, returned to the beaker previously

Precipitation from Alcoholic Solution.

Volume during titration 150 to 250 cm.³.

SrO taken as $Sr(NO_3)_2$. grm.	Ammonium oxalate. grm.	Volume at precipitation. cm.³	Proportion of 85 per cent alcohol.	Acid present during titration.	SrO found. grm.	Error. grm.
0 0974	0.4	100	$\frac{1}{5}$	HCl	0.0973	−0.0001
0.0974	0.4	100	$\frac{1}{5}$	HCl	0.0983	+0 0009
0.0974	0 4	100	$\frac{1}{5}$	HCl	0.0975	+0 0001
0.0974	0.8	100	$\frac{1}{5}$	HCl	0.0981	+0.0007
0.1948	0.4	200	$\frac{1}{5}$	HCl	0.1943	−0.0005
0.1948	0.8	200	$\frac{1}{5}$	HCl	0 1942	−0.0006

dried, treated with 5 cm.³–10 cm.³ of hydrochloric acid and 0.5 grm.–1.0 grm. of a manganous salt, and the liberated oxalic acid is titrated by permanganate. The results obtained by this method are accurate.

Precipitation from Water Solution.

SrO taken as $SrCl_2$. grm.	Ammonium oxalate. grm.	Volume at precipitation. cm.³	Acid present during titration.	SrO found. grm.	Error. grm.
0.0974	0.8	100	HCl	0.0971	−0.0003
0.0974	0.8	100	HCl	0.0980	+0.0006
0.0974	0.8	100	HCl	0.0975	+0.0001
0.0974	0.8	100	HCl	0.0980	+0.0006
0.0974	0.8	100	HCl	0.0973	−0.0001
0.0974	0.8	100	HCl	0.0978	+0 0004

By precipitating in the water solution in presence of a considerable excess of ammonium oxalate and washing with small amounts of water (30 cm.³–40 cm.³) applied judiciously, the loss by solubility may be made practically inappreciable, and in this case, there being no alcohol present to affect the titration, the precipitate need not be dried before treatment with permanganate. The results above show that 0.1 grm. of the strontium salt, calculated as the oxide, may be estimated as the oxalate with accuracy when precipitated in 100 cm.³ of water by a sufficient excess of ammonium oxalate.

Precipitation from Water Solution.

SrO taken as $Sr(NO_3)_2$. grm.	Ammonium oxalate. grm.	Volume at precipitation. cm.³	Acid present during titration.	SrO found. grm.	Error. grm.
0.0974	0.5	100	H_2SO_4	0.0966	−0.0008
0.0974	0.5	100	H_2SO_4	0.0985	+0.0011
0.0974	0.5	100	H_2SO_4	0.0977	+0 0003
0.0974	0.5	100	H_2SO_4	0.0963	−0 0011
0.0974	0.8	100	H_2SO_4	0.0981	+0.0007
0.0974	0.8	100	H_2SO_4	0.0966	−0 0008
0.0974	1.0	100	H_2SO_4	0.0965	−0.0009
0 0974	2.0	100	H_2SO_4	0.0963	−0 0011
0.0974	2.0	100	H_2SO_4	0.0970	−0 0004
0.0778	0.5	100	H_2SO_4	0.0792	+0.0014
0.0778	0.5	100	H_2SO_4	0.0767	−0.0011
0.0778	0.5	100	H_2SO_4	0 0776	−0.0002
0.0778	0.5	100	H_2SO_4	0.0776	−0 0002
0 0974	0.8	250	H_2SO_4	0 0973	+0.0001
0.0974	2.0	250	H_2SO_4	0.0975	−0 0001

Amounts of strontium oxalate approximately equivalent to 0.1 grm. of strontium oxide may be successfully precipitated without alcohol and titrated in a volume of 200 cm.^{3}–300 cm.^{3} in presence of sulphuric acid, as shown by the results which are given in the preceding table.

To precipitate barium as the oxalate, ammonium oxalate is added to a solution of a barium salt, containing 30 per cent of its volume of alcohol, and after standing over night the precipitate is filtered on asbestos, washed by decantation with 100 cm.^{3}–200 cm.^{3} of water containing 30 per cent of its volume of alcohol, and dried over a flame to insure the removal of alcohol. The crucible containing the precipitate is returned to the beaker, also previously dried over a flame, 100 cm.^{3}–200 cm.^{3} of water, 5 cm.^{3}–10 cm.^{3} of strong hydrochloric acid, and 0.5 grm.–1.0 grm. of manganous chloride are added, and the solution is titrated at 35°–45° with permanganate. The results of the experiments given show that barium, either as the nitrate or chloride, may be estimated in the manner described with a fair degree of accuracy.

Precipitation from Alcoholic Solution.

BaO taken as $Ba(NO_3)_2$. grm.	Ammonium oxalate. grm.	Volume at precipitation. cm.^{3}	Acid present during titration.	BaO found. grm.	Error. grm.
0.1165	0.2	100	HCl	0.1177	+0.0012
0.1165	0.2	100	HCl	0.1170	+0.0005
0.1165	0.2	100	HCl	0.1164	−0.0001
0.1165	0.2	100	HCl	0.1151	−0.0014
0.1165	0.2	100	HCl	0.1165	0.0000
0.1165	0.2	100	HCl	0.1176	+0.0011
0.1165	0.2	100	HCl	0.1164	−0.0001
0.2330	0.4	100	HCl	0.2319	−0.0011
0.2330	0.4	100	HCl	0.2335	+0.0005
0.2330	0.4	100	HCl	0.2342	+0.0012
BaO taken as $BaCl_2$.					
0.0942	0.4	100	HCl	0.0952	+0.0010
0.0942	0.4	100	HCl	0.0939	−0.0003
0.0942	0.4	100	HCl	0.0941	−0.0001
0.1884	0.4	100	HCl	0.1893	+0.0009
0.1884	0.4	100	HCl	0.1892	+0.0008

Barium oxalate cannot be successfully titrated in presence of sulphuric acid on account of the great insolubility of barium sulphate and its protecting influence upon undecomposed barium oxalate.

CHAPTER V.

ZINC; CADMIUM; MERCURY.

ZINC.

The Estimation of Zinc as the Pyrophosphate.

IN studying the determination of zinc by the method which involves precipitation as ammonium zinc phosphate and weighing as the ignited pyrophosphate, Austin * has shown that, as in the case of the similar precipitation of manganese,† the presence of a definite excess of ammonium salt during the precipitation is essential to the formation of the ideal salt, NH_4ZnPO_4, uncontaminated by the tribasic phosphate, $Zn_3(PO_4)_2$, while too much tends to produce a double salt too rich in ammonia. The condition of the ammonium zinc phosphate most nearly approximating to the ideal is obtained by precipitating in presence of ammonium chloride in large amount. Microcosmic salt is added until the solution (100 $cm.^3$ to 200 $cm.^3$) containing the ammonium salt is alkaline, and the whole is heated until the mass subsides in crystalline condition. The amount of ammonium chloride should be 20 grm. if the filtration is to be made as soon as the solution cools. One-half this amount will do if the liquid stands a number of hours. Larger amounts tend to give a salt too rich in ammonia. The time of standing seems to be a less important factor than either the excess of microcosmic salt or the amount of ammonium chloride. When the solutions are made finally faintly acid with acetic acid, according to the method of Langmuir,‡ the results are low. Following are the results obtained by the method described.

In a subsequent article§ it is made plain that Dakin's ‖ proposal to wash with a 1 per cent solution of ammonium phosphate, followed by alcohol, leads to erroneous results.

* Martha Austin, Am. Jour. Sci., [4], viii, 210.

† See page 483.

‡ Jour. Am. Chem. Soc., xxi, 115.

§ Am. Jour. Sci., [4], xiv, 156.

‖ Chem. News, lxxxii, 101; lxxxiii, 37.

Estimation as Zinc Pyrophosphate.

$Zn_2P_2O_7$ correspond-ing to $ZnSO_4$ taken. grm.	Found. grm.	Error. grm.	Error in terms of zinc. grm.	$Zn_2P_2O_7$ correspond-ing to Zn left in the filtrate. grm.	$HNaNH_4PO_4.4H_2O$. grm.	NH_4Cl. grm.	Time of standing. hrs.
0.6355	0 6335	−0.0020	−0.0008	none	4.47	10	16
0.6355	0.6381	+0.0026	+0.0010	none	4.47	20	½
0.6355	0 6379	+0 0024	+0.0009	none	4 47	20	2
0 6355	0 6386	+0.0031	+0.0012	none	4.47	20	½
0.6355	0.6393	+0.0038	+0.0014	none	4.47	20	½
0.6367	0 6355	+0.0012	+0.0005	none	4.47	30	16

The Conversion of Zinc Chloride to Zinc Oxide.

Havens* has shown that zinc chloride may be quantitatively converted to zinc oxide by treatment with nitric acid, evaporation of the excess of acid, and ignition of the residue. The solution of zinc chloride is evaporated in porcelain, best with a gentle current of air playing upon the surface of the liquid to avoid spattering, and treated repeatedly with nitric acid, added in small portions with intermediate evaporations. The residue is finally ignited to convert the nitrate to oxide. Results are given in the table.

Conversion of Chloride to Oxide.

ZnO taken as chloride. grm.	ZnO found. grm.	Error. grm.
0.1019	0.1016	−0.0003
0.1010	0 1007	−0.0003
0 1100	0.1095	−0.0005

The Electrolytic Determination of Zinc.

The deposition of zinc upon the rotating crucible† succeeds best, according to Medway,‡ when the zinc salt — preferably the sulphate — is dissolved in 50 cm.^{3} of water to which 4 grm. of potassium oxalate are added. The presence of ammonium salts appears to retard the complete deposition of the metal.

* F. S. Havens, Am. Jour. Sci., [4], vi, 45.

† See Fig. 13, page 12.

‡ H. E. Medway, Am. Jour. Sci., [4], xviii, 56.

Deposition on the Rotating Cathode.

Zinc taken. grm.	Zinc found. grm.	Error. grm.	Current. amp	N. D.$_{100}$.	Time. min.
0.0553	0.0556	+0.0003	2.5	8.3	25
0.0553	0.0553	0.0000	2.5	8.3	25
0.0553	0.0552	−0.0001	2.5	8.3	25
0.0993	0.0995	+0.0002	2.5	8.3	30
0.0993	0.0994	+0.0001	2	6.6	25
0.0993	0.0991	−0.0002	2	6.6	25

In determining zinc by electrolysis with stationary electrodes, it has been found that, when the attempt is made to remove the zinc from the platinum upon which it has been deposited, a coating of platinum black is left, some of the zinc having amalgamated with the platinum. Only by dissolving the zinc, heating the crucible to redness and finally making another application of acid can this black coating be conveniently removed. In order to avoid this formation of platinum black it has been found necessary to coat the platinum with copper and deposit the zinc upon this. The zinc and copper may then be easily removed together by acid. In depositing the zinc upon a rotating cathode, however, it is found to be unnecessary to coat the platinum with copper, since the zinc can be removed without any appearance of platinum black.

In depositing zinc upon the rotating cathode from an acetate solution containing a salt of iron, Moody* has found iron deposited with the zinc.†

The Estimation of Zinc by Precipitation as the Oxalate and Titration with Potassium Permanganate.

Ward‡ has shown that zinc may be accurately estimated by precipitation as oxalate and titration with potassium permanganate. To the boiling water solution of the zinc salt oxalic acid is added, followed by acetic acid in large amount. The precipitate is filtered upon asbestos in the perforated crucible, and washed with small amounts of water. Crucible and precipitate

* Seth E. Moody, Am. Jour. Sci., [4], xxii, 484.

† See page 67.

‡ H. L. Ward, Am. Jour. Sci., [4], xxxiii, 334.

are treated with dilute sulphuric acid, the solution heated to boiling, and the free oxalic acid titrated by permanganate.

Results are given in the table.

Determination of Zinc.

Zinc taken as acetate. grm.	Volume at precipitation cm.³	Oxalic acid. grm.	Acetic acid added. cm.³	Zinc found. grm.	Error. grm.
0 0055	100	2	100	0 0056	+0.0001
0 0274	100	2	100	0 0276	+0 0002
0 0548	50	2	50	0 0553	+0 0005
0 0548	100	2	100	0 0550	+0 0002
0.1370	100	2	100	0 1372	+0 0002

Cadmium.

The Estimation of Cadmium as the Oxide.

Precipitation as Carbonate. Various objections have been made to that method for estimating cadmium which involves precipitation as carbonate, ignition, and weighing as oxide. Browning and Jones* have shown that when the carbonate is filtered upon an asbestos felt in a perforated crucible previously ignited and weighed, danger of reduction is obviated and the process is simplified and placed in the category of good analytical methods. To the solution of the cadmium salt in about 300 cm.³ of hot water is added a 10 per cent solution of potassium carbonate, drop by drop and with constant stirring. The solution, with the precipitate in suspension, is boiled for about fifteen minutes. The precipitated cadmium carbonate becomes granular and settles quickly, and is then filtered upon asbestos in the perforated crucible, ignited and weighed. Results obtained by this method, and quoted in A of the following table, show a slight plus error due, as was shown experimentally, to inclusion of alkali salt, but indicate that the carbonate method can be successfully and simply applied to the quantitative estimation of cadmium.

The work of Flora † fully substantiates these results, giving by similar procedure the analytical figures in B.

* Philip E. Browning and L. C. Jones, Am. Jour. Sci., [4], ii, 269.

† Charles P. Flora, Am. Jour. Sci., [4], xx, 456.

Precipitation as Cadmium Carbonate.

CdO taken. grm.	CdO found. grm.	Error. grm.
	A.	
0 1140	0 1143	+0 0003
0 1142	0.1137	−0 0005
0.1141	0 1148	+0.0007
0 1141	0 1148	+0.0007
0 1142	0 1146	+0.0004
0.1143	0 1147	+0.0004
0.1143	0 1144	+0.0001
0 1139	0 1146	+0 0007
0 1270	0 1272	+0.0002
0 1279	0.1283	+0 0004
0 1272	0 1281	+0 0009
0 1278	0 1281	+0.0003
0 2556	0 2561	+0 0005
0 2550	0 2547	−0.0003
0.1272	0 1279	+0 0007
0 1281	0 1288	+0 0007
0 1274	0 1278	+0 0004
0.1284	0 1290	+0.0006
0 1271	0 1277	+0 0006
0.1278	0.1285	+0.0007
0.2555	0.2555	0 0000
	B.	
0 1277	0.1275	−0 0002
0.1277	0.1280	+0 0003
0.1277	0.1272	−0 0005
0.1399	0.1391	−0 0008
0.1399	0.1399	0 0000
0 1703	0.1700	−0 0003
0.1703	0.1700	−0 0003
0.2129	0.2128	−0 0001
0.2129	0.2128	−0 0001
0.2554	0.2554	0 0000

Precipitation as Hydroxide. Flora* has also tested the similar method by which cadmium is precipitated as hydroxide and weighed as oxide. To the boiling solution of the cadmium salt, about 300 cm.^{3} in volume, a 10 per cent solution of potassium hydroxide is added drop by drop. After boiling about fifteen minutes the precipitate settles quickly in a semigranular state and is filtered on a weighed asbestos felt in the perforated crucible, washed, ignited and weighed. The results are lower than those of the carbonate method.

* Loc. cit.

Precipitation as Cadmium Hydroxide.

CdO taken. grm.	CdO found. grm.	Error. grm.
0 1277	0.1277	0.0000
0 1277	0.1270	−0 0007
0 1277	0.1260	−0 0017
0.1277	0.1286	+0.0009
0 1362	0 1350	−0 0012
0 1399	0 1389	−0 0010
0.1703	0.1697	−0.0006
0 1703	0 1693	−0 0010
0 1703	0.1699	−0 0004
0 1788	0 1802	+0 0014
0 2129	0 2139	+0.0010
0 2129	0.2128	−0 0001

While these figures show that fair results may be obtained, the hydroxide method is not comparable with the carbonate method as to accuracy or convenience. The precipitate does not take the same granular form; it is hard to filter, difficult to wash, and is removed with difficulty from the beaker in which precipitation takes place.

The Estimation of Cadmium as the Pyrophosphate.

It has been shown by Austin* that cadmium may be estimated with accuracy as the pyrophosphate. The precipitate obtained by making alkaline with microcosmic salt the nearly neutral solution, containing ammonium chloride in the proportion of ten grams to one hundred cubic centimeters, is allowed to stand several hours, then filtered off on asbestos in the filtering crucible, dried, ignited and weighed. The cadmium separates out from the solution as a beautiful crystalline mass of cadmium ammonium phosphate of ideal constitution. The conditions must, however, be preserved with care; there must be no excess of ammonia, no free acid, and no excess of ammonium salt beyond the quantity indicated, while that amount is necessary.

A criticism of this method by Miller and Page† was shown in a later article‡ to be without foundation.

* Martha Austin, Am. Jour. Sci., [4], viii, 214.
† School of Mines Quarterly, xxii, 391.
‡ Am. Jour. Sci., [4], xiv, 156.

Estimation as Cadmium Pyrophosphate.

$Cd_2P_2O_7$ corresponding to $CdCl_2$ taken. grm.	Found. grm.	Error. grm.	Error in terms of cadmium. grm.	$Cd_2P_2O_7$ corresponding to Cd found in the filtrate. grm.	$HNaNH_4PO_4.4H_2O$. grm.	NH_4Cl. grm.	Time of standing. hrs.
0.6972	0.6976	+0.0004	+0 0002	trace	4.5	10	16
0.6972	0 6969	−0 0003	−0.0002	trace	4 5	10	18
0.6972	0 6962	−0 0010	−0 0006	trace	4.5	10	16

The Electrolytic Determination of Cadmium.

The deposition of cadmium as the metal, upon the rotating cathode,* has been studied by Medway† for the sulphate solution, and by Flora‡ for solutions containing sulphuric acid, acetates, cyanides, pyrophosphates, phosphates, oxalates, formates, tartrates, free nitric acid, urea, formaldehyde or acetaldehyde.§

Deposition from the Sulphuric Acid Solution. Cadmium taken as the sulphate to the amount of 0.2 grm. approximately and dissolved in 50 $cm.^3$ of water containing 10 drops of dilute sulphuric acid may be successfully deposited upon the crucible rotating at the rate of 650–700 revolutions per minute. To avoid solvent action after stopping the current, dilute ammonia is added drop by drop to faint alkalinity, while the current is still passing and after complete deposition of the metal. That this procedure is satisfactory the following results of Medway‖ show.

Deposition from the Solution of the Sulphate.

Cadmium taken. grm.	Cadmium found. grm.	Error. grm.	Current. amp.	$N. D._{100}$.	Time. min.
0 1088	0 1083	−0.0005	2	6 6	15
0.1088	0.1085	−0.0003	2	6.6	15
0.1088	0.1092	+0.0004	1.5	5	15
0.1088	0.1090	+0 0002	2	6.6	15
0.1088	0.1093	+0.0005	1.5	5	12
0.1088	0.1093	+0.0005	2	6.6	10
0.1088	0.1087	−0.0001	2	6.6	10

* See Fig. 13, page 12.

† H. E. Medway, Am. Jour. Sci., [4], xviii, 56.

‡ Charles P. Flora, Am. Jour. Sci., [4], xx, 268 et seq., and 292 et seq.

§ For descriptions and results, see pages 191 to 195.

‖ Am. Jour. Sci., [4], xviii, 56.

In a detailed study of the use of the rotating cathode for the estimation of cadmium, Flora * calls attention to the fact that the dilution of the solution submitted to electrolysis is of great importance. It is advisable, in order to avoid mechanical loss, to deposit not more than 0.25 grm. of the metal upon the cathode, while even smaller quantities are to be preferred. The current density must also be kept within limits; for otherwise a spongy deposit may result. Cadmium seems to be especially apt to form these spongy, unweighable deposits, and the greatest difficulties come from this behavior of the metal. The best conditions for the deposition in presence of sulphuric acid may be briefly summarized as follows: Cadmium sulphate, equivalent to not more than 0.25 grm. of the metal, is dissolved in 45 $cm.^3$ to 50 $cm.^3$ of water; ten to fifteen drops of dilute sulphuric acid are added; and the solution subjected to electrolysis at a normal current density ($N.D._{100}$) ranging between 3 amp. and 9 amp. per 100 $cm.^2$ of surface. It is not necessary to heat the liquid, as the passage of the current soon heats it sufficiently. When electrolysis is complete, the excess of sulphuric acid may be destroyed with a slight excess of ammonia water, the current broken, and the cathode removed, thoroughly rinsed with water and alcohol, and dried by waving over a free flame. If the deposit is not spongy the drying is a matter of only a few moments, and there is no danger of oxidizing the metallic deposit. If it is preferred, the current may be reduced by interposed resistance, the rotation stopped, and the liquid readily siphoned without danger of injuring the metallic coating.

This process is also available when the cadmium is taken as the chloride if the volume of the solution does not exceed 45 $cm.^3$.†

Deposition from Solutions Containing Acetates. In a volume of 60 $cm.^3$ to 65 $cm.^3$, containing 0.5 grm. to 2 grm. of sodium acetate and a small amount of potassium sulphate to so regulate the conductivity that the normal current density shall not much exceed 3 amp., the deposition of not more than 0.15 grm. of cadmium, taken as the sulphate, proceeds rapidly and satisfactorily. The deposit under the conditions is rather crystalline, fairly compact, and easily washed, so that the method forms one of the very best

* Charles P. Flora, Am. Jour. Sci., [4], xx, 268–276, 392–396, 454–455.

† Flora, loc. cit., page 392.

where the cadmium is taken in the form of the sulphate. At greater concentrations the precipitate shows a tendency to sponginess, and it fails absolutely when the cadmium is introduced as the chloride.*

Deposition from Solution Containing Alkali Acetate.

Cadmium taken. grm.	$NaOC_2H_3O$. grm.	K_2SO_4. grm.	Current. amp.	N. D.$_{100}$ amp.	E.M.F. volts.	Time. min.	Cadmium found. grm.	Error. grm.
0 1118	0 5	1 0	1.0	3 0	8 0	20	0 1121	+0 0003
0 1491	1 5	0.5	0.9	2 7	8 0	15	0 1494	+0 0003
0 1491	1.5	0 5	0.9	2.7	8.0	15	0.1496	+0 0005

Cadmium taken. grm.	NaOH. grm.	K_2SO_4. grm.	Current. amp.	N. D.$_{100}$ amp.	E.M F volts.	Time. min.	Cadmium found. grm.	Error. grm.
0 1491	. .	0 5	1 25	3 75	8 0	10	0 1496	+0.0005
0.1491	0.2*	0.5	0 8	2 4	8 0	15	0.1491	0.0000
0.1491	0.2*	0 5	0 8	2 4	8 0	15	0 1493	+0.0002
0 1223	0 5*	0.2	1 0	3 0	12 0	20	0.1223	0 0000
0 1223	0 5*	0.2	1.0	3 0	12 0	20	0.1223	0 0000

* Neutralized by a slight excess of acetic acid.

Deposition from Solutions Containing Cyanides.

The deposition of cadmium from a solution of the double cyanide has proved to be very satisfactory, and the results with the rotating cathode are in complete accordance with previous work on this method. The range of conditions of current and quantity of electrolyte is broad; the deposit is a beautiful silvery plate which dries very quickly, and is so compact as to be rubbed off only with difficulty; and although the complete deposition of the metal is slower than it is from solutions containing sulphates or acetates, it is sufficiently rapid. Care should be taken to avoid foaming of the solution, as this retards somewhat the deposition of the final traces of cadmium. Generally, a volume of 65 $cm.^3$ to 70 $cm.^3$ is found most satisfactory. In experiments to test the method the cadmium sulphate was treated with sodium hydroxide and the precipitate was dissolved in potassium cyanide. The following results were obtained.†

* Flora, loc. cit., page 392.
† Ibid., page 272.

Deposition from Solution Containing Alkali Cyanide.

Cadmium taken. grm.	NaOH. grm.	KCN. grm.	Current. amp.	N. D.$_{100}$. amp.	E.M.F. volts.	Time. min.	Cadmium found. grm.	Error. grm.
0.1491	1.5	0.5	2.5	7.5	8	35	0.1498	+0 0007
0 1491	1 0	0 5	2.5–4.5	7.5–13.5	8	30	0.1490	−0 0001
0.1223	1.5	1.0	2.5	7 5	8	35	0.1225	+0 0002

Satisfactory results are also found by this method when the cadmium is taken as the nitrate or as the chloride * the best dilution being 60 cm.^{3} to 65 cm.^{3} The time required is a trifle longer than in the estimation of cadmium sulphate by this method, and with the current density necessary to hasten the deposition a considerable tendency to foam is manifest.

Deposition from Solutions Containing Pyrophosphates or Orthophosphates. Brand † has recommended the use of a solution containing sodium pyrophosphate for the electrolytic estimation of metals, among others, cadmium; and the fitness of this solution for use with the rotating cathode has also been studied by Flora.‡

When the cadmium, taken as sulphate or chloride, is precipitated by sodium pyrophosphate, the precipitate dissolved in an excess of ammonium hydroxide, phosphoric acid, sulphuric acid, or hydrochloric acid, and the solution submitted to electrolysis at a volume of 60 cm.^{3}, fairly accurate results may be obtained; but neither is the method so accurate as those previously described, nor are the conditions so flexible. Practically the same thing may be said of the use of the rotating cathode in the electrolysis of cadmium orthophosphate dissolved in phosphoric acid according to the method recommended by Smith.§ Flora‖ has also studied the behavior of solutions containing oxalates, formates, tartrates, urea, formaldehyde and acetaldehyde, nitrates and free nitric acid. The results of this study and of the work previously described may be summarized as follows:

Summary. — Cadmium taken in the form of the sulphate may be very accurately and satisfactorily estimated by deposition

* Flora, loc. cit., page 393, and page 454.
† Zeit. anal. Chem., xxviii, 581.
‡ Loc. cit., page 273.
§ Am. Chem. Jour., xii, 329.
‖ Loc. cit.

upon the rotating cathode from solutions containing sulphuric acid, sodium acetate and acetic acid, or potassium cyanide; but little less satisfactorily from solutions containing urea, formaldehyde or acetaldehyde; and also, with proper precautions, from solutions containing pyrophosphates, phosphates, tartaric acid or formic acid. From solutions containing oxalates or oxalic acid, ammonium tartrate or potassium formate satisfactory deposits are not obtained.

When taken as the chloride, cadmium does not permit such a wide range of conditions. Nevertheless, from solutions of the chloride containing sulphuric acid or potassium cyanide, or the pyrophosphates, the metal is deposited in a form comparable with that obtained when cadmium sulphate is taken. Solutions of the chloride of cadmium to which is added hydrogen disodic phosphate give less desirable results; while solutions containing urea, formaldehyde or acetaldehyde give deposits free from sponginess only after careful regulation of the conditions. In solutions containing the oxalates, oxalic acid, the formates and the tartrates, acetates, formic acid and tartaric acid the results were negative. Cadmium nitrate is in general ill-fitted for electrolytic estimation, the cyanide solution being the only one from which satisfactory results were obtained. From solutions containing one per cent or more of free nitric acid, the cadmium is not deposited by the current.

MERCURY.

The Gravimetric Determination of Mercury as Mercurous Oxalate.

Mercury taken in the form of mercurous nitrate may be estimated as mercurous oxalate precipitated by ammonium oxalate and dried over sulphuric acid, as has been shown by Peters.* It is necessary, however, to control the acidity, dilution, and presence of mercuric salts. It appears that 5 $cm.^3$ of dilute nitric acid, sp. gr. 1.15, may be present in a volume of 100 $cm.^3$, and that 5 $cm.^3$ of the acid will prevent precipitation of small amounts of mercuric salt,—0.0100 grm. to 0.0200 grm., calculated as mercury, depending upon the amount of ammonium oxalate present in excess. According to the procedure recommended by Peters, mercurous nitrate dissolved in 100 $cm.^3$ of water containing 2 per

* C. A. Peters, Am. Jour. Sci., [4], ix, 405.

cent to 5 per cent of dilute nitric acid, sp. gr. 1.15, is precipitated by the addition of ammonium oxalate in slight excess with stirring. It is an easy matter to keep the excess of the precipitant within the limits of 1 cm.³ to 2 cm.³ of the $n/10$ solution, because the mercurous oxalate, when properly stirred, settles rapidly. The precipitate is collected on asbestos in a perforated crucible, washed two or three times with cold water, and dried to constant weight over sulphuric acid at the ordinary temperature, since mercurous oxalate is slowly decomposed at temperatures in the vicinity of 100°. Mercury in the mercuric form may be safely present to the amount of 10 per cent of the mercurous salt.

Results of experiments according to this procedure, in which the drying was effected by exposure for fifteen hours or less at ordinary temperatures over sulphuric acid, are given in the accompanying table.

Mercurous Oxalate Dried over Sulphuric Acid.

Hg taken as $Hg_2(NO_3)_2$.	Hg present as $Hg(NO_3)_2$.	Excess of $n/10$ ammonium oxalate.	HNO_3 (sp. gr. 1.15.)	Volume at precipitation.	Hg found.	Error.
grm.	grm.	cm.³	cm.³	cm.³	grm.	grm.
0.1217	.....	2–4		100	0.1217	0 0000
0 1217	. .	2–4	. .	100	0.1217	0.0000
0.1122	. . .	2–4	.	100	0 1124	+0.0002
0.1122	0 0067	0.93	2	100	0 1130	+0 0008
0.1122	0 0067	0.93	2	100	0 1112	−0.0010
0.1122	0 0067	4.40	2	100	0 1124	+0.0002
0.1122	0.0135	0.72	4	100	0 1125	+0.0003
0.2244	0 0071	1.68	4	100	0 2253	+0 0009
0.2244	0 0071	2.46	4	100	0 2241	−0 0003
0.2244	0.0048	0.54	4	200	0.2248	+0.0004
0.2244	0 0048	2.44	4	200	0.2245	+0.0001

The Determination of Mercury by Titration with Sodium Thiosulphate.

The method proposed by Scherer* for the determination of mercury by titration of mercury salts with sodium thiosulphate has been studied by Norton.† It is shown in the estimation of mercury taken in the form of mercuric chloride the reaction proceeds definitely according to the equation

$$3\,HgCl_2+2\,Na_2S_2O_3+3\,H_2O=2\,HgS.HgCl_2+2\,Na_2SO_4+4\,HCl.$$

* Lehrbuch der Chemie, i, 513.

† John T. Norton, Am. Jour. Sci., [4], x, 48.

This method yields accurate results if carried out under certain fixed conditions. These conditions, which must be closely adhered to, are as follows: The solution containing the mercury in the form of mercuric chloride is diluted to 100 cm.³ and heated to a temperature of 60° C. The sodium thiosulphate in $\frac{1}{20}$th normal solution is run in from a burette until the white precipitate, $2\,HgS.HgCl_2$, first formed begins to take on a brownish tinge due to incipient formation of the sulphide. The solution is then diluted with cold water, some asbestos fiber added to coagulate the precipitate, and the whole is quickly thrown on the filter. After careful washing, potassium iodide is added to the filtrate and the excess thiosulphate is titrated with iodine in presence of starch. The duration of the process need not exceed fifteen minutes. It is worthy of note that there is no necessity of using any hydrochloric acid in addition to that formed in the reaction.

Results obtained by this procedure are given in the table.

Titration of Mercuric Chloride.

$HgCl_2$ taken as Hg. grm.	Volume at beginning. cm.³	Temperature. C.	$Na_2S_2O_3$ in excess. cm.³	$HgCl_2$ found as Hg. grm.	Error. grm.
0 0759	100	60°	3.06	0 0766	+0.0007
0 0384	100	60°	2 81	0.0387	+0.0003
0.1498	100	60°	1 1	0.1500	+0 0008
0 1503	100	60°	1 63	0.1506	+0.0003
0 1479	100	60°	2 41	0.1480	+0.0001
0 1489	100	60°	2 12	0.1503	+0 0014
0 2244	100	60°	2 63	0.2259	+0.0015
·0 1490	100	60°	2.33	0.1484	−0 0006
0 0758	100	60°	2.	0.0762	+0.0004
0 0383	100	60°	2 58	0 0379	−0.0004

In applying Scherer's process to mercurous nitrate and mercuric nitrate, Norton was unable to discover conditions of definite action.

The Estimation of Mercury by Precipitation as Mercurous Oxalate and Titration of the Excess of Precipitant with Permanganate.

Peters* has shown that mercury taken as mercurous nitrate may be estimated by precipitating it as the oxalate and determining by titration with potassium permanganate the excess of

* C. A. Peters, Am. Jour. Sci., [4], ix, 401.

ammonium oxalate used as the precipitant.* The operation is successful in presence of 2 per cent to 5 per cent of dilute nitric acid, sp. gr. 1.15, in 100 cm.³ of the mixture; and mercuric salt to the extent of 10 per cent of the amount of the mercurous salt may be present provided the nitric acid amounts to 2 per cent, and the excess of ammonium oxalate used to effect precipitation is not too great. The precipitate settles well when properly stirred, and the excess of the precipitant need not exceed the safe limit of 1 cm.³ or 2 cm.³. Following are results obtained by the permanganate titration of the filtrate from the precipitated mercurous oxalate.

Precipitation by Ammonium Oxalate and Titration of the Excess.

Hg taken as $Hg_2(NO_3)_2$. grm.	Hg present as $Hg(NO_3)_2$. grm.	Excess of n/10 ammonium oxalate. cm.³	HNO_3 (sp. gr. 1.15). cm.³	Volume at precipitation. cm.³	Hg found. grm.	Error.
0.1217	0.0067	0.86	2	100	0.1232	−0.0015
0.1217	0.0067	0.92	2	100	0.1220	+0.0003
0.1217	0.0067	0.97	2	100	0.1218	+0.0001
0.1217	0.0067	0.90	2	100	0.1224	+0.0007
0.1217	0.0067	0.91	2	100	0.1213	−0.0004
0.1221	0.0067	3.92	2	100	0.1237	+0.0016
0.1217	0.0067	3.93	2	100	0.1235	+0.0018
0.1242	0.0067	8.75	2	100	0.1263	+0.0021
0.1217	0.0134	0.87	2	100	0.1230	+0.0013
0.1217	0.0134	0.86	2	100	0.1232	+0.0015
0.1217	0.0134	3.93	4	100	0.1218	+0.0001
0.1217	0.0134	3.90	4	100	0.1211	−0.0006
0.2244	0.0067	1.88	4	115	0.2244	0.0000
0.2244	0.0067	1.91	4	130	0.2240	−0.0004
0.2244	0.0141	2.98	4	100	0.2230	−0.0014
0.2244	0.0141	2.94	4	100	0.2241	−0.0003

The Titration of Mercurous Salts with Potassium Permanganate.

If a solution of a mercurous salt, such as mercurous sulphate, in dilute sulphuric acid, is titrated with potassium permanganate in the usual manner, the bleaching of the color is rapid at first, but long before the oxidation is complete the solution assumes a golden-yellow color and on standing the brown oxides of manganese are precipitated. For this reason no definite end reaction is obtainable. This difficulty may, however, be avoided, as Randall† has shown, if the permanganate solution is added in

* See page 195.

† D. L. Randall, Am. Jour. Sci., [4], xxiii, 137.

excess, the color then bleached with a standard ferrous sulphate solution, and the end-point finally reached by a few drops of permanganate. Under these conditions the end reaction is perfectly sharp and the oxidation of the mercurous salt complete.

According to Randall's procedure, the mercurous sulphate or nitrate in solution in water containing a suitable excess of sulphuric acid, or nitric acid free from oxides of nitrogen, is diluted and treated with *n*/20 potassium permanganate until the mixture, colored brown by the oxides of manganese, takes on a distinctly red tint. Ferrous sulphate in *n*/10 solution is added in amount sufficient to clear the solution, and the titration is immediately completed with permanganate. The *n*/20 solutions are preferred to the usual *n*/10 solutions on account of the high equivalent of the mercurous salt. Even with *n*/20 solution, 0.1 cm.^{3} of the permanganate is equivalent to about 0.0010 grm. of mercury. For the same reason the titrations are made with all possible care and accuracy. Following are experimental results obtained with a mercurous sulphate solution prepared by shaking up an excess of the salt in water acidified with sulphuric acid, allowing the mixture to stand twenty-four hours, and filtering through asbestos. The solution was standardized by weighing the mercurous chloride precipitated by sodium chloride and dried in a vacuum.

Titration of Mercurous Sulphate.

Hg_2SO_4 sol.	H_2SO_4 [1 : 1].	$KMnO_4$ approx. *n*/20.	$FeSO_4$ =	$KMnO_4$	$KMnO_4$ final.	Hg found.	Hg. (theory).	Error.
cm.^{3}	cm.^{3}	cm.^{3}	cm.^{3}	cm.^{3}	cm.^{3}	grm.	grm.	grm.
			Volume 150 cm.^{3}.					
100	5	14.70	10	10 90	3 80	0.0346	0.0354	−0.0008
100	5	14.72	10	10 90	3 82	0 0347	0.0354	−0.0007
100	5	14.70	10	10.90	3.80	0.0346	0.0354	−0.0008
100	5	14.71	10	10.90	3.81	0 0346	0.0354	−0.0008
Hg_2SO_4 sol. grm.			Volume 500 cm.^{3}.					
505 8	5	29.20	10	10 90	18.30	0.1668	0.1666	+0.0002
500.2	5	32.16	13	14.17	17.99	0.1639	0 1648	−0.0009
510.1	5	29 29	10	10.90	18 39	0.1676	0.1680	−0.0004
499.2	5	28 95	10	10.90	18 05	0.1645	0.1644	+0.0001

For practical purposes the application of the method to mercurous nitrate is of much greater importance. Results obtained with mercurous nitrate in solution — prepared by dissolving the crystallized salt in water containing enough pure nitric acid to prevent the formation of basic salts, and, as an additional precaution, passing a current of hydrogen, washed by alkaline permanganate and alkaline pyrogallol, for twelve hours through the solution to remove nitrous acid — are given below.

Titration of Mercurous Nitrate.

Volume 200 cm.³.

$Hg_2(NO_3)_2$ sol.	HNO_3.	Approx. $n/20$ $KMnO_4$.	$FeSO_4$. = $KMnO_4$		$KMnO_4$ final.	Hg found.	Hg theory.	Error.
cm.³	cm.³	cm.³	cm.³	cm.³	cm.³	grm.	grm.	grm.
25	5	49.99	9.7[illegible]	10 64	39.35	0.3586	0.3594	−0 0008
25	5	49 68	9.50	10 36	39 32	0 3583	0.3594	−0 0011
25	5	49.73	9 50	10.36	39 37	0 3588	0.3594	−0 0006
25	5	49 89	9.67	10.54	39 35	0 3586	0.3594	−0 0008
25	5	49.70	9.50	10.36	39.34	0.3585	0.3594	−0 0009
	H_2SO_4 [1 : 1].							
25	5	50.29	10.00	10.90	39.39	0.3589	0.3594	−0.0005

CHAPTER VI.

BORON; ALUMINIUM; LANTHANUM; THALLIUM.

BORON.

The Gravimetric Determination of Boric Acid.

The Use of Calcium Oxide as a Retainer. THE process of determining boric acid by distillation with methyl alcohol, evaporation of the distillate on calcium oxide, and ignition of the residue,* has been reëxamined by Gooch and Jones† in the light of the experience of many investigators.‡ It is shown that difficulties arising when nitric acid is present in the retort may be obviated by limiting the amount of that acid by the use of phenolphthalein as an indicator at the outset of the distillation. The addition of a drop of the acid and another of the indicator should be repeated once or twice during the distillation to insure the replacement of the acid volatilized from the salt slightly decomposed in the process. The effect of much nitric acid is bad, not only because it neutralizes the calcium oxide when it passes to the distillate, but because when it is used the dried mixture of calcium hydroxide and borate containing nitrate, nitrite and organic matter is likely to puff explosively if ignition is begun as soon as the residue is dry. If the residue is heated gradually and as strongly as possible over a radiator before the flame is actually applied to the crucible, no such action takes place.

That good results may be obtained with small amounts of calcium oxide, provided care as to the use of nitric acid and the conditions of ignition be taken, is shown by the figures of the original description and by the following results of experiments in which phenolphthalein was employed as an indicator and the residue heated strongly over the radiator before actual ignition.

* Gooch, Am. Chem. Jour., ix, 23.

† F. A. Gooch and Louis Cleveland Jones, Am. Jour. Sci., [4], vii, 34.

‡ Penfield, Am. Jour. Sci., [3], xxxiv, 222; Kraut, Zeit. anal. Chem., xxxvi, 3; Moissan, Compt. rend., cxvi, 1084.

Distillation with Nitric Acid.

CaO taken. grm.	B_2O_3 taken. grm.	B_2O_3 found. grm.	Error. grm.
2 3405	0 1788	0.1792	+0.0004
1 7620	0 1790	0.1785	−0 0005
2.1757	0 1824	0.1840	+0 0016
2.5656	0 1788	0.1786	−0.0002

No difficulty exists in respect to explosive effects when acetic acid is used in place of nitric acid, though even in this case it is safer to use the radiator in the first stages of heating, thus avoiding the danger of mechanical loss by too rapid ignition.

Following are determinations made by this method with the use of acetic acid.

Distillation with Acetic Acid.

CaO taken. grm.	B_2O_3 taken. grm.	B_2O_3 found. grm.	Error. grm.
0.9977	0.2065	0 2062	−0 0003
1.0220	0.2067	0 2070	+0.0003
1.3717	0.2077	0.2075	−0 0002
1.1310	0.1791	0.1795	+0.0004

The results of the preceding table, as well as those of the investigators mentioned, are a sufficient answer to the criticism* that acetic acid and nitric acid do not liberate boric acid in the distillation process so that good results may be obtained. In fact, it has been shown† that the prolonged action of carbonic acid is adequate to bring about complete volatility of boric acid with methyl alcohol. The number of distillations required depends, of course, upon the amount of boric acid to be volatilized. To remove 0.2 grm. of the anhydride completely to the distillate, it was shown in the original description of the method‡ that six 10-$cm.^3$ charges of methyl alcohol suffice.

It is also shown that the difficulty§ in bringing calcium oxide to a constant weight before and after absorption of boric acid has been magnified unduly. Thus the following table shows the

* Reischle, Zeit. anal. Chem., xxvi, 512.

† Jones, Am. Jour. Sci., [4], v, 442.

‡ Loc. cit.

§ Thaddeef, Zeit. anal. Chem., xxxvi, 568.

series of weights taken in several experiments in bringing calcium oxide to a constant weight in a 50-cm.[3] platinum crucible ignited over a blast lamp, as well as the weight taken after adding a known amount of standard boric acid solution to the slaked oxide, evaporating, and igniting. The results recorded are those of experiments made on days not moist beyond the average, and with the greatest care to approach the limit of accuracy with which calcium oxide and the boric acid held thereby can be weighed under ordinarily favorable conditions. The first weight of calcium oxide recorded under each experiment was taken after a strong ignition over the blast lamp for about one-half hour. The succeeding weights were taken after similar ignition of five minutes. In all cases the crucible was left to stand a definite period in a sulphuric acid dessicator, and, after the approximate value had once been obtained, the weights of the preceding weighing were replaced on the balance before the crucible was taken from the dessicator. The average of the weights bracketed is the weight taken as constant for the calculations.

Ignition of Calcium Oxide and Calcium Borate.

	CaO taken. grm.		B_2O_3 taken. grm.	$CaO+B_2O_3$ taken. grm.	$CaO+B_2O_3$ found. grm.		Error. grm.
(1)	0.9505 0.9493 0.9493	0.9493	0.2095	1.1588	1.1590 1.1591	1.1591	+0 0003
(2)	1.1319 1.1317 1.1313 1.1315	1.1315	0.2150	1.3465	1 3499 1.3474 1 3475 1.3476	1.3475	+0.0010
(3)	0.8028 0 8025 0 8024	0 8024	0.1184	0.9208	0 9205 0.9206 0.9206	0 9206	+0.0002
(4)	2 6980 2.6975 2.6973 2.6973	2 6973	0.2073	2.9046	2 9043 2 9049 2 9044	2.9048	+0 0002

Obviously calcium oxide may be weighed with accuracy, with or without boric acid; but the fact remains that a less hygroscopic absorbent — one requiring less care in the handling — is to be desired.

The Use of Sodium Tungstate as a Retainer. Gooch and Jones find that sodium tungstate, fused with a slight excess of tungstic acid over that contained in the normal tungstate (to insure its freedom from carbonate), may be used with good results as an absorbent for boric acid. This substance is definite in weight, not hydroscopic, soluble in water, and recoverable in its original weight after evaporation and ignition.

According to the procedure advocated, use is made of the apparatus originally proposed, so modified that the Erlenmeyer flask used as a receiver is fitted tightly to the condenser and trapped with water bulbs.* The retort is made very easily from a 150-cm.^{3} pipette and has the special advantage that particles of the residue spattering during distillation are easily washed from the walls of the vessel by a slight rotary motion of the retort. Special care should be taken to give the tungstate ample time for contact with the distillate before exposing the latter to atmospheric evaporation. The distillate is received, therefore, in a dilute solution of sodium tungstate which is placed in the receiver cooled by ice and trapped with water. The mixture is well stirred, allowed to stand one-half hour after the distillation is complete, evaporated to small volume in a large dish, and transferred to the crucible in which the tungstate was originally weighed. After thorough drying the residue is ignited to fusion and weighed. When acetic acid is employed in the retort, care must be taken in the ignition to expose the fused mass freely to the air (by causing it to flow upon the sides of the crucible) until the color of the cooled tungstate is white, in order that the reducing effect of the acetate may be eliminated. In the experiments recorded in the following table the tungstate was used in the receiver to retain the boric acid, distilled as usual with methyl alcohol from the borates treated with acetic acid, nitric acid or sulphuric acid, in amounts regulated by the use of phenolphthalein as an indicator. Excessive use of acid is disadvantageous, and this is especially true in the case of sulphuric acid; for, if this acid is carried over with the methyl alcohol, as it is at 100° if present in appreciable excess, a part of it, at least, is held permanently by the tungstate to increase the apparent weight of the boric acid estimated. The number of distillations necessary depends upon the amount of boric acid to be volatilized. Six

* See Fig. 2, page 3.

charges of 10 cm.³ each of methyl alcohol are enough to transfer 0.2 grm. of boric anhydride to the distillate.

Distillation of Six 10 cm.³ Portions of Methyl Alcohol.

$Na_2WO_4+WO_3$ taken grm.	B_2O_3 taken. grm.	B_2O_3 found. grm.	Error. grm.
With nitric acid.			
8.5516	0.1582	0 1572	−0 0010
4 9639	0.1329	0 1323	−0 0006
8 0033	0.1267	0.1256	−0 0011
With acetic acid.			
4 9658	0.1434	0 1418	−0.0016
6 0289	0 1431	0 1433	+0.0002
4 6797	0 1589	0.1587	−0.0002
4.0013	0 1433	0 1422	−0.0011
With sulphuric acid.			
6 3439	0 1582	0 1579	−0 0003
8.8227	0 1582	0 1577	−0 0005
10 1516	0 1265	0 1264	−0 0001
6.5738	0 1392	0.1390	−0 0002

The Acidimetric Estimation of Boric Acid.

When boric acid and mannite are mixed in solution, a peculiar compound of strongly acid properties is the result. This compound decomposes carbonates, and its acid taste is comparable to that of citric acid, much stronger than that of boric acid alone. Magnanini * has found that the product of such a mixture of boric acid and mannite solutions shows greater electrical conductivity and a lower freezing point than a similar molecular solution of either substance alone. Other polyatomic alcohols (but all to a less degree than mannite) and some organic acids show this peculiar property of combining chemically with boric acid to increase its acid qualities.† Of the reaction between boric acid and glycerin, Thomson,‡ Barthe,§ and Jörgensen‖ have taken

* Gaz. Chim., xx, 428–440; xxi, 134–145.

† Klein, Compt. rend. lxxxvi, 826; xcix, 144. Lambert, ibid., cviii, 1016–1017.

‡ J. S. C. I., xv, 432.

§ J. Pharm. Chim., xxix, 163.

‖ Zeit. angew. Chem. (1897), 5.

advantage to develop methods for the volumetric estimation of boric acid, glycerin being used to form a combination with boric acid sufficiently acidic to give an acid reaction when used with a sensitive indicator and make possible its titration with an alkali solution; but Hönig and Spitz * show that (in the method of Jörgensen) a very large amount of glycerin must be used to prevent the appearance of the indication of alkalinity with phenolphthalein before all the boric acid is neutralized according to the equation $2\,NaOH + B_2O_3 = 2\,NaOBO + H_2O$. Vadam † uses mannite, which, as he finds, gives sharper indications with litmus. The solution must be boiled to decompose bicarbonates while the volatilization of boric acid is prevented by the use of a return condenser; and silica must be removed by the process of Berzelius, and the solution neutralized to stronger acids, before a titration of the boric acid can be made.

Many indicators said to be insensible to free boric acid have been used to indicate the neutralization of the stronger acids. Hönig and Spitz,‡ and Thomson,§ use methyl orange; Morse and Burton ‖ use tropæolin oo; while Vadam ¶ makes use of litmus.

Neutralization of Stronger Acids.

Finding all these indicators to be more or less affected by boric acid, Jones ** has had recourse to the well-known reaction according to which a stronger acid liberates regularly iodine from a mixture of iodide and iodate, which is the solution of this difficulty. If both the iodide and iodate are in excess of the acid, the entire amount of free acid will be neutralized and the corresponding amount of iodine liberated according to the following equation:

$$5\,KI + KIO_3 + 6\,HCl = 6\,KCl + 3\,H_2O + 3\,I_2.$$

This liberated iodine may be removed by sodium thiosulphate and a solution obtained which is absolutely neutral, containing potassium iodide, iodate and tetrathionate. The statements made by P. Georgevic †† and Furry,‡‡ that boric acid present in

* Zeit. angew. Chem. (1896), 549.
† J. Pharm. Chim. [6], viii, 109–111.
‡ Zeit. anorg. Chem., xviii, 549.
§ J. S. C. I., xv, 432.
‖ Am. Chem. Jour., x, 154.
¶ J. Pharm. Chim. [6], viii, 109–111.
** L. C. Jones, Am. Jour. Sci., [4], vii, 147.
†† J. prakt. Chem., xxxviii, 118.
‡‡ Am. Chem. Jour., vi, 341.

moderate amount in solution has not the slightest action on a mixture of iodide and iodate, have been found to apply to solutions containing not more than 0.1 grm. of B_2O_3 to 25 cm.^{3}. Therefore, when this acid is liberated by an excess of a stronger acid and the iodine set free destroyed by thiosulphate, it remains free in solution to be titrated in any convenient manner. Following along the lines suggested by the above reactions, Jones * has developed a volumetric process for the estimation of boric acid.

The solution in which boric acid is present to an amount not exceeding 0.1 grm. in 25 cm.^{3} is made slightly acid to litmus by hydrochloric acid and treated with 5 cm.^{3} of a solution (10 per cent) of barium chloride. An amount of iodate and iodide of potassium sufficient to liberate iodine at least equivalent to the excess of hydrochloric acid in the acidified solution is mixed with starch in a separate beaker, the iodine which is usually thrown out by this mixture being just bleached by a dilute solution of thiosulphate.

To the now neutral solution of iodide and iodate a single drop of the solution to be analyzed is transferred by a glass rod. If a blue coloration is developed, the solution is known to be acidic with hydrochloric acid, and all the boric acid is in free condition. The amount of iodide and iodate necessary depends upon the acidity of the solution containing boric acid. Usually a mixture of 10 cm.^{3} of a 25 per cent solution of iodide with the same amount of a saturated solution of iodate is sufficient. Any large excess of hydrochloric acid should be neutralized by sodium hydroxide before the iodide and iodate mixture is added. After the addition of the iodide and iodate solution and starch to the boric acid solution, the liberated iodine is carefully bleached by thiosulphate. Excess of thiosulphate in reasonable amount does not seem to be detrimental, but in practice the starch iodide color is clearly bleached, and then no more is added.

Soluble carbonates prevent a definite indication of the neutral point by thiosulphate and starch iodide, therefore the barium chloride was added to transform them to insoluble barium carbonate in the action of the iodide-iodate mixture. The mixture of iodide and iodate is not added to the solution to be analyzed until after it has been made distinctly acidic, for the reason that, when the neutral point is approached in the addition of

* Am. Jour. Sci., [4], viii, 129.

hydrochloric acid, the starch iodide thrown out locally by the acid is not bleached again by the small amount of sodium borate remaining undecomposed and thus obscures the neutral point.

Strengthening of Boric Acid by Mannite. The solution, after the bleaching of iodine by thiosulphate, is colorless and contains only starch, neutral chloride, potassium tetrathionate, iodide and iodate, and all the boric acid present in uncombined condition. The carbonate lies out of the sphere of action in insoluble form as barium carbonate. A few drops of the indicator, phenolphthalein, are now added, and *n*/5 sodium hydroxide is run in until a strong red coloration is produced. A pinch of mannite, 1 grm. or 2 grm., is then added, which bleaches the phenolphthalein coloration, and the alkali solution again run in to a faint indication, which, if permanent on the addition of more mannite, may be taken as the reading point.

The combination of boric acid and mannite liberates immediately in the presence of iodide and iodate about half the amount of iodine required on the theory that B_2O_3 acts with the neutralizing power of metaboric acid, HOBO. If no mannite is present phenolphthalein gives an alkaline indication when only about one-half the amount of alkali theoretically necessary to form the metaborate, NaOBO, has been added. Obviously, then, the starch iodide coloration will not appear on the addition of mannite, if the free boric acid has been neutralized to phenolphthalein by alkali, and the remainder of the alkali is added to complete neutralization immediately after the addition of mannite. The end reaction with phenolphthalein is sharp and the small amount of carbonate present in the standard solution of alkali is precipitated by the barium chloride already in the solution. The calculation must, therefore, be based on the amount of free alkali hydroxide used, according to the following representation:

$$B_2O_3 + 2\,NaOH = 2\,NaOBO + H_2O.$$

The best results and the most definite indications are obtained in cold solution of a volume not greater than 50 $cm.^3$. This fact accords with the observations of Magnanini* that the relative electrical conductivity of the boromannite solution is decreased by dilution and elevation of the temperature. When silicates are

* Gaz. Chim., xx, 428, and xxi, 134.

present in solution, the silicon dioxide is liberated by the excess of hydrochloric acid, and this oxide, whether in hydrous or anhydrous condition, neither affects the indication with iodine nor phenolphthalein, nor does it form with mannite a compound of acidic properties. The presence of fluorides is not detrimental. Ammonium salts do interfere with the indication given by phenolphthalein, but they may be removed by boiling with potassium hydroxide in excess, or an indicator which is not affected by them may be used.

The following table contains the results of a series of analyses in which the boric acid was first drawn into an excess of sodium hydroxide, then estimated according to the method described.

Titration of Boromannite Solution with Standard Alkali.

B_2O_3 sol. taken. cm.^{3}	NaOH sol. required. cm.^{3}	B_2O_3 taken. grm.	B_2O_3 found. grm.	Errors on B_2O_3. grm.
21.95	21.02	0 1571	0.1577	+0 0006
20.68	19.65	0.1479	0 1474	−0 0005
20.73	19 63	0 1483	0 1473	−0 0010
23 05	23.71	0 1776	0 1777	+0 0001
23.10	23 80	0 1780	0 1783	+0 0003
22.76	23.35	0.1754	0 1750	−0 0004
24.08	24.78	0.1855	0.1857	+0 0002
22.00	22.50	0.1695	0.1686	−0 0009
20 78	21.28	0.1601	0.1595	−0 0006

Practical tests of the method upon crude calcium borate and colemanite are given below. The finely ground minerals were dissolved in hydrochloric acid and the analyses proceeded with as above described.

A determination of boric acid by this process can be completed in five minutes and the results are obviously accurate within the limits of ordinary analysis.

Analysis of Crude Borate of Lime.

Ca borate taken. grm.	B_2O_3 found. grm.	B_2O_3. Per cent.
0 4016	0 2289	56.99
0.4044	0.2302	56.92
0 4000	0.2285	57.11

Analysis of Colemanite.

Mineral taken. grm.	B_2O_3 found. grm.	B_2O_3. Per cent.	Average. Per cent.
0 4034	0.2064	51.15	50.99
0 4070	0 2069	50 80	
0.6004	0.3054	50.86	
0.6006	0 3056	50 89	
0.5059	0.2592	51 24	
0.5092	0.2592	50 89	

The usually interfering substances, fluorine, silica and carbon dioxide, have no detrimental influence on the results of this process.

The Iodometric Determination of Boric Acid.

In studying the strongly acidic compound formed when boric acid and mannite are associated in solution, Jones * finds that the acid developed is, under certain definite conditions, sufficiently strong to liberate, quantitatively, from a mixture of potassium iodide and iodate, the amount of iodine required on the supposition that each molecule of metaboric acid (HOBO) acts in a manner similar to a univalent mineral acid under the same conditions.

$$5\,KI + KIO_3 + 6\,HOBO = 3\,I_2 + 6\,KOBO + 3\,H_2O.$$

Obviously, this reaction depends upon the behavior of the acidic boromannite compound as an acid stronger than acetic, tartaric or citric acid; for these acids have been found by Furry † to be incapable of liberating iodine regularly from a mixture of iodide and iodate. Conditions which tend to increase the acidic activity of this compound are high concentrations and moderately low temperatures.‡

It has not been found possible under any conditions to rely upon the immediate liberation of the full amount of iodine; a certain period of time is required for the completion of the reaction. When the solution is of small volume and saturated with mannite, the reaction goes to the end most quickly —

* Am. Jour. Sci., [4], viii, 127.

† Am. Chem. Jour., vi, 341.

‡ Magnanini, Gaz. Chim., xx, 428; xxi, 134, 1016, 1017.

sometimes almost immediately; but if the solution of boric acid is too concentrated — nearly saturated — the boric acid alone throws out some iodine from the iodide-iodate mixture added to destroy other free acid, and on bleaching with thiosulphate a starting point is obtained at which some of the boric acid has already entered into combination. The amount of iodine thus liberated by the boric acid is, however, not large, and if, upon the addition of the iodide and iodate, the iodine thrown out by the free hydrochloric acid present is immediately bleached by thiosulphate and the analysis proceeded with from this as the neutral point, even in concentrated solutions the error is almost inappreciable. If, however, considerable time intervenes between the adding of the iodide and iodate and the determination of the neutral point by thiosulphate, iodine equivalent to as much as several milligrams of boric acid may be liberated. This difficulty was not met with in those experiments in which the iodide and iodate were added to solutions of concentration like that of the standard solution used (7.738 grm. per liter), but in an attempt to estimate the boric acid in colemanite, where the solution was kept as concentrated as possible, hoping in this way to decrease the time required for the complete liberation of iodine, low values were obtained; that is, a false starting point was used. At the time of adding the iodide and iodate the volume should not be less than 25 $cm.^3$ for each decigram of boric anhydride (B_2O_3) present, and should not be much greater than two or three times that amount. At lower concentrations of the boric acid, even though the liquid be saturated with mannite, the necessary time of standing is prolonged and the effect of carbon dioxide upon the iodide and iodate is increased; for carbon dioxide, whether derived from the atmosphere or existing dissolved in the solution, upon standing slowly liberates iodine. The effect of carbon dioxide is, however, small, and in the time required for the completion of the process has never been found equivalent to more than a single drop of the solution of thiosulphate used. Even if the material to be analyzed contains carbonates, after acidifying in concentrated solution and shaking vigorously the small amount of uncombined carbon dioxide remaining has an almost inappreciable effect upon the results. The length of time required for the liberation of the theoretical amount of iodine in a solution of 25 $cm.^3$ to 50 $cm.^3$ to each

0.1 grm. of boric anhydride is 20 to 45 minutes, and at the end of 45 minutes' standing in a solution saturated with mannite the reaction may be considered complete. During this period, however, it is well to keep the solution cool — zero temperature will do no harm — and to insure thorough mixture by occasional shaking. As free iodine would tend to escape upon standing unless kept in a closed flask, it is convenient, immediately after the addition of mannite, to treat with an excess of the standard solution of thiosulphate, — 8 cm.^{3} or 10 cm.^{3} more than the amount required to bleach the iodine liberated, — and at the expiration of 40 to 60 minutes to titrate back with $n/10$ iodine. The strength of the thiosulphate solution found most convenient is $n/5$, while the use of iodine of one-half this strength, $n/10$, diminishes the error of reading correspondingly. In solutions of the volume recommended the addition of starch to give the indication with iodine is unnecessary, since a single drop of one-twentieth normal iodine in excess is sufficient to give a strong lemon coloration.

Procedure in Summary. — The procedure recommended is as follows: The borate is dissolved in as small volume and as little hydrochloric acid as possible, with shaking to remove free carbon dioxide and adjustment of volume so that at the time of adding potassium iodide and iodate there shall be approximately 25 cm.^{3}–50 cm.^{3} of solution for each decigram of boric anhydride present. The greater part of the excess of hydrochloric acid in the solution is destroyed by sodium hydroxide, with the use of litmus paper as an indicator, leaving the solution distinctly acid in reaction. Potassium iodide (3 cm.^{3}–5 cm.^{3} of a 40 per cent solution) and iodate (5 cm.^{3}–10 cm.^{3} of a 5 per cent solution) are added in excess of that required to liberate iodine in an amount corresponding to the hydrochloric acid and the boric acid present. The iodine liberated by the free hydrochloric acid is bleached by a small amount of a strong solution of thiosulphate, and, after agitating to insure thorough mixture, iodine is added to faint coloration. Sufficient mannite is now used to saturate the solution — about 10 grm.–15 grm. for a volume of 50 cm.^{3} — and sodium thiosulphate is added in standard solution 8 cm.^{3}–10 cm.^{3} in excess of that required to bleach the iodine immediately thrown out by the mannite. The solution is again brought to saturation, if necessary, by mannite, and, after standing in a cool place for 40–60 minutes, titrated with decinormal iodine to determine the excess of thiosulphate present.

The results of experiments upon pure boric acid, crude calcium borate and crystallized colemanite are given in the accompanying tables.

Pure Boric Acid.

B_2O_3* taken. cm.³	Thiosulphate† taken. cm.³	Iodine‡ taken. cm.³	Time of standing. min	Volume. cm.³	B_2O_3 taken. grm	B_2O_3 found grm.	Error. grm.
28 00	32 00	1 88	30	28	0 2165	0 2168	+0 0003
27 03	32 00	4 37	27	27	0 2090	0 2081	−0 0009
27.02	31 97	4 04	60	27	0 2089	0 2090	+0 0001
27.06	32 04	3 88	60	50–60	0.2093	0 2101	+0 0008
27 02	32 02	4 40	60	50–60	0 2089	0 2081	−0 0008
27 04	31 72	3 39	60	50–60	0.2091	0 2096	+0 0005
27 01	31 53	2 88	120	50–60	0 2089	0 2100	+0 0011
26.05	31 01	4 01	180	50–60	0.2014	0 2025	+0 0011
27 00	31 00	2 12	30	50–60	0 2088	0 2089	+0 0001
27 00	32 00	4 05	30	50–60	0.2088	0 2092	+0 0004
26.01	32 02	6 20	30	50–60	0 2011	0 2018	+0 0007
27 03	31 01	2 21	48	50–60	0 2090	0 2087	−0 0003
27 05	31 89	3 81	45	50–60	0.2092	0 2093	+0 0001
26 07	31 02	4.14	40	50–60	0 2016	0 2020	+0 0004
27 00	32 04	4 30	40	60	0.2088	0 2086	−0 0002

* 7 773 grm. per liter. † 0 198 normal. ‡ 0.0996 normal.

Calcium Borate.

Mineral. grm.	Thiosulphate taken. cm.³	Iodine taken. cm.³	Time of standing. min	Volume of solutions. cm.³	B_2O_3 found. grm.	Per cent.
0.4015	35 05	4.75	60	40	0.2280	56 92
0 4010	35 34	5.23	120	45	0.2283	56 94

Colemanite.

0.4002	32 00	5.50	90	50	0.2043	51.04
0 2513	32.01	7.36	60	40	0.1279	50.91
0.4007	33 03	7.72	50	65	0.2036	50.81

These results show little variation, and in the case of colemanite correspond closely to the theory 50.97 per cent. The process is convenient, generally applicable, and accurate within the ordinary limits of analysis.

ALUMINIUM.

The Determination of Aluminium by Precipitation with Ether-Hydrochloric Acid.

Crude aluminium chloride may be freed from every trace of a ferric salt by dissolving it in the least possible amount of water, saturating the cooled solution with gaseous hydrochloric acid, filtering upon asbestos in a filtering crucible or cone, and washing the crystalline precipitate with the strongest hydrochloric acid. Prepared in this way the salt gives no trace of color when dissolved in water and tested with potassium sulphocyanate, but the degree of insolubility is not sufficient for the purposes of analysis. Gooch and Havens* have found that in a mixture of hydrochloric acid of highest concentration and ether in equal parts the solubility of aluminium chloride amounts approximately to 5 parts of the hydrous salt, $AlCl_3.6H_2O$, corresponding to 1 part of the oxide, Al_2O_3, in 125,000 parts of the mixture.

Pure aqueous hydrochloric acid of full strength mixes perfectly with its own volume of anhydrous ether, but the addition to this mixture of any very considerable amount of a solution of ferric chloride in strong hydrochloric acid occasions the separation of a greenish etherial solution of the ferric salt upon the surface of the acid. The addition of more aqueous acid does not change the conditions essentially, but more ether renders the acid and the oily solution completely miscible. For the separation of insoluble aluminium chloride from certain small amounts of soluble ferric chloride, the mixture of the strongest aqueous hydrochloric acid and ether in equal parts serves the purpose excellently; when larger amounts of ferric chloride are to be dissolved, ether must be added proportionately in order to prevent the separation of the etherial solution of ferric chloride from the rest of the liquid.

Separation of Aluminium from Iron. The quantitative procedure as finally developed by Gooch and Havens for the separation of aluminium from iron, and the determination of aluminium as the oxide, is as follows: The concentrated aqueous solution of the salts is mixed with a suitable volume of strongest aqueous

* F. A. Gooch and F. S. Havens, Am. Jour. Sci., [4], ii, 416.

hydrochloric acid — enough to make the entire volume approximately 15 cm.^{3} to 25 cm.^{3}. This mixture is saturated with gaseous hydrochloric acid while kept cool by immersing in running water the receptacle containing it. A volume of ether equal to the volume of the liquid is introduced, and the cooled etherial mixture is treated with gaseous hydrochloric acid to saturation. The precipitated crystalline chloride is collected upon asbestos in a perforated crucible, washed with a previously prepared mixture of hydrochloric acid and ether carefully saturated at 15°, dried a half-hour at 150°, covered with a layer of pure mercury oxide * (about 1 grm.), and ignited carefully under a good ventilating flue, finally with the blast lamp.

The gaseous hydrochloric acid is most conveniently produced in regulated current by treating massive ammonium chloride with strong sulphuric acid in the Kipp generator. A platinum dish hung in an inverted bell jar, provided with inlet and outlet tubes through which the current of water for cooling is passed, makes a convenient container for the solution to be saturated with the gas. The filtration is made upon asbestos in a perforated crucible. The filtrate and washings are caught directly in a crucible (placed under the bell jar of the filter pump) in which the subsequent evaporation is to be effected. The heating of the strongly acid solution must be gradual and conducted with care to prevent mechanical loss by a too violent evolution of the gaseous acid.

Results are given in the table.

Precipitation of Aluminium Chloride in Presence of Ferric Chloride.

Al_2O_3 taken in solution as the chloride.	Al_2O_3 found by ignition with HgO.	Fe_2O_3 present as chloride.	Final volume.	Error.
grm.	grm.	grm.	cm.^{3}	grm.
0 0761	0 0758	...	25	−0 0003
0.0761	0.0754	....	25	−0.0007
0.0761	0.0751	....	25	−0.0010
0.0761	0.0757	0.15	25-30	−0.0004
0.0761	0.0756	0.15	25-30	−0.0005
0.0761	0.0755	0.15	25-30	−0.0006
0.0761	0.0755	0.15	25-30	−0.0006

* Loc. cit., p. 419.

Determination of Aluminium and Beryllium.

By similar procedure Havens* has effected the separation of aluminium from beryllium. The beryllium may be recovered in the filtrate from the aluminium chloride by precipitation as hydroxide with ammonia after nearly complete evaporation of the acid, and weighed as oxide after ignition. Or, the filtrate may be evaporated just to dryness on a radiator, care being taken not to heat to the volatilizing point of the beryllium chloride; a few drops of strong nitric acid added; the liquid evaporated, best with a current of air playing on the surface; and the residue heated gently at first, to break up the nitrate, and finally over the blast lamp.

Results of this procedure are given below.

Aluminium and Beryllium.

Al_2O_3 taken in solution as the chloride. grm.	Al_2O_3. grm.	Error. grm.	Final volume. $cm.^3$	BeO taken in solution as the chloride. grm.	BeO found. grm.	Error. grm.
0.1059	0.1058	−0 0001	12	0 0198	0.0204	+0.0006*
0 1053	0.1044	−0 0009	12	0 0194	0.0196	+0.0002*
0 1065	0.1059	−0 0006	12	0 0197	0.0205	+0 0008*
0.1068	0.1060	−0 0008	12	0 0199	0.0207	+0 0008*
0.1049	0.1047	−0 0002	12	0 0198	0.0208	+0 0010*
0 1060	0.1057	−0 0003	12	0.0977	·0 0969	−0 0008*
0 1064	0.1063	−0.0001	12	0 1085	0 1084	−0 0001*
0 1046	0.1038	−0 0008	30	0.1083	0.1087	+0 0004*
0.1051	0 1048	−0 0003	30	0.1071	0.1078	+0.0007*
0.1076	0.1075	−0.0001	30	0.1086	0.1094	+0 0008†

* By the evaporation process.
† By the precipitation process.

Determination of Aluminium and Zinc.

Havens† effected similarly the separation of aluminium from zinc, the zinc chloride in the filtrate being converted to the oxide by evaporation, repeated treatments with nitric acid in porcelain, and ignition. On account of the danger to platinum, the evaporations and treatments with nitric acid are made in porcelain. In these processes, however, the porcelain is somewhat attacked, and correction must be made for a slight contamination of the residual oxide.

Results are given on the following page.

* Franke S. Havens, Am. Jour. Sci. [4], iv, 111.
† Franke Stuart Havens, Am. Jour. Sci., [4], vi, 45.

Aluminium and Zinc.

Al_2O_3 taken as the chloride. grm.	Al_2O_3 found grm.	Error. grm.	ZnO taken. grm.	ZnO found. grm	Error grm.	Error corrected.* grm.	Final volume. cm.³
0 0562	0 0562	0 0000	0 1110	. .	.		. .
0 0580	0.0577	−0 0003	0 1034	.	. .	.	. .
0 0572	0.0572	0 0000	0.1014	0 1027	+0 0013	−0 0007	12
0 0563	0 0550	−0 0013	0 1026	0 1038	+0 0012	−0 0008	16
0 0577	0.0576	−0 0001	0 1000	0 1014	+0 0014	−0 0006	16
0 0559	0.0558	−0 0001	0.1020	0 1035	+0 0015	−0 0005	16
0 0563	0 0556	−0 0007	0.2024	0 2046	+0 0022	+0 0002	20
0 1111	0.1107	−0.0004	0 2092	0 2116	+0 0024	+0 0004	20

* Corrected by the amount of material found in the evaporation of similar amounts of the strong acids in porcelain.

Determination of Aluminium and Copper. The process applies also to the separation of aluminium from copper, as shown by the results given.* The copper in the filtrates of these determinations was converted to the oxide through the sulphate because this operation may be conducted safely in platinum.

Aluminium and Copper.

Al_2O_3 taken as chloride. grm.	Al_2O_3 found. grm.	Error. grm.	CuO taken. grm.	CuO found. grm.	Error. grm.
.	.	.	0 0437	0 0432	−0 0005
. . .	.	. .	0 0359	0 0359	0 0000
. . . .	. . . .	.	0 0345	0 0340	−0.0005
0 0558	0 0545	−0 0013	0.0319	0.0324	+0.0005
0.0538	0 0536	−0 0002	0.0343	0 0356	+0 0013
0 0566	0.0562	−0.0004	0.0337	0 0349	+0.0012
0 0577	0.0575	−0.0002	0.0651	0.0644	−0.0007

Separation of Aluminium from Mercury and Bismuth. Aluminium may be similarly separated from mercury and bismuth, as shown by the results given below.†

Aluminium, Mercury and Bismuth.

Al_2O_3 taken as chloride. grm.	Al_2O_3 found. grm.	Error. grm.	$HgCl_2$ taken. grm.	Bi_2O_3 taken. grm.
0 0570	0.0574	+0 0004	0.1	. . . .
0 0548	0.0557	+0 0009	0.1	. . . .
0 0565	0.0571	+0 0006	. . .	0.1
0 0576	0.0577	+0 0001	. . .	0.2

* Havens, loc. cit.

† Havens, loc. cit.

Lanthanum.

The Estimation of Lanthanum Precipitated as the Oxalate.

Many years ago Stolba* stated that cerium, lanthanum and didymium may be estimated by treating their oxalates with potassium permanganate in the presence of sulphuric acid, but gave no experimental evidence in the form of analytical results. Later this statement was confirmed by Browning and Lynch,† and it was shown that cerium may be estimated by precipitating cerium oxalate with a definite amount of a standard solution of ammonium oxalate used in excess, decomposing the precipitated cerium oxalate by dilute sulphuric acid and estimating the oxalate by permanganate. The ammonium oxalate in excess of the amount required for the precipitation was also estimated by permanganate, and by this process the results were checked.

Lanthanum Precipitated as the Oxalate.

La_2O_3 taken.*	La_2O_3 found.†		Average.	Error.
	Precipitate.	Filtrate.		
grm.	grm.	grm.	grm.	grm.
0.0148	0.0152	0.0144	0.0148	0.0000
0 0148	0.0149	0.0139	0.0144	−0.0004
0.0296	0 0302	0.0291	0.0296	0 0000
0.0296	0.0302	0.0293	0.0297	+0 0001
0 0592	0.0599	0.0586	0 0593	+0.0001
0 0592	0.0598	0.0585	0 0592	0 0000
0.1184	0.1191	0.1179	0.1185	+0.0001
0.1184	0.1191	0.1182	0.1187	+0 0003
0.2368	0.2376	0.2362	0.2369	+0 0001
0.0148	0 0149	0.0145	0 0147	−0 0001
0.0148	0 0150	0.0147	0.0148	0 0000
0.0296	0 0298	0.0293	0 0295	−0 0001
0.0592	0 0596	0 0589	0.0593	+0.0001
0.1184	0.1190	0.1182	0.1186	+0.0002
0.1036	0.1040	0.1029	0.1035	−0.0001

* Taken as ammonium lanthanum nitrate. † La=138.9.

The conditions under which lanthanum may be estimated as the oxalate have been studied by Drushel ‡ and a process of treatment recommended. The procedure is as follows:

* Sitzungsber. d. kgl. böhm. Gesellsch. d. Wissenschaften, v, 4, Juli, 1879.

† See page 248.

‡ W. A. Drushel, Am. Jour. Sci., [4], xxiv, 197.

From a neutral solution of a lanthanum salt, conveniently the nitrate, the oxalate is precipitated by a measured amount of standard $n/10$ oxalic acid, or ammonium oxalate, after the addition of a few drops of acetic acid. The precipitate is thoroughly stirred, allowed to settle, and filtered off in a perforated crucible fitted with an asbestos felt. After thoroughly washing with water the crucible and precipitate are placed in a beaker with 100 $cm.^3$ to 300 $cm.^3$ of water and 10 $cm.^3$ to 30 $cm.^3$ of [1 : 3] sulphuric acid. The contents of the beaker are heated nearly to boiling, and at once titrated to color with standard potassium permanganate. The filtrate is similarly titrated as a check on the titration of the precipitate. The lanthanum is calculated as La_2O_3 from the two titrations. The mean of the two closely agreeing values thus obtained is taken. Results are shown on the preceding page.

THALLIUM.

The Determination of Thallium as the Acid Sulphate and as the Neutral Sulphate.

Crookes* has shown that the salt obtained by heating thallous chloride with sulphuric acid until the excess of the latter is expelled, and then raising the heat to redness, has the constitution of a neutral sulphate and sustains no further appreciable loss of weight on heating, and he has suggested the possibility of applying this treatment in the estimation of thallium. Castanjen† essentially confirms these statements in regard to the sulphate, adding the observation that the salt tends to lose acid on strong ignition in air. Experiments by Browning ‡ show, however, that with proper precautions thallium may be weighed either as the neutral sulphate or as the acid sulphate.

The procedure is as follows: To the solution of the thallium salt of a volatile acid, taken in a crucible, sulphuric acid is added, and the water is removed as far as possible by evaporation on the steam bath. The crucible is then placed within a conical iron radiator and heated at temperatures ranging between 220° and 240° until fuming ceases and the weights after half-hour

* Chem. News, viii, 243.

† Jour. prakt. Chem., cii, 131.

‡ Philip E. Browning, Am. Jour. Sci., [4], ix, 137.

periods of heating remain constant. These weights give the weight of acid thallium sulphate.

Upon further heating of the crucible and contents over a free flame to low redness the weight again becomes constant, after considerable evolution of fumes, and shows a condition closely approximating that of neutral thallium sulphate. Experimental results of this treatment of thallium nitrate are given in the table.

Conversion of Thallium Nitrate to Thallium Sulphates.

$TlHSO_4$ calculated. grm.	$TlHSO_4$ found. grm.	Error. grm.	Tl_2SO_4 calculated. grm	Tl_2SO_4 found. grm.	Error. grm.
0.1605	0 1596	−0 0009	0.1344	0 1346	+0 0002
0.1611	0.1608	−0 0003	0 1349	0.1346	−0 0003
0.1608	0.1608	0 0000	0 1347	0 1352	+0 0005
0.1612	0.1600	−0 0012	0.1350	0.1346	−0.0004
0.1602	0 1596	−0.0006	0.1341	0 1346	+0 0005
0.1608	0 1596	−0 0012	.....	.....	...
0.1617	0.1604	−0 0013	.....	......	..
0.1608	0.1592	−0.0016	0.1347	0 1358	+0 0011
0.1609	0 1590	−0 0019	0.1348	0.1346	−0.0002

The Gravimetric Estimation of Thallium Precipitated as Thallic Hydroxide by Potassium Ferricyanide and Potassium Hydroxide.

The precipitation of thallic hydroxide* by the action of potassium ferricyanide and potassium hydroxide has been recommended by Browning and Palmer,† as satisfactory means for the separation of thallium in easily determinable form according to the following procedure.

To the solution of the thallous salt in 100 cm.^{3} of water are added a solution of potassium ferricyanide, in excess, and potassium hydroxide to complete precipitation of the brown thallic hydroxide. The precipitate is filtered off on asbestos in the perforated crucible, thoroughly washed, best with hot water, and dried over a low flame, at about 200°, to constant weight. Test results are given in the table.

* See page 223.

† Philip E. Browning and Howard E. Palmer, Am. Jour. Sci., [4], xxvii. 380.

Precipitation as Thallic Hydroxide.

Tl_2O_3 taken as thallous nitrate. grm	Tl_2O_3 found. grm.	Error. grm.
0.1305	0 1309	+0 0004
0.1305	0.1314	+0 0009
0.1305	0 1308	+0 0003
0.0870	0 0872	+0 0002
0.1740	0.1741	+0 0001
0.1740	0 1739	−0 0001
0.1740	0.1742	+0 0002
0.1305	0.1307	+0 0002
0.1305	0.1309	+0 0004
0.1305	0.1308	+0 0003
0.0870	0.0872	+0 0002
0.0870	0 0874	+0.0004

The Gravimetric Estimation of Thallium as the Chromate.

Crookes has shown that thallium chromate precipitated by the addition of potassium dichromate to an alkaline solution of a thallous salt has the constitution of a neutral salt, is very insoluble in water, — 100 parts of water at 100° dissolving about 0.2 parts and at 60° about 0.03 parts, — and may be used to effect a rough separation of thallium from cadmium. Browning and Hutchins * have described conditions under which the reaction gives exact quantitative results. To the solution of thallous nitrate, heated to 70° or 80°, are added, with shaking, ammonium hydroxide, or, better, potassium carbonate, to alkalinity, and an excess of potassium dichromate in solution. After cooling and settling, the thallous chromate is filtered off on asbestos in the perforated crucible and dried over a low flame to constant weight. If the precipitation is made in the cold, the chromate does not form well but remains partly in finely divided condition and difficult to filter. The addition of ammonium nitrate before precipitation brings about a better condition even in the cold; but the best results are obtained by precipitating the warm solution made alkaline with potassium carbonate. Results follow in the table.

* Philip E. Browning and George P. Hutchins, Am. Jour. Sci., [4], viii, 460.

Determination of Thallium as Thallous Chromate.

$TlNO_3$ taken. Calculated as Tl_2O. grm.	Tl_2CrO_4 found. Calculated as Tl_2O. grm.	Error. Calculated as Tl_2O. grm.
0 0796	0 0791	−0.0005
0 0792	0 0788	−0 0004
0 0792	0 0786	−0 0006
0 1188	0 1177	−0 0011
0 1192	0 1186	−0 0006
0 1185	0 1178	−0.0007
0 1190	0.1185	−0 0005
0.1189	0 1183	−0 0006
0 1196	0 2000	+0 0004
0 1196	0 2005	+0.0009
0.1173	0 1173	0.0000
0.1171	0 1163	−0.0008

The Iodometric Estimation of Thallium by Precipitation with Potassium Dichromate and Determination of the Excess of the Precipitant.

Browning and Hutchins * have shown that thallium may be estimated by precipitation as chromate and estimation of the excess of precipitant left over from a definite amount taken.

According to this process, a solution of potassium dichromate is standardized † by treatment of the solution, acidified with sulphuric acid, by a definite amount of standard arsenite, setting aside until the change from yellow to bluish green indicates the reduction of the chromate, making alkaline with acid potassium carbonate, and titrating the residual arsenite by standard iodine. To the solution of thallous nitrate, heated to 70° or 80° and made alkaline with acid potassium carbonate, is added an excess of the standardized potassium dichromate and the precipitated thallous chromate is filtered off. After cooling, the filtrate containing the excess of alkali chromate is acidified with sulphuric acid, a definite amount of standard arsenite is added, and the whole allowed to stand a few moments until the change from yellow to bluish green indicates reduction of the chromic acid. Acid potassium carbonate is added to alkaline reaction and the residual arsenite is determined by titration with standard iodine. The difference between the arsenite taken and the arsenite

* Philip E. Browning and George P. Hutchins, Am. Jour. Sci., [4], viii, 461.

† See page 408.

found measures the chromate remaining after precipitating the thallous chromate, and the difference between the chromate remaining and that taken is the chromate equivalent to the thallium. Results of this procedure are given in the table.

Precipitation of Thallous Chromate: Determination of the Excess of Precipitant.

Tl_2O in $TlNO_3$ taken.	Tl_2O in Tl_2CrO_4 found.	Error.
grm.	grm	grm.
0.1192	0 1198	+0 0006
0 1189	0 1205	+0 0016
0.1196	0 1180	−0 0016
0.1196	0 1192	−0 0004
0 1173	0 1182	+0 0009
0.1171	0 1190	+0 0019

The Estimation of Thallium by the Action of Potassium Ferricyanide in Alkaline Solution and of Potassium Permanganate in Acid Solution upon the Ferrocyanide Produced.

The oxidation of a thallous salt with precipitation of thallic hydroxide by the action of potassium ferricyanide and potassium hydroxide and the subsequent oxidation of the potassium ferrocyanide produced, after filtration of the mixture, are the essential steps in a process proposed by Browning and Palmer* for the volumetric estimation of thallium.

The procedure is as follows: To the solution of the thallous salt in about 100 cm.^{3} of water is added a sufficient excess of a solution of potassium ferricyanide and potassium hydroxide to complete precipitation of the brown thallic hydroxide. The precipitate is filtered off on asbestos, generally without settling, and washed thoroughly. The filtrate is acidified with sulphuric acid and titrated with standard permanganate. From the following equations, representing the reactions, the amount of thallium present may be readily calculated:

$$Tl_2O + 4\,K_3FeC_6N_6 + 4\,KOH = Tl_2O_3 + 4\,K_4FeC_6N_6 + 2\,H_2O,$$
$$5\,K_4FeC_6N_6 + KMnO_4 + 4\,H_2SO_4 = 5\,K_3FeC_6N_6 + 3\,K_2SO_4 + MnSO_4 + 4\,H_2O.$$

It is necessary to apply a correction for the amount of permanganate used to give the first tinge of pink color to the amounts

* Philip E. Browning and Howard E. Palmer, Am. Jour. Sci., [4], xxvii, 379.

of ferricyanide used in the determinations, but this seldom exceeds 0.1 $cm.^3$ of the permanganate.

The following table shows the results obtained with different amounts of the thallium salt.

Volumetric Estimation of Thallium.

Tl_2O taken as the nitrate. grm.	Tl_2O found. grm.	Error. grm.
0.0809	0.0809	+0.0000
0.0809	0.0808	−0.0001
0.0809	0.0809	0.0000
0.1213	0.1212	−0.0001
0.1213	0.1216	+0.0003
0.1213	0.1218	+0.0005
0.1213	0.1218	+0.0005
0.1213	0.1212	−0.0001
0.1213	0.1207	−0.0006
0.1618	0.1614	−0.0004
0.1618	0.1613	−0.0005
0.1618	0.1616	−0.0002

CHAPTER VII.

CARBON; SILICON; TITANIUM; ZIRCONIUM; CERIUM; TIN; LEAD.

CARBON.

The Determination of Carbon Dioxide in Carbonates by Loss.

Expulsion of Carbon Dioxide by the Action of Acid. THE determination of carbon dioxide in carbonates, by loss, in the action of acid, is conveniently made with the simple apparatus of Kreider, previously described and figured.*

In carrying out the operation, the carbonate is weighed and placed in the bottom of the test tube, A, which serves as the reaction chamber. The acid, to a volume of 10 cm.³ to 15 cm.³, is drawn into the acid chamber, C, and held there in the manner described. The test tube, A, is slipped over B, and this joint is sealed with paraffin, as has been shown. The apparatus is wiped, placed on the balance and weighed.

Carbon Dioxide in Carbonates.

	Taken. grm.	Found. grm.	Error. grm.
Calcium carbonate	0.2000	0.0879	−0.0001
	0.2000	0.0878	−0.0002
	0.2000	0.0879	−0.0001
	0.2000	0.0879	−0.0001
	0.5000	0.2197	−0.0003
	0.5000	0.2196	−0.0004
	0.5000	0.2194	−0.0006
	0.5000	0.2198	−0.0002
	0.5000	0.2197	−0.0003
	0.5000	0.2197	−0.0003
Barium carbonate	0.5000	0.1134	−0.0011
	0.5000	0.1137	−0.0005
	0.5000	0.1137	−0.0005
	0.5000	0.1136	−0.0006
Strontium carbonate	0.5000	0.1485	−0.0004
	0.5000	0.1486	−0.0003
	0.5000	0.1485	−0.0004

* See Fig. 1, page 1.

Upon removing the cap from the small tube in C, the acid runs from C into A. The carbon dioxide liberated is forced upward through the drying column of calcium chloride and escapes through the annular space between B and C. When the action ceases, a current of dry air is forced through C, to remove the carbon dioxide, the cap is replaced, and the apparatus is weighed. The loss of weight represents the carbon dioxide.

The preceding results show the accuracy which may be expected when carbonates are treated in the apparatus with dilute hydrochloric acid.

Expulsion of Carbon Dioxide by Ignition. From certain carbonates, like those of magnesium, zinc, and cadmium, carbon dioxide may be expelled by simple ignition at a moderate temperature, leaving an oxide in definite and weighable condition. In the case of calcium carbonate, this process of decomposition is completed only at the high heat of the blast lamp, and the reaction, being easily reversible in the atmosphere containing carbon dioxide evolved in the ignition or produced in the source of heat, may leave the oxide not quite pure. Strontium carbonate and barium carbonate are not entirely broken up by simple ignition under the conditions ordinarily available in analysis; nor are the alkali carbonates. For the decomposition of refractory carbonates it is customary to make use of a suitable flux which, by combining with the oxide, will aid in the expulsion of the carbon dioxide. Anhydrous borax,* silicon dioxide,† potassium dichromate,‡ and recently, sodium metaphosphate,§ have been thus used in the analysis of carbonates, and they are applicable similarly to the determination of nitrogen pentoxide in nitrates which leave definite oxides on ignition. Such fluxes, moreover, serve the very essential end of conserving the residual oxides in definite and stable form for weighing under the ordinary atmospheric condition of the balance room. Of those mentioned, the first two, borax and silicon dioxide, require prolonged ignition to bring them to constant weight before making use of them to react with the carbonate or nitrate; and generally they yield in the fusion

* Fresenius, Zeit. anal. Chem., i, 181.

† Rose, Ann. Phys., cxvi, 635; Fresenius, Zeit. anal. Chem., i. 184; Richards and Archibald, Proc. Am. Acad., xxxviii, 443.

‡ Rose, Ann. Phys., cxvi, 131; Fresenius, Zeit. anal. Chem., i, 183.

§ Lutz and Tschischikof, Chem. Zentralblatt, 1905, i, 564; Bottger, Zeit. anal. Chem., xlix, 487.

process a pasty magma, so that prolonged heat at a high temperature is necessary to the complete expulsion of the gaseous product. Sodium metaphosphate, though more fluid in fusion, also demands prolonged care in the preparation. Potassium dichromate is too easily decomposed with loss of oxygen to be employed in exact processes demanding long-continued fusion or heating to temperatures much above its fusing point. These are points which have been sufficiently emphasized in the work to which reference has been made.

It has been shown by Gooch and Kuzirian* that in sodium paratungstate, of composition corresponding approximately to the formulæ $5Na_2O.12WO_3$, or $Na_{10}W_{12}O_{41}$, we have material very easily prepared, stable in fusion, and well suited for use as a flux in the rapid determination of the loss of carbonates and nitrates on ignition. This sodium paratungstate is prepared by dehydrating and fusing over the blast lamp a known weight of normal sodium tungstate, $Na_2WO_4.2H_2O$, adding an equal weight of tungsten trioxide, WO_3 (previously ignited with care to remove all ammonia and to insure complete oxidation), and heating to clear fusion. The cooled mass, which is very easily pulverized, is ground in a mortar and bottled. From this material, kept in a desiccator over sulphuric acid (though not more than ordinarily hygroscopic), portions are weighed for the analytical determinations. Approximately half the weight of the paratungstate is tungsten trioxide (molecular weight 232), and this should be capable of expelling carbon dioxide (molecular weight 44) to the amount of one-fifth its own weight. In practice, the weights of paratungstate used should exceed ten times the weight of carbon dioxide. It is best to weigh a platinum crucible, introduce the dried carbonate and weigh again, add a suitable amount of the prepared sodium paratungstate, stir carefully with a platinum wire with care to avoid mechanical loss, and weigh again. The crucible is then heated over a Bunsen burner, first at very low heat and then to fusion of the mixture for five minutes, cooled in a desiccator over sulphuric acid, weighed, and reignited to test the constancy of weight. The constant weight is usually got in the first ignition. In the following tables are given the results of the estimation of carbon dioxide in calcite, and in the precipitated carbonates of strontium and barium.

* F. A. Gooch and S. B. Kuzirian, Am. Jour. Sci., [4], xxxi, 497.

Analysis of Calcium Carbonate (Calcite).

$CaCO_3$ taken. grm.	$Na_{10}W_{12}O_{41}$ taken. grm.	Loss on ignition. grm.	Theory for CO_2. grm.	Error. grm.
0 5000	2.5	0 2195	0.2198	−0 0003
0 5000	2 5	0.2206	0.2198	−0 0008
0 5000	2.5	0.2200	0.2198	−0 0002
0 5000	2 5	0.2203	0.2198	−0 0005
0 5000	2.5	0 2200	0.2198	−0 0002
0.5000	2 5	0.2204	0.2198	−0.0006
0.5000	2.5	0.2190	0.2198	−0 0002
0.5000	2.5	0.2200	0.2198	−0.0002

*Analysis of Specially Precipitated Strontium Carbonate.**

$SrCO_3$ taken. grm.	$Na_{10}W_{12}O_{41}$ taken (approx.). grm.	Loss on ignition grm.	Theory for CO_2. grm.	Error. grm.
0.5000	2.5	0.1488	0.1490	−0 0002
0.5000	2 5	0.1494	0.1490	+0 0004
0.5000	2.5	0.1494	0.1490	+0 0004
0.5000	2.5	0.1490	0.1490	0 0000
0.5000	2.5	0.1496	0.1490	+0 0006
0.5000	2.5	0.1486	0.1490	−0.0004

*Analysis of Specially Precipitated Barium Carbonate.**

$BaCO_3$ taken. grm.	$Na_{10}W_{12}O_{41}$ taken (approx.). grm.	Loss on ignition. grm.	Theory for CO_2. grm.	Error. grm.
0.5000	2.5	0.1120	0.1115	+0 0005
0.5000	2.5	0.1125	0.1115	+0.0010
0.5000	2.5	0.1109	0.1115	−0.0004
0 5000	2.5	0.1113	0.1115	−0.0002
0.5000	2.5	0.1123	0.1115	−0.0008

* Prepared from the pure chloride by partially precipitating and washing with hydrochloric acid, dissolving the precipitated chloride in water, adding the solution drop by drop to a hot saturated solution of ammonium carbonate, washing the precipitated carbonate and drying below red heat.

The Precipitation and Gravimetric Determination of Carbon Dioxide.

The estimation of carbon dioxide in carbonates, by liberation of that gas and absorption in weighed potash bulbs, demands the careful observance of precautions and the expenditure of much time and attention. According to a method proposed by

Gooch and Phelps,* the rapid absorption of carbon dioxide evolved by the action of acids on carbonates is effected by barium hydroxide contained in a specially devised apparatus; the precipitated barium carbonate is filtered off and washed under a protecting layer of xylene; the washed carbonate on the filter or adhering to the receiver is dissolved in hydrochloric acid; the barium is precipitated from the solution as sulphate, ignited and weighed; and from the weight of barium sulphate is calculated the equivalent amount of carbon dioxide originally liberated.

The apparatus, shown in the figure, consists of an evolution flask (F) of about 50 cm.^{3} capacity; an absorption cylinder (C) made of wide glass tubing; a toy rubber balloon; rubber stoppers, and glass tubes for connection. The tube (A) which connects flask and cylinder is wide enough (about 0.7 cm. bore) to prevent formation of bubbles as the liquid distils, and for further protection is expanded to a bulb. The stopper at the lower end of the cylinder, placed vertically, carries a short tube, about 1.5 cm. in bore, to which the rubber balloon is securely bound. The stopper at the upper end of the cylinder is perforated with two holes, through one of which passes the tube of a glass stopcock, while through the other passes a long tube reaching to the interior of the balloon, and provided with a Bunsen valve (preferably of Kreider's pattern†).

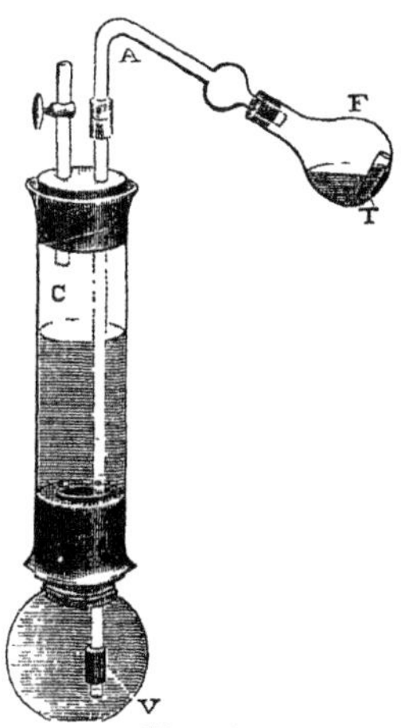

Fig. 18.

In using this apparatus, a saturated solution of barium hydroxide (which is filtered into a siphon bottle, and preserved from atmospheric action by a floating layer of kerosene) is introduced by pressure upon the air in the siphon bottle or by suction applied to the stopcock of the cylinder. Such a solution contains about 5 per cent of its weight of the hydroxide. It is best to use of it in every case an amount at least a fourth in excess of the quantity theoretically required to absorb the

* F. A. Gooch and I. K. Phelps, Am. Jour. Sci., [3], l, 101.
† See page 7.

carbon dioxide, and to fill the cylinder and balloon nearly full of liquid. The carbonate is weighed, introduced into the flask, and washed down with 15 or 20 cubic centimeters of boiled water, which is protected from carbon dioxide of the breath by a balloon attached to the inlet tube within the wash bottle, so that upon blowing into the inlet tube the breath distends the balloon and thus creates the necessary pressure. A small tube, holding enough hydrochloric acid to effect the decomposition of the carbonate to be analyzed, is placed in upright position in the evolution flask. The stopper is inserted in the flask and connections are made as shown in the figure, the little tube containing the acid is overturned by inclining the flask, the acid mixes with the water, and effervescence begins. Heat is applied and the liquid in the flask is boiled until that in the cylinder is heated by the steam nearly to the boiling point, in order that the precipitated barium carbonate may become as granular as possible. The carbon dioxide evolved and the air in the flask are transferred in the process to the absorption cylinder, the valve serving to prevent the back-flow of the liquid while the balloon expands to give room to the air and condensed steam. When the boiling is done the flask and tube are disconnected at the rubber joint, the cylinder is shaken to insure the absorption of the carbon dioxide, and the liquid carrying the greater part of the precipitate is transferred through the stopcock to a paper filter moistened with water, and containing about 5 $cm.^3$ of xylene. The function of the xylene, which was found to be preferable to benzene, kerosene or amyl alcohol, is to rise to the surface when the aqueous solution is added, so as to protect the barium hydroxide from the action of the carbon dioxide of the air. By manipulating the balloon and the stopcock (to which a little funnel may be attached by a piece of rubber tubing for convenience in introducing wash water) the cylinder may be emptied and washed out with hot boiled water, though, of course, a very considerable portion of the precipitate remains adhering to the walls of the absorption apparatus.

The filter is prepared for use with the suction pump, but in the early stages of filtration and washing very little suction should be applied. When the barium hydroxide has been nearly washed out of the precipitate, the xylene is dissolved in a little hot alcohol, the suction is applied and the washing is completed with hot water. The emulsion of xylene and water found in the filtrate is readily cleared up by alcohol. Finally, the barium carbonate in the absorption apparatus and upon the filter is dissolved in hydrochloric acid and precipitated in hot solution by sulphuric acid, the resulting barium sulphate is filtered, washed, and ignited upon asbestos in a perforated crucible, and from its weight the carbon dioxide which originally precipitated the barium, now in the form of the sulphate, is calculated. The results of a series of determinations made in this manner are recorded in the following table.

Carbon Dioxide by Precipitation.

$CaCO_3$ taken. grm.	$BaSO_4$ found. grm.	CO_2 actually present. grm.	CO_2 calculated from $BaSO_4$. grm.	Error in CO_2. grm.
0.0500	0 1180	0.0220	0.0222	+0.0002
0.0500	0.1183	0 0220	0.0223	+0.0003
0 1000	0.2329	0.0440	0.0439	−0.0001
1.1000	0.2347	0.0440	0 0442	+0.0002
0.2000	0.4660	0 0880	0.0878	−0.0002
0.2000	0.4653	0 0880	0.0876	−0.0004
0.5000	1 1650	0.2200	0 2196	−0 0004
0.5000	1 1657	0.2200	0.2197	−0.0003
1.0000	2.3323	0 4400	0.4396	−0.0004
1.0000	2.3309	0.4400	0.4394	−0.0006

The process is fairly rapid and accurate.

The Iodometric Determination of Carbon Dioxide.

When the solution of a metallic hydroxide is acted on by iodine at a temperature high enough to decompose the small amounts of hypoiodites that might otherwise be present, the final action results in the formation of an exactly neutral mixture of iodate and iodide, according to the equation:

$$6\,ROH + 3\,I_2 = RIO_3 + 5\,RI + 3\,H_2O.$$

Upon the assumption that in the case of barium hydroxide this reaction is regular under the conditions of analysis and independent of the excess of iodine which remains in the neutral mixture unacted upon, and that iodine may be estimated by directly titrating with arsenious acid, Phelps * has elaborated a differential method for determining carbon dioxide. In this

* Am. Jour. Sci., [4], ii, 70.

method the liberated gas is run into a measured amount of barium hydroxide, the final excess of which is estimated by treating with iodine in the presence of the precipitated barium carbonate.

The solution of barium hydroxide is standardized by drawing 80–90 cm.^{3} of decinormal iodine into a glass flask, provided with a ground-glass stopper carrying an inlet tube reaching nearly to the bottom of the flask and an outlet tube to which is sealed a Will and Varrentrapp absorption apparatus, and then introducing an appropriate amount of the barium hydroxide solution either from a burette or from a stoppered funnel which is weighed before and after. A glass-stoppered wash bottle answers for a standardizing flask, and, with the glass stopper and its attachments replaced by a rubber stopper, answers the purpose of the absorption flask described later. The glass stopper is introduced, the inlet being closed by a rubber cap, and the absorption apparatus is charged with a solution of potassium iodide, to hinder the escape of iodine. The solution is heated to boiling to break up traces of hypoiodite, then cooled and the excess of iodine determined by decinormal arsenite. It is assumed that the iodine lost has acted on the barium hydroxide according to the equation:

$$6\,BaO_2H_2 + 6\,I_2 = Ba(IO_3)_2 + 5\,BaI_2 + 6\,H_2O.$$

In a precisely similar manner is determined the quantity of barium hydroxide remaining after a measured amount of it has been submitted to the action of carbon dioxide. The difference between the quantity thus formed and that measured out is equivalent to the carbon dioxide entering into action.

Carbon Dioxide in Carbonates. A convenient apparatus for evolving the carbon dioxide from carbonates consists of a wide-mouthed flask of about 75 cm.^{3} capacity, furnished with a doubly perforated stopper carrying a separating funnel for the introduction of acid into the flask, and a tube of 0.7 cm. internal diameter, which is expanded to a small bulb just above the stopper, to carry off the gas. This exit tube is joined by a rubber connector to a tube which passes through the rubber stopper, closing the absorption flask* (the glass-stoppered wash bottle used in standardizing the barium hydroxide solution), and which ends in a valve, preferably

* See left-hand flask in Fig. 19, page 237.

of the Kreider pattern.* This valve is inclosed in a larger tube reaching nearly to the bottom of the absorption flask. Through a second hole in the stopper of the absorption flask passes a glass tube closed by a rubber connector and screw pinchcock.

In making a determination of carbon dioxide in carbonates, the weighed substance is introduced into the boiling flask. Barium hydroxide solution, in amount 7 cm.^{3} to 10 cm.^{3} more than actually necessary to precipitate the carbon dioxide, is drawn into the absorption flask, which is then connected with the boiling flask, as described above. The stopcock of the separating funnel is shut off and the flasks evacuated by connecting the exit tube of the absorption flask with a filter flask previously pumped out by the water pump. Phosphoric acid (chosen as a nonvolatile acid) is introduced into the stoppered funnel with about 50 cm.^{3} of water, previously purified from carbon dioxide by boiling down one-third, and kept in full, stoppered flasks until used. The acid is allowed to enter the boiling flask and the carbon dioxide driven over completely to the absorption flask by boiling for five minutes — the latter being shaken frequently during the passage of the gas into it and kept cool by standing in a dish of water. The atmospheric pressure is restored by admitting purified air through the funnel of the boiling flask. The inlet tube of the absorption flask is then closed by a rubber cap, the exit tube attached to potash bulbs and the flask heated to the boiling point of the liquid (to granulate the precipitated carbonate) and then cooled in a stream of water. The exit tube is removed, a capillary tube long enough to reach below the surface of the liquid introduced and decinormal iodine run in until the liquid is yellow. Then the glass stopper of the absorption flask is introduced, with a rubber cap on the inlet tube and potassium iodide solution in the trap, as in standardizing, and the emulsion brought to a boil. Iodine is again run into the hot solution through the inlet tube until the color remains distinctly red. After cooling, the excess of iodine is determined by standard arsenious acid.

Results obtained by treatment of calcite, essentially as described, are given on the next page, correction being made for the deficiency in carbon dioxide (0.0014 grm. for each gram of the mineral) found by the ignition analysis.

* See page 7.

Carbon Dioxide by Absorption in Barium Hydroxide and Iodometric Determination of the Excess.

Calcite taken. grm	BaO_2H_2 taken. grm.	BaO_2H_2 found. grm.	CO_2 found. grm.	Error on CO_2. grm.	Error corrected. grm.
0 0501	0 2484	0 1604	0 0227	+0 0006	+0 0007
0 0500	0 2381	0 1508	0 0224	+0 0004	+0 0005
0 1022	0 3416	0 1675	0 0447	−0 0003	−0 0001
0 1026	0 3105	0 0351	0 0450	−0 0001	0 0000
0 2032	0 6181	0 2692	0 0896	+0 0002	+0 0004
0 2049	0 5761	0 2223	0 0908	+0 0006	+0 0008
0 5088	1.1301	0 2606	0 2232	−0 0007	0 0000
0.5015	1 0804	0 2245	0 2197	−0 0010	−0 0003
1 0032	2.0125	0.3004	0 4394	−0 0020	−0 0006
1 0064	2.0702	0.3538	0.4405	−0.0023	−0 0009

The Combustion of Organic Substances in the Wet Way.

Phelps * has studied the combustion of substances in the wet way, and, with apparatus like that previously described for the determination of the carbon dioxide of carbonates, has shown that the carbon of many organic compounds may be determined by oxidation and estimation of carbon dioxide.

Carbon Content by the Permanganate Process. Certain organic material may be oxidized by potassium permanganate in acid or alkaline solution, and the carbon content of the organic matter may be estimated by absorbing in a measured amount of standardized barium hydroxide the carbon dioxide produced, acting upon the residual hydroxide with standard iodine in excess, and determining by decinormal arsenite the excess of iodine. According to this process, the organic material is weighed out and introduced into an evolution flask with 10 cm.^{3} to 15 cm.^{3} of pure water (freed from carbon dioxide by boiling down one-third and kept in stoppered bottles), and the evolution flask is connected with an absorption flask charged with a volume of standardized barium hydroxide 3 cm.^{3} to 5 cm.^{3} in excess of the amount required to precipitate the carbon dioxide to be determined.†

The whole system is then evacuated with the water pump to a pressure of 200 mm. to 225 mm. and the boiling flask warmed.

* I. K. Phelps, Am. Jour. Sci., [4], iv, 372.

† See, for the general arrangement, the more elaborate apparatus of Fig. 19, page 237.

An excess of potassium permanganate solution (prepared by dissolving in water, acidulating with sulphuric acid and boiling until free from carbon dioxide) is then run in through the funnel tube and the mixture warmed again. The carbon dioxide is set free by sulphuric acid [1 : 3], either at once or after introducing an excess of pure sodium hydroxide and boiling, and driven completely to the absorption flask by boiling for five minutes. During the passage of the gas the absorption flask is shaken frequently and kept cool by standing in a dish of water and by pouring cold water over it from time to time. During the boiling, the vacuum in the flasks may be tested by opening momentarily the stopcock of the funnel tube and noting the direction of the flow of water contained in the funnel. After the boiling, the atmospheric pressure is restored by allowing air, purified from carbon dioxide by passage through potash bulbs, to enter through the funnel tube of the boiling flask. The flasks are disconnected, the stopper of the absorption flask with its attachments is removed and carefully washed free from barium hydroxide, and a second stopper, provided with a separating funnel and a Will and Varrentrapp absorption apparatus containing water to serve as a trap, is inserted into the mouth of the absorption flask. The contents of the flask are brought to the boiling point, decinormal iodine solution is run in through the funnel tube in sufficient quantity to destroy the larger part of the excess of barium hydroxide, the mixture again heated to boiling, and iodine run in again to permanent red coloration. After cooling, the excess of iodine is determined by titration with decinormal arsenite. The difference between the barium hydroxide taken and that equivalent to the iodine which has disappeared is the

Oxidation with Permanganate in Acid Solution.

Ammonium oxalate taken. grm.	BaO_2H_2 taken. grm.	BaO_2H_2 remaining. grm.	CO_2 found. grm.	CO_2 calculated. grm.	Error on CO_2. grm.
0.2522	0 7267	0.1170	0.1565	0.1561	+0 0004
0.2542	0 7267	0.1113	0.1579	0.1574	+0.0005
0.5020	1 4535	0.2417	0.3110	0.3108	+0.0002
0.5058	1 3954	0.1753	0.3131	0.3131	0.0000
1.0033	2 6163	0.1955	0.6213	0.6211	+0.0002
1 0003	2.5951	0.1836	0.6189	0.6192	−0.0003
1.0010	2.6163	0.2037	0.6192	0 6197	−0 0005

Oxidation with Permanganate in Alkaline Solution.

Barium formate taken. grm.	BaO_2H_2 taken grm	BaO_2H_2 remaining. grm	CO_2 found. grm.	CO_2 calculated grm.	Error on CO_2. grm.
0 5001	0 9302	0 1745	0 1939	0.1935	+0.0004
0 5033	0 9012	0 1402	0 1953	0.1947	+0.0006
1 0002	1 6861	0.1793	0 3867	0 3870	−0 0003
1 0059	1.6279	0.1093	0 3897	0 3892	+0 0005
1 3750	2 2529	0 1820	0 5315	0 5320	−0.0005
1 5028	2 4419	0 1754	0 5816	0.5814	+0 0002

Tartar emetic taken.* grm.	BaO_2H_2 taken. grm.	BaO_2H_2 remaining. grm.	CO_2 found grm.	CO_2 calculated. grm.	Error on CO_2. grm.
0 5051	1 2450	0.1709	0 2756	0.2751	+0.0005
0.5030	1.2226	0 1536	0.2743	0.2739	+0 0004
0 7509	1.7355	0.1401	0.4094	0.4091	+0.0003
0.7541	1.7430	0.1410	0 4111	0 4107	+0.0004
1 0018	2 3456	0.2187	0 5458	0.5456	+0.0002
1.0005	2.2435	0 1196	0.5451	0.5450	+0.0001

* Dried at 100°.

measure of the carbon dioxide. The figures given show the result of the permanganate oxidation when applied to ammonium oxalate in acid solution, and to barium formate and tartar emetic in alkaline solution.

Obviously, certain organic substances oxidizable by permanganate may be analyzed by the process outlined above, but, as noted by Wanklyn and Cooper* and by others, potassium permanganate, whether in acid or alkaline solution, fails to oxidize completely certain organic substances (acetates and carbohydrates, for example) even at the boiling temperature.

Carbon Content by Oxidation with Chromic Acid. It is well known, however, that a mixture of concentrated sulphuric and chromic acids has a much wider field of action in oxidizing organic compounds, and Phelps† has studied the application of this method also. The apparatus recommended for the process, and shown in the accompanying figure, may be described as follows:

A thick-walled liter flask with round bottom, serving as an oxidizing chamber, is closed by a rubber stopper with two per-

* Phil. Mag., [5], vii, 138.

† Loc. cit.

forations. Through one of these passes the tube of a separating funnel reaching nearly to the bottom of the flask and drawn out at the lower end. A disk of platinum foil, through which the tube of the funnel passes, is hung in the neck of the flask, nearly closing it, and held in place by an attached platinum wire, the end of which is squeezed under the rubber stopper. Through the second hole of the stopper passes an exit tube 0.7 cm.[3] in bore. This tube, expanded just above the stopper to a small bulb which serves to prevent mechanical loss of the solid contents of the flask during the boiling, is joined by means of a rubber connector (provided with a screw pinchcock) to the inlet tube of the absorption flask, which is an ordinary 500 cm.[3] round-bottomed flask. This flask is also closed by a rubber stopper with two perforations, through one of which passes the inlet tube and through the other the exit tube, which is also enlarged to a small bulb just above the stopper and is closed by a rubber connector and screw pinchcock. The ground-glass stopper of the funnel tube is carefully cleaned and lubricated with a thick solution of metaphosphoric acid.

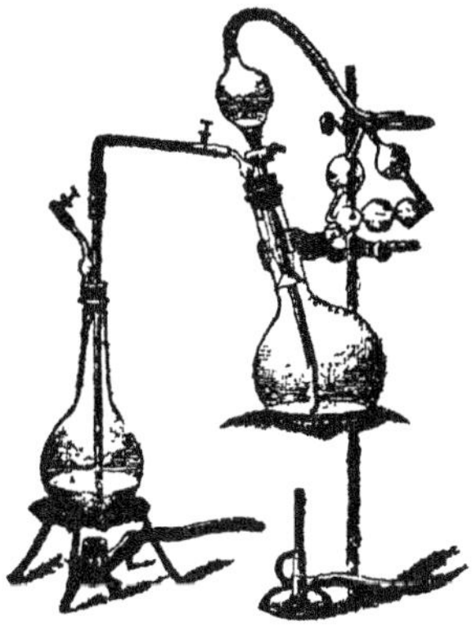

Fig. 19.

A partial vacuum is easily obtained by boiling water in the evolution flask and the barium hydroxide solution in the absorption flask at the same time, closing the flasks, and cooling—both flasks being connected and ready for a determination.

In making a determination, the organic substance is weighed out in a counterbalanced bulb, so thin that it may be easily broken later and made with a wide mouth for convenience in introducing the solid substance. After the substance is weighed, the mouth of the bulb is sealed by heating in a small blowpipe flame and the tube introduced into the evolution flask, together with an amount of pure potassium dichromate, known to be in excess of that required to oxidize the organic substance. The flasks are connected, as already described, with an appropriate amount of barium hydroxide solution in the absorption flask and 10 cm.[3] of pure water in the evolution flask, and the vacuum obtained (as described above) by boiling the contents of both flasks until the water in the evolution flask has decreased to 2 cm.^{3} or 3 cm.^{3}, and cooling. The tube containing the organic substance is then broken by shaking the flask, and 20 cm.^{3} of concentrated sulphuric acid, previously purified from organic material by heating to the fuming point with a few crystals of potassium dichromate, are run in through the funnel tube. While still hot, the acid is shaken in the flask violently, the platinum foil hung in the neck serving to protect the rubber stopper. The flask is warmed to approximately 105°, the highest temperature to which, as shown by Cross and Bevan,* such a mixture of chromic and sulphuric acids may be safely heated without the disengagement of oxygen gas. Water is then run in until the crystals of chromic anhydride have disappeared and the danger of the evolution of oxygen is past. The solution is heated to its boiling point, with care to keep the outward pressure less than the inward pressure — the relation being easily observed by opening momentarily the stopcock of the funnel tube and noting the direction of the flow of water contained in the funnel. The flask is shaken and heated alternately for five minutes to bring about the oxidation of small amounts of carbon monoxide, originally produced. Then more water (60 cm.^{3} to 70 cm.^{3}) is introduced through the funnel and the stopcock between the

Oxidation with Chromic Acid.

Substance taken. grm.	BaO_2H_2 taken. grm.	BaO_2H_2 found. grm.	CO_2 found. grm.	CO_2 calculated. grm.	Error on CO_2. grm.
Analysis of ammonium oxalate.					
0.5009	1.3534	0 1469	0.3097	0.3101	−0.0004
0.5006	1.3400	0.1308	0.3103	0.3099	+0 0004
0.5005	1.3400	0.1343	0.3094	0.3098	−0.0004
1 0002	2.5460	0.1347	0.6188	0.6192	−0.0004
1.0010	2.5192	0.1094	0.6185	0.6197	−0.0012
Analysis of cane sugar.					
0.2001	1.3926	0.1905	0.3085	0.3088	−0.0003
0.2000	1.3926	0.1936	0.3077	0 3086	−0.0009
0.2001	1.3926	0.1857	0.3097	0.3088	+0.0009
0.2014	1.3400	0.1279	0.3111	0.3108	+0.0003

* Jour. Chem. Soc., liii, 889.

boiling and absorption flasks is opened to admit the carbon dioxide to the latter, which is kept cool and shaken as before. The contents of the evolution flask are then heated to boiling and a slow current of air, freed from carbon dioxide by passage through potash bulbs, is allowed to enter through the funnel tube to keep the liquid from undue bumping. The boiling is continued for fifteen minutes, after which the excess of barium hydroxide is determined iodometrically and the carbon dioxide calculated.

The results of experiments with ammonium oxalate and with cane sugar, one of the more difficultly oxidizable substances, are given in the preceding table. The results are evidently very satisfactory.

Carbon Dioxide Evolved and Oxygen Used. Taking advantage of the fact that at 105° the mixture of chromic and sulphuric acids does not evolve oxygen, Phelps* has been able to determine the oxygen used in the combustion of certain organic substances. The knowledge of this amount of oxygen and of that contained in the products of oxidation gives, by difference, the oxygen content of the original substance.

In this operation a known weight of the substance is first treated with a known weight of pure potassium dichromate and 20 cm.^{3} of concentrated and purified sulphuric acid, according to the process described above for the determination of carbon as carbon dioxide. The residual liquid, which, after dilution and boiling for the removal and estimation of the carbon dioxide, should have a volume of 60 cm.^{3} to 80 cm.^{3}, is washed into the Voit flask of the distillation and absorption apparatus figured and previously described.† The absorption chamber is charged with a solution of arsenious oxide in sodium hydroxide, the former being in slight excess of the amount required to take up the chlorine to be evolved by the chromate and the latter in quantity more than sufficient to neutralize the acid which may be volatilized to the receiver in the later operation. The apparatus is connected, hydrochloric acid (35 cm.^{3} of the strongest acid) introduced through the stoppered funnel, a slow current of carbon dioxide started through the system, and the liquid in the flask slowly boiled down, for a period of five or six hours,

* Am. Jour. Sci., [4], ii, 379.

† See Fig. 3, page 4.

until the volume is 30 cm.^{3} to 40 cm.^{3}. After cooling and disconnecting the apparatus, the solution in the receiver is made acid with sulphuric acid and then alkaline with acid potassium carbonate. The residual arsenite is determined by titration with decinormal iodine.

Determination of Carbon Dioxide Evolved and Oxygen Used.

Substance taken.	CO_2 found.	Error on CO_2.	$K_2Cr_2O_7$ taken.	As_2O_3 taken.	As_2O_3 found.	Oxygen used.	Oxygen required by theory.	Error on oxygen.
grm.	grm.	grm.	grm.	grm.	grm.	grm.	grm.	grm.
Analysis of ammonium oxalate.								
1.0122	0 6265	−0.0001	2.0009	1 3002	0 0000	0 1160	0 1139	+0.0021
1.0019	0 6212	+0 0010	2.0002	1.3517	0.0440	0.1147	0.1128	+0.0019
Analysis of phthalic acid.								
0.1002	0.2138	+0.0014	2 0012	1.2004	0 0814	0.1456	0.1448	+0.0008
0 1093	0 2324	+0.0007	2.0000	1.1031	0.0634	0 1582	0.1580	+0 0002
Analysis of cane sugar.								
0.2025	0.3117	−0.0008	3.0000	1.7002	0 0796	0.2275	0.2273	+0.0002
0.4012	0.6166	−0.0024	5.0000	2 3022	0.0366	0.4495	0 4502	−0 0007
Analysis of paper.								
0.3034	0.4932	−0.0010	3.5015	1 4017	0.0879	0.3589	0 3598	−0.0005
0.4523	0.7334	−0.0033	5.0035	1.8000	0 0710	0.5368	0 5358	+0.0010
Analysis of tartar emetic.								
0.5057	0.2671	−0.0009	2.5018	1 7000	0.0766	0.1459	0.1462	−0.0003
1.0099	0.5321	−0.0030	3.5003	1.7520	0.0198	0.2911	0.2919	−0.0008
Analysis of barium formate.								
1.0079	0.3906	+0.0006	3.0026	2 2002	0 0496	0.1423	0.1422	+0.0001
1.5014	0.5814	+0.0005	3.0010	1.8080	0.0890	0 2118	0 2118	0.0000

When the boiling begins the chromate is gradually reduced with evolution of chlorine, but if the evaporation of water is pushed too rapidly the sulphuric acid may reach a degree of concentration such that oxygen is evolved instead of chlorine in the process of reduction. Sometimes, during the reduction, chlorochromic anhydride is visibly volatilized to the receiver, but inasmuch as it is there reduced and registered by the arsenite this transfer is of no moment.

The results of experiments in which the carbon of various organic substances was first determined by oxidation according to the process previously described,* the amount of potassium dichromate employed being accurately known, and the residual dichromate found by reduction with hydrochloric acid by the process just outlined, are given in the tabular statement.

From these results, it will be seen that the process works with accuracy for a great variety of organic substances. It was found impossible, however, to determine the elements in compounds which, like ether and naphthalene, are at the same time volatile and hard to oxidize completely.

SILICON.

The Detection of Silicon in Silicates and Fluosilicates.

The formation of silicon fluoride by the action of hydrofluoric acid or a fluoride and sulphuric acid upon a silicate is often applied to the detection of silica, the silicon fluoride giving with water a white precipitate of silicic acid. The usual procedure in making this test is to allow the gaseous silicon fluoride to come in contact with a moistened glass rod, but the condensation of steam or sulphuric acid on the rod often makes the results uncertain. Browning† has found that when moistened black paper is brought in contact with the fumes of the gaseous fluoride the deposit of silica is very easily detected. According to the procedure recommended, the reaction is made to take place in a small lead cup about one centimeter in diameter and depth, made by running the melted metal into a mold, covered by a flat piece of lead with a small hole in the center, as shown in Fig. 20. Into this cup is put the fluosilicate, or the silicate with a small amount of finely powdered calcium fluoride, generally about 0.1 grm., and a few drops of concentrated sulphuric acid are added. Upon the

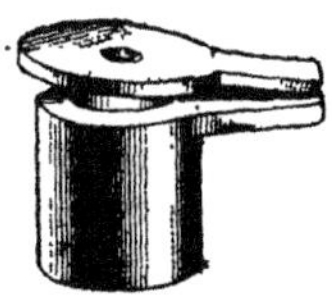

Fig. 20.

* See page 236.

† Philip E. Browning, Am. Jour. Sci., [4], xxxii, 249.

upper side of the cover a piece of moistened black filter paper is placed and upon this a small moistened pad of ordinary filter paper to keep the black paper moist during heating upon a steam bath. At the conclusion of the heating of about ten minutes, a white deposit is found on the under side of the black paper, over the opening in the cover, if silica is present in appreciable amount.

Tests of this procedure are given in the table.

Silicate Tests.

Material tested. grm.	Approximate per cent of SiO_2.	CaF_2 present. grm.	Result.
0 1000 SiO_2	100	0 1000	Nothing.
0.1000 SiO_2	100	0 1000	Very good.
0.0100 SiO_2	100	0 1000	Very good.
0 0050 SiO_2	100	0 1000	Very good.
0 0010 SiO_2	100	0.1000	Trace.
0 0100 Kaolinite	46	0 1000	Very good.
0 0050 Kaolinite	46	0.1000	Very good.
0 0010 Kaolinite	46	0 1000	Trace.
0.0100 Gadolinite	24	0.1000	Very good.
0.0050 Gadolinite	24	0.1000	Trace.
0.0100 Lepidolite	50	0 2000	Good.

Fluosilicate Tests.

0 0050 Na_2SiF_6	...	....	Very good.
0 0010 Na_2SiF_6	...	....	Very good.

Titanium.

The Determination of Titanic Acid by Reduction and Titration with Potassium Permanganate.

The estimation of titanic acid by reduction with zinc and direct titration leads to low results even when precautions are taken to avoid atmospheric oxidation during titration.* Newton† has shown, however, that it is possible to determine titanic acid successfully by reducing with zinc in an atmosphere of hydrogen, adding an excess of ferric sulphate, and titrating the resulting and equivalent ferrous salt with potassium perman-

* Cf. Pisani, Compt. rend., lix, 298; Marignac, Zeit. anal. Chem., vii, 112; Wells and Mitchell, Jour. Am. Chem. Soc., xvii, 878.

† H. D. Newton, Am. Jour. Sci., [4], xxv, 130.

ganate; the reaction between the salts of titanium and iron taking place according to the equation:

$$Ti_2(SO_4)_3 + Fe_2(SO_4)_3 = 2\,Ti(SO_4)_2 + 2\,FeSO_4.$$

According to this procedure, titanic acid dissolved in concentrated sulphuric acid is introduced into a 100-cm.^{3} flask and the solution is diluted with water until it contains 10 per cent of sulphuric acid, this strength being sufficient to hold the titanic acid in solution, while insufficient to reoxidize reduced titanium oxide. Zinc is added in suitable amount, and a rubber stopper carrying a delivery tube and a small separating funnel is inserted in the neck of the flask. After the air has been driven from the flask by the hydrogen evolved, the delivery tube is dipped under water and the stopcock of the funnel is closed. Gentle heat is applied until all the zinc is dissolved, and the solution is cooled. An excess of ferric sulphate is passed into the flask through the separating funnel and followed at once by cold, freshly distilled water until the flask is filled to the neck. The contents of the flask are poured into more cold distilled water and the ferrous salt produced by action of the ferric salt upon the reduced titanium oxide is titrated by $n/10$ permanganate.

Test results are given in the table.

Permanganate Titration of Titanic Acid Reduced by Zinc.

$KMnO_4$. cm.^{3}	TiO_2 taken. grm.	TiO_2 found. grm.	Error. grm.
6.50	0.0520	0 0523	+0 0003
6.52	0.0520	0 0524	+0 0004
6.45	0 0520	0 0519	−0 0001
6 50	0 0520	0 0523	+0 0003
6.48	0 0520	0 0521	+0 0001
6 42	0 0520	0 0518	−0 0002
6.50	0 0520	0 0523	+0 0003
19 95	0 1596	0 1599	+0.0003
19.93	0.1596	0 1598	+0.0002
19.95	0.1596	0 1599	+0.0003
19 90	0 1596	0 1595	−0.0001
19.95	0.1596	0 1599	+0.0003
19.95	0.1596	0 1599	+0.0003
19.90	0.1596	0.1595	−0.0001
19.85	0.1596	0 1591	−0.0005
19.95	0 1596	0 1599	+0.0003
19 88	0 1596	0 1594	−0.0002
19.95	0 1596	0.1599	+0.0003
19 95	0.1596	0.1599	+0.0003

Zirconium.

The Separation of Zirconium from Iron by Volatilization of the Latter in Hydrogen Chloride.

It has been shown by Havens and Way* that zirconium oxide may be separated from ferric oxide by the volatilization of ferric chloride in an atmosphere of hydrogen chloride, containing a little chlorine, at a temperature of 200° to 300°.†

Cerium.

The Separation of Cerium from Other Cerium Earths by the Action of Bromine upon the Mixed Hydroxides in Presence of an Alkali Hydroxide.

One of the best known processes for the separation of cerium from lanthanum and didymium is that of Mosander.‡ This process consists in passing chlorine gas into a mixture of the hydroxides suspended in a distinct excess of a fixed alkali hydroxide, until the solution is saturated and the reaction of the liquid is no longer alkaline to litmus. Under these conditions nearly all the cerium remains undissolved as the ceric hydroxide, while the other cerium earths go largely into solution. In treating mixed material, the residue of ceric hydroxide generally retains some of the cerium earths, so that the treatment with chlorine must be repeated. Two disadvantages associated with this method, therefore, are the preparation and use of chlorine gas, and the solvent action of the hydrochloric acid formed in the reaction upon the ceric hydroxide

$$2\,Ce(OH)_3 + Cl_2 = 2\,CeO_2 + 2\,HCl + 2\,H_2O.$$

Browning and Roberts § have shown that by substituting bromine for chlorine in the Mosander process about 50 per cent of the other cerium earths can be separated from ceric hydroxide in one treatment, and that after three treatments practically all the other cerium earths are removed without any solvent action

* Franke Stuart Havens and Arthur Fitch Way, Am. Jour. Sci., [4], viii, 217.
† See page 508.
‡ Jour. prakt. Chem., xxx, 267.
§ Philip E. Browning and Edwin J. Roberts, Am. Jour. Sci, [4], xxix, 45.

upon the ceric hydroxide. The advantages of the method are, the convenience in the use of the bromine, and the apparent lack of tendency of the hydrobromic acid to dissolve the ceric hydroxide.

The procedure is as follows: The mixed hydroxides are precipitated with a slight excess of sodium hydroxide or potassium hydroxide, and, suspended in the alkaline solution, are treated with liquid bromine or bromine water in distinct excess, and the mixture is placed upon a steam bath until the greater part of the free bromine is expelled. The residue is then filtered off, washed, and treated as before. This process is repeated.

In the test experiments upon a mixture composed of about 50 per cent of cerium oxide and 50 per cent of cerium earth oxides other than cerium oxide, the filtrate after each treatment was found to contain the amounts of cerium earth oxides, free from cerium, indicated in the table. The residue from the last treatment on being dissolved in acid showed only faint didymium bands. In another experiment a larger amount of material, 10 grm., was subjected to a fourth and fifth treatment with bromine, the fourth treatment yielding a small fraction of a gram of oxides other than cerium oxide, and the fifth only a few milligrams. In both cases these oxides were free from cerium. The oxides from the first filtrates were much lighter in color than those obtained from the last, which, of course, indicates that the lanthanum is dissolved by the action of the bromine more readily than the didymium. The quantitative results follow in the table.

Separations by Bromine and Alkali Hydroxide.

Mixed oxides taken. grm.	Oxides found in first filtrate. grm.	Oxides found in second filtrate. grm.	Oxides found in third filtrate. grm.	Total oxides found. grm.
1.0000	0 3310	0 0720	0 0190	0.4420
1.0000	0.2900	0 1010	0.0420	0.4330
1 0000	0 2250	0.1290	0.0640	0.4180
1 0000	0.2750	0.0860	0.0740	0.4350
10 0000	3.1360	1 0050	0 5930	4.7340
10.0000	3.4590	0.5240	0 8560	4.8390

The action of iodine is similar to that of chlorine and bromine, but it is too incomplete to be of practical value.

The Iodometric Estimation of Cerium.

Anhydrous cerium dioxide, prepared by the ignition of the oxalate or hydroxide, is very slowly acted on by acids, especially when pure. For this reason the method which Bunsen described has remained the only one adapted to the satisfactory volumetric estimation of the ignited dioxide.* According to this method, the substance to be determined is weighed out in a glass flask of 10 to 15 cubic centimeters' capacity, a few crystals of potassium iodide are added, and the neck of the flask is drawn out by the aid of a blowpipe to a narrow opening. The flask is filled almost to the narrowing of the neck with hydrochloric acid which is free from chlorine or iron chloride, and a little sodium carbonate is added in order to displace the last trace of air by carbon dioxide. The flask is then closed by sealing off the neck in the blowpipe and warmed in a water bath until the cerium compound is completely dissolved, and the quantity of iodine set free is determined by iodometric analysis.

Browning (with Hanford and Hall)† has shown that good results may be obtained by digesting cerium dioxide with potassium iodide and hydrochloric acid in a glass-stoppered bottle and determining by sodium thiosulphate the iodine set free, or by distilling and estimating the iodine passing to the receiver, in accordance with the reaction:

$$2\,CeO_2 + 8\,HCl + 2\,KI = 2\,CeCl_3 + 2\,KCl + 4\,H_2O + I_2.$$

Digestion Process. According to the first procedure, weighed portions of the pure cerium dioxide are placed in small glass-stoppered bottles of about 100 $cm.^3$ capacity, together with 1 grm. of potassium iodide free from iodate, and a few drops of water to dissolve the iodide. A current of carbon dioxide is passed into the bottle for about five minutes to expel the air, 10 $cm.^3$ of pure strong hydrochloric acid are added, the stopper is inserted and the bottle heated gently, upon a steam radiator, for about one hour, until the dioxide dissolves completely and the iodine is set free. After cooling the bottle, to prevent loss of iodine upon removing the stopper, the contents are carefully washed into about 400 $cm.^3$ of water and titrated with standard

* Ann. Chem., cv, 49.

† Philip E. Browning, with G. A. Hanford and F. J. Hall, Am. Jour. Sci., [4], viii, 452.

sodium thiosulphate to determine the amount of iodine liberated. Results of this procedure, corrected by 0.04 cm.^{3} of $n/10$ iodine, which is the amount of iodine set free in blank determinations, are given in the accompanying table.

Digestion Process.

CeO_2 taken. grm	CeO_2 found grm	Error. grm.
0 1000	0 0994	−0 0006
0 1032	0 1034	+0 0002
0 1016	0 1017	+0 0001
0 1054	0 1041	−0 0013
0 2010	0 2021	+0 0011
0 1104	0 1109	+0 0005
0 1914	0 1907	−0 0007
0 1604	0 1603	−0 0001
0 2146	0 2145	−0 0001
0 1108	0 1099	−0 0009
0 1346	0 1347	+0 0001
0 1540	0 1534	−0 0006
0 1976	0 1968	−0 0008
0 1230	0 1240	+0 0010
0 1199	0 1201	+0 0003
0 1524	0 1528	+0 0004
0 1212	0 1211	−0 0001
0 1528	0 1543	+0 0015

Distillation Process. According to a second procedure, portions of cerium dioxide were weighed out into the retort of a distillation apparatus* consisting of a Voit flask, serving as the retort, sealed to the inlet tube of a Drexel wash bottle, used as a receiver, the outlet tube of which was trapped by sealing on Will and Varrentrapp absorption bulbs. In the retort are placed the cerium dioxide, 15 cm.^{3} of water, 1 grm. of potassium iodide, and 10 cm.^{3} of pure strong hydrochloric acid. In the receiver are 100 cm.^{3} of water and 2 grm. to 3 grm. of potassium iodide, and in the bulbs a dilute solution of potassium iodide. Before adding the hydrochloric acid a current of carbon dioxide is passed through the apparatus for some minutes. After adding the acid, the liquid is boiled in the current of carbon dioxide to a volume of 15 cm.^{3}, and when the free iodine has almost completely left the retort and passed into the receiver the apparatus is allowed to cool. The iodine in the receiver is titrated directly

* See Fig. 3, page 4.

with sodium thiosulphate, and that in the retort after dilution of the residue to about 400 cm.³, the amount in the retort seldom exceeding the equivalent of a few drops of $n/10$ iodine solution. Results are given in the table.

Distillation Process.

CeO_2 taken grm	CeO_2 found. grm.	Error. grm.
0 1028	0 1013	−0 0015
0 2060	0 2055	−0.0005
0 2014	0 2012	−0 0002
0 1716	0 1711	−0 0005
0.0974	0 0972	−0.0002
0.1600	0 1587	−0 0013
0.1268	0 1254	−0 0014
0.1276	0 1268	−0.0008
0.1620	0 1612	−0 0008
0 1016	0 1011	−0 0005
0.1548	0 1543	−0 0005
0 1352	0 1342	−0 0010

The Estimation of Cerium Oxalate by Potassium Permanganate.

Stolba* has stated that cerium oxalate may be estimated volumetrically in the same manner as calcium oxalate by treating the washed precipitate, suspended in warm water, to which a moderate amount of sulphuric acid has been added, with potassium permanganate; that as the titration proceeds the precipitate disappears and the end reaction is sharp; and that the permanganate does not oxidize the cerium from the lower to the higher condition.

Browning and Lynch† have presented experimental proof of the correctness of this statement, and have shown that the cerium may be determined either by titration of the precipitated cerium oxalate or by titration of the excess of ammonium oxalate left over in the precipitation.

To the cerium salt in solution in 100 cm.³ to 200 cm.³ of water is added a definite amount of a standardized solution of ammonium oxalate in excess; the whole is warmed to induce a crys-

* Sitzungsber. d. kgl. böhm. Gesellsch. d. Wissenschaften v. 4 Juli, 1879; Zeit. anal. Chem., xix, 194.

† Philip E. Browning and Leo A. Lynch, Am. Jour. Sci., [4], viii, 457.

talline condition of the precipitate. The precipitate is filtered off on paper, carefully washed, and dissolved by passing 10 $cm.^3$ of hot [1 : 3] sulphuric acid repeatedly through the filter. The filtrate and washings are made up to about 500 $cm.^3$ and warmed to 70° or 80°, and the oxalic acid in solution is titrated with permanganate. The filtrate from the cerium oxalate, containing the excess of ammonium oxalate, is diluted to 500 $cm.^3$, acidified with 10 $cm.^3$ of dilute [1 : 3] sulphuric acid, 1 grm. of manganous sulphate is added to insure regularity of action in presence of hydrochloric acid, and the oxalic acid is titrated by permanganate.

Results obtained in this manner are given in the table.

Permanganate Titration of Precipitate and of Excess of Precipitant.

Amount taken, calculated as $CeCl_3$.	Treatment of precipitate.		Treatment of filtrate.	
	Amount found, calculated as $CeCl_3$.	Error, calculated as $CeCl_3$.	Amount found, calculated as $CeCl_3$.	Error, calculated as $CeCl_3$.
grm.	grm	grm.	grm.	grm.
Precipitation in neutral solution.				
0.1091	0.1103	+0 0012	......	......
0.1091	0.1087	−0 0004	0.1087	−0.0004
0 1364	0.1373	+0 0009	0.1391	+0 0027
0.1364	0.1367	+0 0003	0.1367	+0.0003
0.2182	0.2202	+0.0020	0.2206	+0.0024
Precipitation in acid solution.				
0.1091	.. .	..... .	0 1087	−0.0004
0.1519	0.1535	+0 0016	0 1535	+0.0016
0.1364	0.1367	+0 0003	0 1367	+0.0003
0.2182	0.2183	+0.0001	0.2183	+0.0000

The Estimation of Cerium in the Presence of Other Rare Earths by the Action of Potassium Ferricyanide in Alkaline Solution and Potassium Permanganate in Acid Solution.

Browning and Palmer* have shown that the oxidation of cerium from the cerous to the ceric condition may be effected by potassium ferricyanide in alkaline solution, registered in the

* Philip E. Browning and Howard E. Palmer, Am. Jour. Sci., [4], xxvi, 83.

amount of potassium ferrocyanide formed according to the following equation,

$$2\,K_3FeC_6N_6 + Ce_2O_3 + 2\,KOH = 2\,K_4FeC_6N_6 + H_2O + 2\,CeO_2,$$

and subsequently determined by titration with potassium permanganate in acid solution.

The procedure is as follows: To the cerous sulphate in solution 20 cm.^{3} of a solution containing 20 grm. of potassium ferricyanide to the liter are added, and potassium hydroxide to complete precipitation. The precipitated hydroxide is filtered off, and the filtrate and washings, amounting in volume to from 200 cm.^{3} to 250 cm.^{3}, after being made distinctly acid with dilute sulphuric acid, are titrated with a standard solution of potassium permanganate until the presence of the permanganate color shows the oxidation of all the ferrocyanide to ferricyanide* according to the equation:

$$5\,K_4FeC_6N_6 + KMnO_4 + 4\,H_2SO_4 = 5\,K_3FeC_6N_6 + 3\,K_2SO_4 + MnSO_4 + 4\,H_2O.$$

From this equation and the preceding one the amount of cerium present may be readily calculated.

Each day before the ferricyanide is used a portion of 20 cm.^{3} of the solution is acidified and titrated with the permanganate, and the correction indicated, generally from one to three drops, is subtracted from the amount of the permanganate used in actual determinations.

Oxidation by Ferricyanide and Titration of the Reduced Ferricyanide.

Ce taken, calculated as Ce_2O_3. grm.	Ce found, calculated as Ce_2O_3. grm.	Error. grm.
0.1834	0.1819	−0.0015
0.1376	0.1380	+0 0004
0.1834	0.1829	−0.0005
0.1834	0.1829	−0 0005
0.1834	0.1834	0.0000
0.1376	0.1385	+0 0009
0.1376	0.1371	−0.0005
0.1376	0.1374	−0 0002
0.1376	0.1380	+0 0004
0 1834	0.1824	−0.0010
0.1326	0.1335	+0.0009
0.1326	0.1328	+0.0002

* Sutton's Vol. Anal. 9th ed., page 209.

Oxidation by Ferricyanide and Titration of the Reduced Ferricyanide.

Ce taken, calculated as Ce_2O_3. grm.	Ce found, calculated as Ce_2O_3. grm.	Error. grm	Other rare earths present, calculated as oxides. grm.
0 1328	0 1335	+0 0007	0 1 ThO_2.
0.1327	0 1322	−0 0005	0.1 ThO_2.
0 0266	0 0275	+0 0009	0.1 ThO_2.
0 0267	0 0272	+0 0005	0.1 ThO_2.
0 1324	0.1326	+0 0002	0.1 Y_2O_3.
0 1326	0.1323	−0.0003	0.1 Y_2O_3.
0 0266	0.0264	−0 0002	0.1 Y_2O_3.
0 0264	0.0271	+0 0005	0.1 Y_2O_3.
0 1376	0.1370	−0 0006	0.15 $La_2O_3+Di_2O_3$.
0.1101	0.1091	−0.0010	0.15 $La_2O_3+Di_2O_3$.
0.1324	0 1332	+0.0008	0 03 ZrO_2.

All the various operations in this process are carried on without warming the solution. The filtrations and washings are made under gentle pressure, and require on an average not more than fifteen to thirty minutes. In the preceding tables are results obtained with cerium alone and in presence of salts of other rare earths.

This method presents no difficulties in manipulation and is especially adapted to the rapid estimation of cerium in rare earth mixtures.

TIN.

The Electrolytic Determination of Tin.

From a solution of stannous ammonium chloride in a saturated solution of ammonium oxalate,* Medway has precipitated the tin successfully upon the rotating crucible.†

Deposition of Tin on the Rotating Cathode.

Tin taken. grm.	Tin found. grm.	Error. grm.	Current. Amp.	N. $D._{100}$	Time. min.
0.0804	0.0802	−0 0002	2.5	8 3	20
0.0804	0 0800	−0 0004	2	6 6	20
0.1607	0.1610	+0 0003	2.5	8 3	20
0 1607	0.1603	−0 0004	2.5	8.3	20
0.1607	0.1607	0 0000	3.5	11 6	15

* H. E. Medway, Am. Jour. Sci., [4], xviii, 56.

† See Fig. 13, page 12.

Lead.

The Detection of Lead.

It has been shown by Browning and Blumenthal* that lead may be separated as the sulphate, in association with the alkali earth sulphates, and tested for in the solution obtained by treatment of these insoluble sulphates with ammonium acetate.†

The Electrolytic Determination of Lead as the Dioxide.

In depositing lead dioxide electrolytically, solutions containing nitric acid are employed; precautions must be taken in regard to concentration of acid, strength of current and temperature; and the liquid is siphoned off before interruption of the current.‡ With the rotating cathode making 600 revolutions a minute and a sand-blasted platinum dish for the anode, Exner§ obtained in ten to fifteen minutes adherent deposits with a current $N. D._{100} = 10$ amp. and 4.5 volts acting upon 125 $cm.^3$ of solution containing 20 $cm.^3$ of concentrated nitric acid.

To obviate the necessity of large and expensive apparatus of platinum, Gooch and Beyer ‖ have experimented with the filtering crucible used as a cathode in the devices previously described.¶

In preliminary trials of electrolysis in the closed cell with subsequent filtration** it was found that when the concentration of nitric acid amounted to 10 $cm.^3$ in 60 $cm.^3$ of liquid, with a current of 4 amperes ($N. D._{100} = 10$ amp.) and 6 volts, two sources of error appeared. In the first place, the deposition of metallic lead upon the cathode was often noticeable; and secondly, it appeared to be impossible to make the precipitation of lead dioxide complete so long as that substance was allowed to float in the liquid. Similar results were obtained in experiments in which urea was added to the liquid for the purpose of obviating the solvent action

* Philip E. Browning and Philip L. Blumenthal, Am. Jour. Sci., [4], xxxii, 246.

† See page 442.

‡ Smith, Electro-analysis, page 105, edition of 1911.

§ Jour. Am. Chem. Soc., xxv, 904.

‖ F. A. Gooch and F. B. Beyer, Am. Jour. Sci., [4], xxvii, 59.

¶ See pages 13 to 20.

** See Fig. 15, page 15.

of dissolved oxides of nitrogen upon lead dioxide. In the experiments with this form of apparatus, the stirring of the asbestos felt by gas evolved upon the bottom of the crucible used as an anode, as well as the deposition of oxide on the outer surface of the crucible, was prevented by taking the precaution to moisten the asbestos, from the outside, with a drop of nitrobenzene which, being insoluble in water, prevents the contact of the aqueous electrolyte with the electrode surface underneath the asbestos. An increase of nitric acid to the proportion of 30 cm.^{3} in 100 cm.^{3} of solution served to prevent the deposition of lead upon the cathode, but to prevent the re-solution of lead dioxide it was found to be necessary to use the process of continuous filtration, so that the deposit might be compacted upon the felt, and after deposition was complete to replace the acid liquid by a solution of ammonium nitrate without interruption of the current. After washing out the nitric acid with the solution of ammonium nitrate, the final washing was completed with water. The form of apparatus employed, shown in Figure 16, and the manner of using are described on page 17. In the table are given the results of experiments following this procedure, and, for comparison, the result of an experiment in which it was found that, though electrolysis was continued by the circulating process until the filtrate contained no lead, traces of lead dioxide went into solution after the current had been diminished by the gradual dilution with water used to replace the electrolyte. Tests for lead in filtrates and washings were made by neutralizing with ammonium hydroxide and adding ammonium sulphide, or acetic acid and potassium chromate.

From the results of the experiments described, it appears that good analytical results in the deposition of lead dioxide may be obtained with the filtering crucible used as an electrolytic cell if nitric acid be present to the proportion of 30 cm.^{3} of the concentrated acid in 100 cm.^{3} of solution, the liquid kept in continuous filtration until the electrolysis of the lead salt is complete, the acidic liquid replaced by a solution of ammonium nitrate so that the electric current passing shall not fall off until the nitric acid has been removed, the final washings made with water, and the deposit weighed after drying at 200°. The time required for the complete deposition of 0.15 grm. of lead dioxide under the conditions described is about two hours.

Electrolysis with Continuous Filtration.

$Pb(NO_3)_2$.	Volume.	HNO_3 conc.	Current.			Time.	PbO_2 found.	Theory for PbO_2.	Error.
			Amp.	$N.D._{100}$.	Volt.				
grm.	cm.³	cm.³				min.	grm.	grm.	grm.
A. With no ammonium nitrate in electrolyte or in wash-water.									
0.2023	50	15	2	5	4	5			
			4	10	5	130	0.1460	0.1436	−0 0024
B. With ammonium nitrate in electrolyte and in wash-water.									
0 2022	50	15	2	5	4	40			
			4	10	5	100	0 1459	0.1462	+0 0003
0 2014	50	15	2	5	4	5			
			4	10	5	115	0 1454	0 1458	+0 0004
0.2001	50	15	2	5	4	5			
			4	10	5	115	0 1444	0.1442	−0 0002
0.2006	50	15	2	5	4	5			
			4	10	5	115	0.1448	0 1446	−0 0002
0.2046	50	15	2	5	4	5			
			4	10	5	115	0.1477	0.1472	−0 0003
C. With ammonium nitrate in wash-water only.									
0.2020	50	15	2	5	4	5			
			4	10	5	115	0.1458	0.1460	+0 0003
0.2037	50	15	2	5	4	5			
		4	4	10	5	115	0.1470	0 1473	+0 0003

The Estimation of Lead by Precipitation as Oxalate and Titration with Potassium Permanganate.

Many investigators have made use of precipitation as the oxalate for the estimation of lead. Ward* has shown that the addition of considerable amounts of acetic acid favors completeness of precipitation, whether ammonium oxalate or oxalic acid is used as the precipitant. Precipitation is effected in the boiling solution, the precipitated oxalate is collected on asbestos in the perforated crucible and washed with small amounts of water. The oxalic acid is set free by treatment of the washed precipitate with warm dilute sulphuric acid and titrated with permanganate. When the acetic acid present does not exceed one-fourth of the volume, precipitation is not quite complete, but if half the volume at precipitation is made up of glacial acetic acid, even

* H. L. Ward, Am. Jour. Sci., [4], xxxiii, 334.

in presence of moderate amounts of acetates, the results of titration are accurate. The details of the preferred treatment are given in the table.

Precipitation and Titration of Lead Oxalate.

Lead taken as nitrate. grm.	Volume at precipitation. cm.³	Acetic acid. cm.³	Ammonium oxalate. grm.	Lead found. grm.	Error. grm.
Precipitation by ammonium oxalate.					
0.0050	100	50	4	0.0048	−0 0002
0.0050	100	50	4	0.0045	−0.0005
0 0250	100	50	4	0.0256	+0.0006
0 0250	100	50	4	0.0250	0.0000
0 0500	100	50	4	0.0505	+0 0005
0.1000	200	100	8	0.1002	+0.0002
Precipitation by oxalic acid.					
0.0050	50	25	1	0.0050	0 0000
0.0250	50	25	1	0 0256	+0 0006
0.1000	100	50	2	0 1002	+0.0002
In presence of 2 grm. of ammonium acetate or potassium acetate.					
0.1000	100	50	2	0 1000	0 0000
0.1000	100	50	2	0.0997	−0 0003
0.1000	100	50	2	0 1000	0 0000

CHAPTER VIII.

NITROGEN; PHOSPHORUS; ARSENIC; ANTIMONY; BISMUTH; VANADIUM.

NITROGEN.

The Determination of Nitrogen Liberated by Action of Sodium Hypobromite upon Ammonia Compounds and Derivatives.

THE apparatus designed by Kreider* for the determination of volatile products by loss is well suited to the determination of the nitrogen liberated from urea, ammonium oxalate, ammonium chloride, etc., by the action of sodium hypobromite. Results obtained by the use of this apparatus, described and figured elsewhere,† are given in the accompanying table.

Determination of Nitrogen.

	Taken. grm.	Found. grm.	Error. grm.
Urea	0.1000	0.0469	+0.0003
	0.1000	0.0467	+0.0001
	0.1000	0.0467	+0.0001
	0.1000	0.0468	+0 0002
	0.1000	0.0467	+0.0001
Ammonium oxalate	0.1000	0.0204	+0.0007
	0.1000	0.0197	0.0000
	0.1000	0.0198	+0.0001
	0 1000	0.0198	+0.0001
	0.1000	0.0196	−0.0001
Ammonium chloride	0.1000	0 0264	+0.0002
	0.1000	0 0265	+0.0003
	0.1000	0.0261	−0.0001
	0.1000	0.0263	+0.0001
	0.1000	0.0261	−0.0001

The Estimation of Nitrates by Expulsion of Nitrogen Pentoxide on Ignition.

For the determination of the nitrogen pentoxide combined in nitrates which leave definite oxides on ignition, and estimation of the containing nitrates, various fluxes have been used to aid

* J. Lehn Kreider, Am. Jour. Sci., [4], xix, 188.

† See page 1.

in the expulsion of the volatile oxide and to conserve the residual oxide in definite form for weighing. Borax, silicon dioxide, potassium dichromate and sodium metaphosphate, which have been employed thus, as well as in the similar determination of carbon dioxide in carbonates* — all present certain disadvantages. In sodium paratungstate of composition corresponding approximately to the formula $5Na_2O.12WO_3$, or $Na_{10}W_{12}O_{41}$, Gooch and Kuzirian† find a material very easily prepared, stable in fusion, and well suited for use as a flux in the rapid determination of nitrates by loss on ignition. The sodium paratungstate is prepared by dehydrating and fusing over the blast lamp a known weight of normal sodium tungstate, $Na_2WO_4.2H_2O$, adding an equal weight of tungsten trioxide, WO_3 (previously ignited with care to remove all ammonia and to insure complete oxidation), and heating to clear fusion. The cooled mass, which is very easily pulverized, is ground and bottled. From this material, kept over sulphuric acid (though not more than ordinarily hygroscopic), portions are weighed for the analytical determinations. Approximately half the weight of the paratungstate is tungsten trioxide (molecular weight 232), and this should be capable of expelling nitrogen pentoxide (molecular weight 108.02) to an amount one-half its own weight. The weights of paratungstate to be used are approximately four times the weight of nitrogen pentoxide to be expelled. It is a good practice to weigh a plati-

Analysis of C. P. Nitrates of Commerce, after Drying.

Nitrate taken. grm.	$Na_{10}W_{12}O_{41}$ taken. grm.	Loss on ignition. grm.	Theory for N_2O_5 grm.	Error. grm.
KNO_3.				
0.5000	1 5	0 2668	0 2670	−0 0002
0.5000	1.5	0 2678	0 2670	+0 0008
0.5000	1.5	0 2674	0 2670	+0.0004
0 5000	1.5	0 2672	0 2670	+0 0002
0.5000	1.5	0 2675	0 2670	+0 0005
$Sr(NO_3)_2$				
0.5000	2	0 2544	0 2543	+0 0001
0 5000	3	0 2546	0 2543	+0 0003
$Ba(NO_3)_2$.				
0 5000	3	0 2073	0 2067	+0.0006
0 5000	3	0 2076	0 2067	+0 0009

* See page 226.

† F. A. Gooch and S. B. Kuzirian, Am. Jour. Sci., [4], xxxi, 497.

num crucible, introduce the dried nitrate and weigh again, add a suitable amount of the prepared sodium paratungstate, stir carefully with a platinum wire with care to avoid mechanical loss, and weigh again. The crucible is then heated over a Bunsen burner, first at very low heat and then to fusion of the mixture for five minutes, cooled in a desiccator over sulphuric acid, weighed, and reignited to test the constancy of weight. The constant weight is usually got in the first ignition. In the table are given the results of the estimation of the nitrogen pentoxide in nitrates according to the procedure described.

The Estimation of Nitrates by Reduction with a Ferrous Salt and Titration of the Residual Unoxidized Salt.

In the methods for the quantitative estimation of nitrates which depend upon the reduction in presence of acid by a ferrous salt and the determination of the amount of oxidation produced, scrupulous care is necessary that the atmosphere in contact with the ferrous salt while the nitrogen dioxide is present shall be free from oxygen. This fact was recognized by Fresenius,* who modified the original process of Pelouze† by filling the flask with carbon dioxide or hydrogen at the outset. Eder‡ used carbon dioxide similarly. Holland's method,§ roughly described, consists in boiling, until the air is expelled, the solution of the nitrate in a flask provided with a doubly bent exit tube, rubber-jointed and fitted with a pinchcock; then admitting through the tube as the flask cools a mixture of ferrous salt and strong hydrochloric acid, heating the mixture on a water bath, and, finally, titrating the resulting ferric salt with stannous chloride.

All of these methods give high results, either on account of the oxygen invariably present in carbon dioxide as produced in the laboratory, or because of slow leakage through the rubber connections during the long heating, or because the nitrogen dioxide is not driven out completely from the solution of the iron salts, and acts with atmospheric oxygen to oxidize the ferrous salt during the titration.

* Zeit. anal. Chem., i, 32.
† Ann. Chim., [3], xx, 120.
‡ Zeit. anal. Chem., xvi, 267.
§ Chem. News, xvii, 219.

Phelps* has shown, however, that the oxidation of ferrous sulphate by a nitrate in presence of fairly strong hydrochloric acid may be accomplished quantitatively by the aid of the apparatus to be described, the standard of the solution of the ferrous salt and the amount remaining after the action being determined either iodometrically or by titration with potassium permanganate.

This apparatus consists of a 250-cm.^{3} boiling flask, closed with a rubber stopper carrying a stoppered funnel of 50 cm.^{3} capacity to serve as an inlet tube and a glass tube of 0.8 cm. bore to serve as an outlet tube. The inlet tube is constricted at the lower end, and the outlet tube is enlarged just above the stopper to a small bulb (to prevent mechanical loss during the boiling) and bent twice at right angles. The flask is supported above a Bunsen burner and the outlet tube dips under mercury contained in a test tube.

The nitrate to be analyzed is introduced with water into the flask, the stem of the separating funnel being left full of water, the outlet tube is adjusted to just touch the surface of the mercury in the trap, and air is expelled from the flask by boiling the solution to small volume. An amount of standardized ferrous sulphate solution known to be in excess is introduced into the separating funnel, the outlet tube plunged a centimeter or two deep into the mercury (which is readily accomplished by changing the position of the flask on the wire gauze, provided that the gauze is depressed well at the center and the flask is set well up on the higher part at the beginning of the operation), and then the flame withdrawn until diminution of pressure sufficient to draw the ferrous solution into the flask is made evident by the rise of the mercury in the outlet tube. By applying and withdrawing the flame and by regulating the rate of inflow of the solution, the ferrous salt may be introduced without admitting air, and the funnel washed carefully with an amount of concentrated hydrochloric acid nearly enough to equal the total volume of the liquid in the flask. After the pressure has been restored in the apparatus by heating the flask, the exit tube is again raised to the surface of the mercury and the solution in the flask boiled to a volume of 10 cm.^{3} to 15 cm.^{3}. The excess of acid is then nearly neutralized by introducing sodium carbonate

* I. K. Phelps, Am. Jour. Sci., [4], xiv, 440.

in solution, the carbon dioxide evolved assisting in maintaining the pressure in the apparatus so that the condensed liquid in the test tube which may contain oxidized nitrogen dioxide shall not be returned to the iron solution. The flask is cooled and the ferrous salt remaining determined by titration with potassium permanganate after dilution with 600 cm.^{3} of water and addition of 2 grm. to 3 grm. of crystallized manganous chloride, or iodometrically by first introducing Rochelle salt (3 grm.) in solution and then neutralizing with acid potassium carbonate, after which are added in succession a saturated solution of acid potassium carbonate, iodine, starch paste, standard arsenious acid solution to the bleaching of the starch blue, and, finally, iodine to coloration.

It was shown experimentally that prolonged boiling after the dark compound of nitrogen dioxide with the ferrous salt is broken up is essential and that ammonium salts must be absent if the highest accuracy is desired. Experimental tests of the method as outlined are given in the table.

Reduction of Nitrate by Ferrous Sulphate: Titration of Excess.

KNO_3 taken. grm.	Oxygen value of ferrous salt taken. grm.	Oxygen value of ferrous salt found. grm.	Error on oxygen. grm.	KNO_3 found. grm.	Error on KNO_3. grm.
0.0500	0 01823	0 00621	+0 00015	0 0506	+0 0006
0.0500	0 01865	0 00681	−0.00003	0 0499	−0.0001
0.0500	0 01954	0.00768	0.00000	0 0500	0 0000
0 1000	0 02881	0 00507	+0.00001	0.1000	0 0000
0.1000	0.02822	0.00441	+0 00008	0 1003	+0 0003
0 2500	0.06453	0.00512	+0 00009	0 2503	+0 0003
0 5000	0.13394	0.01524	+0.00005	0.5002	+0.0002
0.5000	0.12210	0.00340	+0 00005	0 5002	+0.0002

The Estimation of Nitrates by Reduction with Ferrous Chloride and Measurement of the Nitrogen Dioxide Evolved.

In consequence of the fact that analytical methods in which nitrates are estimated by the amount of nitrogen dioxide evolved in the reaction with ferrous salts in presence of acid give unexpectedly low results, the form of apparatus to be used in such processes and the conditions affecting accuracy have been made the object of study by Roberts.* As the result of such

* Charlotte F. Roberts, Am. Jour. Sci., [3], xlvi, 126.

study, it is pointed out that to insure rapid action, the hydrochloric acid used should be fairly strong, and that in the measurement of the volume of nitrogen dioxide (NO), swept along by carbon dioxide and collected over sodium hydroxide, the best analytical results are obtained when the gas is passed through a solution of potassium iodide to break up, before the measurement is made, any higher, soluble oxides of nitrogen which may have been formed in the process, notably when the ferrous salt is present in only small excess and when the reaction of decomposition takes place in the hot solution. Note is made of the fact that the action of traces of intermixed air upon the nitrogen dioxide is to form nitrogen trioxide which is soluble, and that the nitrogen of the air is added in exactly the proportion according to which the nitrogen dioxide is removed with the disappearing oxygen. When, however, the gas carrying nitrogen trioxide meets with potassium iodide before measurement, the original amount of nitrogen dioxide is regenerated. Nevertheless, experience shows that the danger of error introduced by the use of potassium iodide when traces of air are present is small in comparison with the danger of error due to the presence of higher oxides of nitrogen produced in the main reaction. Moreover, there is always a small counterbalancing error due to the solubility of nitrogen dioxide in the solution of sodium hydroxide over which it is measured.

The apparatus found to be most satisfactory for this work consists of a small tubulated retort, upon the neck of which is fitted a small condenser to prevent loss of liquid during the distillation. Into the tubulature of this retort is fitted tightly, by a carefully ground joint, a tube drawn out so as to dip below the surface of the liquid, and fitted with carefully ground stopcocks, — as shown in the figure, — and so branched above as to make it possible to transmit carbon dioxide through the apparatus, or to admit any liquid without introducing air. The condenser is joined to a Will and Varrentrapp bulb containing a solution of potassium iodide as a trap, and this in turn is connected by thick vacuum tubing with a Hempel gas burette charged with a strong solution of sodium hydroxide. Carbon dioxide is generated in a Kipp's apparatus by action of boiled hydrochloric acid upon boiled marble, and the liquid is charged with cuprous chloride, following Warrington's device, to take up traces of dissolved oxygen. Notwithstanding all precautions, however, the gas from the generator is never so pure that a hundred cubic centimeters of it will not leave a tiny bubble when shaken with a solution of caustic soda.

In using this apparatus, the nitrate (about 0.1 grm. of potassium nitrate) is introduced into the retort, generally in the dry condition, carbon dioxide is passed through the apparatus until the gas collected over sodium hydroxide leaves only the minute bubble which the gas from the generator alone has been found to

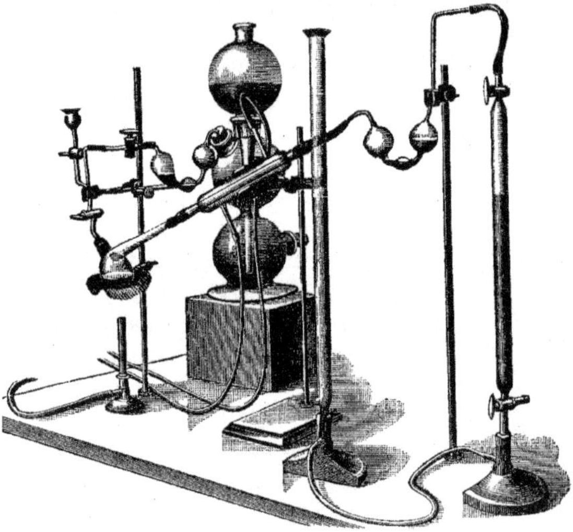

Fig. 21.

give, and 40 cubic centimeters of a boiled solution of ferrous chloride in hydrochloric acid are admitted through the funnel tube, after shutting off the carbon dioxide and lowering the leveling tube of the Hempel burette. With the stopcocks arranged as in sketch, the liquid is then slowly heated to boiling and the process continued until the reaction of the ferrous salt upon the nitrate is apparently complete, when the carbon dioxide is again passed through the apparatus to secure complete removal of the nitrogen dioxide, the absorption of the carbon dioxide being hastened by inclining and shaking the burette at intervals. The volume of the gas under existing barometric and thermometric conditions is noted and from this the weight of the nitrate may be calculated. Results obtained in this manner are given in the table.

Reduction by Ferrous Chloride: Measurement of Nitrogen Dioxide.

KNO_3 taken. grm.	KNO_3 found. grm.	Error. grm.
0.1000	0.0990	−0.0010
0.1000	0.1005	+0.0005
0.1000	0.0992	−0.0008
0.1000	0 0994	−0.0006
0.1000	0.1008	+0.0008
0.1000	0.0989	−0.0011

The Iodometric Determination of Nitrates.

Action of Manganous Chloride in Hydrochloric Acid.

Noting that a saturated solution of manganous chloride in concentrated hydrochloric acid acts like the solution of ferrous chloride in hydrochloric acid in inducing the easy decomposition of nitrates, with the difference, however, that all products of oxidation may be distilled, while the metal chloride reverts to its original form, Gooch and Gruener* have applied this reagent to the quantitative estimation of nitric acid as well as in the qualitative test.

The solution of manganous chloride in hydrochloric acid acts but slowly upon nitrates at the ordinary temperature, but upon warming the decomposition of the nitrate begins at once with the formation of a higher chloride of manganese and liberation of nitrogen dioxide. Ultimately, if heating is continued, chlorine of the higher chloride is evolved and manganous chloride remains. During the process of heating the color of the solution passes from the original characteristic green through darker shades to black, and returns by the reverse changes to the original tint. The decomposition of the nitrate extends under the conditions to the last traces, but the breaking up of the nitrates, with the formation of the higher chloride, does not take place completely in the presence of water amounting to more than a half of the volume of strong acid, and an action already established in strong acid is reversed by the addition of a large amount of water. Chlorates, peroxides and other substances which liberate oxygen

* F. A. Gooch and H. W. Gruener, Am. Jour. Sci., [3], xliv, 117.

or chlorine when in contact with strong hydrochloric acid induce similar phenomena, but in the absence of such other substances the reaction serves to detect nitrates when present in fairly small amounts (perhaps one part in sixty thousand), as shown in the accompanying table:

KNO_3 taken. grm.	$MnCl_2.4H_2O$ in strong HCl. cm.^{3}	Color developed.
0 01000	10	Black.
0.00500	5	Black.
0 00100	5	Dark brown.
0.00050	5	Dark green.
0.00025	5	Deepened tint.
0 00015	5	Deepened tint.
0 00005	5	None.
0 00000	5	None.

In applying this reaction to the quantitative estimation of nitrates, the nitrate to be estimated is treated, in an atmosphere of carbon dioxide, with a saturated solution of crystallized

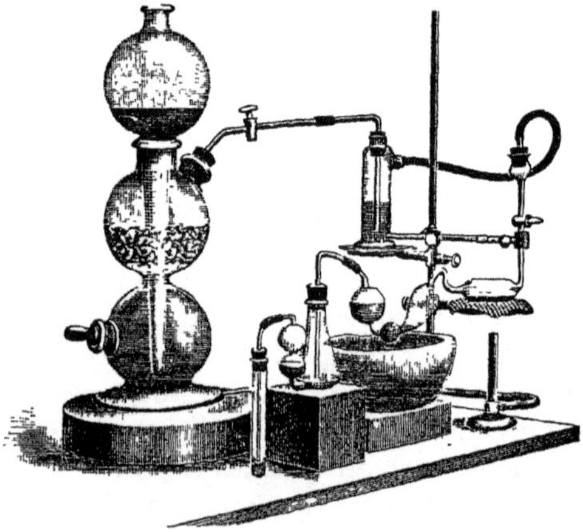

Fig. 22.

manganous chloride in concentrated hydrochloric acid, the volatile products of action—chlorine, nitrogen dioxide and perhaps nitrosyl chloride—are passed into a solution of potassium iodide, and the iodine set free is titrated by sodium thiosulphate. The operation is conducted in an apparatus made wholly of glass where, by any possibility, rubber connections might be acted upon. The retort used was a pipette bent and fitted as shown in Fig. 22. To the retort are sealed Will and Varrentrapp nitrogen bulbs, the outlet tube of which is drawn out so that it may be pushed well within the inlet tube of the second receiver — a Will and Varrentrapp absorption flask — and held in place by an outside rubber connector. The third receiver acts simply as a trap to exclude air from the absorption apparatus proper. In conducting the experiment the receivers were charged with solutions of potassium iodide, the first containing three grams, the second one gram, and the third only a fraction of a gram for every tenth of a gram of nitrate used. The first receiver was cooled in water during the subsequent process of distillation.

Decomposition of Nitrate by Hydrochloric Acid and Manganese Chloride: Titration of Iodine Set Free by Volatile Products.

KNO_3 taken. grm.	$MnCl_2$ mixture. cm.^{3}	KNO_3 found. grm.	Error in terms of KNO_3. grm.	Error in terms of HNO_3. grm.
0.2038	20	0.2047	+0.0009	+0.0005
0.2053	20	0.2057	+0.0004	+0.0003
0.1032	10	0.1035	+0.0003	+0.0002
0.1017	10	0.1004	−0.0013	−0.0008
0.1049	10	0.1049	0.0000	0.0000
0.1027	10	0.1023	−0.0004	−0.0003
0.0524	10	0.0526	+0.0002	+0.0001
0.0513	10	0.0512	−0.0001	−0.0001
0.0354	10	0.0350	−0.0004	−0.0003
0.0232	10	0.0230	−0.0002	−0.0001
0.0107	5	0.0106	−0.0004	−0.0001
0.0127	5	0.0130	+0.0003	+0.0002
0.0145	5	0.0143	−0.0002	−0.0001
0.0053	5	0.0052	−0.0001	−0.0001
0.0043	5	0.0047	+0.0004	+0.0003
0.0014	5	0.0018	+0.0004	+0.0003
0.0000	5	0.0000	0.0000	0.0000

The nitrate and the manganous mixture following it are introduced by applying gentle suction to the end of the absorption train. The current of carbon dioxide is started immediately after putting in the manganous mixture. After a suitable time has elapsed for the removal of air, heat is applied to the retort and the distillation is continued until nearly all the liquid has passed over. Finally, the contents of the receivers are united, the washing of the bulbs was effected easily and expeditiously by passing the wash-water directly through retort and receiver, the introduction of the manganese chloride into the distillate being not at all prejudicial to the accuracy of the titration. The estimation of free iodine is made by titration with sodium thiosulphate as soon as may be after admitting air to the distillate, in order that traces of dissolved nitric oxide may not be reoxidized and again react upon the iodide present to liberate more iodine. The results of the experiments conducted in this manner are given in the table.

Distillation with Phosphoric Acid and Potassium Iodide and Determination of Iodine in the Distillate.

Various attempts to utilize the reduction of arsenic acid brought about upon heating a mixture of standard potassium iodide, potassium arsenate and sulphuric acid with the nitrate, as in the similar process for estimating chlorates,* have proved to be futile. The decomposition of the last traces of nitrates by the action of potassium iodide and sulphuric acid does not occur except at concentrations so great that the sulphuric acid itself liberates iodine from iodides. Gruener† has shown, however, that when phosphoric acid is substituted for sulphuric acid the iodine evolved in the distillation of such a mixture may be taken as the measure of nitrates present in small amounts, provided the concentration of the residue is not carried so far as to bring about reduction of nitric acid to ammonia‡ nor the temperature so raised by removal of water and elevation of the boiling point of the phosphoric acid as to cause dissociation of hydriodic acid.

In Gruener's experiments a small retort was used, the neck of which was bent downward about two inches from the body, so that the retort itself might be tipped backward, allowing the unbent portion of the retort to run upward, thus guarding against loss from spattering. Into the tubulature of the retort was ground a glass tube drawn out at both ends to serve as a perforated stopper for the entrance of carbon dioxide. The neck was passed through a rubber stopper into a side-neck Erlenmeyer flask, the exit tube of which was prolonged and dropped into a side-neck test tube used as a trap. The retort was covered with a

* See page 463.

† Hippolyte Gruener, Am. Jour. Sci., [3], xlvi, 42.

‡ Chapman, Jour. Chem. Soc., xx, 166 (1867).

simply contrived hood which kept the upper parts warm and prevented the iodine from settling anywhere. In the retort was placed the nitrate with an excess of potassium iodide, and in the receiver a known amount of decinormal solution of arsenious oxide strongly alkaline with hydrogen sodium carbonate and diluted to a convenient bulk. The trap contained nothing but water.

The method, so far as it is applicable, may be summed up as follows: The nitrate, not to exceed the equivalent of 0.05 grm. of potassium nitrate, is introduced into the retort, with ten times its weight of potassium iodide, and 17 cm.^{3} to 20 cm.^{3} of phosphoric acid, of specific gravity 1.43. All water used should be recently boiled. Carbon dioxide is passed from a generator set up with materials carefully boiled and containing cuprous chloride to take up the oxygen from any traces of air. The neck of the retort passes into a receiver containing a known amount of decinormal arsenious oxide, alkaline with a good excess of hydrogen sodium carbonate and diluted to a convenient bulk. To this flask is attached for additional safety a simple trap containing water. The solution in the retort is boiled until it is clear that no more iodine remains, when the receiver, after proper washing and addition of the liquid in the trap, is titrated with iodine to find the amount of arsenious oxide still left. This gives the measure of the iodine evolved and consequently of the nitrate present, according to the equation:

$$2\,HNO_3 + 6\,HI = 4\,H_2O + 2\,NO + 3\,I_2.$$

The details of test determinations are given in the table:

Decomposition by Phosphoric Acid and Iodide: Estimation of Iodine Set Free.

KNO_3 taken. grm.	KI taken. grm.	Found. grm.	Specific gravity of solution of phosphoric acid.	Amount of solution used. cm.^{3}	Error, KNO_3. grm.	Error, HNO_3. grm.
0 0500	1	0.0500	1.43	17	0.0000	0 0000
0 0200	0.5	0 0201	1.43	17	+0.0001	+0 0001
0.0200	1	0.0198	1.43	17	−0 0002	−0 0001
0 0250	1	0.0250	1.43	17	0.0000	0 0000
0 0300	1	0.0307	1 43	17	+0.0007	+0 0004
0.0300	1	0.0312	1.43	17	+0.0012	+0 0007
0.0350	1	0.0353	1.43	17	+0 0003	+0 0002
0 0400	1	0 0409	1.35	20	+0.0009	+0 0006
0 0450	1	0.0444	1.35	20	−0 0006	−0.0004
0 0500	1	0 0499	1.37	20	−0.0001	−0 0001

The process is good for estimating nitrates in quantities not exceeding the equivalent of 0.04 grm. or 0.05 grm. of potassium nitrate. With quantities of nitrate above 0.05 grm. it is not safe, inasmuch as with a moderate amount of water present some nitric acid distils over undecomposed and with little water present other complications arise.

Decomposition by Antimony Trichloride: Determination of Oxidation in Residue and of Iodine in Distillate.

To register the action of nitrates, Gruener* tried the effect of antimony trichloride in hydrochloric acid and showed that the reaction proceeds mainly according to the equation,

$$3SbCl_3 + 2HNO_3 + 6HCl = 3SbCl_5 + 2NO + 5H_2O,$$

and in smaller degree according to the equation,

$$SbCl_5 + 2NO = SbCl_3 + 2NOCl.$$

Nitrosyl chloride fails to oxidize arsenious oxide in alkaline solution, breaking up hydrolytically into hydrochloric acid and nitrous acid, but from acidulated potassium iodide out of contact with air it sets free quantitatively an amount of iodine corresponding to the chlorine. These reactions may, therefore, be applied together to the estimation of the nitrate, by noting both the iodine evolved in the action of nitrosyl chloride upon potassium iodide in the distillate and the degree of oxidation of the previously standardized antimony salt in the residue. The procedure is as follows:

Into a diminutive retort — made from a pipette, shaped like a Liebig's drier† and connected by a sliding joint covered by rubber with a Kjeldahl tube used as a receiver, and so placed that carbon dioxide passing through the apparatus shall enter from below — the dry nitrate is introduced, and washed down with a few drops of recently boiled water, or, if more liquid is required, with hydrochloric acid. From a burette a definite amount of antimonious chloride solution, somewhat in excess of the nitrate taken, is added.

The receiver is charged with potassium iodide in recently boiled water and is joined to a trap filled with water. After carbon dioxide has been passed through the apparatus for about ten minutes, the solution is warmed in a bath at even temperature (103°–107°) to insure the safety of the retort, to keep the

* Am. Jour. Sci., [3], xlvi, 47.

† See Fig. 5, page 5.

antimony pentachloride from breaking up, to retain the bulk of the acid in the retort, and to prevent mechanical loss. After fifteen minutes' digestion the contents of the receiver and trap are washed out and at once titrated with sodium thiosulphate or neutralized and titrated with standard arsenite. The residue in the retort is treated exactly as was the antimonious chloride when it was standardized, viz., by dissolving in hydrochloric acid, adding tartaric acid, diluting, nearly neutralizing with sodium hydroxide with careful cooling to prevent action of the tartaric acid upon antimony pentachloride, treating with an excess of acid sodium carbonate, and titrating with decinormal iodine in presence of starch. Below are given the results of experimental tests.

Decomposition by Antimony Trichloride: Determination of Oxidation in Residue and of Iodine in Distillate.

KNO_3 taken. grm.	KNO_3 from $SbCl_5$ in residue. grm.	KNO_3 from I in receiver. grm.	Entire KNO_3 found. grm.	Error in KNO_3. grm.	Error in HNO_3. grm.
0.0222	0 0213	0 0020*	0 0233	+0 0011	+0.0007
0.0336	0 0307	0 0026*	0 0333	−0.0003	−0.0002
0 0470	0.0436	0 0045*	0 0471	+0 0001	+0.0001
0.0553	0.0497	0.0057*	0 0554	+0 0001	+0.0001
0.0664	0.0673	0 0076*	0 0679	+0 0015	+0.0009
0 0759	0.0670	0.0082*	0 0752	−0 0007	−0.0004
0 0837	0.0739	0 0103*	0.0841	+0 0004	+0 0002
0.0934	0.0842	0.0113*	0.0955	+0.0021	+0.0013
0.1034	0 0902	0.0134*	0.1036	+0.0002	+0.0001
0 0262	0.0235	0.0024*	0.0259	−0 0003	−0.0002
0.0127	0.0123	0.0007*	0.0130	+0 0003	+0.0002
0 0065	0.0064	0 0003*	0.0067	+0.0002	+0.0001
0.0026	0.0022	0 0001*	0.0023	−0.0003	−0.0002
0.1232	0.1129	0 0098*	0.1227	−0 0005	−0 0003
0.1540	0.1394	0.0146*	0.1540	0 0000	0.0000
0.1878	0.1655	0.0210*	0.1865	−0.0013	−0.0008
0.0530	0.0481	0.0052†	0 0533	+0.0003	+0.0002
0.0547	0.0484	0 0065†	0.0549	+0 0002	+0.0001
0.0541	0.0474	0.0063†	0 0537	−0 0004	−0.0002

* Found by thiosulphate. † Found by arsenite after neutralization.

The Iodometric Determination of Nitrites.

The apparatus (consisting of a boiling flask fitted with a stopper which carries a stoppered funnel and outlet tube dipping in a mercury trap) previously used by Phelps* for the determination of nitric acid has been applied by him in the determination

* See page 259.

of nitrites.* In this method the nitrite is reduced by the action of potassium iodide and arsenious acid in acid solution and measured by titration of the arsenite remaining, after neutralization.

An amount of standard arsenite solution, slightly in excess of that required to take up the iodine to be set free later by the nitrous acid, and 25 cm.^{3} of a concentrated solution of sodium carbonate, are placed in the flask. The stem of the stoppered funnel is completely filled with water, the rubber stopper inserted tightly and the contents of the flask boiled until all air is expelled, a process requiring an active boiling of 5–8 minutes. The flame is then removed, the outlet tube is plunged deep into the mercury, the flask is cooled with ice water, and enough sulphuric acid [1 : 3] (7 cm.^{3}) is sucked in through the funnel tube to nearly decompose the sodium carbonate previously added and liberate carbon dioxide to balance the atmospheric pressure. When the inward and outward pressures have been equalized the outlet tube is raised so that the end shall dip in the water layer condensed above the mercury in the trap, the acid on the walls of the funnel and in the tube is washed into the flask, and the nitrite solution to be analyzed is run in through the funnel with 2 grm. of potassium iodide. Sulphuric acid [1 : 3] is then added in amount (5 cm.^{3}) sufficient to acidify the contents of the flask, and potassium carbonate is then added, in solution, to alkalinity or until free iodine has been taken up. The mixture is boiled for five minutes to expel nitrogen dioxide and then cooled, and the residual arsenite is titrated with decinormal iodine in presence of starch. In making the various additions of liquid to the flask, care is of course taken to avoid all introduction of air.

When the sulphuric acid is added to the alkaline solution containing the arsenite, iodide and nitrite, iodine is set free locally, but this is at once acted upon by the alkaline arsenite, so that finally, when the acid reaction is reached, there is only a small amount still free, and the possibility of a loss of iodine by volatilization is reduced to a minimum.

The table gives the record of experiments made in this manner upon a solution of commercial sodium nitrite, standardized by treatment with potassium permanganate and oxalic acid in acid solution, according to the procedure of Kinnicut and Nef,† the

* I. K. Phelps, Am. Jour. Sci., [4], xvii, 198.

† Am. Chem. Jour., v, 388.

natural error of which is one of deficiency, as is evidenced by the odor of nitrogen oxides observed when even a very dilute solution of a nitrite is acidified.

Decomposition by Iodide and Arsenite in Acid Solution: Titration of Residual Arsenite in Alkaline Solution.

$NaNO_2$ taken. grm.	Oxygen value of As_2O_3 taken. grm.	Oxygen value of As_2O_3 found. grm.	Error on oxygen. grm.	Error on $NaNO_2$. grm.
0 0958	0 01200	0 00064	+0 00025	+0 0011
0 0958	0.01200	0 00066	+0.00024	+0 0010
0.1916	0 03200	0 00965	+0.00017	+0 0007
0 1916	0 03200	0 00965	+0 00017	+0 0007
0.3832	0.05600	0 01120	+0.00043	+0 0018
0.3832	0 05600	0 01118	+0.00045	+0 0019
0.6716	0 08000	0 00160	+0.00076	+0 0033
0 6716	0 08000	0 00158	+0.00078	+0.0034
0.1916	0 03280	0.01003	+0.00062	+0.0027

The Estimation of Nitrites, and of Nitrites and Nitrates in One Operation.

Determination of Nitrites. By the action of manganous chloride in hydrochloric acid upon a nitrite, passing the products of action into potassium iodide, and collecting the residual nitrogen dioxide, Roberts* has been able to estimate the nitrite both from the iodine set free and from the volume of nitrogen dioxide evolved.

The operation is conducted in a slightly modified form of the apparatus employed in the estimation of nitrates.† This consists of a retort, an absorption system charged with potassium iodide, a Hempel burette used for the collection and measurement of the residual gas over sodium hydroxide,‡ and a carbon dioxide generator for sweeping the gas to the burette. In making the analysis, the air must be thoroughly driven out of the apparatus before the nitrite is introduced, as the carbon dioxide, passing over the solution, decomposes it. Accordingly, carbon dioxide is first passed through the apparatus for some time, then the nitrite is introduced through the funnel tube and rinsed in with

* Charlotte F. Roberts, Am. Jour. Sci., [3], xlvi, 231.

† See page 260.

‡ See page 262.

a little water, followed by the manganous chloride solution, care being taken that the water shall not exceed one-third of the total volume of the liquid, according to the precaution shown to be necessary by Gooch and Gruener.* Working in this way with a solution of specially prepared sodium nitrite, the following results were obtained:

Decomposition by Hydrochloric Acid and Manganese Chloride: Titration of Iodine Set Free: Measurement of Nitrogen Dioxide Liberated.

Volume taken.	$NaNO_2$ determined by $KMnO_4$.*	$NaNO_2$ reckoned from NO.	$NaNO_2$ reckoned from iodine.
cm.^{3}	grm.	grm.	grm.
10	0.0463	0.0456	0.0450
10	0.0460	0.0460	0.0470
15	0.0704	0.0708	0.0722
15	0.0701	0.0704	0.0722
15	0.0688	0.0696	0.0695

* Process of Kinnicut and Nef.

Determination of Nitrites and Nitrates. Roberts† has also shown that in treating a mixture of nitrite and nitrate according to the method just described for the determination of nitrites, the measure of the nitrogen dioxide and the estimation of liberated iodine afford data for the calculation of the nitrite and nitrate in the mixture.

Representing the weight of nitric oxide found by a, and the weight of iodine found by b, and letting x equal the amount of nitric acid operated upon, and y the amount of nitrous acid,

$$\frac{30.01}{63.02}x + \frac{30.01}{47.02}y = a,$$

and

$$\frac{380.76}{63.02}x + \frac{126.92}{47.02}y = b;$$

whence

$$x = 0.248\,b - 1.051\,a,$$

$$y = 2.35\,a - 0.185\,b.$$

Results calculated from data furnished by experiments made in the manner described are given below.

* See page 263.

† Loc. cit.

Nitrite and Nitrate in One Operation.

$NaNO_2$ taken. grm.	$NaNO_2$ found. grm.	Error. grm.	KNO_3 taken. grm.	KNO_3 found. grm.	Error. grm.
0.0702	0.0718	+0.0016	0.1000	0.1000	0.0000
0.0702	0.0712	+0.0010	0.1000	0.0999	−0.0001
0.0702	0.0710	+0.0008	0.1000	0.1004	+0.0004
0.0702	0.0698	−0.0004	0.1000	0.1012	+0.0012
0.0468	0.0453	−0.0015	0.1000	0.0994	−0.0006
0.0468	0.0444	−0.0022	0.0500	0.0513	+0.0013

In calculating these results atomic weights were used, which differ somewhat from those now in vogue, but the differences thus introduced are not significant where the inevitable irregularities are so considerable.

The Estimation of Nitrates and Chlorates in One Operation.

A method for the determination of chlorates which has long been in common use consists in the treatment of those compounds with hydrochloric acid, the passing of evolved chlorine into potassium iodide, and the determination of liberated iodine by titration with sodium thiosulphate. This method is analogous to the method proposed by Gooch and Gruener* for the determination of nitrates, excepting that in the latter case the presence of manganese chloride is essential. In the case of the nitrate, however, there is a second product, nitrogen dioxide, which may be collected and measured, as in the process of treating a nitrate with ferrous chloride in hydrochloric acid, described by Roberts.† By combining the determination of the iodine evolved by the action of the products of decomposition upon potassium iodide with the measurement of nitrogen dioxide evolved, Roberts‡ has been able to effect the estimation of chlorates and nitrates in a single operation involving distillation with a solution of manganous chloride in hydrochloric acid, the amount of nitrogen dioxide found giving the amount of nitrate, and the iodine liberated measuring both nitrate and chlorate.

The operation is carried out with a slightly modified form of

* See page 263.

† See page 260.

‡ Charlotte F. Roberts, Am. Jour. Sci., [3], xlvi, 231.

the apparatus employed in the process* for the estimation of nitrates to which reference has been made.

In this apparatus, Fig. 21, a small retort fitted with a hollow ground-glass stopper prolonged beneath in a tube, and joined above with two branching tubes, one for the admission of carbon dioxide, and the other, attached to a funnel tube with stopcock, for the admission of liquids without introduction of air, is connected with a small condenser, which in turn is attached to an absorption apparatus containing potassium iodide, and this with a Hempel burette containing a strong solution of sodium hydroxide. In treating the mixture of chlorate and nitrate, two Will and Varrentrapp bulbs and generally a Geissler bulb containing potassium iodide are employed as the absorption system to make sure that no chlorine shall escape.

The mixture of chlorate and nitrate is introduced into the retort, the air is driven out by carbon dioxide, and then the solution of manganous chloride in hydrochloric acid is added through the funnel tube. The liquid becomes dark at once, but a short heating suffices to restore it to its original clear, light-green color. When this has been accomplished, a current of carbon dioxide is passed through the apparatus, the bulbed tubes are disconnected, and their contents titrated with sodium thiosulphate. The volume of the gas collected in the burette is noted, and the existing barometric and thermometric conditions, from which the weight of nitrate may be calculated.

Following are the results obtained in tests of the method.

Nitrates and Chlorates in One Operation.

$KClO_3$ taken. grm.	$KClO_3$ found. grm.	Error. grm.	KNO_3 taken. grm.	KNO_3 found. grm.	Error. grm.
0 1000	0.0990	−0.0010			
0.1000	0.0995	−0.0005			
0 0500	0.0484	−0 0016			
0 0500	0 0498	−0 0002			
0.0500	0.0496	−0 0004			
0.0500	0 0515	+0.0015	0.0500	0.0494	−0.0006
0.0500	0.0508	+0.0008	0.0500	0.0493	−0 0007
0.1000	0.0987	−0.0013	0.1000	0.0995	−0.0005
0.1000	0 1007	+0.0007	0 1000	0 0980	−0 0020
0.0300	0 0305	+0.0005	0.1000	0.0990	−0.0010
0.1000	0.1006	+0.0006	0.0300	0.0293	−0 0007

* See page 261.

The Qualitative Separation and Detection of Ferrocyanides, Ferricyanides and Sulphocyanates.

The ordinary method of testing for ferrocyanides, ferricyanides and sulphocyanates by means of ferric and ferrous salts leaves little to be desired in point of delicacy when the substances are not present together. When, however, a sulphocyanate and ferrocyanide occur together the colors tend to mask each other, and various methods to obviate the difficulty have been suggested, such as bleaching the red ferric sulphocyanate by mercuric chloride, and distilling the sulphocyanic acid before testing for that acid. In testing for a ferrocyanide in the presence of a ferricyanide, the formation of the deep-blue color with the ferric salt or ferrous salt has generally been considered of sufficient delicacy for all practical purposes.

Browning and Palmer* have attempted the separation of these substances from one another, as well as their detection.

The Ferrocyanogen Ion. Potassium ferrocyanide has long been mentioned as a precipitant of the ferrocyanides of the rare earth elements, cerium, thorium, yttrium, zirconium, etc., while it is also known that ferricyanides of these elements are soluble. These facts suggested the use of some member of the above-mentioned group as a precipitant of the ferrocyanogen ion, and selection was made of a soluble salt of thorium as perhaps the most satisfactory and available. Experience shows that upon adding a few drops of a 10 per cent solution of thorium nitrate to the solution of a ferrocyanide faintly acidified with acetic acid it is possible to detect by the cloudiness produced so little as 1 part of the ferrocyanide in 500,000 parts of solution. Alkali acetates tend to decompose the thorium ferrocyanide into soluble products, but the difficulty may be overcome by addition of thorium salt or hydrochloric acid. Neither potassium ferricyanide nor potassium sulphocyanate to the amount of 0.1 grm. in 10 $cm.^3$ interferes with this test.

The Ferricyanogen Ion. In making the choice of a precipitant for the ferricyanogen ion with a view to subsequent testing for the sulphocyanogen ion by the ferric salt, some reagent giving a colorless solution is preferable. Salts of the elements zinc and cadmium meet this condition, and cadmium salts prove to be the

* Philip E. Browning and Howard E. Palmer, Am. Jour. Sci., [4], xxiii, 448.

more delicate. It is found that 0.0001 grm. of the ferricyanide may be readily detected in from 5 cm.^{3} to 10 cm.^{3} of water acidified with acetic acid even when 0.1 grm. of potassium sulphocyanate is present.

Both thorium ferrocyanide and cadmium ferricyanide present difficulty in filtering on account of finely divided condition; but this difficulty is met by mixing with the precipitate fine-shredded asbestos and shaking.

The Ferrocyanogen Ion, the Ferricyanogen Ion, and the Sulphocyanogen Ion in Mixtures.

The method recommended for the separation and detection of the ferrocyanogen, ferricyanogen and sulphocyanogen ions is as follows:

I. The solution to be tested, preferably dilute and about 5 cm.^{3} to 10 cm.^{3} in volume, is acidified faintly with acetic acid or hydrochloric acid and treated with a soluble thorium salt to complete precipitation. To the liquid and suspended thorium ferrocyanide finely shredded asbestos is added. The whole is agitated and thrown on a filter, and the precipitate is washed with a little water. The washed precipitate is decomposed by strong sodium hydroxide on the filter, the clear filtrate is acidified with hydrochloric acid and the test for the ferrocyanogen ion is made with ferric chloride.

II. The filtrate from the thorium ferrocyanide is treated with a soluble cadmium salt to complete precipitation of the cadmium ferricyanide, which, after the addition of the asbestos, is filtered and washed. The cadmium ferricyanide on the filter is decomposed by sodium or potassium hydroxide, and the solution is filtered and tested with a ferrous salt.

III. The filtrate from the cadmium ferricyanide is acidified with hydrochloric acid and treated with ferric chloride, which gives the red ferric sulphocyanate.

The table on page 277 gives results of practical tests of this procedure.

The Gravimetric Determination of Sulphocyanates.

Van Name* has shown that while the sulphocyanate of silver, unlike that of copper, is readily soluble in an excess of ammonium or alkali sulphocyanates, which for this reason may not be used to precipitate silver for gravimetric estimation, the reverse process, the precipitation of a soluble sulphocyanate by an excess

* R. G. Van Name, Am. Jour. Sci., [4], x, 454.

of silver nitrate, furnishes a convenient means of standardizing sulphocyanate solutions and in general for estimating sulphocyanic acid.

$K_4FeC_6N_6$ present. grm.	$K_3FeC_6N_6$ present grm.	KSCN present. grm.	Indication.
			Tests for $K_4FeC_6N_6$ only.
0 0010	0 1	0 1	Distinct.
0 0005	0 1	0.1	Distinct.
0.0002	0 1	0 1	Distinct.
0 0001	0 1	0 1	Distinct.
			Tests for $K_3FeC_6N_6$ only.
0 1	0 0010	0 1	Distinct.
0.1	0 0005	0 1	Fairly distinct.
0 1	0 0002	0 1	Faint.
0 1	0 0001	0 1	Very faint.
			Tests for KSCN only.
0 1	0 1	0 0010	Distinct.
0 1	0.1	0 0005	Distinct.
0 1	0.1	0 0002	Distinct.
0 1	0 1	0 0001	Distinct.
			Tests for $K_4FeC_6N_6$, $K_3FeC_6N_6$ and KSCN.
0 0100	0.0100	0 0100	Good tests for $K_4FeC_6N_6$, $K_3FeC_6N_6$, KSCN.
0 0050	0 0050	0 0050	Good tests for $K_4FeC_6N_6$, $K_3FeC_6N_6$, KSCN.
0.0010	0 0010	0 0010	Good tests for $K_4FeC_6N_6$, $K_3FeC_6N_6$, KSCN.
			Tests of mixtures unknown to analyst.
0.0010	0.0010	. .	Found $K_4FeC_6N_6$, $K_3FeC_6N_6$.
0.0010	.... ..	0 0010	Found $K_4FeC_6N_6$, KSCN.
0 0010	0 0010	0.0010	Found $K_4FeC_6N_6$, $K_3FeC_6N_6$, KSCN.

When freshly precipitated the sulphocyanate of silver resembles the chloride in appearance, but when allowed to stand a few hours becomes finely granular and is very easily filtered and washed. It may be safely dried upon an asbestos filter at 110° to 120° to a constant weight corresponding to the theoretical constitution; but at a somewhat higher temperature is decomposed, leaving a residue of silver sulphide.

To the neutral solution of the sulphocyanate in approximately 100 cm.[3] of water, silver nitrate in solution is added in excess. The precipitate is collected upon asbestos in a platinum crucible, washed with cold water and dried to a constant weight at 115°, the drying requiring usually between two and three hours. The filtering is facilitated by allowing a few hours for the precipitate to settle; but this is by no means essential, as it is easy with a little care to obtain a clear filtrate even when the filtering is performed at once.

In the following table are results obtained by this procedure with like volumes of a solution of pure ammonium sulphocyanate free from chloride.

Gravimetric Determination of Sulphocyanates.

Final Volume of Liquid 150 cm.[3].

NH_4SCN. cm.[3]	$AgNO_3$. cm.[3]	Excess of $AgNO_3$. cm.[3]	AgSCN found. grm.
25	25 3	0 15	0.4372
25	25.3	0.15	0.4376
25	25 4	0.25	0 4373
25	25 4	0.25	0.4375
25	30 4	5.25	0.4382
25	Rough excess.		0.4366
25	Rough excess.		0.4381
25	Rough excess.		0.4373
25	Rough excess.		0.4372
25	Rough excess.		0.4369
			Mean 0 4374

The mean of the weights of silver sulphocyanate, 0.4374 grm., is equivalent to 0.2006 grm. of ammonium sulphocyanate for every 25 cm.[3] of solution. Four titrations of the same solution by Volhard's method, against a silver nitrate solution whose standard had been fixed by gravimetric determination as silver chloride, gave as a mean result 0.2003 grm. of ammonium sulphocyanate for 25 cm.[3] of solution. The agreement between these two values is within the possible error of the Volhard standard.

It is, therefore, evident that the standard of a sulphocyanate solution, free from chloride, obtained in the above way may safely be employed for the estimation of unknown amounts of silver by Volhard's method, as well as for other purposes.

Volumetric Estimation of Sulphocyanates by Potassium Permanganate.

When a solution of a sulphocyanate is acidified with sulphuric acid and titrated with potassium permanganate in the usual manner, the end-point is sharp, but the results, calculated from the equation

$$HSCN + 3\,O + H_2O = HCN + H_2SO_4,$$

are invariably low, the magnitude of the error varying greatly, as the following tables show, with the time occupied in titrating, temperature, dilution, and amount of shaking. The principal

Taken for each experiment: 50 cm.³ of approximately *n*/65 NH_4SCN, equivalent to 46.61 cm.³ of the $KMnO_4$ solution.

H_2SO_4 [1 : 1]. cm.³	Concentration of NH_4SCN before titration.	$KMnO_4$ used. cm.³	Error. cm.³	Mode of adding $KMnO_4$.
5	*n*/70	42.75	−3.86	Slow; little shaking except at the end.
5	*n*/70	44.43	−2.18	Rapid; little shaking except at the end.
5	*n*/70	41.41	−5.20	Very slow (50–100 drops per minute) with much shaking.
5	*n*/70	42.78	−3.83	Very rapidly to 42 cm.³ with much shaking; last 0.78 cm.³ slowly.
5	*n*/70	44.63	−1.98	Very rapidly to 44 cm.³ without shaking; last 0.63 cm.³ with shaking.
5	*n*/70	42.96	−3.65	Fairly slow, moderate shaking throughout.

Temperature before titrating 60°–65°.

Taken for each experiment: 15 cm.³ of approximately *n*/10 KSCN, equivalent to 87.30 cm.³ of the $KMnO_4$ solution.

H_2SO_4 [1 : 5].				
15	*n*/20	83.92	−3.38	Very rapid; little shaking except at the end.
15	*n*/20	82.20	−5.10	Slow, much shaking.
30	*n*/30	84.82	−2.48	Very rapid; little shaking except at the end.
30	*n*/30	83.02	−4.28	Medium rate (averaging about 15 cm.³ per minute until near the end), with shaking.

cause of the deficiencies is probably the fact noted by Klason * and others, that sulphocyanic acid undergoes decomposition in water solution in the presence of an inorganic acid.

It has, however, been shown by Van Name † that if the neutral solution of the sulphocyanate be slowly added to a measured and sufficiently large excess of permanganate, previously acidified with sulphuric acid, and the excess of permanganate then estimated with oxalic acid, the results are close to the theory. Under these conditions the sulphocyanic acid is not liberated until in actual contact with the permanganate, and side reactions are thus avoided. To insure this result there must at all times be a sufficient excess of permanganate present to give the solution a deep-red color. Moreover, since manganous salts reduce permanganate,‡ no considerable amount of manganous salt can be produced by the reaction with the sulphocyanate until after all the permanganate has been converted into the brown oxide. The excess of permanganate originally taken must therefore be at least five-thirds of the quantity actually required for the oxidation of the sulphocyanate, in order to insure the presence of unchanged permanganate throughout. In practice a larger excess is desirable, at least twice — or better three times — the theory, and amounts as large as ten times the theory may safely be used.

It is important, however, to remove the excess of permanganate without delay, since the instantaneous oxidation of the sulphocyanate is followed by a slow loss of permanganate, probably due to an oxidation beyond the cyanide stage, the rate of loss being greatest when the solution is concentrated and the excess of permanganate large. This error can be avoided by having the oxalate solution in readiness for rapid addition immediately after the sulphocyanate, or by at once precipitating the permanganate by adding an excess of a manganous salt, thus removing the necessity for further haste.

The following procedure is recommended: A volume of standard *n*/10 permanganate solution, sufficient to oxidize at least two and one-half times the quantity of sulphocyanate actually

* Jour. prakt. Chem. (N. F)., xxxvi, 74.

† R. G. Van Name, Dissertation, Yale University, 1902. The details of the procedure as here given and the three tables are taken from an unpublished portion of this thesis, on file since 1902 in the Yale University Library.

‡ The so-called Guyard reaction, $Mn_2O_7 + 3\,MnO = 5\,MnO_2$.

to be determined, is measured out and acidified with about one-tenth of its volume of [1 : 1] sulphuric acid. To this the sulphocyanate in neutral or faintly alkaline solution, one-fiftieth normal or stronger,* is added rapidly, a few drops at a time, with thorough stirring, and is immediately followed by concentrated manganous sulphate solution in quantity sufficient to precipitate all the residual permanganate. The manganese oxides are then dissolved (with or without previous warming of the liquid to 75°) by a measured amount of standard ammonium oxalate or oxalic acid, and the titration to color completed with the permanganate. It is best to leave for this purpose a few cubic centimeters of permanganate in the burette when the first portion is taken, thus avoiding unnecessary burette readings.

The following analyses were carried out by the method just described, the ratio of permanganate taken to that actually required being about ten to one in the first two experiments, and slightly above two to one in the last four.

Determination of Sulphocyanates by Potassium Permanganate.

$KMnO_4$.	H_2SO_4 [1 : 1].	HSCN equivalent to KSCN taken.	$KMnO_4$ used.	HSCN found.	Error.
cm.³	cm.³	grm.	cm.³	grm.	grm.
45	5	0.00482	4 11	0.00486	+0 00004
45	5	0.00482	4.10	0 00485	+0 00003
45	5	0.01345	12 20	0.01343	−0 00002
45	5	0.02408	20.16	0.02385	−0 00023
45	5	0.02408	20 21	0 02391	−0 00017
95	10	0.04816	40.50	0.04791	−0 00025
195	20	0.09631	81.11	0 09595	−0 00036

When the amount of sulphocyanic acid to be estimated is entirely unknown, the color of the solution must be carefully watched, and if the red color of the permanganate grows weak the addition of the sulphocyanate must be stopped, a further quantity of permanganate added, and the process completed in the usual way.

On account of the large volume of permanganate required, the method is best adapted for the estimation of small amounts of sulphocyanic acid.

* Higher dilutions may be employed, but the excess of permanganate should be proportionately increased.

PHOSPHORUS.

The Determination of Phosphoric Acid by Precipitation as Ammonium Magnesium Phosphate and Weighing as Magnesium Pyrophosphate.

Gooch and Austin* have shown that the precipitation of a soluble phosphate by the magnesia mixture is practically complete in faintly ammonical solutions even when very dilute and charged with large amounts of ammonium chloride, provided the magnesia mixture is present in sufficiently large excess; also that ammonium salts tend to produce an ammonium magnesium phosphate richer in ammonia and phosphoric acid and poorer in magnesium than the normal salt, NH_4MgPO_4, while an excess of the magnesia mixture tends to produce excess of magnesium in the precipitated phosphate. The results of experiment, recorded below, go to show that good results may be expected when the solution of the phosphate, containing a moderate excess of the magnesium salt and not more than 5 to 10 per cent of ammonium chloride, is precipitated by making it slightly ammoniacal, the precipitate being washed in slightly ammoniacal wash-water. In general, however, and especially when more ammonium chloride than this proportion, or more magnesium salt than twice the amount theoretically necessary,

Estimation as Magnesium Pyrophosphate: Effect of Ammonium Salts.

$Mg_2P_2O_7$ corresponding to HNa_2PO_4 taken.	$Mg_2P_2O_7$ found.	Error in terms of $Mg_2P_2O_7$.	Error in terms of P.	Volume.	NH_4Cl in magnesia mixture.	NH_4Cl added.		$MgCl_2.6H_2O$ in magnesia mixture.
						I.	II.	
grm.	grm.	grm.	grm.	cm³.	grm.	grm.	grm.	grm.
Single precipitation.								
0 8615	0.8613	−0 0002	−0 00005	150	1 68	..	..	3.3
0.8615	0.8615	0.0000	0.00000	200	1 68	20	..	3.3
0.8615	0.8602	−0.0013	−0.00036	200	1 68	20	..	3 3
0 8615	0.8561	−0.0054	−0.00151	300	1 68	60	..	3.3
Double precipitation.								
0 8111	0.8114	+0.0003	+0.00008	150 100	1 68	.	.	3 3
0 8615	0 8613	−0.0002	−0.00006	150 100	1.68	..	.	3 3
0 8615	0 8578	−0.0037	−0.00103	200 100	1 68	..	20	3.3
0.8615	0.8487	−0.0128	−0.00358	200 100	1.68	..	60	3.3

* F. A. Gooch and Martha Austin, Am. Jour. Sci., [4], vii, 187.

is present, it is safer to decant the supernatant liquid from the precipitate (through the filter to be used subsequently to hold the phosphate), to dissolve the precipitate in a little hydrochloric acid and reprecipitate by dilute ammonia, washing with faintly ammoniacal wash-water.

The Iodometric Determination of Phosphorus in Iron.

The very careful work of Blair and Whitfield * shows that the ammonium phosphomolybdate, precipitated under the conditions ordinarily prescribed for the determination of phosphorus in iron or iron ores, is of the definite constitution expressed by the symbol $24\ MoO_3, P_2O_5, 3\ (NH_4)_2O, 2\ H_2O$, and contains 1.794 parts of phosphorus to every 100 parts of molybdic anhydride. It has been demonstrated in the work of Fairbanks † that phosphorus may be determined with advantage by finding iodometrically the combined molybdic anhydride. Molybdic acid may be reduced by hydriodic acid to the condition of oxidation represented by the symbol Mo_2O_5 in acid solution,

$$2\ MoO_3 + 2\ HI = Mo_2O_5 + I_2 + H_2O;$$

while in an alkaline solution the reduced product is reoxidized by standard iodine.

$$Mo_2O_5 + I_2 + H_2O = 2\ MoO_3 + 2\ HI.$$

The amount of iodine necessary to reoxidize reduced molybdic acid is large, and the amount of molybdic acid compared with the phosphorus contained in the phosphomolybdate is also large, so that a method of great theoretical accuracy should result from the utilization of these reactions for the determination of phosphorus in iron.

According to the procedure described, the solution, not less than 150 cm.³ nor more than 300 cm.³ in volume, and containing iron as the nitrate and phosphorus as phosphoric acid in presence of not too much free nitric acid, is heated to 85° and shaken for five minutes with 40 cm.³ of filtered ammonium molybdate solution,‡ and filtered on asbestos in the perforated crucible.

* Jour. Am. Chem. Soc., xvii, 747.

† Charlotte Fairbanks, Am. Jour. Sci., [4], ii, 181.

‡ Blair and Whitfield, loc. cit. This solution is made of 100 grm. of MoO_3; 400 cm.³ of water; 80 cm.³ of concentrated NH_4OH added to 300 cm.³ of HNO_3 (sp. gr. 1.42), diluted with 700 cm.³ of water.

The precipitate is washed with 10 per cent nitric acid and then with 1 per cent potassium nitrate. The asbestos felt is transferred to a 150-cm.[3] flask or narrow-based Erlenmeyer. The precipitation flask and cork are thoroughly washed with a mixture of 5 cm.[3] of ammonia and 10 cm.[3] of water, and the washings are allowed to rinse the sides of the perforated crucible — standing on a small funnel — and so to run into the flask. Strong hydrochloric acid is added, 25 cm.[3], and, when the phosphorus does not exceed 0.0060 grm., 0.5 grm. of potassium iodide; but when more phosphorus is present a little more potassium iodide is needed. Experience has shown that the iodide present should not exceed the amount theoretically necessary by more than a half-gram. The flask is trapped loosely with a short bulbed tube hung in the neck.*

The liquid is boiled down from a total volume of 40 cm.[3] to just 25 cm.[3], easily marked by two strips of paper pasted on opposite sides of the flask. If the solution is boiled farther, the molybdic acid is likely to be reduced beyond the degree of oxidation indicated by the symbol Mo_2O_5.

The residue is cooled and transferred to a stoppered bottle (shown in Fig. 2) fitted with a separatory funnel and a trap filled with a solution of potassium iodide. Through the stoppered funnel are added in solution 1 grm. of tartaric acid, enough sodium hydroxide to nearly neutralize the free acid, followed by acid sodium carbonate to complete the neutralization, and a measured amount of iodine in excess of that required for the oxidation.

After neutralization the iodine color in the solution should perceptibly fade within fifteen minutes; but for complete oxidation the bottle should be set aside, out of sunlight, for an hour and a half, and then the excess of the iodine is titrated with a standard solution of arsenious acid.

Since there is a slight tendency on the part of the iodine to form a little iodate during the long digestion, it is wise to acidulate the solution in each case slightly with dilute hydrochloric acid after the titration with the arsenic solution, and then to determine by sodium thiosulphate the trace of iodine which has taken the form of iodate.

In the following table of test experiments the absolute errors in terms of phosphorus are given; and the percentage errors,

* See. Fig. 6, page 6.

between the phosphorus taken and the phosphorus found, referred to 10 grm. of material (the maximum amount of high-grade iron or steel usually taken for analysis), are also added.

Reduction of Phosphomolybdate by Hydriodic Acid: Iodometic Determination of Reduced Molybdic Acid.

Amount of P taken. grm.	Amount of P found. grm.	Error on P. grm.	Error of P. Per cent. grm.	Neutralized by
0 002727	0 002778	+0.000051	+0 0005	$NaHCO_3$.
0 001812	0 001743	−0 000069	−0 0007	$NaOH+NaHCO_3$.
0 000909	0 000914	+0.000005	+0 00005	$NaOH+NaHCO_3$.
0 003508	0.003262	−0 000246	−0 002*	$NaOH+NaHCO_3$.
0.005454	0.005417	−0.000037	−0 0003	$NaHCO_3$.
0 001818	0.001861	+0.000043	+0 0004	$NaOH+NaHCO_3$.
0 003636	0.003716	+0.000080	+0 0008	$NaHCO_3$.
0 000909	0.000988	+0.000079	+0 0008	$NaHCO_3$.
0 000363	0.000289	−0.000074	−0 0007	$NaOH+NaHCO_3$.
0 008180	0.008179	−0 000001	−0.00001	$NaHCO_3$.

* Obviously accidental.

The Estimation of Phosphoric Acid and Phosphorus Precipitated as Ammonium Phosphomolybdate.

The method studied by Randall* for the estimation of molybdic acid by the aid of the zinc reductor, the receiving flask charged with ferric alum, and the permanganate titration, has been applied by him† to the estimation of phosphorus in iron, precipitated in the form of ammonium phosphomolybdate. According to Randall's procedure, the phosphomolybdate, precipitated in a flask and shaken in the usual manner, is allowed to settle, then filtered on asbestos in a perforated crucible, and washed with a solution of ammonium acid sulphate (15 cm.^{3} ammonia, 25 cm.^{3} sulphuric acid, 1 liter water). The flask is washed out with a solution of 20 cm.^{3} of water and 5 cm.^{3} of ammonia, and this is poured on the asbestos in the crucible. The molybdenum solution is acidified with 10 cm.^{3} of strong sulphuric acid and passed through the reductor into the ferric alum solution, preceded by 100 cm.^{3} of hot water and followed by 200 cm.^{3} of the hot dilute acid with 100 cm.^{3} of water, the reduced solution being titrated immediately with approximately tenth normal permanganate. The results are calculated on the assumption that the am-

* See page 424.

† D. L. Randall, Am. Jour. Sci., [4], xxiv, 315.

monium phosphomolybdate contains phosphorus and molybdenum in the proportion given by the symbol $(NH_4)_3 12MoO_3PO_4$, and that the reduction proceeds to the condition represented by the symbol Mo_2O_3. In the following table are shown results obtained by this procedure applied to pure ferric nitrate and a known amount of microcosmic salt.

Determination of Phosphorus by Titration of Reduced Phosphomolybdate.

P taken. grm.	P found. grm.	Error. grm.
0 003645	0 003673	+0 000028
0 003645	0 003697	+0 000052
0 003645	0 003638	−0.000007
0 003645	0 003726	+0.000081
0 003645	0 003630	−0.000015
0 003645	0 003661	+0.000016

The Determination of Phosphoric Acid by Precipitation as Uranyl Phosphate and Estimation of the Uranium Volumetrically.

The method of estimating uranium by reduction in the zinc reductor and oxidation with permanganate* has been applied by Pulman† to the determination of uranic oxide, and so of phosphoric acid, precipitated as ammonium uranyl phosphate. This precipitate is filtered with extreme difficulty, but with an asbestos felt coated with the finer floating particles of the partially settled emulsion of prepared asbestos it is possible to obtain filtrates from the ammonium uranyl phosphate which are perfectly clear, though the process of filtering and washing is slow on account of the compactness of the felt surface and the gelatinous nature of the precipitate.

The process as worked out for the determination of the phosphoric acid is as follows: A measured amount of a standard phosphate solution (containing about 4.7 grm. of microcosmic salt per liter) is drawn into a beaker, and a solution containing 12 grm. of ammonium acetate, formed by neutralizing about 10 cm.^{3} of ammonium hydroxide (0.90 sp. gr.) with acetic acid

* See page 430.

† O. S. Pulman, Jr., Am. Jour. Sci., [4], xvi, 229.

(50 per cent), and from 2 cm.³ to 4 cm.³ of free acetic acid is added. The total volume is made up to about 150 cm.³ and the solution heated nearly to boiling. The ammonium uranyl phosphate is then precipitated by slowly adding an excess of uranium nitrate, with stirring, and the mixture is boiled gently for about twenty minutes, allowed to settle, and filtered on a tight felt of asbestos. The precipitating beaker and the precipitate are washed thoroughly with a dilute solution of ammonium acetate containing a little free acetic acid (to overcome the tendency of the precipitate to pass through the filter), and the crucible containing the precipitate is placed in a glass funnel. Enough dilute sulphuric acid [1:5] is then added to dissolve the precipitate and thoroughly wash out all the soluble uranium salt from the asbestos, the solution being caught below as it passes through the crucible and funnel in the beaker used for the precipitation. The solution is made up to a volume of from 100 cm.³ to 150 cm.³ with dilute sulphuric acid [1:5], heated to boiling. A few cubic centimeters of warm dilute sulphuric acid [1:5] are passed through the reductor and followed by the uranium solution, a few cubic centimeters more of the dilute sulphuric acid, and 250 cm.³ of hot water. The contents of the flask are then poured into a porcelain dish, diluted with 200 cm.³ of hot water, and titrated with a $n/10$ solution of potassium permanganate.

Results obtained by this method are shown in the following table:

Reduction of Uranyl Phosphate and Titration with Permanganate.

P_2O_5 taken.	UO_3 corresponding to P_2O_5 taken	H_2SO_4 (1.84).	Dilution at reduction.	$KMnO_4$.	UO_3 found.	Error on UO_3	Error on P_2O_5.
grm.	grm.	cm.³	cm.³	cm.³	grm.	grm.	grm.
0.0404	0 1630	25	150	11 06	0 1632	+0 0002	+0 00005
0.0404	0 1630	25	150	11 03	0 1628	−0.0002	−0 00005
0.0226	0 0912	20	120	6.14	0.0906	−0 0006	−0 00015
0.0226	0 0912	20	120	6 17	0 0911	−0 0001	−0 00002
0.0719	0.2902	25	150	19 62	0 2896	−0 0006	−0.00015
0.0719	0.2902	25	150	19.61	0 2894	−0 0008	−0.00020

The process for the determination of uranium by the reductor depends upon the fact that any reduction of uranium lower than uranous oxide — and such reduction undoubtedly takes place in the reductor — is corrected by exposure to the air, the lower oxide being rapidly oxidized to exactly the uranous state, while the uranous salts are stable enough to be estimated before they are oxidized appreciably by atmospheric action.

ARSENIC, ANTIMONY AND TIN.

The Determination of Arsenic by Precipitation as Ammonium Magnesium Arsenate and Weighing as Magnesium Pyroarsenate.

The striking analogy between the phosphates and the arsenates led Levol* to undertake the separation of an ammonium arsenate corresponding to the ammonium magnesium phosphate, the composition of which Berzelius had given. Levol states that ammonium magnesium arsenate of the composition $NH_4MgAsO_4.10H_2O$ is obtained by adding a solution of a double ammonium magnesium salt to arsenic acid, that it is a salt possessing about the same degree of solubility in water, in ammoniacal water, and in ammoniacal water containing magnesium salt, as the corresponding phosphate, and that at red heat, after carefully drying, it yields magnesium pyroarsenate. Several sources of error in this process have been pointed out by many investigators. First, the low indications obtained when the pyroarsenate is weighed suggest a loss of arsenic during ignition, in consequence of the reducing action of ammonia evolved in the process.† Another possible source of error is the solubility of ammonium magnesium arsenate in ammoniacal water and solutions of ammonium salts.‡ There is also the possibility that the constitution of ammonium magnesium arsenate as precipitated may not be ideal, in consequence of the action of ammonium salts known to be influential in determining the constitution of the analogous ammonium phosphates of magnesium and other elements.§

* Ann. Chim., [3], xvii, 501.

† Wach and Rose, Schweigger, Jour. Ch. Phys., lix, 297. Reichel, Ann. Phys., lxxvi, 20. Rammelsberg, Ber. Dtsch. chem. Ges., xiv, 279. Kaiser, Zeit. anal. Chem., xiv, 250.

‡ Rose, Zeit. anal. Chem., iii, 206. Wood, Am. Jour. Sci., [3], vi, 368. Brauner, Zeit. anal. Chem., xvi, 57.

§ Neubauer, Zeit. anorg. Chem., ii, 45; Zeit. angew. Chem., 1896, 435; Jour. Am. Chem. Soc., xvi, 289. Gooch and Austin, Am. Jour. Sci., [4], vi, 233; vii, 187; viii, 206.

The conditions to be observed in applying the method to the determination of arsenic have, therefore, been carefully studied by Austin.*

It is shown in the first place by special tests that the presence of ammonium chloride does produce solubility of the precipitate thrown down by magnesia mixture,† but that this solvent effect may be overcome even when the ammonium chloride present amounts to as much as 60 grm. in 300 $cm.^3$ of total volume, by a sufficiency of the magnesia mixture; and, further, that the ammonium magnesium arsenate once precipitated may be safely washed with small amounts (25 $cm.^3$ to 50 $cm.^3$) of faintly ammoniacal water.

It appears also that the ammonium salt induces the formation of an ammonium magnesium arsenate too rich in ammonia to give the pyroarsenate, $Mg_2As_2O_7$, on ignition, even when the precipitation is complete. It is found, however, that a suitable increase in the amount of magnesia mixture present at precipitation may bring about the formation of an ammonium magnesium arsenate of ideal constitution, even in presence of a considerable amount of the ammonium salt.

According to the most favorable procedure, the slightly acid solution of the arsenate containing no ammonium salts is added drop by drop to the distinctly ammoniacal magnesia mixture, and a little more ammonia is added. The precipitate is filtered off as soon as it subsides, on asbestos in the perforated crucible and with use of the filtrate to effect the transfer, and washed with about 25 $cm.^3$ of faintly ammoniacal water applied in small portions. After careful drying, the residue is ignited with caution and weighed as magnesium pyroarsenate, $Mg_2As_2O_7$.

When no ammonium salts are present an excess of about 30 $cm.^3$ of magnesia mixture in a total volume of 200 $cm.^3$ is sufficient to form the arsenate in ideal condition. If ammonium salts are present and conditions prevent their removal, the

* Martha Austin, Am. Jour. Sci., [4], ix, 55.

† The magnesia mixture is prepared by dissolving 110 grm. of the crystallized magnesium chloride in a small volume of water, filtering, and adding to it 58 grm. of ammonium chloride (purified in solution by adding bromine water and bleaching with ammonia), filtering, diluting to a volume of 2 liters, and adding enough ammonia — 10 c.c. — to make the solution smell distinctly of ammonia.

slightly acidulated arsenate should be added gradually to a very large excess (150 cm.^{3}) of the magnesia mixture.

Results obtained by the procedure are given in this table.

Determination of Arsenic as Magnesium Pyroarsenate.

$Mg_2As_2O_7$ corresponding to As_2O_5.			Magnesia mixture.	NH_4Cl.
Taken	Found.	Error.		
grm.	grm.	grm.	cm.^{3}	grm.
In absence of ammonium salts.				
0.7843	0 7830	−0 0013	50	..
0 7843	0 7849	−0 0006	50	..
0 7843	0 7841	−0 0002	50	..
0 7843	0 7843	0 0000	50	..
In presence of ammonium salts.				
0 7843	0 7763	−0 0080	75	10
0.7843	0 7762	−0.0081	75	10
0 7843	0 7832	−0.0011	100	10
0 7843	0.7838	−0.0005	100	10
0 7843	0 7784	−0.0059	100	20
0 7843	0 7810	−0.0033	100	20
0 7843	0 7849	+0.0006	150	60
0 7843	0.7846	+0.0003	150	60

In no one of the many precipitates tested by silver nitrate for included chlorides was more than a trace found.

Precipitation of Small Amounts of the Arsenate.

Magnesia mixture.	NH_4Cl.	$Mg_2As_2O_7$.		
		Taken.	Found.	Error.
cm.^{3}	grm.	grm.	grm.	grm.
25	..	0.0015	0 0014	−0.0001
25	..	0.0015	0 0017	+0 0002
25	..	0.0077	0.0078	+0.0001
50	..	0.0077	0.0076	−0.0001
25	10	0.0015	0.0013	−0.0002
25	10	0.0077	0.0070	−0.0007

To effect the precipitation of amounts of arsenate so small as not to be at once precipitable, recourse may be taken to a process of freezing and melting, in which the magnesium arsenate is made insoluble in the freezing and continues to be insoluble when the medium melts. This is best effected by putting the solution in a platinum dish, surrounding the dish with a mixture of ice and salt until the mass is solid, and then allowing the mass to melt at the room temperature. Results obtained in this manner by Gooch and Phelps* are given in the table.

The Iodometric Estimation of Arsenic Acid.

Holthoff's development of Mohr's suggestion relative to the reduction of arsenic acid to the lower condition of oxidation by the action of sulphurous acid,† with the demonstration that arsenic acid can be evaporated even to dryness in presence of hydrochloric acid without danger of significant volatilization, has placed the analysis of ordinary compounds of arsenic within the scope of Mohr's classical and exact method of determination by titration with iodine. As Holthoff left the method, it is satisfactory so far as regards accuracy, and as modified by McCay,‡ who substitutes for the four hours' digestion heating for one hour in a pressure bottle, is eminently successful. Gooch and Browning§ have still further shortened the process of reduction of arsenic acid by making use of hydriodic acid as the active agent instead of sulphurous acid.

Reduction by Hydriodic Acid and Oxidation by Iodine in Alkaline Solution.

A method for the determination of iodine in haloid salts based upon the action of arsenic acid, in the presence of sulphuric acid, according to the equation,

$$H_3AsO_4 + 2\,HI = H_3AsO_3 + H_2O + I_2,$$

the iodine being completely volatilized, but leaving behind in the arsenious acid produced by the action the record of the amount of hydriodic acid originally present, is described elsewhere.‖ This reaction is in the present case utilized conversely, and potassium iodide in excess, in presence of sulphuric acid, is employed to bring about the reduction of the arsenic acid to arsenious acid, which may be determined, after neutralization, by

* F. A. Gooch and M. A. Phelps, Am. Jour. Sci., [4], xxii, 492.

† Zeit. anal. Chem., xxiii, 378.

‡ Am. Chem. Jour., vii, 373.

§ F. A. Gooch and P. E. Browning, Am. Jour. Sci., [3], xl, 66.

‖ See page 457.

the iodine method. The conditions of the methods are different, in that in the former the hydriodic acid is entirely broken up by the action of the arsenic acid, and the iodine volatilizes easily; while in the latter some hydriodic acid must remain in solution until a very low degree of concentration is reached, and remaining must exhibit its characteristic proneness to retain free iodine.

It is found in practice that when a solution made up to contain sulphuric acid, an arsenate, and potassium iodide to an amount somewhat in excess of that theoretically demanded to effect the conversion of the arsenic acid to arsenious acid, is boiled, iodine is evolved. The color of the liquid passes from the dark red when the iodine is abundant through the various gradations of tint to a canary yellow, and then, as the sulphuric acid reaches a degree of concentration sufficient to determine by its own specific action the liberation of iodine, the color again darkens. If the process of concentration is continued, and much arsenic is present, crystals of arsenious iodide separate and form more abundantly on cooling. If evaporation is pushed still farther arsenious iodide begins to volatilize, and at the point where the sulphuric acid fumes the liquid loses all color and the arsenic has vanished more or less completely. In one experiment conducted in this manner it was found, by the method to be described later, that of 0.3861 grm. of arsenic pentoxide originally present with 1 grm. of potassium iodide and 10 $cm.^3$ of sulphuric acid [1 : 1] the equivalent of 0.1524 grm. remained. In another similar experiment, in which, however, only a few milligrams of arsenic oxide were involved, not a trace of arsenic remained at the end.

It is obvious that two points in this course of action demand attention: First, means must be used for removing the remnant of free iodine which is withheld by the hydriodic acid, or of rendering it harmless in the titration process to follow; and, secondly, the degree to which the solution may be concentrated without loss of arsenic must be fixed. In the converse of this process, the marked influence of the amount of sulphuric acid present upon the degree of concentration necessary to expel the iodine was particularly noted. In the present case, the effect of the proportion of sulphuric acid in solutions containing definite amounts of potassium iodide and potassium arsenate is likewise of first importance. In studying the effects of concentration, the solution was made up to about 100 cm.^{3} and concentrated by boiling until the color was faintest; then, to determine provisionally, and for preliminary purposes, the point at which volatilization of arsenic was likely to occur, the concentration was continued until the arsenious iodide began to separate. The results are tabulated as follows:

KI. grm.	As_2O_5. grm.	H_2SO_4 [1 : 1]. cm.^{3}	Volume when color was lightest. cm.^{3}	Volume when AsI_3 appeared. cm.^{3}
1	0 1900	20	80	33
1	0 1900	15	65	25
1	0 1900	10	40	19
1	0 1900	5	30	11

The amount of sulphuric acid which, considering rapidity in concentrating to the proper point, ease in neutralizing the acid previous to titration, and general convenience in manipulation, seemed to be best is 10 cm.^{3} of the [1: 1] mixture. The most suitable limit of concentration of the solution appears to be 40 cm.^{3}.

It is manifest from the phenomena described that when much hydriodic acid remains in the solution the last portions of free iodine cannot be completely removed by heat without volatilization of the arsenic. It was found that upon adding approximately $n/100$ sulphurous acid drop by drop to the hot concentrated solution the point at which the color vanished could be determined without difficulty, but that if the solution was permitted to stand a single minute the color of iodine returned, developed by the action of air upon the hot hydriodic acid. By diluting the solution with cold water as soon as the sulphurous acid has done its work and immediately neutralizing with potassium carbonate, reversion of arsenious acid to arsenic acid is precluded, magnesia mixture producing in the solution no precipitate of the ammonium magnesium arsenate.

The process as recommended by Gooch and Browning may be summarized briefly as follows: To the arsenate taken in solution in a 200-cm.^{3} Erlenmeyer flask are added potassium iodide in excess of the amount needed according to the equation to complete the reduction, and 10 cm.^{3} of [1:1] sulphuric acid

The liquid is diluted to about 100 cm.³ and boiled rapidly (with the precaution of trapping with a two-bulbed tube hung with the large end downward)* until the volume diminishes to 40 cm.³, shown by a mark upon the flask. The color of free iodine is bleached by cautious additions of sulphurous acid (corresponding roughly to centinormal iodine) and the solution is instantly diluted with water, nearly neutralized with potassium carbonate, and completely with the acid carbonate. The liquid is cooled and titrated as usual with iodine, using starch as an indicator. The whole operation is easily completed in a half-hour.

Results obtained by the procedure are given in the table.

Reduction by Hydriodic Acid: Oxidation by Iodine in Alkaline Solution.

KI taken.	H_2SO_4 [1 : 1] taken.	As_2O_5 taken.	As_2O_5 found.	Error.
grm.	cm.³	grm.	grm.	grm.
1.5	10	0.3861	0.3862	+0.0001
1.5	10	0.3862	0.3856	−0.0006
1.5	10	0 3861	0.3862	+0.0001
1.5	10	0.3860	0.3862	+0.0002
1.5	10	0.3863	0.3862	−0.0001
1.5	10	0.3862	0.3862	0.0000
1	10	0 1927	0.1922	−0 0005
1	10	0 1928	0.1922	−0.0006
1	10	0 1930	0.1925	−0.0005
1	10	0.1930	0.1927	−0 0003
1	10	0.1936	0.1929	−0 0007
1	10	0.1929	0.1928	−0.0001
1	10	0.0383	0 0380	−0.0003
1	10	0 0383	0.0385	+0 0002
0.5	10	0.0383	0 0384	+0 0001
0.4	10	0 0383	0 0385	+0.0002
0.3	10	0 0383	0 0386	+0.0003
0.2	10	0.0383	0 0384	+0.0001
0.2	10	0 0076	0 0074	−0 0002
0.2	10	0 0076	0 0074	−0.0002
0 2	10	0 0038	0 0034	−0.0004
0 2	10	0 0038	0 0034	−0.0004

In subsequent work by Gooch and Morris † it is shown that the process may be shortened by restricting the volume at which heating begins so that the boiling need not be extended beyond five or six minutes.

* See Fig. 6, page 6.

† F. A. Gooch and Julia C. Morris, Am. Jour. Sci., [4], x, 151.

According to this slight modification, the solution of the arsenate is heated in a trapped Erlenmeyer flask* with potassium iodide to an amount about 0.5 grm. in excess of the amount theoretically required and 10 cm.^{3} of sulphuric acid of half strength in a total volume of 50 cm.^{3} to 75 cm.^{3}. The liquid is boiled till the iodine vapors are no longer visible in the flask above the liquid, the iodine color in the still hot liquid is bleached by the cautious addition of sulphurous acid, the whole is diluted with cold water, and cooled quickly. The solution is nearly neutralized with potassium hydroxide and the neutralization is completed with acid potassium carbonate. The reduced acid is titrated with iodine after adding the starch indicator. By this procedure the results of the following table were obtained.

Reduction by Hydriodic Acid and Sulphurous Acid: Oxidation by Iodine in Alkaline Solution.

Volume. cm.^{3}	H_3O_3AsO taken. grm.	H_3O_3AsO found. grm.	Error. grm.
35	0 1559	0.1559	0.0000
35	0.1559	0 1560	+0.0001
40	0 1559	0 1559	0 0000
65	0.1559	0 1559	0 0000
50	0 2495	0.2499	+0.0004
50	0 2557	0.2449	−0 0008
60	0 3119	0 3117	−0.0002
60	0.3119	0 3120	+0 0001
75	0 3119	0 3124	+0 0005
75	0.3119	0 3132	+0 0013
75	0.3119	0.3121	+0 0002
75	0 3119	0 3115	−0 0004
75	0 3119	0 3124	+0 0005

Reduction by Hydriodic Acid: Titration of Iodine Liberated.

The process just described for the reduction and estimation of arsenic acid, depending upon the removal by volatilization of all but the last traces of liberated iodine, and the conversion of this minute residue by sulphurous acid, involves no secondary reactions of a sort likely to influence the main effect. It is exact and rapid.

The method of Williamson,† brought forward more recently, depends upon the conversion of the liberated iodine to hydriodic acid. The interaction at ordinary temperatures of a suit-

* See Fig. 6, page 6.

† Jour. Soc. Dyers and Colorists, 1896, 86–89.

ably strong acid, hydrochloric or sulphuric acid, upon the mixture of the arsenate and iodide, sets free iodine, and the liberated iodine is converted to hydriodic acid by the action of sodium thiosulphate, the end-point being the disappearance of the iodine color.

According to Williamson's directions, 25-cm.^{3} portions of the solution of the arsenate are treated with potassium iodide and mixed with an equal volume of hydrochloric acid of sp. gr. 1.16. The precaution is recommended that the strength of the solution of the arsenate shall not exceed the decinormal value, in order that the dilution consequent upon titration by the thiosulphate may not be too great; the reducing action brought about by the action of the strong acid upon the arsenate and iodide being reversible upon the dilution of liquid with water. This procedure thus limits the process to the determination of about 0.18 grm. of arsenic acid in 25 cm.^{3} of the solution to be treated with an equal volume of hydrochloric acid of sp. gr. 1.16. Obviously, however, the process should, so far as the reduction is concerned, be applicable to larger amounts of arsenic, provided the strength of the acid is kept up proportionately. It is essential that the liquid at the end of the titration should contain approximately 10 per cent of its mass of absolute hydrochloric acid or about one-third of its volume of the aqueous acid of sp. gr. 1.16.

The arsenic acid is measured either by the amount of standard thiosulphate required to bleach the iodine or by the amount of iodine required afterward to reoxidize the arsenious acid, after neutralizing with acid potassium carbonate. If the former alternative is followed, the end reaction must be the disappearance of the yellow color of the iodine, since in solutions so strongly acid it is impossible to place dependence upon the starch indicator; in using the latter alternative, the starch indicator is, of course, permissible and preferable.

In the direct titration of the iodine by thiosulphate two sources of error present themselves as possibilities: first, the excessive liberation of iodine by the action of air upon the strongly acidulated iodide; and second, the liability of the thiosulphate,* if present even in momentary or local excess during the process of titration, to break down under the action of strong acid, thus

* Norton, see page 364.

changing its capacity to convert iodine to hydriodic acid. The latter contingency should be remote in proportion to the caution used in adding the thiosulphate and in keeping the liquid well stirred; the former must of necessity vary with the acidity of the solution containing the iodide, the time of exposure to atmospheric action, and the degree of contact with the air incidental to stirring. How far each of these possibilities is likely to interfere in the practical conduct of an ordinary analysis has been investigated by Gooch and Morris.*

The effects likely to result simply from the strong acidification of the solution containing potassium iodide, and their variation for conditions of dilution representing the beginning and the end of a titration on the lines laid down, are shown in the following table. The solution of potassium iodide was diluted as indicated before the addition of the acid, and the iodine set free was titrated by thiosulphate.

Effect of Concentration of Acid and Time of Action upon Potassium Iodide.

HCl (sp. gr. 1.16) taken.	KI taken.	Total volume.	$Na_2S_2O_3$ added at once. In terms of H_3O_3AsO.	$Na_2S_2O_3$ added after 5 minutes. In terms of H_3O_3AsO.	$Na_2S_2O_3$ added after stirring 5 minutes. In terms of H_3O_3AsO.
cm.³	grm.	cm.³	grm.	grm.	grm.
25	2	50	0.0013	.....	.....
25	2	75	0.0004	...	...
25	2	50	...	0.0035	.....
25	2	75	....	0.0019	....
25	2	50	.....	.....	0.0042
25	2	75	....	......	0.0021
50	2	100	0.0017	.....	...
50	2	150	0.0004	......	....
50	2	100	......	0.0035	.....
50	2	150	......	0.0019	......
50	2	100	......	......	0.0035
50	2	150	......	......	0.0014

The concentration of acid and the time before titration are, obviously, the essential factors. The absolute amount of acid present and the stirring seem to make little difference.

As to the action of the hydrochloric acid on small amounts of the thiosulphate, there is the evidence of the experiments detailed in the following statements, in which 1 cm.³, 2 cm.³ and 5

* F. A. Gooch and Julia C. Morris, Am. Jour. Sci., [4], x, 151.

Effect of Concentration of Acid upon Thiosulphate.

HCl (sp. gr. 1.16). cm.³	Volume before titration. cm.³	$Na_2S_2O_3$ nearly $n/10$. In terms of H_3O_3AsO. cm.³	grm.	Iodine to color without dilution. In terms of H_3O_3AsO. grm.	Error of titration without dilution. In terms of H_3O_3AsO. grm.	Iodine to color after diluting to 75 cm.³ In terms of H_3O_3AsO. grm.	Error of titration after dilution. In terms of H_3O_3AsO. grm.
25	26	1	0.0071	0 0062	−0 0009	0 0071	0 0000
25	50	1	0.0071	0 0071	0 0000	0 0071	0 0000
25*	50	1	0.0071	0 0079	+0 0008	0 0079	+0 0008
25	50	2	0.0141	0 0146	+0.0005	0.0146	+0 0005
25*	50	2	0.0141	0 0157	+0 0016	0 0157	+0 0016
25	30	5	0 0353	0 0336	−0.0017	0 0374	+0.0024
25	50	5	0.0353	0 0359	+0 0006	0 0359	+0.0006
25*	50	5	0.0353	0.0411	+0.0058	0 0411	+0.0058

* In these experiments the acid stood in contact with the thiosulphate 5 minutes before titration.

cm.³ of nearly $n/10$ thiosulphate were exposed to the action of 25 cm.³ hydrochloric acid (sp. gr. 1.16), without dilution or diluted with an equal volume of water, and titrated with nearly $n/10$ iodine. The condition of acidity when the volume of 50 cm.³ contains 25 cm.³ of hydrochloric acid (sp. gr. 1.16) is that of the beginning of titration of Williamson's process. In order that the effect of error due to such action upon the determination of arsenic acid may appear immediately, the thiosulphate and iodine used are expressed in terms of that acid.

Time Effect of Acid upon Thiosulphate.

HCl (sp. gr. 1.16). cm.³	KI. grm.	Volume. cm.³	$Na_2S_2O_3$ nearly $n/10$. In terms of H_3O_3AsO. cm.³	grm.	Iodine in terms of H_3O_3AsO, at once. grm.	Iodine in terms of H_3O_3AsO, after 5 min. grm.	Error in terms of H_3O_3AsO. grm.
25	2	50	1	0.0071	0.0057	......	−0 0014
25	2	75	1	0 0071	0.0071	. ... .	0 0000
25	2	50	2	0 0141	0 0131	.... ..	−0.0010
25	2	75	2	0 0141	0 0143	......	+0.0002
25	2	50	5	0 0353	0 0322	......	−0.0021
25	2	75	5	0 0353	0 0357	. ...	+0.0004
25	2	50	1	0 0071	.	0 0028	−0.0043
25	2	75	1	0 0071	. ....	0.0067	−0.0004
25	2	50	2	0 0141	. ....	0 0116	−0.0025
25	2	75	2	0 0141	......	0 0139	−0.0002
25	2	50	5	0 0353	......	0.0314	−0.0041
25	2	75	5	0 0353	......	0 0361	+0.0008

The two sources of error due to the action of hydrochloric acid, the liberation of iodine and the decomposition of the thiosulphate, naturally tend to neutralize one another, but the completeness of such neutralization must be largely a matter of chance in the varying conditions of actual analysis. The experiments of the preceding table, in which *n*/10 thiosulphate, to the amount of 1 cm.^{3}, 2 cm.^{3} and 5 cm.^{3}, was added to the liquid, 50 cm.^{3} and 75 cm.^{3}, containing 25 cm.^{3} acid, and titrated with iodine at once, and after five minutes, were made to test the effects for the conditions of dilution prevailing at the beginning and at the end of a titration.

It is clear that under the conditions covered by the experiments of the two preceding tables the decomposition of the thiosulphate is likely to occur in greater or less degree, and that when the acid of sp. gr. 1.16 is not much diluted the products of decomposition are not oxidized by the iodine completely. The latter observation is quite in harmony with the fact that sulphur dioxide bleaches iodine in strong hydrochloric acid only slowly and incompletely. In such cases dilution favors further action of the iodine, but results obtained by titration with iodine in the acid solution diluted with an equal amount of water are unmodified by further dilution.

In the following tables are recorded actual determinations of arsenic according to Williamson's process. To each 25 cm.^{3} of the arsenate were added 1, 2 or 3 grm. of potassium iodide and 25 cm.^{3} hydrochloric acid (sp. gr. 1.16). The iodine was bleached

Williamson's Procedure.

HCl.	KI.	Volume at beginning of titration.	Volume at end of titration.	H_2KAsO_4 in terms of H_3O_3AsO.	H_3O_3AsO found.	Error.
cm.^{3}	grm.	cm.^{3}	cm.^{3}	grm.	grm.	grm.
25	2	50	51	0.0062	0 0085	+0 0023
25	2	50	52	0 0125	0 0156	+0 0031
25	2	50	55	0 0312	0 0350	+0 0038
25	2	50	55	0.0624	0.0666	+0 0042
25	2	50	73	0.1559	0.1588	+0 0029
25	2	50	73	0 1559	0.1587	+0 0028
25	2	50	73	0 1559	0.1591	+0 0032
25	2	50	73	0 1559	0 1595	+0 0036
25	3	50	73	0.1559	0.1595	+0.0036
25	1	50	73	0.1559	0.1581	+0 0022
25	2	50	73	0 1559	0.1581	+0.0022
25	2	50	73	0.1559	0.1588	+0.0029

by nearly decinormal thiosulphate without addition of the starch indicator, which loses all delicacy in the presence of strong acid. The time occupied by each titration was about five minutes. The standards of the arsenate were determined by the vaporization process,* the purity of reagents employed in that process having been proved by trying the process in the estimation of a solution of arsenic acid made by oxidizing pure decinormal arsenious acid by iodine.

The range of error in these results is from +0.0023 grm. to +0.0042 grm. with a mean of +0.0031 grm. — not very different from what might be expected from the effect of the interaction of the strong hydrochloric acid and the iodide alone. The counter effect due to the decomposition of the thiosulphate is not large, yet it is probably real, as will appear in the sequel.

In the following series of determinations, made with new solutions and new standards throughout, the arsenic acid was determined in two ways: (I) The iodine set free by 25 cm.^{3} of hydrochloric acid (sp. gr. 1.16) and 3 grm. potassium iodide, the solution having a total volume of 50 cm.^{3} at beginning and of 75 cm.^{3} at the end, was titrated by sodium thiosulphate; (II) The arsenious acid remaining after the first titration by sodium thiosulphate was titrated, after being neutralized with acid potassium carbonate by iodine, in the presence of the starch indicator.

Titration of Iodine Liberated and Titration of Arsenite Produced.

H_2KAsO_4 taken in terms of H_3O_3AsO. grm.	H_3O_3AsO found by the thiosulphate. grm	Error. grm.	H_3O_3AsO found by titration of H_3O_3As with iodine. grm.	Error. grm.
0.1767	0.1798	+0.0031	0.1776	+0.0009
0.1767	0.1798	+0 0031	0.1777	+0.0010
0.1767	0.1795	+0.0028	0.1785	+0.0018
0 1767	0.1793	+0.0026	0.1785	+0 0018
0.1767	0.1794	+0.0027	0.1780	+0.0013
0.1767	0.1798	+0.0031	0.1785	+0 0018

The average error of the first operation is 0.0029 grm., not far from that of the previous series; the error of the second operation, the titration of the arsenious acid, amounts on the average to 0.0014 grm. In the second operation the error due

* See page 291.

to over use of the thiosulphate by iodine set free outside the main reaction is obviously eliminated. The tetrathionate present after neutralization with acid potassium carbonate is unaffected by iodine, as was found by titrating 25 cm.^{3} of $n/10$ iodine mixed with 25 cm.^{3} hydrochloric acid (sp. gr. 1.16) by the thiosulphate, neutralizing with acid potassium carbonate,* adding starch and getting the starch blue with a single drop of $n/10$ iodine. The average error of this process (0.0014), therefore, is probably due to the products of decomposition of the thiosulphate in the first operation.

From the foregoing experiments it is clear that an arbitrary correction of about 0.0030 grm. must be deducted from the indications of Williamson's process of direct titration by thiosulphate, made with the greatest care under the conditions mentioned; and that a correction varying from one-half that amount (0.0015 grm.) to nothing (according to the amount of arsenious acid present), when the determination is made by iodine after neutralization with acid potassium carbonate. After making these arbitrary corrections in the results of the preceding table, the individual variations fall within reasonable limits.

On the other hand, the vaporization process, in which the arsenate is reduced by boiling with sulphuric acid and potassium iodide in the manner described,† gives indications reasonably regular and accurate without the application of an arbitrary correction.

The Detection and Approximative Estimation of Minute Quantities of Arsenic in Copper.

Sanger's successful application of the Berzelius-Marsh process to the quantitative determination of arsenic in wall papers and fabrics,‡ by the comparison of test mirrors with standard mirrors carefully prepared under the conditions of the test, opens the way, naturally, to the similar estimation of minute amounts of

* It is worthy of note that, as was found, it is not possible to substitute an alkali hydroxide for the carbonate in the early stages of the process of neutralization, on account of the decomposing effect of the former reagent upon the tetrathionate. This effect is in proportion to the heating of the solution, but is never wholly absent even when ice is intermixed with the liquid and the greatest care taken to prevent a rise of temperature.

† See pages 291, 294.

‡ Am. Chem. Jour., xiii, 431.

arsenic in any substances which may be submitted to the process immediately or after suitable preparation. Gooch and Moseley * have studied the application of Sanger's process to the determination of traces of arsenic in copper.

It has been shown by Headden and Sadler † that the presence of copper in the Marsh generator is instrumental in holding back the arsenic. It is obvious, therefore, that means must be employed for the complete removal of the copper from the arsenic before the solution of the latter is put into the reduction flask, and there is no method by which arsenic may be removed from copper easily, and without loss, aside from those methods which depend upon the volatility of arsenious chloride from solution in strong hydrochloric acid. Of such methods, on the score of rapidity in execution, accessibility of pure materials, and compactness of apparatus, preference is to be given to that process which is based upon the simultaneous action of strong hydrochloric acid and potassium bromide upon the salt of arsenic.‡

To get the copper into condition for the application of the process of separation from arsenic, it is sufficient to dissolve an amount not exceeding 1 grm. in nitric acid somewhat diluted with water, to add to the solution 2 cm^3 or 3 $cm.^3$ of strong sulphuric acid, and to evaporate the liquid until fumes of the sulphuric acid are disengaged abundantly. A single treatment of this sort serves to remove the nitric acid so completely that no interference with the normal action of the Marsh apparatus is apparent in the subsequent operation. The residue after concentration is diluted with water to about 5 $cm.^3$ and washed into the distillation flask with an amount of the strongest hydrochloric acid (sp. gr. 1.20) equal to that of the remainder of the liquid. It is desirable that the entire volume of the liquid should not much exceed 10 $cm.^3$. The flask, which has a capacity of 40 or 50 $cm.^3$, is inclined at an angle of about 45° and joined by means of a pure rubber stopper to a bent pipette which serves as a distillation tube. The lower end of the vertical limb of the pipette dips beneath the surface of about 5 $cm.^3$ of hydrochloric acid of half-strength contained in a test tube which is cooled and supported by water nearly filling an Erlenmeyer flask. A gram of

* F. A. Gooch and H. P. Moseley, Am. Jour. Sci., [3], xlviii, 292.

† Am. Chem. Jour., vii, 342.

‡ See page 316.

potassium bromide is introduced, and the distillation (which may be completed in three or four minutes) is pushed nearly to dryness. The flask is washed out, another portion of potassium bromide is introduced, and the first distillate is introduced and redistilled as before, excepting that the condensation is this time effected in pure water. This second operation serves merely to hold back traces of copper carried over in the first distillation, but the addition of the potassium bromide in the second distillation is quite as necessary as in the first, since the bromine liberated in the process has the effect of reoxidizing the arsenic in the receiver and so making that element nonvolatile under the conditions until the reducing agent is again introduced. The free bromine in the final distillate must be reconverted to hydrobromic acid before the contents of the receiver may be introduced into the reduction flask, and this effect may be most easily and unobjectionably accomplished by the addition of a little stannous chloride dissolved in hydrochloric acid of half-strength and purified from arsenic by prolonged boiling. Incidentally and simultaneously the arsenic is reduced to the arsenious form, and, though Sanger has shown that minute amounts of arsenic are completely eliminated from the solution in the reduction flask when that element is introduced in the higher form of oxidation, it is our experience that the rapidity of elimination of the arsenic is so increased by the introduction of the small amount of stannous chloride needed to bleach the bromine that the mirror appears in from five to ten minutes and is practically complete in half an hour, especially if the precaution is taken to add a little more stannous chloride, according to Schmidt's suggestion,* after the operation has been in progress about twenty minutes.

Schmidt has shown that the addition of stannous chloride to the Marsh apparatus in action not only does not effect the retention of arsenic, as many other metallic salts do, but actually brings about the final evolution in the form of the hydride of that portion of the arsenic which may have been deposited during the process in elementary form upon the zinc.

The Sanger apparatus is used in form essentially unchanged; but the zinc in the reserve generator is coated with copper by the action of a solution of copper sulphate, since in this way it is made more sensitive to the action of the dilute sulphuric acid,

* Zeit. anorg. Chem., i, 353.

while the presence of copper (which is of course out of the question in the reduction flask) can be of no disadvantage in the reserve generator and may even serve a useful end in fixing traces of arsenic if the zinc and acid employed are not absolutely free from that element. In the formation of the mirror, too, it has proved to be an advantage to inclose the portion of the glass tube to be heated in a short thin tube of iron or nickel slightly larger than the glass tube and kept from contact with it except at the ends, which are notched and bent inward. By keeping the outer tube of metal at a low red heat it is possible to diminish the tendency of the arsenic, particularly when the amounts are fairly large, to form a double mirror corresponding to the allotropic conditions of the arsenic. It is necessary, moreover, to substitute hydrochloric acid for the sulphuric acid usually employed in the reduction flask; but, though the opinion is current that hydrochloric acid introduces difficulties in the Marsh test, no evidence of the formation of a zinc mirror in the ignition tube or of other unfavorable action due to the use of pure hydrochloric acid has been noted in this operation. It is, of course, obvious that the hydrochloric acid used must be arsenic free.

Tests of this process were made with copper-prepared by electrolyzing in ammoniacal solution the purest copper sulphate obtainable and stopping the deposition before the solution had become exhausted. This copper, in which no arsenic was found, was dissolved in nitric acid, arsenic in the higher condition of oxidation was added, and the process of the separation of the arsenic from the copper and conversion to the mirror carried out in the manner described.

Arsenic in Copper.

Copper taken. grm.	Arsenic taken. mgrm.	Mirror estimated (by comparison with standard mirror). mgrm.	Error. mgrm.
None.	None.	None.	None.
0.7	None.	None.	None.
0.5	0 005	0 003	−0 002
0.5	0 011	0 013	+0 002
0.35	0 020	0 015	−0 005
0.3	0 030	0 030	None
0.43	0 040	0 035	−0.005
0.44	0.050	0 040	−0.010

It is plain from the results given above that the method is capable of detecting sharply minute amounts of arsenic in copper and of effecting the estimation of quantities less than 0.05 mgrm. with some approximation to accuracy.

There is, as Sanger has pointed out, a good deal of variation even in standard mirrors made with all possible care and precaution, and in the estimation of mirrors containing as much as 0.05 mgrm. of arsenic the uncertainty of comparison as well as the actual variation of the mirror is considerable.

When a sample of copper is under test which may contain more than 0.05 mgrm. of arsenic, it is desirable to introduce into the reduction flask the measured solution containing the arsenic gradually and in definite portions, and to judge by the formation of the mirror in an interval of ten minutes after the introduction of a portion of this test solution whether it is wiser to add the entire solution or to estimate the arsenic in the entire solution from that found in an aliquot portion.

Results of the analysis of several samples of commercial electrolytic copper are appended. The last two represented, presumably, the very purest electrolytically refined copper obtainable commercially.

Arsenic in Commercial Copper.

	Copper taken. grm.	Arsenic found. mgrm.	Percentage of arsenic.
Sample A.......	0 3	0 015	0 005
Sample B.......	0.3	0 030	0 010
Sample C.......	1	0 018	0.0018
	1	0 015	0.0015
Sample D.......	1	0 005	0 0005
	1	0.005	0.0005

The Separation of Arsenic from Copper by Precipitation as Ammonium Magnesium Arsenate.

Arsenic acid in an alkali salt may be completely precipitated as the ammonium magnesium arsenate, even in presence of ammonium salts, by adding the solution, with stirring, to a sufficient excess of magnesia mixture kept ammoniacal.*

* See page 288.

The fact that ammonium magnesium arsenate is insoluble in presence of an abundance of magnesia mixture, while many salts of copper are soluble in ammonia, suggests that arsenic existing as an arsenate may be separated from copper in ammoniacal solution by magnesia mixture. Obviously, the action of the ammoniacal copper solution on cellulose renders it impossible to make such an estimation by the use of paper filters; but ammonium magnesium arsenate, as has been shown, may be filtered off upon a mat of fine asbestos under pressure, in a perforated platinum crucible.

Gooch and Phelps* have found that, while a single precipitation is sufficient to effect the separation of small amounts of the arsenic acid from copper, copper is held in appreciable amount, apparently in combination, when the amount of the magnesium pyroarsenate exceeds a few milligrams. By dissolving the first precipitate and reprecipitating, reasonably good separations may, however, be effected for amounts not exceeding 0.4 grm. For larger amounts a second dissolving and reprecipitation must be made.

According to procedure outlined, the slightly acid solution of the alkali arsenate is run from a burette with stirring into the slightly ammoniacal solution of copper sulphate and magnesia mixture,† and the mixture is made distinctly ammoniacal. The precipitate is transferred to a weighed crucible, and, after rinsing once with distilled water made faintly ammoniacal, is dissolved in hot hydrochloric acid [1 : 3]. For convenience in handling the solutions the filtrate is received in a beaker under an evacuated bell-jar rather than in the usual filter flask. After cooling and adding ammonia nearly to neutrality, the solution is poured with stirring into an abundance of magnesia mixture kept constantly ammoniacal. The precipitate obtained in this way is collected on the asbestos felt used for the first filtration, the first portion of the filtrate being employed in each case to remove the last portions of the ammonium magnesium arsenate from the platinum dish. For amounts exceeding 0.4 grm. the process of dissolving, reprecipitating and filtering is repeated. After rinsing all traces of reagents from the precipitate with distilled water made faintly ammoniacal (20 $cm.^3$–50 $cm.^3$), the crucible

* F. A. Gooch and M. A. Phelps, Am. Jour. Sci., [4], xxii, 488.

† See page 289.

and contents are dried over a low Bunsen flame until all ammonia is driven off, and then ignited cautiously. Results of this procedure are given in the table.

Separation of Arsenic from Copper.

$CuSO_4$.	H_2KAsO_4.	Magnesia Mixture	$Mg_2As_2O_7$.			Error in terms of arsenic.
			Theory	Found.	Error.	
grm.	cm.³	cm.³	grm.	grm.	grm.	grm.
One precipitation.						
2	0.2	25	0.0015	0.0015	0.0000	0.0000
2	0.2	25	0.0015	0.0015	0.0000	0.0000
2	0.2	25	0.0015	0.0015	0.0000	0.0000
2	1	25	0.0077	0.0086	+0.0009	+0.0004
Two precipitations.						
2	0.2	25–25	0.0015	0.0012	−0.0003	−0.0001
2	0.2	25–25	0.0015	0.0009	−0.0006	−0.0003
2	1	25–25	0.0077	0.0072	−0.0005	−0.0002
2	1	25–25	0.0077	0.0079	+0.0002	+0.0001
2	5	25–25	0.0386	0.0389	+0.0003	+0.0001
2	5	25–25	0.0386	0.0383	−0.0003	−0.0001
2	10	25–25	0.0766	0.0754	−0.0012	−0.0006
2	10	25–25	0.0766	0.0754	−0.0012	−0.0006
2	25	25–25	0.1931	0.1927	−0.0004	−0.0002
2	25	100–100	0.1931	0.1919	−0.0012	−0.0006
2	50	50–50	0.3862	0.3867	+0.0005	+0.0002
2	50	25–25	0.3862	0.3864	+0.0002	+0.0001
Three precipitations.						
2	50	50–50–50	0.3830	0.3822	−0.0008	−0.0004
2	100	50–50–50	0.7660	0.7653	−0.0007	−0.0003
2	100	50–50–50	0.7660	0.7656	−0.0004	−0.0002
2	100	100–100–100	0.7724	0.7726	+0.0002	+0.0001

To recover amounts of the arsenate so small as not to be at once precipitable by magnesia mixture (at the outset or after the solution in hydrochloric acid), the solution, best contained in a platinum dish, is frozen in a mixture of ice and salt. Upon allowing the frozen mass to melt at the ordinary room temperature, insoluble magnesium arsenate remains. Results of the procedure are given in the table below.

Recovery of Small Amounts by Freezing.

$CuSO_4$.	Mg mixture.	NH_4Cl.	$Mg_2As_2O_7$.		
			Theory.	Found.	Error.
grm.	cm.3	grm.	grm.	grm.	grm.
2	25-25	.	0.0015	0.0015	0 0000
2	25-25	. .	0 0015	0 0015	0 0000
2	25-25	10	0 0015	0 0012	-0 0003

The Iodometric Determination of Antimonic Acid and of Antimonic Acid and Arsenic Acid.

Bunsen's method of determining qualitatively the condition of oxidation of salts of antimony, by boiling these substances in solution with potassium iodide and hydrochloric acid and noting whether the liquid takes the color of free iodine, has been applied successfully to the quantitative determination of antimony in its highest condition of oxidation. Weller * distils the iodine from the solution, collects it in the distillate and, determining it volumetrically, calculates from the amount found the antimonic salt which has set it free according to the equation

$$SbCl_5 + 2\,HI = SbCl_3 + 2\,HCl + I_2.$$

The advantage of treating the residue, rather than the distillate, in analytical processes in general which involve distillation is, however, so obvious that Gooch and Gruener † have determined the conditions under which Bunsen's reaction may be applied in such manner that the antimony is held and estimated directly in the residue. The general plan of work was laid down in a similar process elaborated in this laboratory for the reduction of arsenic acid.‡ According to this process, the arsenic to be reduced is taken in a solution of appropriate dilution, and treated with sulphuric acid in adjusted amount and an excess of potassium iodide; the liquid thus prepared is boiled to a definite degree of concentration, the iodine then remaining unexpelled, if any, is bleached by the very careful addition of dilute (centinormal)

* Ann. Chem., ccxiii, 246.

† F. A. Gooch and H. W. Gruener, Am. Jour. Sci., [3], xlii, 213.

‡ See page 291.

sulphurous acid, and the liquid is immediately diluted and neutralized; after cooling, the reduced arsenic is titrated by standard iodine in presence of starch.

It was found in preliminary experimentation that the same general plan of treatment is available in the handling of antimonic compounds, but it is necessary to take precautions to prevent the deposition of the antimony from solution upon the addition of the sulphuric acid. Tartaric acid accomplishes this effect satisfactorily and does not, as the result proved, introduce undesirable complications. It appears that the dilution of the solution at which the crystalline iodide or oxyiodide separates out during the boiling is greater than is the case when similar amounts of arsenic are dealt with, and that concentration to 45 $cm.^3$ is sufficient to cause crystallization and slight sublimation when the amount of antimonious oxide present (with excess of potassium iodide and 10 $cm.^3$ of sulphuric acid, 1 : 1) is approximately 0.2 grm. Otherwise the process as employed in the reduction of arsenic appears to be applicable to the similar treatment of antimony.

According to the procedure developed experimentally, the antimonate, in amount not exceeding the equivalent of 0.2 grm. of antimonious oxide, is treated in a trapped 300-$cm.^3$ Erlenmeyer flask* with tartaric acid (4 grm.) and sulphuric acid to acidity and thereafter with 10 $cm.^3$ of [1 : 1] sulphuric acid and 1 grm. of potassium iodide. The mixture is boiled, after introducing folded platinum foil, to prevent bumping of the liquid, and a trap to prevent mechanical loss. When the liquid has been concentrated to a volume of 45 $cm.^3$ to 55 $cm.^3$ the boiling is stopped, the color bleached by the cautious addition of sulphurous acid (approximately centinormal). The solution is diluted, nearly neutralized with sodium hydroxide, made alkaline by hydrogen sodium carbonate in an excess amounting to about 20 $cm.^3$ of the saturated solution, and titrated with the standard (decinormal) iodine after the addition of starch.

Less concentration in the boiling may result in incomplete reduction and greater concentration when the amount of antimony present is large may result in loss of that element by volatilization.

Results of experiments made under the conditions prescribed are given in the following table.

* See Fig. 6, page 6.

Determination of Antimonic Acid.

Final volume.	Tartar emetic taken.	Sb_2O_3 taken and oxidized.	Iodine used in final oxidation.	Sb_2O_3 found.	Error.
cm.³	grm.	grm.	cm.³	grm.	grm.
55	0.5023	0.2178	0.3827	0.2178	0.0000
55	0.5015	0.2175	0.3806	0.2166	−0.0009
50	0.5007	0.2172	0.3814	0.2171	−0.0001
50	0.5039	0.2185	0.3839	0.2185	0.0000
45	0.5001	0.2169	0.3818	0.2173	+0.0004
45	0.5004	0.2170	0.3825	0.2176	+0.0006

The antimonate used in these experiments was made by titrating tartar emetic in solution in presence of acid sodium carbonate by $n/10$ iodine. This treatment of the tartar emetic served the double purpose of providing a perfectly definite antimonic salt and of adjusting the standard of the iodine to be used subsequently in reoxidizing the antimony after its reduction. The imperfection of the process, whatever it may be, whether in the reduction or elsewhere, becomes apparent and is measured immediately by the difference between the amounts of iodine employed in the two oxidations.*

The method is also applicable to the reduction and estimation of antimony and arsenic in association, as is shown by the following test results.

Determination of Antimonic Acid and Arsenic Acid.

Final volume.	Tartar emetic taken.	Sb_2O_3 taken and oxidized.	As_2O_3 taken and oxidized.	Iodine used in first oxidation.*	Iodine used in final oxidation.	Difference between the amounts of iodine used in the two oxidations.	Error in terms of Sb_2O_3.	Error in terms of As_2O_3.
cm.³	grm.	grm.	grm.	cm.³	cm.³	cm.³	grm.	grm.
50	0.1530	0.0870	0.0500	19.37	19.43	+0.06	+0.0004	−0.0003
50	0.1503	0.0855	0.0495	19.05	19.02	−0.03	−0.0002	−0.0001
50	0.1503	0.0855	0.0544	20.05	19.97	−0.08	−0.0006	−0.0004
50	0.1503	0.0855	0.0495	19.05	19.00	−0.05	−0.0004	+0.0003

* To form the antimonate and arsenate.

* Hale has shown that a variation between the value of the iodine determined against tartar emetic and the value determined against arsenious oxide is due to variation in the composition of the tartar emetic and is not inherent in the process of titration (see page 40).

The Separation of Antimony from Arsenic by the Simultaneous Action of Hydrochloric Acid and Hydriodic Acid, and the Estimation of Antimony in the Residue.

The separation of arsenic from antimony by taking advantage of, the difference in volatility of the lower chlorides is due to Fischer.* This method of treatment consists in the reduction of the chlorides by means of ferrous chloride and the volatilization of the arsenic by repeated distillations of the mixture with hydrochloric acid of 20 per cent strength added in successive portions. The process has been subsequently modified by Hufschmidt† by the substitution of gaseous hydrochloric acid, introduced in a continuous current into the distilling mixture, for the aqueous acid, and later changed further and improved by Classen and Ludwig,‡ who employ ferrous sulphate, or ammonio-ferrous sulphate, in place of the less easily prepared ferrous chloride. In its latest form the method is exceedingly exact, but the conditions are such that the antimony in the residue must be determined gravimetrically. Gooch and Danner § have arranged the process so that the determination of the antimony may be made by a rapid volumetric method, by substituting for the iron salt, which utterly precludes the direct volumetric estimation of the antimony, another reducer — hydriodic acid — which can interfere in no way with the subsequent determination of the antimony by the well-known iodometric method.

It has been shown that arsenic ‖ and antimony** may be reduced by the action of hydriodic acid applied under conditions such that the arsenic shall not volatilize. In the present case the reducing action of hydriodic acid takes place in the presence of strong hydrochloric acid, and at the boiling temperature of the solution, — conditions arranged to bring about the volatilization of the arsenic as rapidly as possible.

The method of proceeding is briefly summarized in the following statement: To the solution of the oxides of arsenic and antimony, taken in amounts not exceeding 0.5 grm. of each, potas-

* Ann. Chem., ccviii, 182.

† Ber. Dtsch. chem. Ges., xvii, 2245.

‡ Ber. Dtsch. chem. Ges., xviii, 1110.

§ F. A. Gooch and E. W. Danner, Am. Jour. Sci., [3], xlii, 308.

‖ See page 291.

** See page 308.

sium iodide is added in a little more than the equivalent quantity, and enough strong hydrochloric acid to raise the entire volume of the solution to 100 cm.^{3}. The solution is saturated with hydrochloric acid gas and submitted to distillation in a current of that gas until the volume decreases to 50 cm.^{3} or a little less. The liquid is cooled rapidly, treated first with an excess of sulphurous acid and then with iodine to the exact oxidation of the former reagent; and, after the addition of 1 grm. of tartaric acid to every 0.2 grm. of antimonious oxide, the acid present is nearly neutralized with sodium hydroxide. The solution is made alkaline with hydrogen sodium carbonate added in excess to an amount corresponding to 10 cm.^{3} of the saturated solution for every 0.1 grm. of antimonious oxide present. Titration with decinormal iodine gives the antimony quickly and with a fair degree of accuracy. The whole process requires about an hour and a half for completion. Results obtained according to the procedure are given below.

Separation of Antimony from Arsenic.

H_2KAsO_4 taken.	KI taken.	Volume.		Color.		Sb_2O_3 taken.	Sb_2O_3 found.	Error.
		Initial.	Final.	On cooling.	With starch.			
grm.	grm.	cm.^{3}	cm.^{3}			grm.	grm.	grm.
0.5	0.5	100	50	Pale yellow.	Faint.	0.2268	0.2265	−0.0003
0.5	0.5	100	50	Pale yellow.	Faint.	0.2306	0.2300	−0.0006
0.5	0.5	100	50	Pale yellow.	Faint.	0.2272	0.2264	−0.0008

When the treatment with sulphurous acid is omitted the final titration indicates apparent deficiency in the antimony oxide amounting sometimes to more than two milligrams. This deficiency appears to be due, at least in part, to the presence, at the time of neutralization, of a small amount of iodine chloride, which might easily be produced by the oxidizing effect of antimonic and arsenic oxides upon the mixed halogen acids.

The Detection of Arsenic, and of Antimony with Tin, in Mixtures Containing Compounds of These Elements.

Upon the well-known fact that hot strong hydrochloric acid dissolves easily the sulphides of antimony and tin, but arsenious sulphide to only a slight degree, is based the simplest and most rapid method in common use for the separation of arsenic from antimony and tin. Unfortunately, however, the forcible treatment necessary to bring about the solution of large amounts of antimony is sufficient * to dissolve small quantities of arsenious sulphide, so that for the purposes of general analysis the method is inadequate. Koehler † has shown that only the arsenic is precipitated, and that very completely, when hydrogen sulphide acts upon the solution of arsenious and antimonious salts in hydrochloric acid of 20 per cent strength, but the adaptability of Koehler's treatment to the detection of arsenic in the ordinary course of analysis is limited by the necessity of so constituting the solution to be tested that hydrogen sulphide shall occasion no deposit of free sulphur to conceal or be mistaken for a precipitation of arsenious sulphide. In the course of analysis the mixed sulphides of arsenic, antimony, and tin, remaining after the removal of the sulphides insoluble in alkaline sulphides and recovered from solution by the action of hydrochloric acid, require for their complete solution the action of an oxidizing agent, which must, of course, interfere with the immediate use of Koehler's method.

Action of Hydrochloric Acid and Potassium Iodide. Gooch and Hodge ‡ have applied the action of hydrochloric acid and potassium iodide to the reduction of the salts of arsenic, antimony, and tin, with the volatilization of the arsenic by repeated distillations. The arsenic may be looked for in the distillate by hydrogen sulphide by Koehler's method after removal of free iodine by means of stannous chloride. To effect the repeated distillations of small portions of concentrated hydrochloric acid upon mixtures of the salts with potassium iodide, an apparatus which is essentially the distillation apparatus of Mohr is employed. This consists of a 25-cm.^{3} flask fitted by means of a rubber stopper to a pipette, bent, drawn out at the lower end, and dipped into a test tube which is at the same time supported and cooled in a flask partly filled with water. The pipette tube is wide enough (about 0.7 cm. in diameter) to prevent the formation of bubbles within it, and the bulb, holding about 20 cm.^{3}, is sufficiently large to retain any liquid which may be momentarily forced back by the accidental cooling of the flask during the distillation.

* Rose-Finkener, Anal. Chem., ii, 423.

† Zeit. anal. Chem., xxix, 192.

‡ F. A. Gooch and B. Hodge, Am. Jour. Sci., [3], xlvii, 382.

The mixture of salts of arsenic, antimony, and tin, in the higher condition of oxidation, is put in the flask with 3 grm. of potassium iodide in 5 cm.^{3} of water and 5 cm.^{3} of the strongest hydrochloric acid (sp. gr. 1.20). The distillation is carried nearly to dryness, and the distillate is condensed in 10 cm.^{3} of a mixture of strong hydrochloric acid and water in equal parts. The iodine evolved during the distillation is bleached by the addition to the distillate of stannous chloride dissolved in hydrochloric acid of half-strength, and hydrogen sulphide is passed to precipitate the arsenic if present. The residue in the flask is treated with 10 cm.^{3} of the strongest hydrochloric acid and the process of distillation repeated, but this time the distillate is condensed in 10 cm.^{3} of water in order that the final acidity of the liquid may be that of acid of half-strength, and so, after bleaching by stannous chloride, immediately available for the test for arsenic by hydrogen sulphide. Subsequent treatments of the residue are carried out similarly until arsenic ceases to appear in the distillate.

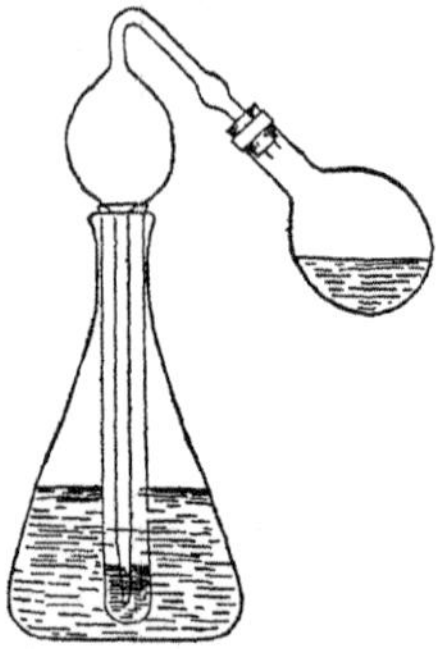

Fig. 23.

Four distillations of 10-cm.^{3} portions of the strongest acid suffice to transfer 0.01 grm. of arsenic completely to the distillate, while a single distillation appears to be sufficient to volatilize anything less than 0.003 grm.

Experiments made with antimony taken in the form of purified tartar emetic and oxidized by iodine in alkaline solution previous to treatment, either alone or with arsenic, show that antimony is discoverable in the residues when even so little as 0.0001 grm. of that element is originally introduced, though it is evident that a portion of the antimony may pass to the distillate when much of it is present in the flask. Indeed, when large quantities of antimony are treated the appearance of the brownish-red fumes of antimonious iodide in the distilling tube may serve as an indication that the concentration should go no further, since the antimonious iodide will, if it reaches the receiver in quantity, impart to the distillate a color which is not discharged by the stannous chloride used to bleach the iodine. In this case it will be necessary to look subsequently for a precipitate of arsenious sulphide in a liquid of its own tint. The amount of antimony volatilized seems to be proportioned to the amount present, and, when the distillation is properly conducted, enough antimony remains in the residue to be found if it was originally present in discoverable quantity.

The results of similar work with tin alone, and with tin and arsenic, go to show that, though tin like antimony may pass to the distillate under the conditions, enough tin always remains to be found in the residue, if the amount was originally appreciable.

Tests with Hydrochloric Acid and Potassium Iodide.

Arsenic taken as H_3O_3AsO. grm.	Antimony taken as H_3O_3SbO. grm.	Tin taken as $SnCl_4$. grm.	Precipitation by H_2S in successive distillates.	Precipitation by H_2S in the residue dissolved in water.
0.0001	......	....	I. Found. II. None.	None.
0.0033	.....	....	I. Found. II. None.	None.
0.0050	.....	...	I–III. Found. IV. None.	None.
0.0100	...	...	I–IV. Found. V. None.	None.
0.1000	....	.....	I–VII. Found. VIII. None.	None.
......	0.0001	......	I. None.	Distinct color.
0.0050	0 0001	.....	I–IV. Found. V. None.	Distinct color.
0.0001	0 4	...	I. Found. II. None.	Large.
0.0100	0.4	....	I–IV. Found. V. None.	Large.
......	......	0 0001	I. None.	Distinct color.
0.0100	......	0.0001	I–IV. Found. V. None.	Distinct color.
0.0001	......	0 0005	I. Found. II. None.	Distinct.
0 0100	......	0.0005	I–IV. Found. V. None.	Distinct.
0.0001	......	0 5	I. Found. II. None.	Large.
0.0100	......	0.5	I–IV. Found. V. None.	Large.

A single distillation, which may easily be completed in five minutes, is sufficient to discover the presence of 0.0001 grm. of arsenic associated with so much as 0.4 grm. or 0.5 grm. of antimony or tin, and to remove from the residue amounts of arsenic not exceeding 0.003 grm. When larger amounts of arsenic are to be removed, so that the tin and antimony may be obtained free from that element, the result may be accomplished by a suitable number of distillations; or, inasmuch as only a little iodine remains after the first distillation, the end may be attained by dissolving the residue in hydrochloric acid of half-strength, bleaching the iodine with exactly the necessary amount of sulphurous acid or sodium thiosulphate (since the use of the stannous chloride is here precluded), and passing hydrogen sulphide.

The details of experimental tests are given in the preceding statement.

Action of Hydrochloric Acid and Potassium Bromide. Gooch and Phelps* have shown that the action of hydrobromic acid upon arsenic acid is so similar to that of hydriodic acid that potassium bromide may with advantage be substituted for the iodide in the process of reduction and distillation. According to the procedure indicated, the mixture to be analyzed is introduced into the flask† with 5 cm.^{3} of water, 3 grm. of potassium bromide and 5 cm.^{3} of hydrochloric acid of full strength (sp. gr. 1.20). The end of the pipette tube is dipped into 5 cm.^{3} of hydrochloric acid of half-strength contained in this test tube used as a receiver, and the distillation is carried on until the liquid in the flask has almost entirely passed to the receiver. The residue is treated with 10 cm.^{3} of the strongest hydrochloric acid, and the distillation is repeated with the modification that this time the condensation is effected by passing the volatile material into 10 cm.^{3} of water, so that the liquid in the receiver at the end of this operation may have the acidity of hydrochloric acid of half-strength. This process of treating the residue with the strongest hydrochloric acid and distilling is continued until arsenic ceases to be discoverable in this distillate. At the beginning of the distillation bromine is liberated and collects in this distillate; but later, as the arsenious chloride volatilizes and condenses again, the color of the bromine in the distillate vanishes with the simultaneous reconversion of the arsenic to the higher form of oxidation. In such a solution, especially if it is not very hot, hydrogen sulphide

* F. A. Gooch and I. K. Phelps, Am. Jour. Sci., [3], xlviii, 216.

† See Fig. 23, page 314.

precipitates the arsenic only slowly, but the addition of a little stannous chloride dissolved in hydrochloric acid of half-strength to the hot solution reduces the arsenic to the lower form of oxidation and prepares the way for the immediate precipitation of arsenious sulphide by hydrogen sulphide. Antimonic acid is likewise reduced under the conditions of the distillation; but, as Koehler has shown,* neither the small amounts of antimony and tin, which if present originally may pass partially to the distillate, nor the tin added later to effect the reduction of the arsenic, will be precipitated by hydrogen sulphide under the existing conditions of temperature and acidity.

The results of experiments are recorded in the accompanying table.

Tests with Hydrochloric Acid and Potassium Bromide.

Arsenic taken as H_3AsO_4. grm.	Antimony taken as H_3SbO_4. grm.	Tin taken as $SnCl_4$. grm.	Precipitation by H_2S in successive distillates after treatment with $SnCl_2$.	Precipitation by H_2S in the residue dissolved in water.
......	......	......	I. None.	None.
......	......	......	I–X. None.	Faint coloration.
0 0001	......	......	I. Found. II. None.	None.
0.0010	......	......	I. Found. II. None.	None.
0 0100	......	......	I–II. Found. III. None.	None.
0.1000	......	......	I–III. Found. IV. None.	Faint coloration.
0.4000	......	......	I–VI. Found. VII. None.	Orange precipitation.*
1 0000	......	......	I–X. Found. XI. None.	Orange precipitation.*
......	0 4000	......	I. None.	Large.
0 0001	0 4000	......	I–II. Found. III. None.	Large.
0 0001	0.0001	......	I. Found. II. None.	Distinct color.
0 0010	0 0001	......	I. Found. II. None.	Distinct color.
0 0100	0.0001	......	I–II. Found. III. None.	Distinct orange.
......	......	0 4000	I. None.	Large.
0.0001	......	0.4000	I–II. Found. III. None.	Large.
0.0001	......	0.0001	I. Found. II. None.	Distinct color.

* Subsequently identified as antimony sulphide by depositing the metal on platinum.

* Zeit. anal. Chem., xxix, 192.

The Iodometric Determination of Arsenic and Antimony, and of Associated Copper.

The determination of arsenic and antimony in the filtrate from cuprous iodide after the titration of free iodine by sodium thiosulphate has been studied by Heath* who has thus made use of processes described elsewhere† for the reduction of arsenic acid and antimonic acid by the action of potassium iodide and sulphuric acid. The reactions involved in these processes may be expressed by the general symbols,

$$M_2O_5 + 4\,HI = M_2O_3 + H_2O + 2\,I_2$$

and

$$M_2O_3 + 2\,I_2 + 2\,K_2O = M_2O_5 + 4\,KI.$$

Trials of the method were made with tartar emetic and potassium arsenate. The tartar emetic was dissolved in water and the antimony oxidized to the higher condition by means of standard iodine solution in presence of sodium or potassium bicarbonate, the amount of iodine required being taken as a measure of the amount of antimony used. The solution of antimony thus obtained, or the solution of arsenic taken as potassium arsenate, was acidified and a known volume of a standard solution of copper nitrate was added. The copper was determined iodometrically with precautions recommended elsewhere.‡

During the determination of copper, mineral acids must not be present on account of their tendency to bring about reduction of the higher salts of arsenic and antimony by action of the excess of potassium iodide used in throwing out the copper. Such action causes high results on copper and low results on arsenic and antimony. A mixture of acetic acid and potassium iodide reduces the higher salts slowly. Tartaric acid is somewhat irregular in action and tends to cause an interfering precipitation of acid potassium tartrate. The action of citric acid is, however, satisfactory if the solution is not allowed to stand. The experimental results show that in the iodometric determination of copper associated with arsenic there must be no delay in titrating the iodine. Antimonic acid is not reduced appreciably, in a reasonable time, under similar conditions.

* F. H. Heath, Am. Jour. Sci., [4], xxv, 513.

† See pages 291, 308.

‡ See pages 118, 123.

In the course of preliminary work the fact developed that tetrathionic acid, which results from titration of free iodine by sodium thiosulphate in the copper determination, makes trouble in the subsequent operation. For, when the solution is boiled, after addition of sulphuric acid, the tetrathionic acid decomposes to give hydrogen sulphide and free sulphur, and sulphides of antimony and arsenic may be precipitated. It was found, however, that if liquid bromine is added to the cold solution in sufficient quantity to decompose all the excess of potassium iodide present, and the solution then boiled, there is very little subsequent trouble on account of tetrathionic acid.

The procedure for the determination of copper and arsenic or of copper and antimony may be outlined as follows:—To the solution containing the copper and also the arsenic or antimony in the higher condition of oxidation, add 1 grm. to 2 grm. of citric acid. To precipitate amounts of copper not exceeding 0.3 grm. in a volume of 50 cm.^{3}, add 3 grm. of potassium iodide; in a volume of 100 cm.^{3}, add 5 grm. of potassium iodide. Titrate the free iodine with $n/10$ sodium thiosulphate. The reactions involved in the copper determination are

$$2\,CuSO_4 + 4\,KI \rightarrow 2\,K_2SO_4 + Cu_2I_2 + I_2$$

and

$$I_2 + 2\,Na_2S_2O_3 = 2\,NaI + Na_2S_4O_6.$$

Filter off the cuprous iodide on asbestos. To the filtrate add 1 cm.^{3} of liquid bromine and boil the solution in an Erlenmeyer flask, using a trap to prevent loss by spattering. If, after boiling for a short time and allowing the large amount of free iodine to volatilize, the solution does not become clear, cool it, add a little more bromine (0.5 cm.^{3}) and boil again. When the solution has become clear, concentrate it somewhat (to about 60 cm.^{3}) to expel excess of bromine. Dilute to about 100 cm.^{3}, add 2 grm. of potassium iodide and boil to a volume of 50 cm.^{3} Cool the solution, bleach the free iodine by adding sulphurous acid,* using starch as indicator. Dilute to 100 cm.^{3}, add $n/10$ iodine to color and just bleach by careful addition of dilute sulphurous acid from a pipette. Neutralize the solution with sodium or potassium bicarbonate and titrate the arsenic or antimony with standard iodine solution in the usual way.

* Compare page 295.

Following are tables showing results obtained by this procedure:

Copper and Arsenic.

KI used.	Volume at end of precipitation	Cu taken.	Cu found.	Error in Cu.	Liquid bromine used.	KI used.	As taken.	As found.	Error in As.
grm.	cm.³	grm.	grm.	grm.	cm.³	grm.	grm.	grm.	grm.
3	50	0 0700	0 0700	0 0000	1 0	2	0.1238	0 1231	−0 0007
3	50	0 0700	0 0693	−0.0007	1 0	2	0.1238	0 1231	−0 0007
3	50	0 0875	0 0869	−0 0006	1 0	2	0.1238	0.1239	+0 0001
4	50	0 0700	0 0698	−0.0002	{0 6 / 0 4}	1	0.1238	0.1235	−0 0003
4	45	0 0700	0 0703	+0 0003	1 0	2	0.1238	0.1247	+0.0009
4	50	0 0700	0 0700	0 0000	1.0	2	0.1238	0.1235	−0.0003
5	70	0 1400	0 1407	+0.0007	{1 0 / 0 5}	2	0.1238	0.1233	+0.0005
5	50	0 0910	0 0907	−0.0003	{1.0 / 0.5}	2	0.1238	0.1239	+0.0001
4	50	0.0700	0.0703	+0.0003	1 0	2	0.1238	0 1237	−0 0001
4	50	0.0875	0 0879	+0.0004	1.0	2	0.1238	0.1234	−0 0004
5	65	0.1400	0 1410	+0 0010	1 0	3	0.1238	0.1239	+0 0001
3	30	0.0875	0 0860	−0 0015	0 8	2	0.0495	0.0493	−0.0002

Copper and Antimony.

KI used.	Volume at end of precipitation.	Cu taken.	Cu found.	Error in Cu.	Liquid bromine used.	KI used.	Sb taken.	Sb found.	Error in Sb.
grm.	cm.³	grm.	grm.	grm.	cm.³	grm.	grm.	grm.	grm.
4	85	0.0700	0 0703	+0.0003	{1.0 / 0.5}	2	0 1417	0.1421	+0.0004
4	80	0.0700	0 0701	+0.0001	{1 0 / 0 5}	2	0 1727	0.1725	−0 0002
4	80	0.0875	0.0869	−0.0006	{1 0 / 0 5}	2	0 1286	0.1289	+0 0003
5	80	0.0735	0.0739	+0.0004	{1 0 / 0 5}	2	0 1641	0.1645	+0.0004
5	90	0.1050	0.1050	0.0000	{1.0 / 0 5}	2	0.1378	0.1372	−0 0006
4	75	0.0875	0.0874	−0.0001	{1.0 / 0.5}	2	0.1329	0.1326	−0.0003
6	115	0.0700	0.0707	+0.0007	{1 3 / 0.5}	2	0.2474	0.2477	+0 0003
8	120	0.1575	0.1571	−0.0004	{1.5 / 0.5}	2	0.1419	0.1413	−0.0006

From the results obtained it seems possible by this method to separate and determine copper and arsenic, or copper and antimony, with errors of only a few tenths of a milligram. It is also possible to determine the sum of arsenic and antimony

Copper. *Copper, Arsenic and Antimony.*

KI used. grm.	Volume at end of pre-cipita-tion. cm.³	Cu taken. grm.	Cu found. grm.	Error in Cu. grm.	As taken as KH_2AsO_4. grm.	Cm.³ of I required by theory to oxidize As.	I used to oxidize Sb. cm.³	Sb atken as found by I used. grm.	Br used. cm.³	KI used in reduc-tion. grm.	Cm.³ of I required by theory to oxidize As and Sb.	I used to oxidize As and Sb. cm.²	Error in terms of I.* grm.
..	...	......	......	......	0.1238	32.54	17.87	0.1088	1.0	2	50.41	50.06	−0 0043
..	...	......	......	......	0.0495	13.01	13.46	0.0820	1.0	2	26.47	26.46	−0 0001
..	...	......	......	......	0.1238	32.54	15 40	0 0938	0.5	2	47.94	47.70	−0 0030
..	...	......	......	......	0.0990	27.21	18 12	0 1054	1.0	2	45 33	44 90	−0 0053
5	90	0.0700	0.0718	+0.0018	0.0495	13 01	17 15	0 1044	{1.0, 0 5}	2	30.16	30 00	−0.0020
5	90	0.0700	0.0710	+0.0010	0.0990	27.21	21.86	0.1272	{1.0, 0 5}	2	49.07	49.00	−0 0009
5	95	0.1222	0.1223	+0.0001	0.0990	27.21	29.68	0.1727	{1 5, 0 3}	2	56.89	56.23	−0 0081
9	115	0.0875	0.0868	−0.0007	0.0990	27 21	20 20	0.1176	2.0	3	47 41	46 90	−0 0063
6	110	0.1050	0.1067	+0.0017	0.0990	27.21	15.55	0.0905	1.3	2	42.76	42.64	−0 0015
7	100	0.1613	0.1611	−0.0002	0.0990	27.21	18.92	0.1101	1.5	2	46.13	45.90	−0 0028

* Weight of iodine × 0.50=error calculated entirely on Sb.
Weight of iodine × 0.30=error calculated entirely on As.

present with a fair degree of accuracy, and to separate and determine copper when associated with both arsenic and antimony. In the latter case the sum of the arsenic and antimony may also be determined, but the values here obtained for copper tend to come a little too high and those for arsenic and antimony a little too low.

The Estimation of Arsenic, Antimony and Tin in the Lower Condition of Oxidation by the Action of Potassium Ferricyanide in Alkaline Solution and Potassium Permanganate in Acid Solution.

In 1892 Quincke * published a method for estimating arsenic and antimony gasometrically, consisting essentially in oxidizing the arsenic or the antimony by a known excess of potassium ferricyanide in the presence of alkali, and determining the excess by measuring in a gasometer the oxygen evolved by the action of hydrogen peroxide on it according to the following equation:

$$2\,K_3FeC_6N_6 + H_2O_2 + 2\,KOH = 2\,K_4FeC_6N_6 + 2\,H_2O + O_2.$$

The same process of oxidation by use of the ferricyanide has been utilized by Palmer † in the estimation of arsenic, antimony and tin; but, to determine the amount of ferricyanide changed to ferrocyanide in the operation, use is made of the process of titration in acid solution by potassium permanganate previously employed by Browning and Palmer ‡ in the estimation of cerium and thallium. In alkaline solution the oxidations of the arsenious, antimonous and stannous salts take place according to the following equations:

$$As_2O_3 + 4\,K_3FeC_6N_6 + 4\,KOH = As_2O_5 + 4\,K_4FeC_6N_6 + 2\,H_2O$$
$$Sb_2O_3 + 4\,K_3FeC_6N_6 + 4\,KOH = Sb_2O_5 + 4\,K_4FeC_6N_6 + 2\,H_2O$$
$$SnO + 2\,K_3FeC_6N_6 + 2\,KOH = SnO_2 + 2\,K_4FeC_6N_6 + H_2O.$$

The ferrocyanide formed is oxidized in acid solution by permanganate according to the equation

$$10\,K_4FeC_6N_6 + 2\,KMnO_4 + 8\,H_2SO_4 = 10\,K_3FeC_6N_6 + 6\,K_2SO_4 + 2\,MnSO_4 + 8\,H_2O.$$

* Zeit. anal. Chem., xxxi, 1.

† Howard E. Palmer, Am. Jour. Sci. [4], xxix, 399.

‡ See pages 223, 249.

Determination of Antimony in Antimonious Condition. To the solution containing antimonious chloride dissolved in enough hydrochloric acid to prevent the formation of a basic salt, are added at least five times as much potassium ferricyanide as is theoretically necessary and about 25 cm.[3] of a 20 per cent solution of potassium hydroxide. After standing a few minutes, the solution is strongly acidified with dilute sulphuric acid and titrated with permanganate without previous removal of the antimonic salt. Results are given in the table.

The Antimonious Salt.

Sb_2O_3 taken. grm.	$K_3FeC_6N_6$. grm.	KOH. grm.	Volume. cm 3	Sb_2O_3 found grm.	Error. grm
0 0986	8	4	100	0 0989	+0 0003
0 0986	4	4	75	0 0984	−0 0002
0 0986	2	4	75	0 0984	−0 0002
0 0986	4	4	150	0 0984	−0 0002
0 0986	4	4	75	0 0984	−0 0002
0 0493	4	4	75	0 0495	+0 0002
0 0493	4	4	75	0 0497	+0 0004
0 0493	4	4	75	0 0495	+0 0002
0 1479	4	4	75	0 1482	+0 0003
0 1479	4	4	75	0 1477	−0.0002
0.1479	4	4	75	0 1476	−0 0003
0 1971	8	8	125	0 1972	+0.0001

Determination of Tin in Stannous Condition. To the solution of stannous chloride free from oxygen and kept under an atmosphere of hydrogen are added in solution at least five times as much potassium ferricyanide as is theoretically necessary and enough

The Stannous Salt.

Sn taken. grm.	$K_3FeC_6N_6$. grm.	KOH. grm.	Volume. cm.[3]	Sn found. grm.	Error. grm.
0 1032	2.5	6	65	0 1033	+0.0001
0 1022	2.5	6	65	0 1016	−0 0006
0 1029	3	7	60	0 1030	+0 0001
0 1009	5	6	85	0 1011	+0 0002
0.1005	5	5	60	0 1010	+0.0005
0.1011	10	5	85	0.1015	+0 0004
0 0995	10	5	85	0.1004	+0 0009
0.2020	4	6	80	0.2019	−0.0001
0.2003	5	6	90	0.1998	−0.0005
0.2021	10	5	85	0.2027	+0.0006

of a solution of potassium hydroxide to completely dissolve the precipitated stannic acid, the two solutions of ferricyanide and potassium hydroxide having been previously mixed. The stannic salt is precipitated by the addition of about 10 grm. of ammonium sulphate, and warming to 50° or 60°. After settling, the precipitate is filtered off on asbestos, under gentle pressure, and washed with a 10 per cent solution of ammonium sulphate. The filtrate is strongly acidified with sulphuric acid, and titrated with permanganate.

The test results are given in the preceding table.

Determination of Arsenic in Arsenious Condition. The essential procedure in estimating arsenic taken in the arsenious form is to oxidize by potassium ferricyanide in presence of a fixed alkali hydroxide, neutralize the last by the addition of ammonium sulphate, precipitate the arsenic by magnesia mixture, filter off the ammonium magnesium arsenate, and titrate the filtrate with permanganate, after acidification with sulphuric acid.

The procedure recommended is as follows: To the solution containing the arsenic in the arsenious condition is added an amount of potassium ferricyanide equal to at least five times the amount theoretically required to oxidize the arsenic to the higher condition of oxidation, and about 25 $cm.^3$ of a 20 per cent solution of potassium hydroxide, keeping the volume of the solution less than 100 $cm.^3$ After standing a few minutes, the free fixed alkali is removed and the solution made ammoniacal by dissolving in it about 10 grm. of ammonium sulphate, and magnesia mixture,* about 100 $cm.^3$, is added.

After settling, the ammonium magnesium arsenate is filtered off on asbestos, and washed with faintly ammoniacal water. The filtrate is strongly acidified with dilute sulphuric acid, and titrated with permanganate.

As Grützner † has shown, during the titration of large amounts of ferrocyanide by permanganate, a precipitate of $K_2MnFeC_6N_6$ often forms by the action of unchanged ferrocyanide on the manganese sulphate reduced from the permanganate. This precipitate slowly clears up as more permanganate is added, disappearing entirely as the end point is reached, but it tends

* About 100 $cm.^3$ of the mixture made up of MgCl .6 H_2O, 55 grm.; NH_4Cl, 29 grm.; NH_4OH, 5 $cm.^3$, in a liter of water.

† Chem. Centralblatt, 1902, i, 500.

to cause high results by obscuring the end point. This difficulty is overcome, however, by titrating in the presence of a large amount of sulphuric acid, the formation of this precipitate being thus prevented. The titration may safely be made in the cold in the presence of 10 per cent of sulphuric acid, and this amount is generally sufficient to prevent the formation of the precipitate.

Experimental results are given in the table.

Arsenious Oxide.

As_2O_3 taken. grm.	$K_3FeC_6N_6$. grm.	KOH grm.	Volume. cm.³	As_2O_3 found grm.	Error. grm.
0 0499	8	1 25	100	0 0502	+0 0003
0 0499	4	1 25	75	0 0499	0 0000
0 0499	3	1 25	75	0 0501	+0 0002
0 0501	4	1 25	75	0 0500	−0 0001
0 0997	8	1 25	100	0 0999	+0 0002
0 0997	6	1 25	90	0 0993	−0 0004
0 0997	6	1 25	90	0 0993	−0 0004
0 1001	8	1 25	100	0 0998	−0 0003
0 1496	9	1 25	110	0 1492	−0 0004
0 1496	10	1 25	110	0 1496	0.0000
0 1502	10	1 25	110	0 1502	0 0000
0 1994	12	1 25	125	0 1994	0 0000
0 1001	4	4.	75	0 0998	−0.0003
0 1001	3	4	75	0 0998	−0.0003

The Estimation of Arsenic Acid and Antimonic Acid Associated with Vanadic Acid.

Methods for the estimation of arsenic acid and of antimonic acid in association with vanadic acid, dependent upon differential reduction followed by reoxidation, are discussed in connection with methods for the determination of vanadium.*

VANADIUM.

The Gravimetric Estimation of Vanadic Acid Based on Liberation of Iodine and Absorption of that Element by Silver.

Perkins has shown that when a soluble vanadate is added to an excess of potassium iodide made acid with hydrochloric and shaken with electrolytic silver in an atmosphere of hydrogen,† iodine is set free and then is absorbed by the silver.‡ From the

* See page 350.

† Claude C. Perkins, Am. Jour. Sci. [4], xxix, 540.

‡ See Fig. 10, pages 27, 444.

increase in weight of the silver taken the amount of vanadium pentoxide may be calculated upon the assumption that one molecule sets free two atoms of iodine according to the equation

$$V_2O_5 + 2\,KI + 2\,HCl = V_2O_4 + H_2O + 2\,KCl + I_2.$$

Results obtained in this manner with ammonium vanadate, the composition of which had been determined by fusion with slightly acidic sodium tungstate, are shown in the following table.

Reduction of Vanadic Acid by Hydriodic Acid: Absorption of Liberated Iodine by Silver.

Ag taken. grm.	V_2O_5 taken. grm.	I found. grm.	Calculated V_2O_5. grm.	Error. grm.
2.0012	0.1946	0.2699	0 1939	−0 0007
2 0012	0.2051	0 2856	0.2051	0.0000
2 0024	0.1964	0.2746	0.1972	+0.0008
2.0024	0.2362	0.3295	0.2367	+0.0005
2 0062	0.2953	0.4106	0.2949	−0 0004
2 0062	0.2681	0.3729	0 2679	−0.0002
2.0527	0.5770	0.8038	0.5774	+0.0004
2.0006	0.2035	0.2829	0.2032	−0.0003

The Precipitation of Ammonium Vanadate by Ammonium Chloride.

Berzelius was the first to point out and utilize in analysis the fact that when a vanadate in concentrated solution is treated with a saturated solution of ammonium chloride, white insoluble ammonium metavanadate is deposited.* Modifications of the method have been proposed by V. Hauer,† Ditte,‡ and Gibbs,§ all of which have been unfavorably criticized.||

Gooch and Gilbert** have shown, however, that the process of Gibbs gives a practically complete precipitation of ammonium vanadate, when to the slightly ammoniacal solution of the soluble vanadate such an excess of ammonium chloride is added that

* Ann. Phys., xcviii, 54, 1831.

† Jour. prakt. Chem., lxix, 388.

‡ Compt. rend., civ, 982.

§ Proc. Am. Acad., x, 242, 249.

|| Milch, Inaug. Diss., Berlin, 1887. Rosenheim, Inaug. Diss., Berlin, 1888. Liebert, Inaug. Diss., Halle, 1891. Euler, Inaug. Diss., Berlin, 1895.

** F. A. Gooch and R. D. Gilbert, Am. Jour. Sci. [4], xiv, 205.

the solution deposits ammonium chloride after concentration and cooling, and the mixture is then allowed to stand twenty-four hours. Should too much ammonium chloride for convenient handling crystallize out on cooling, this is to be redissolved by the careful addition of ammonia; but care should be taken that, after standing, a little solid ammonium chloride and free ammonia still remain in the mixture. The precipitated ammonium vanadate is washed with a cold saturated solution of pure ammonium chloride, and the vanadium in the vanadate determined by any appropriate means. Volumetric processes are to be preferred.

According to this procedure the vanadate is dissolved by digestion upon the steam bath in 25 $cm.^3$ of slightly ammoniacal water. The mixture is allowed to remain upon the steam bath, with addition from time to time of a little ammonia, to keep the metavanadate of normal composition and colorless, until the volume has been reduced to about 25 $cm.^3$. On cooling, a small amount of ammonium chloride crystallizes out, but only a little if the proportion has been properly adjusted. If too large an amount crystallizes out, it is nearly redissolved by the cautious addition of ammonium hydroxide. The mixture is allowed to stand twenty-four hours to insure complete crystallization of the ammonium metavanadate, and is then filtered on a weighed asbestos filter in the perforated crucible, the precipitate being transferred and washed with a cold saturated solution of ammonium chloride. Crucible and precipitate are heated, at first very gently to drive off the ammonium chloride without occasioning mechanical loss of the vanadium, and finally to redness and fusion of the pentoxide remaining.

Without special precaution some trouble is occasionally found in removing from the walls of the beaker the adherent crystals of ammonium vanadate, but the difficulty may be overcome by the device of forming upon the walls of the beaker before using it a film of paraffin of extreme thinness by rinsing the beaker with a dilute solution of paraffin in naphtha (0.5 grm. of paraffin in 300 c.c. of naphtha) and allowing the naphtha to evaporate. Crystals of the vanadate adhering to the walls of the beaker thus previously prepared are easily removed by means of the ordinary rubber or "policeman." The table contains the record of six consecutive experiments made after some preliminary study of the method. Tests of the washings and filtrate were made in several instances by acidifying with hydrochloric acid and testing with hydrogen dioxide, but no indication of the presence of vanadium was obtained.

Precipitation by Ammonium Chloride and Ammonia: Gravimetric Determination.

NH_4VO_3 taken.* grm.	V_2O_5 present. grm.	V_2O_5 found. grm.	Error. grm.
0.5	0.3807	0.3814	+0.0007
0.5	0.3807	0.3818	+0 0011
0 5	0.3807	0.3813	+0 0006
0 5	0 3807	0.3808	+0 0001
0 5	0 3807	0.3808	+0.0001
0 5	0.3807	0 3799	−0.0008

* Containing 76.14% of V_2O_5, according to Holverscheit's method.

These results are sufficient to show that the method of Gibbs is capable of yielding an analytical separation of value, but, as Gibbs pointed out, it is ordinarily preferable to estimate the vanadium volumetrically rather than to go through the tedious and exacting process of ignition to recover the vanadium pentoxide.

Other results obtained in determining by the Holverscheit process the ammonium vanadate precipitated by the Gibbs process confirm the gravimetric results.

Precipitation by Ammonium Chloride and Ammonia: Volumetric Determination.

NH_4VO_3 taken. grm.	V_2O_5 found. grm.	Error. grm.
0.1000	0 0760	−0.0001
0 1000	0 0764	+0 0003
0 1000	0 0763	+0.0002
0 1000	0 0962	+0.0001
0.1000	0.0757	−0 0004
0.1000	0.0751	−0.0010
0.1000	0.0761	0.0000
0 1000	0.0758	−0 0003

The Estimation of Vanadium as Silver Vanadate.

The conditions under which vanadium may be estimated as silver vanadate have been investigated by Browning and Palmer.* The experimental results indicate that the composition of the

* Philip E. Browning and Howard E. Palmer, Am. Jour. Sci. [4], xxx, 220.

precipitate thrown down in solutions acidified with acetic acid is variable, while that of the precipitate from neutral solution has after ignition the composition of the metavanadate $AgVO_3$.

The preferred procedure is as follows: The solution of the vanadate is made acid with nitric acid and then alkaline with ammonia. It is boiled until, when the ammonia is almost expelled, the solution begins to turn yellow. At this point the boiling is stopped because if it is continued further the solution becomes too acidic and the precipitate which forms on addition of silver nitrate is no longer definitely the metavanadate. To the solution, about 200 cm.^{3} in volume and heated to boiling, a solution of silver nitrate is added with vigorous stirring to coagulate the precipitate. The precipitate is then filtered off on an asbestos felt contained in a perforated crucible, washed thoroughly, ignited at a gentle heat below the fusing point of silver vanadate, and weighed as $AgVO_3$.

Following are results of this procedure:

Precipitation as Silver Metavanadate.

V_2O_5 taken. grm.	$AgVO_3$ found. grm.	V_2O_5 found. grm.	Error. grm.
Precipitated from neutral solution by $AgNO_3$.			
0.1139	0.2595	0.1143	+0.0004
0.0569	0.1291	0.0569	+0.0000
0.0569	0.1277	0.0562	+0.0007
0.0569	0.1303	0.0574	+0.0005
0.1066	0.2436	0.1073	+0.0007
Ammoniacal solution boiled to expel ammonia: $AgNO_3$ added to boiling solution.			
0.1066	0.2430	0.1070	+0.0004
0.1066	0.2429	0.1070	+0.0004
0.0533	0.1224	0.0539	+0.0006
0.0533	0.1221	0.0538	+0.0005
0.0569	0.1312	0.0578	+0.0009
Solution acid with HNO_3 made ammoniacal and boiled to expel ammonia: $AgNO_3$ added as soon as the solution began to turn yellow.			
0.0569	0.1293	0.0569	0.0000
0.1066	0.2419	0.1065	−0.0001
0.1066	0.2424	0.1068	+0.0002
0.1066	0.2422	0.1066	+0.0000
0.0533	0.1215	0.0535	+0.0002

The Estimation of Vanadic Acid by the Action of the Halogen Acids.

The Action of Hydrochloric Acid. The estimation of vanadic acid by reduction with concentrated hydrochloric acid followed by the iodometric determination of the chlorine evolved, as proposed by Bunsen * and noted by Mohr,† has been unfavorably criticized by many investigators.‡ On the other hand, Gibbs § has determined the small amounts of vanadium pentoxide found in the vanadio-tungstates and other complex combinations, by boiling with strong hydrochloric acid, collecting in potassium iodide the chlorine evolved, titrating the freed iodine with sodium thiosulphate and calculating from the amount of iodine thus found the vanadium pentoxide corresponding to a change of condition from V_2O_5 to V_2O_4.

Gooch and Stookey || have shown that when the action of hydrochloric acid is sufficiently continued, vanadic acid may be completely reduced to the condition of the tetroxide; and have pointed out that the method of reduction is of special advantage in those cases which call for titration of the tetroxide by permanganate, since in such cases the use of Holverscheit's admirable method of reduction by hydrobromic acid is precluded. It appears also that it is possible to effect the reduction of vanadic acid to within a few per cent of the amount present by a single treatment with concentrated hydrochloric acid, and that the amount of the reduction may be determined by titrating the residue with potassium permanganate.

In the action of the most highly concentrated aqueous hydrochloric acid upon vanadic acid three stages are marked: First, the vanadic acid dissolves to a solution of a deep brown color without perceptible evolution of chlorine; upon warming, the solution evolves chlorine and takes on a deep green color; thereafter the evolution of chlorine becomes weaker, and the solution, giving off a small amount of chlorine on strong boiling, assumes a clear

* Ann. Chem., xcvi, 265.

† Titrirmethode, v[te] Aufl., 314.

‡ Czudnowicz, Ann. Phys., cxx, 17. Milch, Inaug. Diss., Berlin, 1887, 10. Rosenheim, Inaug. Diss., Berlin, 1888; Ann. Chem., ccli, 197. Holverscheit, Inaug. Diss., Berlin, 1890.

§ Proc. Am. Acad., x, 250.

|| F. A. Gooch and L. B. Stookey, Am. Jour. Sci., [4], xiv, 369.

blue color. Observing, however, that the liberation of chlorine actually begins to some extent as soon as the hydrochloric acid and vanadic acid come into contact, Gooch and Stookey have used the form of apparatus shown in Fig. 3, p. 4, so that no chlorine may be evolved before the apparatus is connected and ready. For the retort is used a Voit flask upon the inlet tube of which is sealed a stoppered funnel, while the outlet tube is sealed to a Drexel wash bottle used as a receiver, and this in turn is joined to Will and Varrentrapp bulbs. Connection is provided between the funnel and a carbon dioxide generator, so that by passing carbon dioxide through the apparatus all danger of regurgitation may be avoided. A branched connection with the funnel tube, so arranged that either hydrochloric acid gas or carbon dioxide or both may at pleasure be sent into the apparatus, is a convenience.

According to the procedure described, the receiver and trap are charged with water containing 3 grm. of potassium iodide. The vanadate is introduced into the dry flask and the apparatus is adjusted. Hydrochloric acid is put in the stoppered separating funnel and, when all is ready, allowed to run into the flask and boiled until the liquid takes a clear blue color. Thereafter, either the evaporation is continued to dryness and fresh concentrated hydrochloric acid added, or else the weak acid remaining after the boiling process is cooled and resaturated with gaseous hydrochloric acid; and, in either case, the boiling is repeated.

Several successive treatments with strongest hydrochloric acid, consisting either in the addition of the concentrated aqueous acid to the dry residue in successive portions or in passing gaseous acid through the residual liquid cooled in ice water, finally bring the residue to a condition of reduction in which the recharging of the liquid residue with gaesous acid produces no brown or green color, but a clear blue. When this condition is reached the reduction is complete, the application of the Holverscheit procedure to the residue producing no appreciable further reduction. The holding of the blue color, when the solution cooled in ice water is thoroughly charged with the gaseous acid, appears to be the best evidence of the complete reduction of V_2O_5 to V_2O_4; it is not sufficient that the boiled solution, in which the acid has been weakened, should be blue. The determination of

Action of Hydrochloric Acid on the Vanadate: Determination of Chlorine Liberated.

NH_4VO_3 taken. grm.	HCl taken.	V_2O_5 calculated. grm.	V_2O_5 found. grm.	Error. grm.	Details of treatment.
0.1000	10 cm.³, sp. gr. 1.20	......	.....	. .	Evaporated to dryness.
	5* cm.³, sp. gr. 1.20	......	....	..	Evaporated to dryness.
	5* cm.³, sp. gr. 1.20	0.0752	0.0737	−0.0015	Evaporated to dryness. Residue brown when treated with concentrated HCl.
0.1000	40 cm.³, sp. gr. 1.20	0.0752	0 0700	−0 0052	Boiled 15 minutes.
	....	......	......	......	Evaporated to dryness.
	7* cm.³, sp. gr. 1 20	......	.....	...	Evaporated to dryness.†
	7* cm.³, sp. gr. 1.20	......	......	......	Evaporated to dryness.†
	7† cm.³, sp. gr. 1.20	0.0752	0.0742	−0.0010	Evaporated to dryness.† Residue treated with KBr gave Br equivalent to 0.0001 V_2O_5.
0.1000	40 cm.³, sp. gr. 1.20	0 0752	0.0699	−0.0053	Boiled 20 minutes.*
	Gas.‡	0 0752	0.0733	−0.0029	Boiled 5 minutes †
	Gas.†	0.0752	0.0751	−0.0001	Boiled 5 minutes.† Residue blue with concentrated HCl and with KBr gave Br equivalent to 0.0004 V_2O_5.

* Brown. † Blue. ‡ Green.

the free iodine in the receiver by titration with thiosulphate measures the amount of vanadate present, according to the equations

$$V_2O_5 + 2\,HCl = V_2O_4 + H_2O + Cl_2$$
$$Cl_2 + 2\,KI = 2\,KCl + I_2.$$

Test experiments made with ammonium vanadate analyzed by the Holverscheit* process, are given in the preceding table.

The results of other similar experiments in which the reduced vanadium tetroxide in the residue was determined by titration with potassium permanganate after the addition of a manganous salt are given in the following table; and in some cases the results of the iodometric determination of chlorine set free are brought into comparison with the permanganate determination of the tetroxide.

Action of Hydrochloric Acid upon the Vanadate: Oxidation of the Residue by Permanganate.

NH_4VO_3.	HCl.	V_2O_5 calculated.	V_2O_5 found by chlorine in distillate	Error.	V_2O_5 found by $KMnO_4$ as residue.	Error.
grm.	$cm.^3$	grm.	grm.	grm.	grm.	grm.
0.1000	25, sp. gr. 1.20	0.0765	......	......	0.0740*	−0.0025
0.1000	25, sp. gr. 1.20	0.0765	......	....	0.0736*	−0.0029
0.1000	25, sp. gr. 1.20	0.0765	....	....	0.0744*	−0.0021
0.1000	30, sp. gr. 1.20	0.0765	0.0711	−0.0054	....*	......
	Gas.‡	......	0.0734	−0.0031	0.0758†	−0.0007
0.1000	25, sp. gr. 1.20	0.0765	0.0714	−0.0051	....*	......
	Gas.‡	......	0.0746	−0.0019	......*	......
	Gas.‡	......	0.0765	0.0000	0.0765†	..0.0000
0.3000	25, sp. gr. 1.20	0.2295	......	......	......*	......
	Gas.‡	......	......	......	......*	......
	Gas.‡	......	......	......	0.2284†	−0.0011
0.3000	25, sp. gr. 1.20	0.2295	0.2259	−0.0036	....*	......
	Gas.‡	......	0.2288	−0.0007	0.2302†	+0.0007
0.3000	25, sp. gr. 1.17	0.2295	0.2207	−0.0088	....*	....
	Gas.‡	......	0.2244	−0.0051	....*	.....
	Gas.‡	......	0.2258	−0.0037	....*	......
	Gas.‡	......	0.2269	−0.0026	....*	......
	Gas.‡	......	0.2281	−0.0014	0.2293†	−0.0002

* Residue after boiling was blue.
† Residue blue after boiling was still blue when resaturated.
‡ Residue as cooled in ice water and re-saturated with gaseous HCl.

* Inaug. Diss., Berlin, 1890, p. 48.

When hydrochloric acid of suitable concentration and the vanadate come into contact, the liberation of chlorine is immediate, some of which escapes from the solution while some is retained, and the reaction proceeds to a balance as indicated in the expression

$$V_2O_5 + 2\,HCl \rightleftarrows V_2O_4 + H_2O + Cl_2.$$

To complete the reduction of the higher oxide it is necessary to remove the free chlorine from the system while keeping up the requisite strength of the hydrochloric acid. In removing the chlorine by boiling, the concentration of the hydrochloric acid is diminished below the point at which action upon vanadic acid may take place with liberation of chlorine. This is why, in pushing the action to completion by the boiling process, it is necessary to increase the concentration of the hydrochloric acid from time to time, either by cooling and recharging with gaseous acid or by evaporating off the weak acid and replacing it by strong acid.

In continuing the study of this reaction, Gooch and Curtis* have tried to complete the reaction by bubbling a current of gaseous hydrochloric acid through the cooled residue of a single treatment by boiling. Under these conditions the hydrochloric acid is always at the concentration of activity, but the removal of the chlorine is necessarily slow since the current of gas must not be rapid enough to cause mechanical loss from the mixture.

The apparatus used was that described above and the vanadate was standardized by the method of Holverscheit.

In every experiment approximately 0.1 grm. of ammonium vanadate was first introduced into the reduction flask B. The air was expelled from the apparatus by carbon dioxide from the generator, the receiver C being charged with hydrochloric acid and the trap *g* with water. Concentrated hydrochloric acid (15 $cm.^3$) was admitted through the stoppered funnel A, and the mixture was boiled. The deep red color, produced when the acid was first added, gradually passed through green to blue. The flask was allowed to cool, carbon dioxide being admitted to fill the partial vacuum, and surrounded with ice. Hydrochloric acid gas was passed through the solution in the reduction flask,

* F. A. Gooch and R. W. Curtis, Am. Jour. Sci., [4], xvii, 41.

at the rate of one or two bubbles a second, for periods varying from ½ hour to 112½ hours, the solution turning brown at first and then changing to green or blue, according to the length of the period. For continuing the flow of gas for long periods small Kipp generators set up with massive ammonium chloride and concentrated sulphuric acid were found very convenient, a single charge serving to keep up the flow continuously over night.

At the end of the operation the degree of reduction was determined by titrating the contents of the flask, after dilution, with potassium permanganate in presence of a manganous salt. The data of the experiments are detailed in the following table:

Reduction by Hydrochloric Acid at Room Temperature.

NH_4VO_3 taken.	Time	V_2O_4 per 0.1000 grm. of vanadate taken.		Difference
grm.	hrs.	Calculated.	Found.	
0.1022	½	0.0695	0.0619	0.0076
0.1121	17	0.0695	0.0621	0.0074
0.1010	18½	0.0695	0.0678	0.0017
0.0980	21	0.0695	0.0658	0.0037
0.1044	30	0.0695	0.0600	0.0095
0.1043	112½	0.0695	0.0691	0.0004

The reduction of the vanadic acid to the condition of the tetroxide by the action of hydrochloric acid in the cold is obviously slow, as would be expected, but the results show that it may be made practically complete in this manner. No indication of reduction below the condition of the tetroxide by the agency of hydrochloric acid is apparent.

The Action of Hydrobromic Acid. In Holverscheit's method the reduction of vanadic acid by the action of hydrochloric acid and small amounts of potassium bromide to the condition represented by vanadium tetroxide is ideally complete. Under these conditions the concentration of hydrobromic acid is low. The effect of more concentrated hydrobromic acid upon the course of reduction has been investigated by Gooch and Curtis.*

In the experiments recorded in the following table, weighed portions of ammonium vanadate were introduced into the reduction

* F. A. Gooch and R. W. Curtis, Am. Jour. Sci., [4], xvii, 43.

flask, the receiver and trap were charged with a solution of potassium iodide (3 grm. : 350 cm.^{3}), the apparatus was filled with carbon dioxide, hydrobromic acid (15 cm.^{3}) of sp. gr. 1.68 (made by distilling a mixture of potassium bromide and sirupy phosphoric acid) was introduced through the funnel and the mixture was boiled eight or ten minutes. On the addition of the acid the vanadate dissolved and the solution took on a light green color, which on heating changed to red-brown and finally to a clear deep green. After cooling, the degree to which the vanadic acid had been reduced was estimated in two ways — by determining by means of standard sodium thiosulphate the iodine set free in the receiver by the bromine evolved, and by oxidizing by standard iodine the reduced product in the flask. The latter process followed the lines recommended by Browning* and consisted essentially in neutralizing the acid in the reduction flask by potassium bicarbonate, adding an excess of twentieth normal iodine, allowing the mixture to stand twenty minutes (all without admission of air), then transferring to a larger flask, introducing a slight excess of twentieth normal arsenite, and titrating with iodine in presence of starch.

The results of these experiments are calculated upon the hypotheses (first) that the vanadic acid is reduced to the tetroxide, (second) that it is reduced to the trioxide, and (third) that the trioxide and tetroxide are left in mixture.

Reduction by Hydrobromic Acid.

In 0.1000 NH_4VO_3 taken.		Found; calculated as V_2O_4.		Found; calculated as V_2O_3.		Found; calculated as mixture from figures for receiver.	
V_2O_4. grm.	V_2O_3. grm.	Flask. grm.	Receiver. grm.	Flask. grm.	Receiver. grm.	V_2O_4. grm.	V_2O_3. grm.
0.0699	0 0632	0 0913	0 0877	0.0413	0 0396	0.0521	0 0160
0.0699	0.0632	0.0879	0.0885	0.0397	0 0400	0 0513	0 0168
0 0699	0.0632	0.0849	0.0896	0.0384	0 0405	0.0502	0 0179
0.0699	0 0632	0.0858	0.0849	0.0388	0 0384	0.0549	0 0136
0.0699	0 0632	0.0854	0.0853	0 0386	0 0385	0 0545	0.0139
0.0699	0 0632	0.0841	0.0839	0.0380	0.0379	0.0559	0.0127

In other experiments the aqueous hydrobromic acid remaining after boiling was strengthened by cooling and recharging with the

* See page 345.

gaseous acid liberated by heating strong aqueous hydrobromic acid, and then reboiled. The bromine evolved in each boiling was determined, and finally the degree of reduction of the residue was estimated.

Reduction by Hydrobromic Acid.

In 0.1000 NH_4VO_3 taken.		Found; calculated as V_2O_4.		Found; calculated as V_3O_2.		Found; calculated as mixture from figures for receiver.	
V_2O_4. grm.	V_2O_3. grm.	Flask. grm.	Receiver. grm.	Flask. grm.	Receiver. grm.	V_2O_4. grm.	V_2O_3. grm.
0.0699	0.0632	*	......	......	......	......	.. ..
......	.....	0.0945	0.0943	0.0427	0.0426	0.0445	0.0221
0.0699	0.0632	*	0.0860	.....	0.0389	0.0538	0.0146
......	......	*	0.1104	......	0.0499	0.0295	0.0366
......	....	*	0.1291	.. ..	0.0584	0.0107	0.0535
.....	.....	0.1258	0.1291	.....	0.0569	0.0107	0.0535

* Recharged with HBr.

So it appears that increase in concentration of the hydrobromic acid tends to carry the reduction below the condition of the tetroxide. The highest degree of reduction reached in these experiments corresponds to a mixture of one-sixth tetroxide and five-sixths trioxide.

The Action of Hydriodic Acid. Browning * has shown that in accordance with the theory of reduction given in the following equation,

$$V_2O_5 + 2\,HI = V_2O_4 + H_2O + I_2,$$

good analytical results are obtained when solutions of the vanadate, 1 or 2 grm. of potassium iodide and 10 cm.[3] of sulphuric acid of half strength are boiled to a volume of about 35 cm.[3] and the residual solution is cooled, neutralized with an alkali bicarbonate (after the addition of a tartrate to prevent precipitation), and treated for some time with an excess of iodine which is followed by an excess of arsenious acid, the last being titrated by iodine in presence of starch.

In Browning's process the estimation of the reduced product in the residue is made the measure of action. The determination of the iodine set free by the action of hydriodic acid has been studied by Gooch and Curtis.† The experiments recorded be-

* See page 343.

† F. A. Gooch and R. W. Curtis, Am. Jour. Sci., [4], xvii, 45.

low were conducted, in the apparatus of Figure 3, in an atmosphere of carbon dioxide. The determinations of the iodine collected in potassium iodide in the receiver are compared with determinations of the reduction in the residue. The determinations of reduced vanadic oxide were made either by neutralization with potassium bicarbonate, by addition of a measured amount of standard iodine followed by a measured amount of standard arsenite and back titration with iodine, or by treatment of the residue with iodine before neutralizing with the bicarbonate (to avoid atmospheric oxidation), addition of arsenite and final titration with iodine.

Reduction by Sulphuric Acid and Potassium Iodide in Moderate Concentrations.

V_2O_4 in 0 1000 grm. NH_4VO_3 taken.	KI.	H_2SO_4 1 : 1.	Initial volume.	Final volume.	Reduction flask V_2O_4.		Receiver V_2O_4.	
					Found.	Error.	Found.	Error.
grm.	grm.	cm.^{3}	cm ^{3}	cm.	grm.	grm.	grm.	grm.
The residue neutralized before addition of iodine.								
0 0699	1	10	..	35	0 0668	−0.0031	0.0700	+0.0001
0.0699	1	10	.	35	0 0692	−0.0007	0.0715	+0.0016
0 0699	1	10	..	35	0.0686	−0 0013	0.0718	+0.0019
0.0699	1	10	50	35	0.0696	−0.0003	0.0744	+0.0045
0.0699	1	10	45	35	0 0678	−0.0021	0 0704	+0.0005
0.0699	1	10	50	35	0.0690	−0 0009	0.0710	+0.0011
0 0699	1	10	60	35	0 0681	−0 0018	0 0738	+0.0039
0.0699	1	10	55	35	0 0689	−0.0010	0 0724	+0 0025
0.0699	0.6	6	55	35	0 0679	−0 0020	0.0722	+0 0023
Iodine added before neutralizing the residue.								
0.0699	1	10	55	35	0 0699	0.0000	0.0713	+0 0014
0.0699	1	6	55	35	0.0713	+0.0014	0.0722	+0.0023
0 0699	1	10	80	35	. . . .	. . . .	0 0725	+0.0026
0.0699	1	10	75	35	0 0710	+0 0011	0.0718	+0 0019
0 0699	0.6	4	55	35	0.0701	+0 0002	0.0709	+0.0010
0 0699	0.6	10	55	35	0 0717	+0 0018	0.0745	+0.0046
0.0699	0.6	6	55	35	0 0706	+0 0007	0 0734	+0 0035
0 0699	0.6	6	55	35	0 0703	+0.0004	0 0727	+0 0028
0.0699	0 6	4	55	35	0 0700	+0 0001	0.0731	+0.0032

In those experiments in which the vanadate was treated with dilute sulphuric acid and potassium iodide, the iodine found in the receiver indicates a trifling reduction beyond the condition of the tetroxide V_2O_4, averaging 0.0023 grm.; and the same in general is true in regard to those determinations of the residue in the reduction flask in which the iodine was added before the bicarbonate, the over-reduction averaging 0.0007 grm. The determinations by reduction in the residue, in which the neutralization took place before the addition of the iodine, uniformly show an incomplete reduction — amounting in the average to 0.0015 grm. — an effect which is without doubt due to the action of air upon the sensitive alkaline solution of the reduced vanadate. Although the conditions of concentration are such that in absence of the vanadic acid there is no tendency (barring the insignificant action of dissolved air) toward liberating iodine, a little more iodine is liberated by vanadic acid when acted upon by sulphuric acid and potassium iodide than would correspond to a reduction of vanadic acid to the condition of the tetroxide.

Reduction by Hydrochloric Acid and Potassium Iodide at High Concentrations.

V_2O_5 in 0.1000 NH_4VO_3 taken.	HCl, concentrated.	KI.	Initial volume.	Final volume.	V_2O_5 found.	
grm.	cm.[3]	grm.	cm.[3]	cm.[3]	Flask.	Receiver.
A.						
0.0632	5	0.6	50	*	...	0.0323
				2	0.0613	0.0627
0.0632	5	1	55	*	. . .	0.0327
				2	0.0618	0.0637
0.0632	12.5	1	50	*	.....	0.0378
				2	0.0615	0.0644
0.0632	15	1	45	*	......	0.0372
				2	0.0617	0.0642
0.0632	25	1	50	*	.....	0.0518
				2	0.0618	0.0657

* Approximately 40 cm.[3], the vanishing point of the iodine color.

B.						
0.0632	15	1	16	2	0.0618	0.0630
0.0632	15	1	16	2	0.0612	0.0627
0.0632	15	1	16	2	0.0617	0.0625
0.0632	15	1	16	2	0.0620	0.0630
0.0632	15	1	16	2	0.0616	0.0627
0.0632	15	1	16	2	0.0618	0.0628
0.0632	15	1	16	2	0.0617	0.0630
0.0632	15	1	16	2	0.0617	0.0629
0.0632	15	1	16	2	0.0616	0.0629
0.0632	15	1	16	2	0.0618	0.0630

On the other hand, the degree to which vanadic acid may be reduced by hydrochloric acid and potassium iodide is found to depend upon the concentration. By carrying the distillation to a very low final volume, as may be done without difficulty in the apparatus described, the reduction may be made to correspond very closely to the condition of vanadium trioxide, V_2O_3. This is shown by the results of the preceding table.

The addition of phosphoric acid, as proposed by Friedheim and Euler * is unnecessary and may even work disadvantageously by permitting a dangerous rise in temperature in the still liquid residue when low volumes are reached.

This is shown in the following series of experiments in which 1 grm. of potassium iodide, 2 cm.³ of sirupy phosphoric acid (sp. gr. 1.70) and 0.1 grm. ammonium vanadate were treated, the initial volume being 60 cm.³

Reduction by Phosphoric Acid and Potassium Iodide at High Concentrations.

In 0.1000 NH_4VO_3 taken.		Final volume.	Reduction flask.		Receiver.	
V_2O_4. grm.	V_2O_3. grm.	cm.³	V_2O_4. grm.	V_2O_3. grm.	V_2O_4. grm.	V_2O_3. grm.
0.0699	0.0632	35	0 0693	......	0.0698	......
0.0699	0.0632	25	0.0705	......	0.0711	......
0.0699	0.0632	22	0 0711	......	0.0706	....
0.0699	0.0632	4	......	0.0606	......	0.0623
0.0699	0.0632	2	....	*	....	0.0617
0.0699	0.0632	2	.....	0.0597	....	0.0612
0.0699	0.0632	1.7	....	0.0621	....	0 0613
0 0699	0 0632	1.6	...	*	.....	0 0624
0 0699	0.0632	1.4	......	0.0604	......	0.0629

* Flask broke.

These figures indicate that when the distillation is continued until the volume is about 35 cm.³ the condition of oxidation corresponds nearly to that of the tetroxide and that when the residue is concentrated almost to dryness the figures approach the value for the trioxide. Under the conditions the determinations in the residue are of doubtful value, for more or less spattering occurs, and the temperature may be such that a volatile compound of vanadium begins to distil.

It is evident, therefore, that hydrochloric acid is capable of reducing vanadic acid, even in the cold, to the condition of the tetroxide, and under none of the conditions tried does reduction go further; that hydrobromic acid, which in small concentra-

* Ber. Dtsch. chem. Ges., xxviii, 2071.

tions gives a definite reduction to the condition of the tetroxide,* may easily push the reduction well on toward the condition of the trioxide at higher concentration; and that the reduction by hydriodic acid may be carried at will to either of two stages — that of the trioxide or that of the tetroxide.

The Determination of Vanadic Acid by Reduction in Acid Solution and Reoxidation by Iodine in Alkaline Solution.

Reduction by Organic Acids. A method for the determination of vanadium has been described by Browning,† in which tartaric acid is used to reduce vanadic acid to the condition of the tetroxide which is then reoxidized in alkaline solution by standard iodine and thus estimated. Browning and Goodman ‡ have shown that oxalic acid and citric acid may be similarly used to reduce vanadic acid, and have studied the conditions under which the determination is accurate in presence of tungstic and molybdic acids.

The mode of proceeding in the estimation of vanadium, by the use of either tartaric, oxalic or citric acid to effect reduction in hot solution, may be briefly summarized as follows: To the solution of a vanadate add approximately 1 grm. of the organic acid for every tenth of a gram of substance to be determined. Heat the solution to boiling. To the cooled liquid add about 5 grm. of potassium bicarbonate for every gram of acid used. Add iodine in slight excess and set aside until no further bleaching is noticeable. Destroy the excess of iodine with standard arsenite, add starch, and titrate back with standard iodine. The total amount of iodine used less the equivalent of the arsenious oxide is the measure of oxidation according to the equation

$$V_2O_4 + I_2 + H_2O = V_2O_5 + 2\,HI.$$

Results of this procedure are given on pages 342, 343 and 344.

The process of reduction by means of tartaric acid in boiling solution fails in presence of molybdic acid, which under the con-

* Holverscheit's procedure.

† Zeit. anorg. Chem., vii, 158.

‡ Philip E. Browning and Richard J. Goodman, Am. Jour. Sci., [4], ii, 355.

Reduction by Tartaric Acid in Boiling Solution: Applicable in Presence of Tungstic Acid.

V_2O_5 taken. grm.	V_2O_5 found grm.	Error. grm.	Sodium tungstate. grm.	Tartaric acid. grm.
0 1621	0 1618	−0 0003	..	2
0 1620	0.1624	+0.0004	..	2
0 1614	0 1622	+0.0008	..	2
0 1619	0 1606	−0.0013	..	1
0 1604	0 1597	−0 0007	..	2
0 1618	0 1615	−0.0003	..	3
0.1298	0 1305	+0.0007	..	1
0 1294	0 1297	+0 0003	..	1
0 1618	0 1618	0 0000	..	2
0 2588	0 2575	−0 0013	..	3
0 2722	0 2726	+0 0004	..	2
0 3273	0 3269	−0 0004	..	2
0 1618	0.1615	−0 0003	1	3
0 1615	0 1606	−0.0009	1	3
0 1618	0 1624	+0 0006	1	3
0.1619	0.1624	+0 0005	1	3
0 1627	0.1623	−0 0004	1	3
0.1621	0 1624	+0 0003	1	4
0.2587	0 2574	−0 0013	1	4
0 2587	0 2589	+0 0002	1	4

Reduction by Oxalic Acid in Boiling Solution: Applicable in Presence of Tungstic Acid and Molybdic Acid.

V_2O_5 taken. grm.	V_2O_5 found. grm.	Error. grm.	Oxalic acid. grm.	Ammonium molybdate. grm.	Sodium tungstate. grm.
0 1806	0.1803	−0.0003	1	..	..
0 1950	0.1955	+0 0005	1	..	..
0.1959	0.1955	−0 0004	1	..	..
0.1950	0.1959	+0.0009	1	..	..
0 1954	0.1977	+0.0023	1	..	..
0.1956	0.1960	+0.0004	1	..	..
0.1956	0.1964	+0.0008	1	..	..
0.1956	0.1957	+0.0001	1	..	..
0.3900	0.3899	−0 0001	2	..	..
0.3897	0 3917	+0.0020	2	..	..
0.3903	0.3905	+0.0002	2	..	..
0.1954	0.1959	+0.0005	2	1	..
0.1957	0.1960	+0.0003	2	1	..
0.1954	0 1961	+0.0007	2	1	..
0.1806	0.1818	+0.0012	3	..	..
0.1807	0.1827	+0.0020	3	..	..
0.1809	0.1803	−0.0006	3	1	..
0.1956	0.1961	+0.0005	3	..	1
0.3611	0.3617	+0.0006	5	..	..
0.3616	0.3626	+0.0010	5	1	..

Reduction by Citric Acid in Boiling Solution: Applicable in presence of Tungstic Acid and Molybdic Acid.

V_2O_5 taken grm.	V_2O_5 found. grm.	Error. grm.	Citric acid. grm.	Ammonium molybdate. grm.	Sodium tungstate grm.
0 1956	0 1956	0 0000	1	..	.
0 3905	0 3921	+0 0016	2	...	. .
0 1960	0 1960	0 0000	1	...	..
0 1953	0 1060	+0 0007	1	...	.
0 2088	0 2082	−0 0006	2	...	
0 2100	0 2098	−0 0002	2	...	...
0 2092	0 2107	+0 0015	1	...	.
0 2092	0 2107	+0 0015	2	..	
0 2096	0 2082	−0 0014	2	0.5	.
0.2099	0 2116	+0 0017	3	0.5	
0 2005	0 2101	+0 0006	2	...	0 5
0 2099	0 2095	−0 0004	3	...	0 5

ditions is also somewhat reduced. Browning and Goodman have shown, however, that a period of long standing — twenty-four hours — may be substituted in this process for the period of boiling and that then the process, otherwise the same, is effective in presence of molybdic acid as well as tungstic acid.

Reduction by Hydriodic Acid. Browning * has shown that vanadic acid may be reduced by the action of potassium iodide† and sulphuric acid, and the residue of vanadium tetroxide titrated by iodine in alkaline solution ‡ with sufficient exactness to form the basis of a rapid and fairly accurate method for the determination of the vanadium.

According to this procedure a vanadate solution is placed in a trapped Erlenmeyer flask § with potassium iodide (1 grm. to 2 grm.) and 10 cm.³ of sulphuric acid [1 : 1]. The mixture is boiled until fumes of iodine are no longer visible and the escaping steam gives no indication of free iodine with red litmus paper.‖ This point is reached, if the potassium iodide is kept within the above limits, when the volume has diminished from approximately 100 cm.³ to 35 cm.³ The cooled solution is nearly neu-

* Philip E. Browning, Am. Jour. Sci., [4], ii, 185.

† See page 337.

‡ Browning, Zeit. anorg. Chem., vii, 158.

§ See Fig. 6, page 6.

‖ See page 450.

Reduction by Tartaric Acid on Standing 24 Hours or More: Applicable in Presence of Tungstic Acid and Molybdic Acid.

V_2O_5 taken. grm.	V_2O_5 found. grm.	Error. grm.	Time in. days.	Tartaric acid. grm.	Total volume. cm.^{3}
0 1646	0 1649	+0.0003	1	4	25
0 1640	0 1606	−0.0034	1	4	65
0 1293	0 1264	−0 0029	2	3	55
0 1633	0.1628	−0 0005	2	4	65
0.1293	0.1288	−0.0005	3	2.5	50
0 1298	0.1299	+0.0001	3	2.5	50
0 1295	0.1279	−0.0016	3	3	55
0 1617	0.1597	−0.0020	4	2	70
0 1623	0 1622	−0.0001	4	3	80

V_2O_5 taken. grm.	V_2O_5 found. grm.	Error. grm.	Ammonium molybdate. grm.	Sodium tungstate. grm.	Time in days.	Total volume. cm.^{3}	Tartaric acid. grm.
0 1552	0.1558	+0.0006	..	..	1	25	5
0.1289	0.1301	+0.0012	...	..	1	25	5
0.2583	0.2587	+0.0004	...	..	1	50	5
0.1293	0.1299	0.0006	1	..	1	25	6
0 2582	0.2591	+0.0009	1	..	1	50	6
0.2582	0.2588	+0.0006	1	..	1	50	5
0.1297	0.1308	+0.0011	1	..	1	25	5
0.1291	0.1289	−0.0002	...	1	1	25	6
0.2582	0.2568	−0.0014	...	1	1	50	5
0 1293	0.1299	+0.0006	1	1	1	25	8
0 2582	0.2579	−0.0003	1	1	1	50	5
0.1550	0.1538	−0.0012	...	..	2	25	5
0 1556	0.1545	−0.0011	...	..	2	25	5
0.1289	0.1296	+0.0007	...	..	2	25	5
0.1549	0.1527	−0.0022	0.5	..	2	25	5
0.1553	0.1548	−0.0005	1	..	2	25	5
0 1556	0.1554	−0.0002	1	..	2	25	5
0.1293	0.1310	+0.0017	1	..	2	25	6
0.1295	0.1299	+0.0004	...	1	2	25	6
0.1293	0.1289	−0.0004	1	1	2	25	7
0.1293	0.1301	+0.0008	...	..	3	25	5
0.1289	0.1299	+0 0010	0.5	..	3	25	5
0.1293	0.1292	−0 0001	1	..	3	25	7
0 1556	0.1567	+0.0011	1	..	3	30	5
0.1291	0.1289	−0.0002	1	1	3	25	7
0.1550	0.1557	+0.0007	...	..	4	25	5
0.1554	0.1557	+0.0003	1	..	4	25	5
0 1556	0.1557	+0.0001	0.5	..	4	25	5

tralized by sodium hydroxide,* a little tartaric acid is added to prevent precipitation of the tetroxide, and neutralization is completed by acid potassium carbonate. To the cooled solution n/10 iodine is added in slight excess, the flask is closed with a paraffined cork, and the mixture is allowed to stand about half an hour. The excess of iodine is taken up by n/10 arsenite, and the excess of the last is titrated with n/10 iodine in presence of starch. The difference between the amount of n/10 arsenite used, expressed in terms of iodine, and the whole amount of n/10 iodine employed gives the amount of iodine necessary to oxidize the vanadium tetroxide to vanadium pentoxide.

Results of this procedure are given in the table.

Reduction by Sulphuric Acid and Potassium Iodide: Reoxidation by Iodine in Alkaline Solution: Final Titration with Standard Arsenite.

V_2O_5 taken. grm.	V_2O_5 found. grm.	Error. grm.	KI. grm.	H_2SO_4(1 : 1). cm^3
0 1699	0.1690	−0 0009	1	10
0 1704	0 1699	−0 0005	1	10
0 1706	0 1700	−0 0006	1	10
0.1702	0 1692	−0 0010	1	10
0 3613	0 3620	+0 0007	1.5	10
0.1805	0 1803	+0 0002	1	10
0 3614	0 3620	+0 0006	1.5	10
0 1811	0.1814	+0.0003	1	10
0 1807	0 1815	+0.0008	1	10
0 3613	0 3620	+0 0007	1 5	10
0.3679	0 3674	−0 0005	1 5	10
0.3612	0 3608	−0 0004	1 5	10
0 2893	0 2907	−0 0014	1 5	10
0 3456	0.3448	−0 0008	1.5	10
0 3453	0.3448	−0 0005	1 5	10
0 3907	0 3912	+0 0005	2	10
0 3908	0.3898	−0 0010	1	10
0.3906	0.3921	+0 0015	2	10
0 3909	0.3912	+0 0003	1.5	10

Reduction by Hydrobromic Acid. For the estimation of the bromine set free by the action of potassium bromide and sulphuric acid upon a vanadate according to Holverscheit, Browning † has substituted the reoxidation of the reduced vanadium tetroxide by iodine in presence of an acid carbonate.

* The potassium or sodium hydroxide for this work must be free from alcohol, as the solution is allowed to stand with iodine after neutralization. It is conveniently prepared by mixing potassium or sodium carbonate in proper proportions with calcium oxide and filtering off the calcium carbonate.

† Philip E. Browning, Am. Jour. Sci., [4], ii, 185.

According to this method, reduction is effected by boiling in a trapped Erlenmeyer flask * a mixture of the vanadate in solution with 1 grm. to 2.5 grm. of potassium bromide and 10 cm.³ of [1 : 1] sulphuric acid until bromine is no longer present in the steam, a point ordinarily reached when the original volume of about 100 cm.³ has diminished to a volume of about 25 cm.³. At this point the solution is nearly neutralized with alkali hydroxide free from alcohol † and completely neutralized by an acid alkali carbonate, a little tartaric acid having been first added to prevent precipitation of the vanadium tetroxide. An excess of $n/10$ iodine is added and the mixture is allowed to stand half an hour with the flask closed by a paraffined stopper. The excess of iodine is taken up by $n/10$ arsenite, and the slight excess of the last is determined by titration with $n/10$ iodine. The difference between the entire amount of iodine used and the iodine equivalent of the arsenite employed measures the oxidation of vanadium tetroxide to vanadium pentoxide.

Results of this procedure are given in the table.

Reduction by Sulphuric Acid and Potassium Bromide: Reoxidation by Iodine in Alkaline Solution: Titration with Arsenite.

V_2O_5 taken. grm.	V_2O_5 found. grm.	Error. grm.	KBr. grm.	H_2SO_4 [1 : 1]. cm.³
0.1800	0 1876	−0 0014	1	10
0.1886	0.1886	0 0000	2	10
0.1885	0.1882	−0 0003	1	10
0.1885	0.1886	+0 0001	1.5	10
0.1881	0.1873	−0 0008	1.5	10
0.1886	0 1882	−0 0004	2	10
0.3907	0 3894	−0 0013	2	10
0 3907	0.3903	−0.0004	2	10
0.3907	0.3894	−0 0013	2	10
0.3909	0.3889	−0 0020	2	10
0.3911	0.3903	−0 0008	1.5	10
0.3902	0.3900	−0.0002	2.5	10

The Use of the Jones Reductor in the Estimation of Vanadic Acid.

The reduction of vanadic acid to the condition of vanadium tetroxide preparatory to estimating it by titration with potassium permanganate may be accomplished by the use of sulphurous acid or hydrogen sulphide. The more convenient method of

* See Fig. 6, page 6. † See note, page 345.

treatment by zinc and free acid, followed by direct titration, is not applicable on account of the irregularities in reduction found under the conditions. Gooch and Gilbert * have developed a reliable method for bringing the product of reduction of vanadic acid by zinc and acid definitely to the condition of the tetroxide, in order that the Jones reductor so useful in the preparation of salts of iron for titration by potassium permanganate may be made similarly applicable in the estimation of vanadic acid and its salts.

A convenient form of the Jones reductor † is made as follows: The contracted end of a piece of glass tubing, 2 cm. in inside diameter and 50 cm. long, is sealed to a stopcock prolonged in a smaller tube, 0.5 cm. in inside diameter, to a length of 24 cm. At the point of contraction in the larger tube is placed a piece of platinum gauze, next to this a mat of fine glass wool 2 cm. in thickness, and upon the last a column 40 cm. long of amalgamated zinc of a size to pass a sieve of eight meshes to the centimeter. The smaller tube is passed through a rubber stopper fitted to a vacuum flask and the last is connected through a pressure regulator with the vacuum pump. In using this apparatus the pump is started, the regulator set to give a pressure in the flask less than the outside pressure by an amount equal to 20 cm. of water, the reducing column of zinc warmed by passing through it hot distilled water followed by 100 $cm.^3$ of hot 1 per cent sulphuric acid, and then the solution of the salt of vanadic acid to be reduced is gradually drawn through the zinc in small portions, alternating with portions of the 1 per cent sulphuric acid amounting to 100 $cm.^3$ Finally, the column is washed down with 100 $cm.^3$ more of the dilute sulphuric acid and about 250 $cm.^3$ of hot distilled water. Throughout the entire series of operations care is taken to keep the zinc covered with liquid and so out of contact with the air.

The lavender solution collected in the flask contains vanadium dioxide, which, when exposed to the action of a current of air takes, as Roscoe has shown,‡ the blue color of the tetroxide; but the action of the molecular oxygen of the air proves to be insufficient to bring about the complete oxidation of the lower

* F. A. Gooch and R. D. Gilbert, Am. Jour. Sci., [4], xv, 389.

† Described in Blair's Chemical Analysis of Iron.

‡ Ann. Pharm. Suppl., vi, 98 (1868).

vanadium oxides in acid solution to the condition of the tetroxide within a reasonable time. Ordinary oxidizers carry the action too far, oxidizing the tetroxide as well as the lower oxides. It is found, however, that silver oxide and silver salts supply oxygen sufficiently to affect the lower oxides easily while leaving the tetroxide intact. Silver sulphate appears to be the most convenient form in which to use the silver compound.

Regulation of Reduction by Use of Silver Sulphate. In treating the solution, therefore, in the reductor as described, the receiving flask is charged with a saturated solution of silver sulphate, and when the reduced solution issuing from the reductor meets the silver sulphate a muddy deposition of finely divided silver begins. The mixture is boiled and filtered upon asbestos in a perforated crucible. The solution, now about 700 cm.^{3} in volume, is heated to the boiling point and titrated with $n/10$ potassium permanganate. Upon boiling the mixture, the metallic silver gathers into a single spongy mass and leaves the solution so clear that, were it not that spongy silver is easily acted upon by the permanganate, the titration might be made without previous filtration.

For small amounts of vanadium the determinations are accordant and exact. Wider variations, due to the difficulty of catching the pink end reaction in presence of the reddish yellow color which appears as the vanadic acid is formed in considerable amount, are inherent in the permanganate process of titration when large amounts of vanadic acid are involved.

The results of test experiments follow in the table:

Reduction by Zinc: Partial Oxidation by Silver Sulphate: Titration with Permanganate.

V_2O_5 taken.	$KMnO_4$ required nearly $n/20$.	V_2O_5 found.	Error.
grm.	cm.^{3}	grm.	grm.
0 0767	17	0.0770	+0 0003
0.0767	17.04	0.0771	+0 0004
0.0767	17.05	0.0772	+0 0005
0.0767	16.96	0.0768	+0.0001
0.0767	16.98	0.0769	+0.0002
0.0767	17	0.0770	+0 0003
0.1918	42.9	0.1942	+0 0024
0.1918	42.7	0.1933	+0 0015

Registration of Reduction by Use of Ferric Sulphate.

In a study of the effects of zinc and magnesium in the reduction of vanadic acid, Gooch and Edgar,* have shown that the degree in which vanadic acid may be reduced by magnesium in the presence of hydrochloric acid or sulphuric acid is irregular and dependent upon conditions not easily controlled. With magnesium amalgam the reduction proceeds most readily and approximates more or less to the condition of V_2O_2, but under none of the conditions tried is the reduction by magnesium sufficiently definite to be applied in a good analytical process. With zinc the reduction is more regular, but the condition of oxidation of the product of the reduction of vanadic acid by zinc and hydrochloric acid in the flask, by zinc and sulphuric acid in the flask, or by amalgamated zinc in the reductor, never corresponds exactly to V_2O_2 when titration is made in air. It is shown that the use of the zinc reductor carries the reduction easily and rapidly to the condition of V_2O_2, and that by anticipating the oxidizing action of the air by means of a ferric salt in the receiver † the solution is made less sensitive to the action of the air, while the highest degree of reduction is registered by the ferrous salt formed; but a solution containing vanadium in the condition of V_2O_2 cannot be exposed to air, even momentarily, without undergoing oxidation.

Reduction by Zinc: Treatment with Ferric Salt: Titration by Permanganate.

Ferric alum, 10 per cent solution. cm.^{3}	Phosphoric acid syrup. cm.^{3}	$KMnO_4$ $n/10 \times 1.052$. cm.^{3}	V_2O_5 taken as $NaVO_3$. grm.	V_2O_5 calculated from oxidation from V_2O_2. grm.	Error in terms of V_2O_5. grm.
25	5	43.10	0.1381	0.1378	−0 0003
25	5	43.20	0.1381	0 1381	0.0000
25	5	43 30	0.1381	0.1384	+0.0003
25	5	43.28	0.1381	0 1384	+0.0003
25	5	43.32	0.1381	0.1385	+0 0004
15	3	21.60	0.0691	0 0690	−0.0001
15	3	21.62	0.0691	0 0691	0 0000
15	3	21.80	0.0691	0.0696	+0.0005
40	8	64 90	0.2072	0.2075	+0.0003

According to the recommended procedure, the receiver attached to the zinc column is charged with a solution of ferric

* F. A. Gooch and Graham Edgar, Am. Jour. Sci., [4], xxv, 233.

† See page 426.

alum, gentle suction is applied, and through the column of amalgamated zinc are passed in succession hot water (100 cm.^{3}), 2.5 per cent sulphuric acid (100 cm.^{3}), the solution of vanadic acid in 2.5 per cent sulphuric acid (125 cm.^{3}), and finally hot water (200 cm.^{3}). To the receiver is added sirupy phosphoric acid (4 cm.^{3}), to decolorize the solution, and the titration of the hot solution is made in the usual manner with potassium permanganate. Correction should be made for the action of the zinc column upon the reagents without the vanadic acid. Experimental tests of the method are given in the table.

The Estimation of Vanadic and Arsenic Acids and of Vanadic and Antimonic Acids in the Presence of One Another.

If a solution containing arsenic and vanadic acids or antimonic and vanadic acids is boiled with tartaric or oxalic acids, the vanadic acid is reduced to tetroxide and may be reoxidized in alkaline solution by iodine * according to the equation

$$V_2O_4 + I_2 + H_2O = V_2O_5 + 2\,HI.$$

If a solution containing arsenic and vanadic acids, or antimonic and vanadic acids, is reduced with sulphur dioxide under proper conditions, the arsenic acid is reduced to arsenious acid, or the antimonic acid to antimonious acid, and the vanadic acid to tetroxide, so that after boiling off the excess of reagent the reoxidation by iodine in alkaline solution should proceed according to one of the equations

$$As_2O_3 + V_2O_4 + 3\,I_2 + 3\,H_2O = As_2O_5 + V_2O_5 + 6\,HI$$
$$Sb_2O_3 + V_2O_4 + 3\,I_2 + 3\,H_2O = Sb_2O_5 + V_2O_5 + 6\,HI,$$

and the iodine thus used should in each case correspond to the sum of the two oxides. Edgar † has shown that if aliquot parts of the same solution are treated according to these processes the titration of the solution reduced by tartaric acid determines the vanadium present and that the titration of the solution reduced by sulphur dioxide determines the sum of the oxides of arsenic and vanadium, or of antimony and vanadium, the difference in the number of cubic centimeters of iodine used in the two

* Browning, Zeit. anorg. Chem., vii, 158; Browning and Goodman, Am. Jour. Sci., [4], ii, 355.

† Graham Edgar, Am. Jour. Sci., [4], xxvii, 299.

cases being the amount required to oxidize the arsenic oxide or antimony oxide.

The method of treatment in detail is as follows: The solution of arsenate and vanadate, or of antimonate and vanadate, is divided into two portions.

(I) One portion is boiled with 1 to 2 grm. of tartaric or oxalic acid until the blue color of the vanadium tetroxide indicates complete reduction. The solution is then cooled, nearly neutralized with potassium bicarbonate, and an excess of standard iodine solution is added. Neutralization is completed, an excess of bicarbonate added, and the solution allowed to stand for from fifteen to thirty minutes. The excess of iodine is then removed with standard arsenious acid and the solution titrated to color after the addition of starch.

(II) The second portion of the solution is placed in a small pressure flask and slightly acidified with sulphuric acid. A strong solution of sulphurous acid (25 cm.^{3}) is then added and the flask is closed and heated for one hour on the steam bath.* After cooling, the flask is opened and the solution transferred to an Erlenmeyer flask and boiled to remove the excess of sulphur dioxide, a current of carbon dioxide being passed into the liquid to facilitate the removal of the last traces. The solution is then cooled, nearly neutralized with potassium bicarbonate and an

Arsenic and Vanadium.

V_2O_5 taken. grm.	V_2O_5 found. grm.	V_2O_5 error. grm.	As_2O_5 taken. grm.	As_2O_5 found. grm.	As_2O_5 error. grm.	(I) n/10 iodine. cm.^{3}	(II) n/10 iodine. cm.^{3}
0.1183	0.1181	−0.0002	0.0960	0 0961	+0.0001	12 95	29 65
0.1183	0.1183	0 0000	0.0960	0.0962	+0 0002	12 97	29 70
0.1183	0 1182	−0 0001	0 0960	0 0962	+0.0002	12 96	29 70
0.0591	0 0593	+0.0002	0.0480	0.0480	0 0000	6 50	14 85
0 0591	0 0594	+0 0003	0 0480	0 0482	+0.0002	6 52	14 90
0.0591	0.0589	−0 0002	0 0480	0.0483	+0 0003	6 45	14.86
0.1774	0.1779	+0.0005	0 1440	0 1438	−0 0002	19.50	44.50
0.1774	0.1774	0.0000	0.1440	0 1440	0.0000	19.45	44 50
0.1774	0.1776	+0.0002	0.1440	0.1442	+0 0002	19.47	44.45
0.2366	0.2371	+0 0005	0 0480	0 0478	−0.0002	26.00	34.31
0 2366	0.2366	0 0000	0.0480	0.0480	0.0000	25.95	34.30
0.0591	0 0593	+0.0002	0.1440	0.1439	−0.0001	6.50	38 90

* McCay, Am. Chem. Jour., vii, 273; Von Knorre, Zeit. angew. Chem. (1888), 155.

Antimony and Vanadium.

V_2O_5 taken. grm.	V_2O_5 found. grm.	V_2O_5 error. grm.	Sb_2O_5 taken. grm.	Sb_2O_5 found. grm.	Sb_2O_5 error. grm.	(I) $n/10\times$ 0 9375 iodine. cm³.	(II) $n/10\times$ 0 9375 iodine. cm³.
0.1183	0.1185	+0 0002	0 0757	0.0759	+0.0002	13 85	23 69
0.1183	0 1186	+0 0003	0.0757	0 0764	+0.0007	13 87	24 05
0.1183	0.1183	0 0000	0.0757	0 0760	+0.0003	13 83	23 95
0.1774	0.1777	+0 0003	0.1261	0 1258	−0.0003	20 80	37.55
0.1774	0.1777	+0 0003	0.1261	0 1258	−0.0003	20 80	37 55
0.1774	0.1773	−0 0001	0.1261	0 1260	−0 0001	20 75	37 50
0.2366	0.2376	+0 0010	0.1261	0 1257	−0.0004	27 80	44.52
0 2366	0.2369	+0.0003	0.1261	0.1263	−0.0002	27 70	44 50
0.2366	0.2369	+0.0003	0 1261	0 1260	−0.0001	27 70	44.45

excess of iodine added as before. After completing the neutralization, adding an excess of bicarbonate and allowing to stand for one-half hour, the excess of iodine was determined by arsenious acid as before.

As before stated, the titration figures of (I) determine the vanadium, and the subtraction of these figures from those of (II) gives figures for the arsenic or the antimony, as the case may be.

Results of the method are given in the table.

The Estimation of Vanadic Acid Associated with Chromium, with Molybdenum, and with Iron.

Methods for the estimation of vanadic acid associated with chromic acid, with a ferric salt, with chromic acid and a ferric salt, or with molybdic acid, are described elsewhere.*

The Estimation of Vanadium in the Tetroxide Condition by the Action of Potassium Ferricyanide in Alkaline Solution and Potassium Permanganate in Acid Solution.

To the estimation of vanadium in the tetroxide condition of oxidation, Palmer † has applied the oxidizing action of potassium ferricyanide in alkaline solution with reoxidation of the resulting ferrocyanide in acid solution, following the general plan formerly employed by Browning and Palmer for the estimation of cerium ‡

* See page 510.
† Howard E. Palmer, Am. Jour. Sci., [4], **xxx, 141.**
‡ See page 249.

and thallium,* and by Palmer for the estimation of arsenic, antimony and tin †. The reactions involved may be represented by the following equations:

$$V_2O_4 + 2\,K_3FeC_6N_6 + 2\,KOH = V_2O_5 + 2\,K_4FeC_6N_6 + H_2O,$$

$$5\,K_4FeC_6N_6 + KMnO_4 + 4\,H_2SO_4 = 5\,K_3FeC_6N_6 + 3\,K_2SO_4 + MnSO_4 + 4\,H_2O.$$

Vanadium was prepared in the tetroxide condition by treating portions of a solution of ammonium vanadate, made slightly acid with hydrochloric acid, with a current of sulphur dioxide until the clear blue color indicated complete reduction of the vanadic acid to the condition of the tetroxide, V_2O_4. The solution was then boiled in a current of carbon dioxide until the last traces of sulphur dioxide had been removed. To determine the vanadium in such a solution the following procedure was adopted.

To the cooled solution are added in solution potassium ferricyanide to at least tenfold the amount theoretically necessary for the oxidation and about 6 grm. of potassium hydroxide, the concentrations being such that the total volume amounts to 100 cm.^{3} or 125 cm.^{3}. Barium hydroxide is added to precipitate the vanadate, which otherwise would form with the ferrocyanide ‡ an insoluble compound refractory in the subsequent titration by the permanganate. The precipitate is settled and filtered off on an asbestos felt. The filtrate and washings are acidified with hydrochloric acid and titrated with permanganate. The results obtained are recorded in the table.

Estimation of Vanadium Tetroxide.

V_2O_5 taken.* grm.	$K_3FeC_6N_6$ used. grm.	KOH used. grm.	V_2O_5 found. grm.	Error. grm.
0 0960	4	6	0 0959	−0.0001
0 0960	4	6	0 0954	−0 0006
0.0960	4	6	0 0956	−0.0004
0 0960	4	6	0 0962	+0.0002
0 0960	4	6	0 0956	−0.0004
0.0960	4	6	0 0959	−0 0001
0 0960	4	6	0 0961	+0.0001
0.0960	4	6	0 0962	+0.0001
0 0960	4	6	0 0961	0.0000
0.0960	4	6	0 0961	+0.0001

* Reduced by sulphur dioxide as described.

* See page 223. † See page 322.

‡ Grützner, Chem. Centralblatt, 1902, i, 500.

In this process the large proportions of ferricyanide and potassium hydroxide are necessary to insure complete oxidation of the vanadium at the concentrations employed. If the dilution is greater, more ferricyanide is required. Titration by permanganate in a solution acidified with sulphuric acid is unsatisfactory on account of the difficulty in noting the end point in the presence of barium sulphate precipitated by the action of the excess of barium hydroxide. The ferrocyanide, however, may be safely titrated by permanganate in the cold in dilute hydrochloric acid solution.

CHAPTER IX.

OXYGEN; SULPHUR; SELENIUM; TELLURIUM.

OXYGEN.

The Iodometric Determination of Oxygen in Air and in Aqueous Solution.

THE reactions by which a satisfactory iodometric determination of the oxygen of perchlorates may be accomplished * have also been applied by Kreider † to the determination of the oxygen of the air or of oxygen dissolved in water.

Determination of Oxygen in Air. The method, in brief, consists in conducting a known volume of air through a strong solution of hydriodic acid in the presence of nitrogen dioxide, subsequently neutralizing the acid with potassium bicarbonate, titrating the liberated iodine with standard decinormal arsenic solution, and calculating the equivalent weight and then the volume of oxygen. By several simple devices, to be described, all calculations may be done away with and the percentage of oxygen seen immediately by the volume of arsenic solution required for titration.

The volume of oxygen indicated under the standard conditions of temperature and pressure (0° and 760 mm.) must either be calculated to that which it would occupy under the conditions of the experiment, or the volume of air taken must be reduced to the standard conditions of temperature and pressure. The latter plan is the more satisfactory, since by Lunge's ingenious device ‡ the reduction can be readily effected without any calculation, and independently of changing temperature and pressure. For this purpose the following arrangement of two burettes answered admirably. One burette graduated to 120 cm.^{3} contains over mercury the same volume of moist air which 100 cm.^{3} of moist air at 0° and 760 mm. would occupy under the given conditions, this standard being very carefully de-

* See page 467.

† D. Albert Kreider, Am. Jour. Sci., [4], ii, 361.

‡ Zeit. angew. Chem., 1890, 139.

termined. By means of a T-tube this standard burette is made to connect with the burette in which the volume of air to be analyzed is measured and with a movable reservoir of mercury. Both burettes are firmly fastened to a movable iron rod and the zero marks accurately adjusted to the same level. By drawing into the measuring burette a volume of moistened air greater than that required, and then, by raising the reservoir of mercury, compressing the air in the standard tube to the 100 cm.^{3} mark, at the same time allowing the excess of air to escape from the measuring burette, exactly 100 cm.^{3} of air under the standard conditions of temperature and pressure may be obtained. To facilitate the adjustment, two strips of wood are fastened to the rubber connection by means of screw pinchcocks in such a way that by closing one pinchcock the flow of mercury from the reservoir is shut off, while by tightening the other pinchcock mercury is gradually forced out of the rubber.

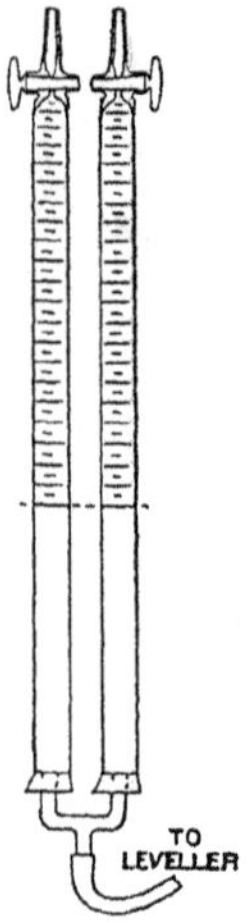

Fig. 24.

The action of the oxygen upon hydriodic acid is effected in a 300 cm.^{3} bulb pipette, both ends of which are cut off short and sealed to glass stopcocks. The tube from one of the stopcocks is cut off short after being tapered and constricted so as to hold a rubber connector tightly, while the tube from the other stopcock is left sufficiently long to reach to the bottom of a 500 cm.^{3} Erlenmeyer beaker. These tubes are preferably of about 3 mm. bore, since for the several connections all air may be expelled from tubes of this size by displacement with water. The bulb is filled with water to expel all air. The water is then displaced by pure carbon dioxide (prepared as described below) and the flask is exhausted by connecting it with a large bottle kept vacuous by a water pump. The required amounts of potassium iodide solution, hydrochloric acid and nitrogen dioxide are introduced in the order named, after which the measured volume of air is gradually admitted while the bulb is constantly agitated to keep the hydriodic acid continually renewed upon the surface of the bulb. The shaking is continued for a minute or two until

Fig. 25.

the action is completed, when a dilute solution of potassium bicarbonate is admitted. Upon opening the lower stopcock, the pressure of the carbon dioxide forces the liquid from the bulb into a solution of the bicarbonate in amount sufficient, as previously determined, to neutralize all the acid taken. Bicarbonate is again admitted through the upper stopcock, and after neutralization has been completed the bulb may be washed out without any danger from the admission of air.

All the water employed, both for the solution of potassium iodide and for the various connections, must be free from oxygen. It is prepared by filling a three liter flask with distilled water and boiling until the volume of the liquid is reduced about one-third, when the flask is closed by a doubly perforated rubber stopper fitted with tubes like a wash bottle. By means of the tube which reaches below the surface of the water, pure carbon dioxide is passed through while the water is still boiling. Then the source of heat is removed and the escape tube is closed by a piece of rubber tubing and screw pinchcock. As the water cools it is well shaken while carbon dioxide is admitted from the generator, and finally pumped in under considerable pressure by the little hand pump shown in Fig. 9. From this supply the water may be drawn as needed without the introduction of any air, the escape tube being provided with a rubber tube and screw pinchcock and a long, slender nozzle which may be inserted into the tubes of the absorption apparatus. A bottle thus charged suffices for many determinations and requires only occasional recharging with carbon dioxide.

The potassium iodide solution is made up to contain 1 grm. of the salt in 30 $cm.^3$ of water, and is contained in an ordinary wide-mouthed bottle, fitted as a wash bottle, and graduated approximately to volumes of 30 $cm.^3$ — the amount usually taken. In making the solution, the potassium iodide is weighed into the bottle, which is then closed. All air is expelled by a current of carbon dioxide, the desired amount of water, free from oxygen, is drawn in, and attachment is again made with the carbon dioxide generator. After allowing the gas to pass for several minutes the exit is closed and the gas is pumped in by the little hand pump. Inasmuch as this solution is drawn when used into an exhausted bulb, the bottle may be emptied without exposing its contents to the air.

Nitrogen dioxide is generated by the action of nitric acid upon granulated shot copper in a Kipp generator. When the nitric acid is diluted with an equal volume of water the evolution of the gas is sufficiently rapid without the application of heat, but contamination by the higher oxide is more likely. However, by passing the gas, as it issues from the generator, through a set of Geissler bulbs containing an acidified solution of potassium iodide and washing with potassium iodide solution, perfectly purified gas is obtained. Theoretically, only a small amount of the nitrogen dioxide is required for the transference of the oxygen to the hydriodic acid, but when too little is taken the action is very slow. On the other hand, too large an amount relieves the vacuum to such an extent as to interfere with the introduction of the air. A little device to measure the volume of gas taken is therefore attached to the generator. It consists of a tube filled with water and roughly graduated for every five cubic centimeters, so attached to the generator that the gas enters with displacement of the water to a lower bulb, and as it is withdrawn is replaced by water. A volume of fifteen cubic centimeters of the gas is a convenient and satisfactory amount for an analysis.

Carbon dioxide is generated in a Kipp generator, which is charged with previously boiled acid and marble and a little cuprous chloride. To remove traces of reducing matter, it is first passed through a solution of iodine and then washed with potassium iodide.

Relation of Arsenic to Oxygen.

$n/10$ As_2O_3. cm.^{3}	Oxygen equivalent at 0° and 760 mm. cm.^{3}	Correction for 0.01 cm.^{3} $n/10$ As_2O_3.
37.0	20.714	0.005
37.1	20.770	.. ..
37.2	20.826	.. .
37.3	20.882	...
37.4	20 938	....
37.5	20.994	.....
37.6	21.050	.....
37.7	21.106	. ..
37.8	21.162	.. ..
37.9	21.218	. ..
38.0	21.274	.....

For the titration a decinormal solution of arsenious oxide (4.95 grm. to the liter) was employed, one cubic centimeter being equal to 0.55985 cm.[3] of oxygen at 0° and 760 mm. when the weight of a liter of oxygen at 0° and 760 mm. is taken as 1.42895 grm. When the volume of air taken is 100 cm.[3] under standard conditions of temperature and pressure, as obtained by Lunge's device, the preceding table, calculated for the volume of oxygen equivalent to the volume of arsenic solution, shows directly the percentage of oxygen corresponding to the reading of the burette. The correction necessary for the fraction of a tenth of a cubic centimeter of the arsenic solution is obtained with sufficient accuracy by simply multiplying the figure expressing hundredths of a cubic centimeter by 0.005.

In the table are shown results obtained upon portions of a sample of air either measured in the adaptation of Lunge's device to make 100 cm.[3] under standard conditions or measured in an ordinary gas burette and calculated to standard conditions.

Oxygen in Air.

Volume of air reduced to 0° and 760 mm. cm.^{3}	n/10 As_2O_3 required. cm.^{3}	Volume of oxygen found at 0° and 760 mm. cm.^{3}	Per cent of oxygen in air.
Measured to standard conditions.			
100.00	37 44	20 96	20.96
100.00	37 54	21.01	21.01
100.00	37.50	20 99	20.99
100.00	37 57	21.03	21.03
100 00	37.47	20 97	20.97
100.00	37.50	20 99	20.99
Calculated to standard conditions.			
91.18	34 06	19 07	20 91
91 73	34.47	19.30	21.04
90 84	34.25	19 17	21.11
90.60	34.20	19 16	21.13
86.06	32 55	18.22	21.17
85.96	32.40	18 14	21.10
86.49	32.53	18.21	21.06
87.85	33 00	18.47	21.03
44 17	16.60	9 29	21.04
44.11	16.70	9.35	21.10
44.54	16 80	9.41	21.12

As is evident from the table, the determinations according to this method are not reliable beyond 0.05 per cent, but for practical purposes this is sufficiently accurate.

Determination of Dissolved Oxygen. A determination of oxygen dissolved in water may be completed in about ten minutes by means of the apparatus illustrated by the accompanying figure. The apparatus consists of a flask of about 300 cm.^{3} capacity, into the bottom of which is sealed a stopcock with a long exit tube. Upon the neck is cut the fiducial circle *c* and above the stopcock *e* is sealed on, as shown. The neck of the flask is drawn out and sealed to the stopcock *d* and the bulb *a*, of about 30 cm.^{3} capacity blown in it. The capacity of the apparatus between stopcock *b*, and the fiducial mark *c*, is carefully determined.

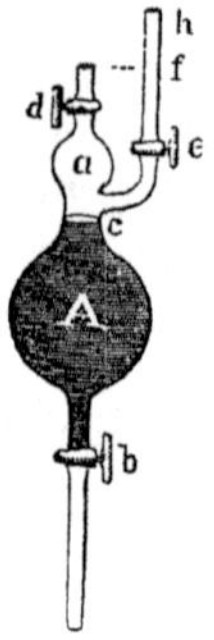

Fig. 26.

The manipulation for the determination of dissolved oxygen is as follows: The flask is held in position by a clamp fastened to a movable support. Stopcock *b* being closed, the water is admitted through *e* and the air allowed to escape through *d* until the level of water is that indicated by the line *f*. (When the water to be examined is not saturated with air, the flask must first be filled with carbon dioxide and the water introduced by replacement of that gas.) With *d* closed, sufficient water is allowed to escape through *b* to bring the surface to *e*, which is then closed. The nitrogen dioxide generator is then attached to *d*, and by opening *b* the gas is allowed to replace the water until the meniscus coincides with *c*, when *d* is closed and the generator disconnected. Two cubic centimeters of strong hydrochloric acid are introduced through *e* by expelling nitrogen dioxide through *d*, in which a drop of water forms an effective trap to prevent the entrance of air. Then the potassium iodide is admitted in the same way. The solution of iodide for this purpose is free from oxygen and contains 1 grm. in 3 cm.^{3}. It is kept under pressure of carbon dioxide in the bottle previously described, and by means of a long nozzle may be conducted to the bottom of *eh* and thus be admitted with but momentary and slight contact with the air. The tube *eh* contains approximately 3 cm.^{3}. With all the stopcocks closed, the flask is inverted several times and thoroughly shaken, and the ends of the stopcocks are washed out with distilled water. After again placing the apparatus in its position, enough potassium bicarbonate solution is admitted through *e* to expel all the nitrogen dioxide through *d*, the bulb *a* holding sufficient bicarbonate to neutralize all the acid taken. The bicarbonate being heavier quickly diffuses through the contents of the flask and neutralizes the acid; *d* and *e* are kept closed for a minute with *b* open so as to allow sufficient of the liquid to escape into a beaker containing some bicarbonate to provide space for the carbon dioxide evolved. Then the flask is washed out into the beaker and its contents titrated with standard arsenite.

The bleaching, by the aid of starch for the final reaction, may be accurately read to a single drop and the reading verified by adding a drop of $n/10$ iodine solution to characteristic coloration.

The table gives the results of a series of determinations.

Oxygen in Water.

Volume of water taken cm.³	Temperature. C°.	As_2O_3 required. cm.³	Volume of oxygen dissolved in 1000 cm.³ of water at 760 mm.
314 63	20	3 42	6 04
314 63	20	3 45	6 09
314 63	20	3 40	6 00
314.63	20	3 41	6 02
314 63	20	3 43	6 05
314 63	20	3.40	6.00
314 63	20	3.36	5.93
314 63	20	3.40	6 00
314.63	20	3.40	6 00
314.63	20	3.50	6.18
314.63	20	3 38	5.96
314.63	20	3.40	6.60

The mean of these determinations gives 6.022 cm.³ of oxygen as the amount dissolved in distilled water at 20° and 760 mm., and while some of the determinations vary considerably from this mean, as a whole they are fairly accordant. This method, moreover, is applicable to carbonated water.

The Estimation of Oxidizers by the Gravimetric Determination of Iodine Set Free in Reaction.

The process developed by Perkins for the gravimetric estimation of iodine by absorption in metallic silver may be utilized for the determination of oxidizers.* In determining available oxygen by this process a definite amount of the oxidizer is added

* Claude C. Perkins, Am. Jour. Sci., [4], xxix, 339.

to a solution of potassium iodide acidified with hydrochloric acid and the mixture is shaken with a weighed amount of silver in an atmosphere of hydrogen. The value of the oxidizer is then calculated from the increase in weight of silver which represents the amount of liberated iodine. The table shows figures obtained in the determination of *potassium permanganate, hydrogen dioxide, potassium dichromate*, and *ferric chloride* taken in standardized solutions. In all of the results the error is well within reasonable experimental variation.

Iodine Liberated by Oxidizers.

Silver taken. grm.	Oxygen available in oxidizer. grm.	Iodine found. grm.	Oxygen found. grm.	Error. grm.
Determination of potassium permanganate.				
3.0000	0.0123	0.1956	0.0123	0.0000
3.0000	0.0123	0.1936	0.0122	−0 0001
3.0100	0.0123	0.1964	0.0124	+0.0001
3.0100	0.0123	0.1968	0.0124	+0.0001
4.0101	0.0185	0.2926	0.0184	−0 0001
4.0101	0.0247	0.3886	0.0245	−0.0002
Determination of hydrogen peroxide.				
3.0000	0.0202	0.3200	0.0202	0 0000
3.0000	0 0202	0.3214	0.0203	+0.0001
3.0100	0.0404	0.6427	0.0405	+0 0001
3 0000	0.0311	0.4937	0 0311	0.0000
3.0000	0.0322	0.5128	0.0323	+0 0001
3 0000	0 0606	0.9590	0.0604	−0.0002
Determination of potassium dichromate.				
3 0000	0.0080	0.1272	0.0080	0.0000
3.0000	0.0160	0.2552	0.0161	+0.0001
3.0000	0.0201	0.3141	0.0198	−0.0003
3.0000	0.0402	0.6390	0.0403	+0.0001
3.0000	0.0160	0.2552	0.0161	+0.0001
3 0000	0.0160	0.2571	0.0162	+0.0002
Determination of ferric chloride.				
3.0000	0 0218	0.3470	0.0219	+0.0001
3.0000	0.2018	0.3476	0.0219	+0.0001
3.0000	0.0437	0.6922	0.0436	−0.0001
3.0000	0.0218	0.3489	0 0220	+0.0002
3.0000	0.0262	0.4183	0 0264	+0 0002
3.0000	0.0218	0.3516	0.0222	+0.0004

SULPHUR.

The Detection of Sulphides, Sulphates, Sulphites and Thiosulphates in the Presence of One Another.

A method by Browning and Howe * presents improved procedure upon lines suggested by the method of R. Grieg Smith,† for the detection of sulphides, sulphates, sulphites and thiosulphates. The method as modified may be thus summarized: To about 0.1 grm. of the substance to be analyzed dissolved in 10 cm.^{3} of water or more, add sodium, potassium, or ammonium hydroxide to distinct but faintly alkaline reaction. The solution should be neutral or alkaline rather than even faintly acid, owing to the readiness with which sulphur separates.

(I) To the alkaline solution add zinc acetate in distinct excess and filter. The precipitate is treated with acid and a test made for hydrogen sulphide, to indicate presence or absence of a sulphide.

(II) To the filtrate add acetic acid, a few drops in excess of the amount necessary to neutralize, and barium chloride. A precipitate of barium sulphate indicates a sulphate.

(III) Filter through a double filter. To the filtrate add iodine until the solution takes on a permanent yellow tinge, and then bleach with stannous chloride, best after adding a few drops of hydrochloric acid to prevent the precipitation of a basic salt of tin. A precipitate of barium sulphate indicates a sulphite.

(IV) Filter, add bromine water in faint excess to the filtrate, bleaching again with stannous chloride. A precipitate of barium sulphate indicates a thiosulphate originally present.

The results of test experiments are given below.

Tests for Sulphite after Iodine Treatment (III).

K_2SO_3 taken. grm.	Volume of water. cm.^{3}	$BaSO_4$ precipitated after oxidation with iodine.	Remarks.
0.1	10	Very abundant.	Plainly visible before adding $SnCl_2$.
0.01	10	Abundant.	Plainly visible before adding $SnCl_2$.
0 001	10	Distinct.	More distinct after adding $SnCl_2$.
0.0005	10	Fair.	Hardly visible before adding $SnCl_2$.
0.0001	10	Faint.	Invisible before adding $SnCl_2$.

* Philip E. Browning and Ernest Howe, Am. Jour. Sci., [4], vi, 317.

† Chem. News, lxxii, 39.

Test for Thiosulphate After Iodine and Bromine Treatment (III) and (IV).

$Na_2S_2O_3$ taken. grm.	Volume of water. cm³	$BaSO_4$ precipitated by action of iodine.	$BaSO_4$ precipitated by action of bromine.	Remarks.
0.1	10	Faint.	Abundant.	No sulphur separated in 1 minute.
0.01	10	None.	Abundant.	No sulphur separated in several minutes.
0.001	10	None.	Distinct.	No sulphur.
0.0005	10	None.	Faint.	No sulphur; $SnCl_2$ necessary.
0 0001	10	None.	Very faint.	No sulphur; $SnCl_2$ necessary.

Test for Sulphate (III) and Thiosulphate (IV), After Removal of Sulphide (I) and Sulphate (II).

K_2SO_4 taken. grm.	$Na_2S_2O_3$ taken. grm.	$BaSO_4$ precipitated after oxidation with iodine.	$BaSO_4$ precipitated after oxidation with bromine.
0.1	0.01	Abundant.	Good.
0.1	0.001	Abundant.	Distinct.
0.01	0 1	Good.	Abundant.
0.001	0 1	Faint.	Abundant.
0.001	0.001	Fair.	Fair.

The Iodometric Determination of Thiosulphates.

In an investigation of the reaction between iodine and sodium thiosulphate, Pickering* has shown that more iodine is required to oxidize the thiosulphate as the proportion of hydrochloric acid increases. This effect is by him ascribed to the formation of sulphate, apparently by the increased activity of the iodine; but the more rational explanation is that, although some sulphate is ultimately formed, the thiosulphate is first partially decomposed into free sulphur and sulphur dioxide. Finkener † and Mohr ‡ also mention the decomposing effect of free acid upon sodium thiosulphate.

Norton § has studied carefully the effect of acidity in the iodine titration of thiosulphate, with results given in the following account.

* Jour. Chem. Soc., vol. xxxvii, pages 135.
† Anal. Chem., 6 Aufl., pages 620.
‡ Titrirmethode, 6 Aufl., pages 279.
§ John T. Norton, Jr., Am. Jour. Sci., [4], vii, 287.

The sodium thiosulphate used in the following experiments, taken in nearly decinormal solution, was standardized by running it into an approximately decinormal solution of iodine, the value of which had been determined by comparison with decinormal arsenious acid made from carefully resublimed arsenious oxide. In the experiments recorded below the solutions were stirred continuously while the thiosulphate ran into the acidified liquid. The results show clearly that the amount of iodine used in the titration of a fixed amount of sodium thiosulphate is dependent to a very marked degree upon the concentration of the thiosulphate in presence of the acid, the concentration of the acid, the time and the temperature.

Varying Concentrations of Thiosulphate and Acid; Temperature 0°–5°; Titrations made Rapidly.

Volume of liquid at beginning of titration.	$Na_2S_2O_3$ approximately $n/10$ taken.	Volume of $n/10$ iodine used in titration.				
		HCl=none.		=1 cm.³	=5 cm.³	=10 cm.³
cm.³	cm.³	cm.³		cm.³	cm.³	cm.³
100	30	30.25	mean=30.22	30.75	30.76	31.2
200	30	30 22		30.21	30.56	31.4
300	30	30 20		30.22	31.03	30.9
400	30	30.21		30 20	30.20	30.55
500	30	30.20		30.20	30 21	30.55
100	25	25.29	mean=25.27	25.32	25.98	25.70
200	25	25.28		25.34	25 40	25.45
300	25	25 29		25.41	25.38	25.83
400	25	25.27		25.24	25.30	25.63
500	25	25.22		25.23	25.40	25.30
100	20	20.15	mean=20.15	20.17	20.33	20.23
200	20	20.20		20 13	20.27	20.23
300	20	20.21		20.15	20.20	20.17
400	20	20.20		20.10	20.27	20.07
500	20	20.10		20.10	20.17	20.13

Varying Concentrations of Thiosulphate and Times; Temperature 0°–5°.

Volume of the liquid at beginning of titration.	HCl sp. gr. 1.12 present.	$Na_2S_2O_3$ approximately $n/10$ taken.	Volume of $n/10$ iodine used in titration after standing.		
			5 minutes.	10 minutes.	15 minutes.
cm.³	cm.³	cm.³	cm.³	cm.³	cm.³
200	10	30	30.80	31.30	32.32
200	10	25	25.50	26.00	26.30
200	10	20	20.30	20.70	20.68

Varying Temperatures.

Volume of liquid at beginning of titration. cm.³	HCl sp. gr. 1.12 taken. cm.³	Temperature.	$Na_2S_2O_3$ approximately $n/10$ taken. cm.³	Volume of $n/10$ iodine used in titrations at different temperatures. cm.³
400	10	6°	25	23.52
400	10	22°	25	23.73
400	10	34°	25	24.35
400	10	42°	25	24.5
400	10	54°	25	25
400	10	64°	25	26.1

From these results it is plain that the conditions under which considerable amounts of sodium thiosulphate are titrated in presence of hydrochloric acid must be carefully guarded when accuracy is a consideration. The amount of acid should be restricted, the temperature should be reduced as nearly to 0° as possible, and titration by the iodine should be made promptly. So long as the thiosulphate present does not exceed 20 cm.³ of the $n/10$ solution in a volume of 200 cm.³, rapid titration in cold solution proceeds with fair regularity in presence of hydrochloric acid up to 10 cm.³ of the acid of sp. gr. 1.12.

Fortunately, in most analytical processes involving the use of the thiosulphate as a reagent it is possible to add that reagent from the burette to the solution to be acted upon, so that it is destroyed normally as fast as it is introduced and the danger of interaction with the acid does not occur.

The Iodometric Determination of Sulphites in Alkaline Solution.

According to Bunsen, Dupasquier's method of oxidizing sulphurous acid by iodine in an acid solution, proceeds to completion when the concentration of the sulphur dioxide does not exceed 0.5 per cent of the solution. When, on the other hand, the proportion of sulphur dioxide exceeds this value, there is a secondary reaction, which, according to Volhard, involves the reduction of the sulphur dioxide by the hydriodic acid produced. The reaction proceeds normally in dilute solutions according to the equation

$$2\,SO_2 + 2\,I_2 + 4\,H_2O = 4\,HI + 2\,H_2SO_4.$$

In solutions too concentrated, however, the secondary reaction,

$$SO_2 + 4\,HI = 2\,I_2 + 2\,H_2O + S,$$

takes place, as Volhard has shown,* and vitiates the indications. The difficulty may be obviated, however, as has been shown by Volhard,† if the solution of the sulphurous acid or a sulphite is run with stirring into a solution of iodine in potassium iodide, acidified with hydrochloric acid, to the bleaching of color, using starch as an indicator. Volhard's method is accurate and reliable, but it involves the inconvenience of making up every solution to be examined accurately to a standard volume of which portions are to be drawn from a burette and made to react with definite amounts of a standardized solution of iodine. It rests upon the facts that the oxidation of sulphite is brought about in the acidified solution and that no more than a small proportion of hydriodic acid is present at the point at which the oxidation of sulphur dioxide takes place.

To avoid the inconvenience of the Volhard method it has been proposed by Rupp ‡ to bring about the oxidation of sulphites by treatment with an excess of standardized iodine in a solution made alkaline by acid sodium carbonate, and then, after fifteen minutes, to titrate the excess of iodine by sodium thiosulphate. This procedure, however, as has been shown by Ruff and Jaroch § and by Ashley,‖ is faulty in principle and practice, and gives correct results only by a chance balancing of opposing errors. It has been further pointed out by Ashley** that it is possible to overcome the difficulties by treating the alkaline mixture with acid before attempting to titrate by sodium thiosulphate the excess of iodine. According to Ashley's procedure, the practical estimation of sulphurous acid or of a soluble sulphite may be accomplished with a reasonable degree of accuracy by adding to the solution of the substance, not exceeding 100 $cm.^3$ in volume and containing a gram of acid sodium carbonate, at least twice as much iodine as is theoretically necessary to effect oxidation, acidifying cautiously with

* Ann. Chem. 242, 93.

† Loc. cit.

‡ Ber. Dtsch chem. Ges., xxxv, 3694.

§ Ber. Dtsch. chem. Ges., xxxviii, 409.

‖ Am. Jour. Sci., [4], xix, 237.

** R. Harmon Ashley, Am. Jour. Sci., [4], xx, 13.

hydrochloric acid, and determining with standard sodium thiosulphate the excess of iodine remaining in the acidified solution.

Results of this procedure are given in the table.

Determination of Sulphite.

Iodine value of SO_2 taken.	Iodine taken.	Iodine value of $Na_2S_2O_3$ used.	Error.		Excess of HCl [1 : 3.]	Volume at titration.
			In terms of iodine.	In terms of SO_2.		
grm.	grm.	grm.	grm.	grm.	cm.³	cm.³
0.1143	0.3143	0.1990	+0.0010	+0.0003	5.0	125
0.1143	0.3143	0.1982	+0.0018	+0.0004	5.0	125
0.1143	0.3143	0 1992	+0.0008	+0.0002	5.0	125
0.1143	0.3143	0.1986	+0.0014	+0.0003	5.0	125
0.1482	0.3187	0.1708	−0.0003	−0.0001	7.5	125
0.1576	0.3187	0.1586	+0.0025	+0.0006	7.5	125
0.1576	0.3187	0.1643	−0.0032	−0.0008	7.5	125
0.1576	0.3187	0.1598	+0.0013	+0.0003	7.5	125
0.1576	0.3187	0.1606	+0.0005	+0.0001	7.5	125
0.1576	0.3187	0.1602	+0.0009	+0.0002	7.5	125
0.1576	0.3187	0.1622	−0.0011	−0.0003	7.5	125
0.1560	0.3195	0.1660	−0.0025	−0.0006	7.5	125
0.1992	0.4460	0.2482	−0.0014	−0.0003	7.5	125
0.1915	0.3825	0.1919	+0.0009	−0.0002	7.5	125
0.2056	0.3771	0.1701	+0.0014	+0.0003	7.5	125
0.2056	0.3771	0.1697	+0.0018	+0.0004	7.5	125
0.2056	0.3771	0.1707	+0.0008	+0.0002	7.5	125
0.2056	0.3771	0.1709	+0 0006	+0.0002	7.5	125
[0.2131	0.4470	0.2412	−0.0073	−0 0018]	7.5	125
0.2354	0.3825	0.1490	−0.0019	−0.0005	7.5	125
0.2597	0.4463	0.1869	−0.0003	−0.0001	7.5	125
0.2638	0.4463	0.1847	−0.0022	−0.0005	7.5	125
0.2908	0.6375	0.3505	−0.0038	−0.0009	7.5	125
0.3187	0.4463	0 1326	+0.0050	+0.0012	7.5	125
0.3395	0.6275	0.2842	+0.0038	+0.0009	7.5	125
0.3395	0.6275	0 2852	+0 0028	+0.0007	7.5	125
0.3395	0.6275	0.2844	+0.0036	+0.0009	7.5	125
0.3395	0.6275	0.2855	+0.0025	+0.0006	7.5	125

It is probable that the formation of a small amount of dithionate instead of sulphate is the occasion of the deficient expenditure of iodine noted when the concentration of this element is low, and that the dithionate is not formed appreciably when the iodine concentration is high. The dithionate once formed is but slowly attacked by iodine, and that is apparently the reason why long standing of the mixtures containing a small proportion of iodine does not result in complete oxidation of the sulphite to sulphate.

The Determination of Dithionic Acid and Dithionates.

That dithionic acid may be best liberated from combination by the action of sulphuric acid, rather than hydrochloric acid, has been shown by Ashley.* In studying these reactions, Ashley treated a specially prepared barium dithionate with the acids in the distillation apparatus previously described and figured,† and estimated the resulting sulphur dioxide by absorbing it in standard iodine the excess of which was determined by titration with thiosulphate. The successful procedure is as follows: A weighed amount of dithionate is introduced into the Voit flask and there dissolved in water. Sulphuric acid is run in through the separating funnel and the mixture then boiled, the sulphur dioxide being collected in the Drexel receiver charged with a measured amount of standard iodine in potassium iodide and trapped with potassium iodide. A slow current of carbon dioxide is driven through the system to sweep the sulphur dioxide into the iodine and to prevent any sucking back. When boiling has been carried so far that fumes of sulphuric acid begin to appear the operation is stopped, and the excess of iodine remaining is determined by means of sodium thiosulphate, starch iodide being used as an indicator. Results of experiments carried out in this manner with barium dithionate are given in the table.

Decomposition of Barium Dithionate by Boiling with Sulphuric Acid.

S_2O_5 taken. grm.	I value of S_2O_5 taken. grm.	I taken. grm.	I value of $Na_2S_2O_3$ required. grm.	S_2O_5 found. grm.	Errors in I. grm.	Errors in S_2O_5. grm.	Time. min.
0.1039	0.2310	0.5759	0.3435	0.1045	+0.0014	+0 0006	20
0 1046	0.2326	0 5708	0.3372	0.1051	+0.0010	+0 0005	28
0.1039	0.2311	0 5740	0.3435	0.1037	−0.0006	−0 0002	34
0 1033	0.2297	0 5701	0.3387	0.1041	+0.0017	+0 0008	45
0 1721	0.3827	0 5712	0.1876	0.1726	+0.0009	+0 0005	35
0 1719	0.3820	0 5702	0.1894	0.1713	−0.0012	−0 0006	50
0 1726	0.3838	0 5734	0.1898	0 1726	−0.0002	0.0000	12
0 1724	0.3832	0 5727	0.1885	0 1728	+0.0010	+0.0004	10
0 1721	0.3826	0 3727	0.1886	0.1728	+0.0015	+0.0007	10
0 0692	0.1539	0 3130	0.1599	0 0689	−0.0008	−0.0003	12
0.0350	0.0777	0 3109	0.2323	0 0354	+0 0009	+0.0004	4
0 2061	0.4582	0 6205	0.1632	0.2057	−0.0009	−0.0004	15
0 2402	0.5340	0 6215	0 0862	0.2408	+0.0013	+0.0006	15

* R. Harmon Ashley, Am. Jour. Sci., [4], xxii, 259

† See Fig. 3, page 4.

Similar experiments in which hydrochloric acid was substituted for sulphuric acid were not satisfactory, chiefly because the decomposition of the dithionate was not complete even after long and repeated treatments, and the large amount of the hydrochloric acid carried to the receiver tends, as is well known, to render titration of the residual iodine by thiosulphate less exact, and the starch iodide less delicate as an indicator.

There are reasons why sulphuric acid should work better than hydrochloric acid in this process:

First, when sulphuric acid is added to the solution of barium dithionate, barium sulphate is precipitated and dithionic acid is left in free condition, this reaction proceeding at once to completion because the barium sulphate formed is removed from the system. When hydrochloric acid is used the dithionic acid is completely liberated only by a gradual change in the conditions of equilibrium.

Second, when the solution containing sulphuric acid is boiled, the water is driven off, the concentration of the solution increases and the temperature of the sulphuric acid reaches a high point. Under such conditions the decomposition of the dithionic acid is rapid and complete, the time being dependent upon the original dilution of the solution.

Third, no appreciable amount of acid distils over from the Voit flask into the receiver containing the iodine, to interfere with the back titration with sodium thiosulphate, the only acid present being that produced by the oxidation of the sulphur dioxide. Under these conditions the starch indicator acts sharply, which is not the case when hydrochloric acid is used.

The Determination of Persulphates.

Arsenate-Iodide Method. The estimation of persulphates is accomplished by Peters and Moody * by procedure similar to that of Gooch and Smith † for the estimation of chlorates. According to this process a mixture containing the persulphate, potassium iodide, 2 grm. to 3 grm. of hydrogen potassium arsenate, 20 cm.^{3} of [1 : 1] sulphuric acid, and water enough to make a total volume of about 100 cm.^{3}, is boiled in a trapped Erlen-

* Charles A. Peters and Seth E. Moody, Am. Jour. Sci., [4], xii, 367.

† See page 463.

meyer beaker * until the volume is diminished to 35 cm.³, when the solution is made alkaline with potassium bicarbonate and the arsenite present is estimated by standard iodine in presence of starch. The results obtained are in close agreement, as shown by the record below:

Arsenate-Iodide Method.

Ammonium persulphate solution. cm.³	KI present. grm.	Iodine required for oxidation of of arsenite. grm.	Iodine liberated by persulphate. grm.	$(NH_4)_2S_2O_8$ equivalent to iodine liberated. grm.	$(NH_4)_2S_2O_8$ average. grm.
12.5	0 1875	0.0514	0 1361	0.1225	
12.5	0 1875	0.0522	0.1352	0.1217	0 1222
12.5	0.1875	0.0514	0 1361	0.1225	
12.5	0 1875	0.0516	0.1359	0.1222	

Making use of the arsenate-iodide method as a control, Peters and Moody have brought into comparison the results of other and simpler methods, viz., the methods of LeBlanc and Eckardt, Grützner, Mondolfo and Namias.

Method of LeBlanc and Eckardt. According to LeBlanc and Eckardt,† when a mixture containing a persulphate, a sufficient excess of ferrous salt, and sulphuric acid is heated at 60°–80°, or allowed to stand ten or twelve hours, the persulphate is reduced, and the amount of ferrous salt oxidized is the measure of the amount of persulphate originally present in solution. The experiments of Peters and Moody show results agreeing quite closely with one another and in fair agreement with the average of the arsenate-iodide method.

Experiments made in blank to discover the amount of ferrous salt oxidized in eleven hours by other agencies than the persulphate showed an amount of oxidation in that time too small to be detected; but, when the solution was allowed to stand thirty-six hours, a slight oxidation was noticed, which, if calculated as persulphate, would be equivalent to 0.0006 grm. The correction for eleven hours' standing, upon this basis, therefore, would be 0.0002 grm., which is obviously so small that it may be disregarded. This absence of any significant oxidation of the ferrous

* See Fig. 6, page 6.

† Zeit. Elektrochem., 5, 355–7.

salt was undoubtedly due to the fact that the solution of ammonio-ferrous sulphate had been standing some time before standardizing and was devoid of dissolved oxygen.

Ferrous Sulphate Method.

Volume of liquid, 100 cm.[3]

Ammonium persulphate, 10 grm. to liter. cm.[3]	$(NH_4)_2Fe(SO_4)_2.6H_2O$ solution, 20 grm. to liter. cm.[3]	$KMnO_4$ approx. $n/10$, to oxidize the excess of ferrous salt. cm.[3]	$(NH_4)_2S_2O_8$ calculated from ferrous salt oxidized. grm.	$(NH_4)_2S_2O_8$ average. grm.	Variation from results 80° of arsenate-iodide method. grm.
Heated 10 minutes at 60°–80°.					
12.5	25	2.11	0.1216	0.1217	−0 0005
12.5	25	2.18	0.1209		
12.5	25	2.11	0.1216		
12.5	25	2.11	0.1216		
12.5	50	15.00	0.1218		
25.0	50	4.21	0.2426		
Stood 11 hours at 21°–25°.					
12.5	26	1.89	0.1214	0.1220	−0.0002
12.5	26	1.80	0 1224		
12.5	40	8.75	0.1221		
12.5	40	8.75	0.1221		
Carbon dioxide above the liquid: stood 11 hours at 21°–25°.					
12.5	40	8.75	0.1221	0.1219	−0 0003
12.5	40	8.78	0.1218		

Method of Grützner. Grützner * has stated that arsenious oxide is well suited to the determination of the oxidizing power of persulphates, the reaction being hastened by heat and the presence of an alkali hydroxide.

In testing the value of the persulphate solution according to the method of Grützner, the mixture containing the persulphate, standard arsenite solution, and sodium hydroxide, was raised to the boiling point, cooled, faintly acidified with sulphuric acid, then made alkaline with potassium bicarbonate, and, after adding a large excess of the last, the arsenite remaining unoxidized was

* Chem. Centralblatt, 1900, i, p. 835.

titrated by iodine. Experiments made in blank, — that is, experiments in which the mixture of arsenite and alkaline hydroxide was raised to the boiling point, cooled, made acid with sulphuric acid, then alkaline with potassium bicarbonate, and titrated with iodine, — showed a deficiency of arsenious acid as compared with the experiments, otherwise similarly conducted, in which the mixture was not subjected to heat. The effects of various specimens of sodium and potassium hydroxides were tested in the process — sodium and potassium hydroxides made by the alcohol process, potassium hydroxide by the barium hydroxide process, sodium hydroxide especially prepared from sodium carbonate and calcium hydroxide — as well as a specimen of potassium carbonate. With all excepting the potassium carbonate more or less oxidation was observed, sometimes trifling, sometimes considerable. The same specimens of potassium and sodium hydroxides and potassium carbonate were tested in the absence of the arsenite solution, by dissolving them in water, heating the solution to the boiling point, cooling, adding sulphuric acid in faint excess and then potassium bicarbonate, and titrating with iodine to a color without starch. In all cases, save that of the potassium carbonate, the color of the first drop of iodine solution added was destroyed, and from 0.06 cm.^{3} to 0.19 cm.^{3} were necessary to bring the permanent iodine color. Potassium carbonate, however, proves to be inefficient as a substitute for an alkali hydroxide. By using definite amounts of the hydroxide and introducing a correction for the oxidation determined by the blank tests the real oxidizing effect of the persulphate upon the arsenite may be deduced. The table contains the values found by Grützner's process, both corrected and uncorrected.

Arsenious Oxide Method.

Ammonium persulphate solution taken. cm.^{3}	Arsenite solution. cm.^{3}	NaOH by CaO_2H_2. grm.	Calculated from iodine actually used. grm.	$(NH_4)_2S_2O_8$ corrected for oxidation in blank. grm.	$(NH_4)_2S_2O_8$ corrected, average. grm.
12.5	25	2	0.1223	0.1218	
12.5	25	2	0.1223	0.1218	0.1219
12.5	25	2	0.1225	0.1220	
12.5	25	2	0.1229	0.1220	

Method of Mondolfo. Marshall * is authority for the statement that persulphates liberate iodine from potassium iodide, and that the action is hastened by heat and affected little by the addition of dilute sulphuric acid. Upon this reaction Mondolfo † has based a method for the estimation of persulphates, which consists in heating a mixture containing the persulphate and potassium iodide in a stoppered bottle for ten minutes at 60° to 80°, and titrating by thiosulphate the iodine set free. According to Peters and Moody,‡ however, the reduction of persulphate is incomplete even when the volume of the mixture is restricted to 25 $cm.^3$, and the digestion continued thirty minutes with the addition of small amounts of sulphuric acid. Under the best conditions the amount of persulphate found was 0.1207 grm. in an amount of solution for which the arsenate-iodide process indicated 0.1222 grm., and the LeBlanc and Eckardt process 0.1218 grm. in the average.

Method of Namias. The process proposed by Namias, without knowledge of Mondolfo's method — differs from the method of Mondolfo in the particular that the reaction is carried out at the ordinary temperature. The mixture of persulphate and potassium iodide in solution is allowed to stand eleven hours in a stoppered bottle and the iodine set free is titrated with thiosulphate. In experiments conducted by Peters and Moody § the reduction of the persulphate under the conditions described was incomplete, the color of iodine returning upon longer standing after the titration. Under the best conditions the value of 0.1208 grm. was found for the persulphate contained in an amount of solution for which the arsenate-iodide method indicated 0.1222 grm. and the LeBlanc and Eckardt process 0.1218 grm. in the average.

Comparison of Methods. To compare the values obtained for the persulphate solution the averages of the results obtained by the different methods, together with the average of all the experiments, are given in the table.

The process of Mondolfo and the process of Namias, both of which involve the liberation of iodine from potassium iodide and

* Jour. Chem. Soc. 59, 771.

† Chem. Ztg., 23, 699.

‡ Loc. cit.

§ Loc. cit.

the titration of that iodine by thiosulphate, give results which are practically identical and lower than those obtained by the other three methods.

Process.	Number of experiments.	Average of results. grm.
Mondolfo	6	0.1207
Namias	8	0.1208
LeBlanc and Eckardt	12	0.1217
Grutzner (corrected)	4	0.1219
Arsenate-iodide method	4	0.1222

The process of LeBlanc and Eckardt in which the persulphate is reduced by a ferrous salt, the process of Grützner, in which an arsenite solution is the reducing agent, and the arsenate-iodide method, in which the persulphate is determined by the difference between the amount of iodine in an iodide added and the amount necessary to oxidize the arsenite remaining after boiling the solution, are all in close agreement and all higher than those obtained by the process of Namias or that of Mondolfo.

The process of LeBlanc and Eckardt is simple, rapid and convenient. The method of Grützner is advantageous in that the ordinary arsenite solution is the standard for the process, though requiring the application of a correction. The arsenate-iodide method, introduced as a control, is accurate but less simple than the other methods.

Selenium.

The Gravimetric Estimation of Selenious Acid by Liberation of Iodine and Absorption of that Element by Silver.

When selenious acid is treated in acidulated solution with potassium iodide and shaken with silver in an atmosphere of hydrogen,* the reaction proceeds to the complete reduction of the element, according to the equation

$$SeO_2 + 4\,KI + 4\,HCl = 4\,KCl + 2\,H_2O + Se + 2\,I_2.$$

The increase in weight of the insoluble material represents the iodine evolved plus the selenium of the selenium oxide.† Tests

* See page 444.

† Claude C. Perkins, Am. Jour. Sci., [4], xxix, 338.

made in the application of this process to selenium dioxide twice crystallized from nitric acid and resublimed over manganese dioxide are given in the table.

Weighing of Silver Iodide and Selenium.

Ag taken. grm.	Se taken. grm.	Increase. grm.	Calculated Se. grm.	Error. grm.
2 0133	0 0050	0.0365	0.0049	−0.0001
2 0133	0 0075	0 0529	0 0071	−0 0004
2.5639	0.0126	0.0894	0 0121	−0 0005
2.5639	0 0428	0 3178	0.0429	+0 0001
3.0018	0.0504	0.3799	0.0501	−0 0003

The Gravimetric Determination of Selenious Acid by Precipitation of Selenium.

For the gravimetric determination of selenious acid it is usual to precipitate the selenium with sulphurous acid in presence of hydrochloric acid and to weigh the elementary selenium. Precipitation by this method, however, is slow and incomplete in many cases, so that it is always necessary to treat the filtrate a second time with sulphurous acid and to digest for some time. Adopting the idea from volumetric methods for the determination of selenium * in which an iodide in acid solution is used to reduce the selenious acid, Peirce † effects the reduction by the use of potassium iodide in large excess.

When potassium iodide is added in slight excess to the solution of selenious acid acidified with hydrochloric acid the selenium is precipitated in the form of a red powder. Boiling for ten minutes removes most of the liberated iodine and changes the selenium into the black modification. This may be collected upon an asbestos felt, washed, dried at 100° to a constant weight, and weighed. If the selenium amounts to less than 0.1 grm. the results accord well with theory. When the amount is larger, the selenium is apt to assume on boiling a pasty, molten condition which makes filtering and washing impossible. In this condition the selenium holds iodine. By using the potassium iodide in large excess, however, the pasty condition may be modified, and by effecting the reduction in large volumes of solution the danger of inclusion is lessened.

* See pages 377, 379.

† A. W. Peirce, Am. Jour. Sci., [4], i, 416.

According to this procedure, it is sufficient to dilute the solution containing selenious acid or a selenite to 400 $cm.^3$ before acidifying with hydrochloric acid and then to add potassium iodide to an amount about three grams in excess of that actually required. Boiling from ten to twenty minutes will change the selenium to the black modification and remove most of the iodine. The process of precipitation and filtering can be completed in half an hour. The selenium is dried at 100° to a constant weight.

When the selenium occurs in the higher form of oxidation the reduction follows the same course, though iodine is not liberated until the solution is quite warm; but at the end of the usual time of boiling the action is complete.

Results obtained by this procedure are given in the table.

Reduction of Selenium.

Se taken as SeO_2. grm.	Se found. grm.	KI. grm.	Volume. $cm.^3$	Error. grm.
0.2853	0.2861	7	900	+0 0008
0.3189	0.3192	8	400	+0.0003
0.3318	0 3324	7	500	+0 0006
0 3798	0.3805	7	500	+0 0007
0 4252	0.4259	7	350	+0 0007
0.4430	0.4434	10	450	+0 0004
Se taken as SeO_3.				
0.1063	0.1065	5	500	+0 0002
0.1063	0 1062	5	375	−0 0001
0 2010	0.2017	5	350	+0.0007
0 3115	0.3126	6	500	+0.0011

The Iodometric Determination of Selenious Acid by Methods Based upon the Action of Potassium Iodide in Presence of Acid.

The Contact Method. According to the method of Muthman and Schaefer,* when selenious acid is brought into contact with potassium iodide in an acidulated solution, iodine and selenium are liberated in elementary condition, the former being directly determinable by titration with sodium thiosulphate after addition of starch. On account of the difficulty in determining the exact point in the titration at which the starch blue disappears

* Ber. Dtsch. chem. Ges., xxvi, 1008.

from the liquid in which the finely divided and opalescent selenium is held in suspension, the process was recommended by the authors for use only when great accuracy is not essential.

Evidently if the reaction between the acidulated iodide and selenious acid is simple and complete, the process should be capable of improvement by removing the selenium before the titration is attempted. This has been done by Gooch and Reynolds * by filtration by means of the vacuum pump upon a thick felt of asbestos in a perforated crucible or cone of large filtering surface. With a properly prepared filter of this description there is no difficulty in separating the selenium in a very few moments so completely that it is possible to determine the iodide remaining dissolved in the excess of potassium iodide with all the accuracy characteristic of this most exact of titration processes. When, however, the difficulty of determining the end-reaction in the titration of the iodine by the thiosulphate is overcome, it becomes apparent that the reaction itself is likely to be incomplete. Even when the potassium iodide is used in moderate excess in presence of a large proportion of hydrochloric acid, and with addition of the thiosulphate previous to the filtration in order that there may be every opportunity for the iodine and thiosulphate to interact, the results show marked deficiency in the reduction of the selenious acid. Either the reaction according to the equation

$$SeO_2 + 4\,HI = Se + 2\,H_2O + 2I_2$$

is incomplete or else there is formed between the selenium and iodine some combination, such as was noted by Hautefeuille † in the interaction between iodine and hydrogen selenide.

In further study of this reaction, and following the suggestion of Peirce's experience in the gravimetric determination of selenium, Norton ‡ has found that the accuracy of the process is very much increased for small amounts of selenium by the use of relatively very large amounts of potassium iodide. The process is still inaccurate when large amounts of selenium are present. This is shown by the results of the accompanying table.

* F. A. Gooch and W. G. Reynolds, Am. Jour. Sci., [3], l, 255.

† Compt. rend., lxviii, 1554.

‡ J. T. Norton, Jr., Am. Jour. Sci., [4], vii, 292.

Contact Method.

SeO_2 used grm.	KI. grm.	Volume of solution cm.³	HCl sp. gr. 1.12. cm ³	SeO_2 found. grm.	Error grm
0.0553	10	150	10	0 0558	+0.0005
0 0574	5	150	10	0 0567	−0 0007
0.0683	5	150	10	0.0683	0 0000
0.0487	5	150	10	0 0484	−0.0003
0 2617	10	150	10	0 2589	−0.0028

The Distillation Method. The reaction between selenious acid, potassium iodide, and hydrochloric acid, in the sense of the equation given above, may be pushed further toward completion by submitting the mixture to distillation. For this purpose Gooch and Reynolds * make use of the apparatus previously described and figured.† The distillation flask is a Voit gas-washing flask, and this is sealed to the inlet tube of a Drexel wash-bottle used as a receiver, to the outlet tube of which is sealed a Will and Varrentrapp absorption apparatus to serve as a trap. The mixture to be distilled, containing not more than 0.2 grm. of selenium dioxide, is introduced into the flask, a solution of 3 grm. of potassium iodide in 100 cm.³ of water is put into the receiver and trap, and during the distillation a slow current of carbon dioxide is passed through the apparatus to keep the boiling regular. Naturally, the acidified solution of the iodide in the flask retains with great tenacity traces of dissolved iodine, so that, in order to determine all the iodine liberated in the reaction, the residue in the flask as well as the distillate in the receiver and trap must be titrated with sodium thiosulphate. Results are fairly good, though a little deficient, for amounts of selenium dioxide up to 0.2 grm.; but when the amount of the dioxide reaches 0.5 grm. it is found that the sum total of iodine in the distillate and in solution in the residue falls far below the theory based upon the assumption that the products are selenium, iodine, and water. The details of treatment and the results for the smaller amounts of selenium are recorded in the table.

The selenium in the residue is, for the smaller amounts, left, after the boiling, in fine dense crystalline condition, so that it does not interfere with the titration of the free iodine; but for the larger amount, 0.5 grm., it is in pasty form adhering to the flask.

* Am. Jour. Sci., [3], i, 256. † See Fig. 3, page 4.

Examination proves that the selenium holds iodine, which is liberated slowly to water and more rapidly to an aqueous solution of potassium iodide; but the error thus introduced is allowable up to the limit of 0.2 grm. of selenium dioxide.

Distillation Method.

SeO_2 taken.	KI in flask.	HCl in flask (sp. gr. 1.20).	Total volume boiled.	Time in minutes.	SeO_2 found.	Error.
grm.	grm.	cm.³	cm.³		grm.	grm.
0.0499	1	5	60	5	0 0497	−0 0002
0.0499	1	5	60	5	0 0497	−0 0002
0.0499	1	5	60	10	0.0496	−0 0003
0.2000	3	5	60	10	0.1995	−0 0005
0 2000	3	5	60	10	0.1991	−0 0009
0.2023	3	5	60	10	0.2018	−0 0005

Peirce states * that the range of the process may be much extended by the use of very large amounts of potassium iodide. In this case, however, more iodine is retained in the residue and the difficulty resulting from atmospheric action when the acidulated residue is exposed becomes magnified.

Differential Method: Treatment of the Residue. When a solution of an arsenate, potassium iodide, and sulphuric acid is boiled under defined conditions† arsenic acid is reduced to arsenious acid with liberation of iodine. When the arsenic acid is in excess the whole of the iodine is evolved and the arsenious acid produced is its exact measure. Upon making the solution alkaline with acid potassium carbonate, the arsenious acid may be reoxidized by standard iodine, and the amount of iodine thus used is the exact equivalent of that set free in the reduction process. Gooch and Peirce ‡ have shown that when selenious acid is present during the reaction between arsenic acid and the iodide, selenium is reduced, and the subsequent estimation of arsenious acid in the residue will be less than should be produced by the iodide by an amount equivalent to the selenious acid present, the reduction taking place according to the equation

$$SeO_2 + 4\,HI = Se + 2\,H_2O + 2\,I_2.$$

* See page 376.

† See page 457.

‡ F. A. Gooch and A. W. Peirce, Am. Jour. Sci., [4], i, 31.

According to the procedure worked out, the selenious acid to be determined is put into an Erlenmeyer flask of 300 cm.[3] capacity; a known amount of standardized potassium iodide (somewhat in excess of that theoretically required) is added; and a solution containing about 2 grm. of pure di-hydrogen potassium arsenate with 20 cm.[3] of sulphuric acid of half-strength is introduced. During the boiling the mixture is protected from ordinary mechanical loss by a trap* (consisting of a two-bulbed drying tube, cut short and hung loosely with the wide end downward in the mouth of the flask) and by the introduction of a few bits of porcelain. The liquid is boiled until the volume decreases, according to indicating marks on the flask, from 100 cm.[3] or more to 35 cm.[3], concentration to about this lower limit having been found to be necessary for the completion of the reaction. The residue is cooled, the acid is nearly neutralized with potassium hydroxide, acid potassium carbonate is added until it is present to the amount of 20 cm.[3] of its saturated solution in excess of the quantity needed for complete neutralization, and, after the addition of starch, standard iodine is introduced until the starch-blue appears.

Differential Method.

Initial volume.	Final volume.	H_2SO_4 half-strength.	Di-hydrogen-potassium arsenate.	KI taken.	SeO_2 taken.	SeO_2 found.	Error.
cm.[3]	cm.[3]	cm.[3]	grm.	grm.	grm.	grm.	grm.
100	35	20	2	1.3277	0.1280	0 1275	−0.0005
100	35	20	2	1.0429	0.0998	0.0994	−0.0004
100	35	20	2	1.0887	0.1024	0.1028	+0.0004
100	35	20	2	1.0405	0.1036	0.1028	−0.0008
100	35	20	2	1.0721	0 1030	0 1029	−0.0001
100	35	20	2	0.9958	0.1273	0 1272	−0.0001
125	35	20	2	2.0828	0 1997	0 2000	+0.0003
125	35	20	2	2.2272	0.2110	0.2113	+0.0003
125	35	20	2	2.1535	0.2067	0.2069	+0.0002
150	40	20	2	2.6554	0.2560	0.2549	−0.0011
175	35	20	2	3.2428	0.3110	0.3118	+0.0008
175	35	20	2	3.2428	0.3085	0.3083	−0.0002

The iodine introduced measures the arsenious acid (and so the quantity of iodine set free by the arsenic acid), and the difference between it and the iodine originally present in the

* See Fig. 6, page 6.

form of the iodide represents the amount set free by the selenious acid.

The preceding table comprises the details and results of a series of determinations made in the manner outlined.

The Determination of Selenious Acid by Potassium Permanganate.

In the action of potassium permanganate upon selenious acid, whether in a solution acidified with sulphuric acid or made alkaline by caustic soda, the reduction of the permanganate does not proceed to the lowest degree of oxidation of the manganese, the selenious acid being unable to reduce the higher hydroxides which separate. When the permanganate is introduced into the acidified solution the color vanishes, leaving a clear colorless liquid, but as more is added the solution becomes yellow and deepens gradually in color to a reddish brown, until turbidity due to the deposition of a brown hydroxide of manganese ensues, and finally the characteristic color of the permanganate is plainly distinguishable. The exact point at which precipitation of the manganic hydroxide begins depends upon the dilution, acidity, and temperature of the solution. In employing the reaction quantitatively it is necessary, according to Gooch and Clemons,* to add the permanganate in distinct excess, to destroy the surplus by means of standard oxalic acid added to the solution acidified with sulphuric acid, and then to determine the excess of oxalic acid in the warmed solution by addition of more permanganate. The difference between the amount of permanganate actually used and that required to oxidize the known amount of oxalic acid introduced is the measure of the selenious acid acted upon, provided the amount of sulphuric acid present in the final titration and the temperature are adjusted, to prevent on the one hand interference with the end-reaction by precipitation of manganese hydroxide, according to Guyard's reaction, and on the other hand to obviate evolution of oxygen outside the main reaction.† According to the method of treatment prescribed, the solution of selenium dioxide in 100 $cm.^3$ of water containing 10 $cm.^3$ of sulphuric acid of half-strength is heated to 75°, an approximately decinormal standardized solution of potassium permanganate is added until the characteristic color predominates over that of the

* F. A. Gooch and C. F. Clemons, Am. Jour. Sci., [3], 1, 51.

† See page 47.

brown hydroxide deposited during the oxidation, oxalic acid in solution of known strength is introduced until the excess of permanganate has been destroyed and the insoluble hydroxide dissolved, and, finally, at a temperature of 50° or less, permanganate is added to coloration. The final volume varies between 250 cm.[3] and 350 cm.[3], and the sulphuric acid (absolute) between about five per cent at the start and one and a half or two per cent at the end.

The determination of large amounts of selenious acid by this method is somewhat less advantageous than would be the case if the reduction of the permanganate proceeded further in the first action. One hundred cubic centimeters of a standard solution is as much as can be conveniently handled in a single process of titration, and that volume of decinormal permanganate (which is about as strong as the standard solution should be when accurate work is expected) is capable of oxidizing about 0.25 grm. of selenium dioxide. In the table are given the results of practical tests of this method.

Permanganate Oxidation.

SeO_2 taken. grm.	Oxygen equivalent to permanganate used. grm.	Oxygen equivalent to oxalic acid used. grm.	SeO_2 found. grm.	Error. grm.
0 1000	0 03506	0 02065	0 1001	+0.0001
0.1000	0 03519	0 02073	0 1004	+0.0004
0.1000	0 03706	0 02255	0 1007	+0 0007
0.1000	0 03853	0.02422	0 0994	−0.0006
0.1000	0 03512	0 02065	0 1005	+0.0005
0 2000	0 06124	0 03256	0.1994	−0.0006
0.2011	0 06069	0.03177	0.2008	−0.0003
0.2004	0.06072	0 03177	0 2010	+0 0006
0.2020	0.06083	0 03185	0 2012	−0.0008
0.2038	0.06106	0 03185	0 2028	−0.0010

The Determination of Selenious Acid by the Direct Action of Sodium Thiosulphate, According to the Method of Norris and Fay.

In the method of Norris and Fay * for the iodometric determination of selenious acid, advantage is taken of a direct and unique action of sodium thiosulphate upon selenium dioxide in the presence of hydrochloric acid, four molecules of sodium

* Am. Chem. Jour., xviii, 703.

thiosulphate acting upon one molecule of selenious acid.* The method, which consists in treating the solution of selenious acid in ice water, in the presence of hydrochloric acid, with an excess of a $n/10$ solution of sodium thiosulphate and titrating back the excess of the thiosulphate with iodine, involves the addition of an excess of the thiosulphate to the solution of selenious and hydrochloric acids, and thus establishes conditions which demand care as to the relation of the acid, thiosulphate, degree of dilution, and temperature.

Norton † shows that with precautions noted the process of Norris and Fay is simple, rapid and accurate; without them, as the experimental results given below indicate, errors of considerable amount may enter.

Reduction by Thiosulphate and Titration of Excess.

Amount of SeO_2 taken. grm.	HCl (sp. gr. 1.12). cm.³	Excess of $Na_2S_2O_3$. cm.³	SeO_2 taken. grm.	Error. grm.
Volume at beginning, 200 cm.³				
0.1042	5	24.16	0.1041	−0.0001
0.0611	10	13.3	0 0611	0.0000
0.0850	10	21.9	0 0828	−0 0022
0.0757	25	13.07	0.0749	−0.0008
0.0540	25	21.02	0.0522	−0.0018
Volume at beginning, 400 cm.³				
0.0616	10	2.28	0 0625	+0.0009
0.0628	10	7 11	0 0631	+0 0003
0.0508	10	11.4	0 0511	+0.0003
0.0587	10	12.8	0 0594	+0.0007
0.0807	10	15.3	0 0813	+0.0006
0.0633	10	20 85	0.0638	+0 0005
0.0682	25	1 11	0.0685	+0 0003
0.0779	25	1 35	0.0788	+0 0009
0.0465	25	18.93	0 0469	+0 0004

It is recommended to so adjust conditions that no more than 20 cm.³ of $n/10$ thiosulphate shall be present in excess. If this limit be placed upon the thiosulphate, 5 cm.³ of hydrochloric acid (sp. gr. 1.12) may safely be present in a volume of 200 cm.³ at the beginning, or 10 cm.³ of the acid in a volume of 400 cm.³ The presence of 5 cm.³ of the acid in 400 cm.³ of solution is really sufficient to bring about the reaction.

* The complete reaction is not stated.

† J. T. Norton, Jr., Am. Jour. Sci., [4], vii, 287.

The Iodometric Determination of Selenic Acid by the Action of the Halogen Acids.

Reduction by Hydrochloric Acid, with Distillation. It has long been known that selenic acid is reducible by hydrochloric acid with evolution of chlorine, but the reaction was regarded as more or less uncertain until Petterson showed * that conditions of action may be secured under which the reduction proceeds regularly according to the equation

$$SeO_3 + 2\ HCl = SeO_2 + H_2O + Cl_2.$$

The chlorine evolved may be estimated iodometrically and taken as the measure of the selenic acid originally present or of the selenious acid produced. According to this method of determination, it is only necessary to boil a solution of selenic acid in hydrochloric acid of moderate concentration, and if the solution is not too dilute the reduction is obtained in a few moments.

Gooch and Evans † have determined the limits within which a successful determination of the selenic acid may be expected. It is shown that so long as the volume of the hydrochloric acid, sp. gr. 1.20, does not amount to more than 10 per cent of the entire liquid no chlorine whatever is evolved, and that only when the percentage of this acid rises as high as thirty does the chlorine evolved during boiling for five minutes approach the theoretical yield. Care must be taken, however, not to prolong the boiling after the solution reaches a concentration corresponding to hydrochloric acid of half-strength; for under such conditions — easily attained in boiling down mixtures of selenious acid and hydrochloric acid — over-reduction may take place and selenium appear visibly in the distillate. Obviously it is advantageous, in attempting the practical reduction of selenic acid, to begin the distillation with acid of strength sufficient to insure the evolution of chlorine in quantity at the outset, and it has been found best to start with a mixture one-third of which is the strongest aqueous hydrochloric acid, sp. gr. 1.20. With solutions so constituted the reduction goes on rapidly. Good results may be expected when the mixture, containing one-third of its volume of the strongest aqueous hydrochloric acid at the beginning, is boiled until the

* Zeit. anal. Chem. xii, 287.

† F. A. Gooch and P. S. Evans, Jr., Am. Jour. Sci., [3], l, 400.

chlorine is expelled, care being taken that the volume of the liquid shall not become less than two-thirds of the original volume.

The apparatus * made by sealing to the outlet tube of a Voit wash bottle (used as a retort) to the inlet tube of a Drexel wash bottle (charged with potassium iodide and used as a receiver) with a set of Will and Varrentrapp bulbs (sealed to the receiver, to serve as a trap) is convenient for the operation. A current of carbon dioxide aids in carrying the chlorine to the receiver and in promoting quiet boiling.

From solutions having a total volume of 75 cm.[3] at the outset and containing 25 cm.[3] of the strongest aqueous hydrochloric acid (sp. gr. 1.20), the entire amount of chlorine corresponding to the reduction of 0.2 grm. of selenic acid to selenious acid is liberated in ten minutes. The iodine in the receiver is estimated by thiosulphate. The details of experiments made under this procedure with selenic acid obtained by oxidizing with permanganate pure selenium dioxide† are given in the table.

Reduction by Hydrochloric Acid.

SeO_3 taken.	Total volume at the outset. cm.[3]	HCl (sp. gr. 1.20) present. cm.[3]	Time in minutes.	SeO_3 found. grm.	Error.
0 0572	75	25	10	0 0568	−0 0004
0 0572	75	25	10	0.0569	−0.0003
0 1144	75	25	10	0 1143	−0 0001
0 1144	75	25	10	0.1137	−0 0007
0 1144	75	25	10	0.1147	+0 0003
0 2288	75	25	10	0 2233	−0 0005
0.2288	75	25	10	0 2279	−0 0009

Reduction by Hydrobromic Acid, with Distillation. When acted upon by sulphuric acid and potassium bromide in solution, selenic acid liberates bromine in proportion to the excess of acid, the amount of bromide, and the temperature.

$$SeO_3 + 2\ HBr = SeO_2 + H_2O + Br_2.$$

When such a solution is boiled the bromine is evolved and may be collected in potassium iodide, and the iodine thus set free may be determined by standard sodium thiosulphate and taken as the measure of the bromine distilled.

* See Fig. 3, page 4.

† See page 382.

Gooch and Scoville * have shown that the applicability of the reaction to quantitative purposes turns upon the adjustment of the proportions of the reagents used. The apparatus shown in Fig. 3 † is convenient for the distillation process.

When the proportions of sulphuric acid, potassium bromide, and selenic acid are favorable, the bromine liberated is removed rapidly to the distillate, leaving the residue perfectly colorless, but as the distillation is continued the liquid residue again takes on color and more iodine is set free by the action of the distillate upon potassium iodide, while selenium is plainly visible in the receiver. When the amount of potassium bromide is large, its effect is to retain bromine in the liquid so obstinately that no period of colorlessness intervenes before the second stage of color arrives; when its amount is small, while that of the sulphuric acid is also small, the reduction of the selenic acid and the evolution of the bromine progress slowly; and the interval of colorlessness is prolonged when the amount of bromide is small, while that of the acid is comparatively large. The proportions found best in handling 0.25 grm. of selenic acid, or less, are an initial volume of 60 $cm.^3$ containing 20 $cm.^3$ of sulphuric acid of half-strength, with 1 grm. of potassium bromide. Under these conditions it is found that the reduction is almost theoretically exact when the distillation is continued until the recoloration of the boiling liquid is distinctly recognizable; and this point corresponds in practice very closely to a concentration of volume to 35 $cm.^3$. In the following table are given the results of experiments made under these conditions of action.

Reduction by Hydrobromic Acid.

SeO_3 taken as H_2SeO_4.	H_2SO_4 of half-strength.	KBr taken.	Initial volume.	Final volume.	SeO_3 calculated.	Error.
grm.	$cm.^3$	grm.	$cm.^3$	$cm.^3$	grm.	grm.
0 0590	20	1	60	35	0.0588	−0.0002
0 0590	20	1	60	35	0.0591	+0.0001
0.0614	20	1	60	35	0 0616	+0.0002
0 0614	20	1	60	35	0 0607	−0 0007
0 1180	20	1	60	35	0 1177	−0.0003
0 1180	20	1	60	35	0 1180	0.0000
0.1534	20	1	60	35	0 1527	−0.0007
0 2349	20	1	60	35	0 2350	+0.0001

* F. A. Gooch and W. S. Scoville, Am. Jour. Sci., [3], l, 402.

† See page 4.

Reduction by Hydriodic Acid, with Distillation. The determination of selenic acid by acting with hydrochloric acid and potassium iodide and estimating the iodine liberated has been studied by Gooch and Reynolds.* While the simple contact of selenic acid and potassium iodide in solution acidified with hydrochloric acid does not produce a regular liberation of iodine, it is possible by submitting such mixtures to distillation, when the amounts of selenic acid present are not too large, to bring about a definite reaction in which the products are selenium, water and iodine, according to the equation

$$SeO_3 + 6\,HI = Se + 3\,H_2O + 3\,I_2.$$

In applying this reaction to analytical purposes it is convenient to make use of an apparatus and procedure previously described† and to treat, preferably, not more than 0.2 grm. of the selenic oxide with 1 grm. to 3 grm. of potassium iodide, 5 cm.^{3} of concentrated hydrochloric acid in a total volume of 60 cm.^{3}, and to continue the boiling for ten minutes.

The results of experiments made in this manner with selenic acid obtained in solution by oxidizing known amounts of selenium dioxide by potassium permanganate‡ are given in the table.

Reduction by Hydriodic Acid.

SeO_3 taken.	KI in flask.	HCl in flask (sp. gr. 1.20).	Total volume boiled.	Time in minutes.	SeO_3 found.	Error.
grm.	grm.	cm.^{3}	cm.^{3}		grm.	grm.
0.0593	1	5	60	5	0 0593	0.0000
0.0593	1	5	60	5	0.0591	−0 0002
0.0593	3	5	60	10	0.0596	+0 0003
0.1779	3	5	60	10	0 1769	−0.0010
0.1779	3	5	60	10	0.1780	+0.0001
0.1779	3	5	60	10	0.1764	−0.0015

Reduction by Hydriodic Acid, Differential Method. In a mixture made up of a selenate, an arsenate, potassium iodide, and sulphuric acid, the arsenic acid attacks the hydriodic acid before all of the selenic acid is reduced. In order to apply to selenic acid the differential method of determination, the selenic acid must first be reduced to

* F. A. Gooch and W. G. Reynolds, Am. Jour. Sci., [3], l, 258.

† See page 379.

‡ See page 382.

the condition of selenious acid. Ordinarily, the simplest mode of reducing selenic acid is by boiling it in solution with hydrochloric acid of definite strength,* but in this case the presence of hydrochloric acid is precluded on account of the consequent volatilization of arsenious chloride during the process of concentration in the subsequent treatment with the iodide. It is possible, however, to make use of reduction by hydrobromic acid, since arsenious bromide is not appreciably volatile under the conditions. Gooch and Peirce† have shown that the determination of selenic acid may therefore be accomplished by first reducing it to selenious acid by the bromide process and then treating the residue by the differential method for the determination of selenious acid.‡

Differential Method.

SeO_2 taken as H_2SeO_4. grm.	KI used in second reduction. grm.	SeO_2 found. grm.	Error. grm.
0.0378	0.6306	0.0380	+0.0002
0.0378	0.5643	0.0374	−0.0004
0.0516	0.7136	0.0517	+0.0001
0.0503	0.7302	0.0508	+0.0005
0.0541	0.6671	0.0544	+0.0003
0.1007	1.3277	0.1011	+0.0004
0.1008	1.3277	0.1011	+0.0003
0.1007	1.2082	0.1005	−0.0002
0.1007	1.1684	0.1016	+0.0009
0.1007	1.0522	0.0999	−0.0008
0.1009	1.2679	0.1005	−0.0004
0.1031	1.1119	0.1032	+0.0001
0.1870	1.8720	0.1879	+0.0009
0.2014	1.9915	0.2020	+0.0006
0.2016	2.0745	0.2025	+0.0009
0.2059	1.8687	0.2064	+0.0005

The procedure is as follows: The solution containing the selenate to be determined is put in an Erlenmeyer flask of 300 cm.³ capacity with 1 grm. of potassium bromide and sulphuric acid amounting to 20 cm.³ of the acid of half-strength. The solution amounting to 60 cm.³ or 100 cm.³ is boiled until the colorless solution left when the bromine vanishes begins to color again. Experience shows that the reappearance of the brownish color is very easily seen and that it is not safe to conclude that the free

* See page 385.

† F. A. Gooch and A. W. Peirce, Am. Jour. Sci., [4], i, 33.

‡ See page 380.

bromine has been eliminated, under the conditions of dilution and proportion, until this stage of concentration — which corresponds to a volume of about 35 cm.[3] — has been reached; but the distillation should not be pushed beyond the point at which the returning color is noted. When this condition has been reached the solution is cooled and treated exactly in the manner described for the reduction of selenious acid. The neutralization by acid potassium carbonate, after the final boiling, generally occasions the precipitation of manganous carbonate, but the precipitate does not interfere in the slightest with the titration which follows.

The preceding table comprises determinations made to test the accuracy of the iodometric determination of selenic acid by the combined processes of reduction.

The Separation of Selenium from Tellurium by Procedure Based upon the Difference in Volatility of the Bromides.

When small amounts of selenic acid are boiled in aqueous solution with potassium iodide and hydrochloric acid, selenium is precipitated, while the iodine set free simultaneously may be estimated in the distillate and residue, and taken as the measure of the selenic acid originally present.* If the iodide is omitted from the mixture, so that the hydrochloric acid alone shall be the reducer, the reduction proceeds only to the point of formation of selenious acid, provided the boiling is not continued after the hydrochloric acid has reached the condition of half-strength at which it boils unchanged under normal atmospheric pressure. A solution of selenic acid, potassium bromide, and sulphuric acid, of regulated dilution and proportions, also yields under defined conditions selenious acid as the product of reduction. When, however, the ebullition of a solution of selenious acid in hydrochloric acid is continued after the acid has reached the condition of half-strength, traces of selenium appear in the receiver and connecting tubes, the distillate sets free iodine from potassium iodide, and it is evident that the selenious acid is undergoing further reduction; and the same effects are produced when the boiling of the mixture of sulphuric acid, potassium bromide, and selenious acid is pressed beyond the point at which the solution begins to be colored. Obviously, under certain conditions

* See page 388.

of concentration, selenium tetrachloride and selenium tetrabromide, respectively, are forming from the acid; and the appearance of the elementary selenium is due to partial decomposition of the halogen salts. Phenomena of a similar nature are seen when an aqueous solution of selenious acid, phosphoric acid, and sodium chloride is submitted to distillation: that is to say, there comes a time in the process of boiling such mixtures when the appearance of elementary selenium and the action of the distillate upon potassium iodide make evident the volatilization and partial decomposition of the selenium compounds of the halogens, and the further continuance of the treatment results in the more or less complete removal of the selenium compounds to the distillate. From the mixture containing the phosphoric acid, selenious acid and sodium chloride only a partial volatilization of the selenium chloride takes place. The volatilization of selenium bromide, however, produced by the reaction between phosphoric acid, selenious acid and potassium bromide, may be made complete. Upon this reaction Gooch and Peirce * have based a process for the separation of selenium and tellurium, taking advantage of the volatility of selenium tetrabromide and the non-volatility of tellurium tetrabromide under definite conditions.

The distillation apparatus used in the process of separation is shown in Fig. 4.† It consists of two Voit flasks, a Drexel bottle, and Will and Varrentrapp bulbs, connected by sealed or ground joints, as shown in the figure.

The operation is conducted as follows: In the first Voit flask, V^1, selenium dioxide and tellurium dioxide are dissolved in potassium hydroxide, the alkali is neutralized and the precipitate thus formed is redissolved by phosphoric acid, added in excess to the amount of 20 $cm.^3$ of the acid of sp. gr. 1.70. To the solution is added 1 grm. of potassium bromide with enough water to make the entire volume of the solution 50 $cm.^3$ The second flask, V^2, contains 10 $cm.^3$ of water, and the Drexel bottle and trap are charged with a solution of potassium iodide. Carbon dioxide is passed through the apparatus and the solution in V^1 is boiled until the volume has diminished to 15 $cm.^3$, the flask and connecting tube being cloaked with a mantle of asbestos board and gently

* F. A. Gooch and A. W. Peirce, Am. Jour. Sci., [4], i, 181.

† See page 5.

flamed toward the last to remove traces of the selenium bromides held back mechanically by the oily tellurium compound which collects. After cooling, the first flask V^1 is removed; 1 grm. of potassium iodide and 5 cm.^{3} of hydrochloric acid are added to the contents of the second flask, V^2; the current of carbon dioxide is again started through the apparatus; the mixture is boiled ten minutes; and the iodine in the flask, receiver and trap, determined by titration with sodium thiosulphate, is taken as the measure of the selenium dioxide.*

Results of this procedure are given in the table.

Double Distillation.

TeO_2 taken.	KBr taken.	H_3PO_4 (sp. gr. 1.70) taken.	Final volume.	SeO_2 taken.	SeO_2 found.	Error.
grm.	grm.	cm.^{3}	cm.^{3}	grm.	grm.	grm.
..	1	20	15	0 0366	0 0372	+0 0006
...	1	20	15	0 0366	0 0377	+0 0011
..	1	20	15	0 1098	0.1090	−0 0008
...	1	20	15	0 1098	0.1101	+0.0003
0 1	1	20	15	0 0733	0 0735	+0 0002
0.1	1	20	15	0 0997	0.0995	−0.0002
0 1	1	20	15	0.1004	0.1003	−0 0001
0 1	1	20	15	0 0916	0 0914	−0.0002
0 1	1	20	15	0.0997	0.0995	−0 0002
0 1	1	20	15	0.1010	0.1014	+0 0004
0 1	1	20	15	0.1015	0.1008	−0.0007
0 1	1	20	15	0 1019	0 1022	+0.0003
0 1	1	20	15	0 1010	0.1012	+0.0002
0 1	1	20	15	0.1002	0.1000	−0.0002
0.1	1	20	15	0 1006	0.1004	−0.0002
0 1	1	20	15	0.1006	0.1001	−0.0005

The phenomena of the distillation are very characteristic. When selenious acid is present without tellurous acid, the solution boils quietly in the first flask until the volume of liquid has decreased to about 30 cm.^{3}, when traces of red selenium begin to deposit in the tube joining the first and second flask. When the volume has further diminished to about 25 cm.^{3} the liquid begins to take on color, darkens rapidly, and evolves bromine, which at once attacks the selenium previously deposited. The greater part of the bromine is absorbed in the second flask, V^2, but a trace finds its way to the Drexel bottle, in which it sets free a slight amount of iodine from the iodide. As the operation

* See page 379.

progresses, an orange-yellow crystalline solid, presumably selenium tetrabromide for the most part, appears in the tube where the selenium has been, while a dark oily liquid, consisting largely, no doubt, of the monobromide, condenses in drops upon the walls of the flask and returns to form a floating layer upon the hot liquid. Finally, when the volume has diminished to 15 $cm.^3$, the liquid has become perfectly clear and colorless, white fumes of hydrobromic acid are evolved, and the tube between the two flasks has been cleared. At this point the second flask, V^2, contains (besides a trace of selenium corresponding to the slight amount of bromine which has escaped to the Drexel bottle) the colorless selenious acid regenerated by the action of the water and free bromine upon the mixed selenium bromides. The contents of this flask may now be treated with potassium iodide and hydrochloric acid as directed above * and the iodine in the receiver, including, of course, the small amount set free by the bromine which reaches the receiver in the first stage of the process, and the small amount remaining in the flask, measure the selenium dioxide acted upon.

When tellurium dioxide is subjected without the selenium dioxide to similar treatment the phenomena are different. The solution containing the tellurous acid, potassium bromide, and phosphoric acid, in the proportions used in the experiments with selenious acid, colors at about the same degree of concentration at which the solution containing the selenious acid began to darken. As the concentration progresses, the color deepens, ruby red crystals (probably hydrated tellurium tetrabromide) form, which accumulate upon the walls of the flask and turn yellow, and when the volume of the solution is diminished to 15 $cm.^3$ a green vapor begins to distil. During the process no iodine is set free in the Drexel bottle, and upon stopping the boiling and adding potassium iodide to V^2 no iodine is liberated, even when the boiling has gone so far that a trace of the green vapor has condensed and run into the water in the flask.

When the tellurium dioxide and selenium dioxide are both present the characteristic phenomena occur together — the evolution of bromine, coloring of the liquid, distillation of selenium bromides, crystallization of tellurium tetrabromide, and volatilization of the selenium compounds.

* See page 379.

TELLURIUM.

The Gravimetric Estimation of Tellurous Acid by Liberation of Iodine and Absorption of that Element by Silver.

Tellurous acid reacts with potassium iodide in presence of hydrochloric acid and silver * in an atmosphere of hydrogen according to the reaction

$$TeO_2 + 4\,KI + 4\,HCl = 4\,KCl + 2\,H_2O + Te + 2\,I_2.$$

The increase in weight of insoluble material represents both iodine and tellurium. Tests made by Perkins † with tellurium dioxide prepared from the basic nitrate gave the following results.

Weighing of Silver Iodide and Tellurium.

Ag taken. grm	Te taken grm	Increase grm.	Calculated Te. grm	Error grm.
2 0152	0 0330	0 1654	0 0332	+0 0002
2 0152	0 0990	0 4931	0 0989	−0 0001
2 0815	0 0528	0 2635	0 0529	+0 0001
2 0815	0 0660	0 3294	0 0661	+0 0001
2 0815	0 0990	0 4948	0 0993	+0 0003
2 1693	0 1650	0 8240	0 1654	+0 0004
3 0126	0 1650	0 8258	0 1657	+0 0007
3 0126	0 0660	0 3302	0 0663	+0 0003

The Determination of Tellurous Acid by Oxidation with Potassium Permanganate.

The estimation of tellurous acid by oxidation with excess of potassium permanganate (either in acid or alkaline solution), destruction of the higher oxides of manganese or the manganate by standard oxalic acid in presence of sulphuric acid, and titration of the residual oxalic acid by more permanganate has been shown by Brauner ‡ to be feasible, but the tendency of the permanganate to throw off too much oxygen when the oxidation is made in solutions strongly acidified with sulphuric acid (as must be the case if the tellurous oxide is to be held permanently in solution by sulphuric acid) necessitates the application of a considerable correction. Gooch and Danner § have shown,

* See page 444.
† Claude C. Perkins, Am. Jour. Sci., [4], xxix, 540.
‡ Jour. Chem. Soc., 1891, 238.
§ F. A. Gooch and E. W. Danner, Am. Jour. Sci., [3], xliv, 301.

however, that when the tellurous oxide is first dissolved in an alkali hydroxide and the solution is made acid to a limited degree with sulphuric acid, either before or after oxidation by the permanganate, no correction appears to be necessary.

According to the first procedure, the alkaline solution of the oxide is diluted to 100 cm.³, a measured amount of standardized permanganate is added in excess, sulphuric acid [1 : 1] is introduced to an amount not exceeding by more than 5 cm.³ that needed for neutralization, standardized oxalic acid is measured in to an amount more than sufficient to destroy the manganic oxide and permanganate left after the oxidation, and the surplus of oxalic acid is titrated by permanganate.

According to the second procedure, the alkaline solution of the oxide is treated with sulphuric acid [1 : 1] until the precipitate first thrown down is just redissolved, and 1 cm.³ more of the [1 : 1] acid is added. To the solution thus acidulated is measured in potassium permanganate in excess, and then standardized oxalic acid to an amount a little more than sufficient to destroy the permanganate remaining; the liquid is warmed to 80°, and the excess of oxalic acid is titrated by permanganate.

The results of experiments made according to these procedures are given below.

Oxidation in Alkaline Solution.

TeO_2 taken. grm.	TeO_2 found. grm.	Error. grm.	Mean error. grm.
0.1200	0.1199	−0.0001	+0.0006
0.0783	0.0783	0.0000	
0.0931	0.0938	+0.0007	
0.1100	0.1116	+0.0016	
0.0904	0.0907	+0.0003	
0.1065	0.1077	+0.0012	

Oxidation in Acid Solution.

TeO_2 taken. grm.	TeO_2 found. grm.	Error. grm.	Mean error. grm.
0.0910	0.0912	+0.0002	+0.0003
0.0910	0.0908	−0.0002	
0.0911	0.0922	+0.0011	
0.0913	0.0913	0.0000	
0.0912	0.0913	+0.0001	
0.0914	0.0921	+0.0007	

Oxidation in Presence of a Chloride.

In the presence of free hydrochloric acid the action of the permanganate upon tellurous acid has been shown by Brauner * to be irregular and excessive, and the irregularity cannot be corrected (as in the titration of ferrous salts in presence of hydrochloric acid) by the addition of a manganous salt according to the well-known procedure of Kessler † and Zimmermann.‡ Gooch and Peters § have pointed out that there should be nothing to prevent the accurate determination of tellurium in tellurous compounds in the presence of chlorides by the permanganate process, provided the first oxidation is made in alkaline solution and the second oxidation is carried out with such precautions as are necessary to a correct determination of oxalic acid by permanganate in presence of hydrochloric acid; for the special danger of over-action on the part of the permanganate cannot exist while the solution is alkaline, and has passed when the tellurite has become a tellurate and before the solution is made acid.

It has been shown that the presence of a manganous salt is necessary and sufficient to secure regularity of action when oxalic acid is titrated in presence of a considerable amount of hydrochloric acid. When the amount is no more than would be formed in the decomposition of a gram or two of halogen salt of tellurium the disturbing effect under ordinary conditions of work is probably inappreciable, but even in such a case it is better to oxidize in the presence of a manganous salt for the reason that the titration of the oxalic acid may then be made at the ordinary atmospheric temperature.

According to the procedure recommended, the alkali hydroxide solution of tellurous oxide containing the alkali chloride is treated with standardized potassium permanganate until the characteristic permanganate color is visible; standardized ammonium oxalate is introduced in excess of the quantity required to reduce the remaining permanganate, manganate, and higher oxides; and enough sulphuric acid [1 : 1] is added to neutralize the alkali hydroxide and leave an excess of about 5 $cm.^3$. Then the final titration with permanganate may be made either after heating

* Jour. Chem. Soc., 1891, 241.

† Ann. Phys., cxviii, 48; cxix, 225, 226.

‡ Ann. Chem., ccxiii, 302.

§ F. A. Gooch and C. A. Peters, Am. Jour. Sci., [4], viii, 122.

the solution to 80° or at the ordinary temperature after the addition of 0.5 grm. to 1 grm. of manganous chloride.

Experimental results of the procedure are given below:

Permanganate Oxidation in Alkaline Solution: Treatment with Oxalic Acid: Permanganate Titration in Acid Solution.

Volume at beginning, 150 cm.³: Te=127.5.

TeO_2 taken. grm.	NaCl. grm.	H_2SO_4 1 : 1. cm.³	$MnCl_2$ $4H_2O$. grm.	TeO_2 found. grm.	Error. grm.
Temperature of titration, 60°–80° C.					
0.1000	0.4	5	...	0.1003	+0.0003
0.1000	0.4	5	...	0.1000	0.0000
0.1000	0.4	5	...	0.1004	+0.0004
0.1000	1.0	5	...	0.1003	+0.0003
0.0650	1.0	5	...	0.0653	+0.0003
Temperature of titration, 20°–26° C.					
0.0700	0.4	5.7	1.0	0.0705	+0.0005
0.0700	0.4	5.7	1.0	0.0698	−0.0002
0.0700	0.4	5.7	0.5	0.0701	+0.0001
0.1000	0.4	5.7	0.5	0.1008	+0.0008

Oxidation in Presence of a Bromide.

Fairly good determinations of tellurous acid may be made similarly in the presence of a bromide, provided the titration is made at the atmospheric temperature in the presence of a sufficiency (0.5 grm. to 1 grm.) of a manganous salt and of an excess of sulphuric acid limited to

Permanganate Oxidation in Alkaline Solution: Treatment with Oxalic Acid: Permanganate Titration in Acid Solution.

Volume at beginning, 150 cm.³: Te=127.5.
Temperature of titration, 24°–26° C.

TeO_2 taken. grm.	NaCl. grm.	KBr. grm.	H_2SO_4 12.5 per cent. cm.³	$MnCl_2$. $4H_2O$. grm.	TeO_2 found. grm.	Error. grm.
0.0650	...	0.5	1	1.0	0.0661	+0.0011
0.0650	...	0.5	1	1.0	0.0647	−0.0003
0.1000	...	0.5	1	1.0	0.1002	+0.0002
0.3000	..	0.5	5	0.5	0.3010	+0.0010
0.0650	0.5	0.5	1	1.0	0.0661	+0.0011

5 cm.^{3} of the 12.5 per cent mixture, in a volume of 150 cm^3. At higher temperatures and higher concentrations of acid, bromine is liberated by the permanganate. The experimental results are given in the table.

The Determination of Tellurous Acid by the Precipitation of Tellurous Iodide.

Hydriodic acid and tellurous acid interact with the formation of tellurium tetraiodide, converted by water into an oxyiodide and by excess of alkali iodides to soluble double salts. Gooch and Morgan * have observed that when potassium iodide is added to a cold solution of tellurous acid containing at least one-fourth of its volume of strong sulphuric acid, no tendency to form a double salt becomes apparent until the potassium iodide amounts to more than enough to convert all the tellurous acid present into tellurium tetraiodide according to the equation

$$H_2TeO_3 + 4\,H_2SO_4 + 4\,KI = TeI_4 + 4\,KHSO_4 + 3\,H_2O.$$

The tellurium tetraiodide thus formed is extremely insoluble in sulphuric acid of one-fourth strength, though soluble in excess of potassium iodide and acted upon by water with formation of tellurium oxyiodide and hydriodic acid. It is produced at first in the condition of a finely divided dark brown precipitate which upon agitation of the liquid gathers in curdy masses and settles, leaving the liquid clear. By taking advantage of this tendency to curd, it is possible to determine without great difficulty the exact point during the gradual addition of potassium iodide when the precipitation of the tellurium iodide is complete. Upon this property Gooch and Morgan have based a simple titrimetric method for the direct determination of small amounts of tellurous acid.

According to the procedure described, tellurous oxide is dissolved in a very little of a strong solution of potassium hydroxide, and dilute sulphuric acid is added carefully until the tellurous acid which is precipitated upon neutralization of the alkali hydroxide is just redissolved. To this solution, contained in an Erlenmeyer flask, sulphuric acid [1 : 1] is added in such amount that the liquid shall contain, after the subsequent addition of potassium iodide in solution, at least one-fourth of its volume of

* F. A. Gooch and W. C. Morgan, Am. Jour. Sci., [4], ii, 271.

strong sulphuric acid. The flask is placed upon a pane of window glass supported upon strips of wood about 1 cm. above the level of a work table covered with white paper. A solution of approximately decinormal potassium iodide, free from iodate and carefully standardized in terms of iodine by a method to be described,* is introduced gradually from a burette into the middle of the Erlenmeyer beaker. As the drops of the potassium iodide touch the liquid the precipitation forms at the center and travels in rings toward the outer walls of the beaker. When the liquid becomes so opaque that the effect of the potassium iodide is distinguished with difficulty, the beaker is rotated and the curded precipitate permitted to settle; and then the process of titration is continued as before until precipitation ceases. With an Erlenmeyer flask 10 cm. in diameter across the bottom and a final volume of liquid amounting to not more than 100 cm.^{3} the precipitation is easily followed.

The results of a series of determinations made according to the method described are recorded in the following table:

Precipitation of Tellurous Iodide.

Te=127 *

Final volume. cm.^{3}	Strongest H_2SO_4 present. cm.^{3}	Iodine value of KI used. grm.	TeO_2 taken. grm.	TeO_2 found. grm.	Error. grm.
50	17	0 0706	0.0223	0 0221	−0.0002
50	17	0.0764	0.0244	0 0239	−0.0005
50	17	0.1591	0.0496	0 0499	+0.0003
60	17	0.1655	0.0517	0 0519	+0 0002
60	17	0.1578	0 0498	0 0494	−0.0004
80	30	0.1591	0.0498	0 0499	+0.0001
100	30	0 3179	0.1001	0 0997	−0.0004
100	30	0.3186	0.1008	0 0999	−0 0009
100	30	0 3208	0 1011	0 1005	−0.0006
100	30	0 3208	0 1010	0 1005	−0.0005

* Determined by permanganate oxidations and reductions by hydrobromic acid (see p. 402).

The Iodometric Estimation of Tellurous Acid.

The determination of tellurous acid by oxidation of the alkali hydroxide solution with potassium permanganate, reduction of residual permanganate and higher oxides of manganese with oxalic acid in presence of sulphuric acid, and titration of the excess

* See page 457.

of oxalic acid by permanganate, is not feasible in presence of an iodide, because upon acidifying the mixture, iodine is at once set free.

Potassium permanganate and the higher oxides of manganese are, however, completely and rapidly reduced by an excess of potassium iodide upon the addition of acid, and the iodine liberated is a measure of the permanganate. Norris and Fay * have utilized this reaction in an excellent iodometric method for the determination of tellurous acid. This method consists in treating the alkaline solution of tellurous oxide with standard permanganate until the meniscus of the liquid shows a deep pink color, then diluting the solution with ice water, adding potassium iodide and sulphuric acid, and titrating with sodium thiosulphate. The difference between the amount of iodine thus found and the amount found by treating similarly the same amount of permanganate, taken by itself, is the measure of the tellurous acid.

It is plain that any agent capable of converting the iodine to hydriodic acid without at the same time reducing telluric acid should be capable of measuring the excess of the permanganate, and so the amount of tellurous acid originally present. Gooch and Peters † make use of the standard arsenite, employed also in standardizing the permanganate,‡ to take up the free iodine.

According to this procedure, the solution of tellurous oxide in alkali hydroxide is added to a solution of potassium iodide; standardized permanganate is run in until the green color of the manganate appears (about 30 $cm.^3$ of the $n/10$ solution for every 0.1 grm. of TeO_2); dilute sulphuric acid is introduced in slight excess, followed, after the iodine has separated, by an excess of acid potassium carbonate; and the iodine is titrated to vanishing color (without starch) by standard arsenite. It is evident that when the solution is acidified more than enough iodide to complete the reduction of the manganese oxides should be present, or else that the arsenious acid should be present in suitable amount before the sulphuric acid is put in. This latter procedure may be used in case, for any reason, it is preferred not to introduce more iodide into the solution than was present originally, as,

* Am. Chem. Jour., xx, 278.

† F. A. Gooch and C. A. Peters, Am. Jour. Sci., [4], viii, 125.

‡ See page 41.

for example, when a direct determination of the iodine present is to follow.

Experimental results follow in the table:

Permanganate Oxidation in Alkaline Solution and Iodometric Determination of the Excess.

Te = 127.5.

TeO_2 taken.	NaCl.	KBr.	KI.	Total volume at end.	NaOH present during oxidation.	TeO_2 found.	Error.
grm.	grm.	grm.	grm.	cm.³	grm.	grm.	grm.
0.1000	...	...	0.5	160	0.1	0.1005	+0.0005
0.1000	...	...	0.5	160	0.1	0.1001	+0.0001
0.1000	...	...	0.5	160	0.1	0.1003	+0.0003
0.1000	...	...	1.0	250	0.1	0.1007	+0.0007
0.2000	...	...	1.0	250	0.2	0.1997	−0.0003
0.1000	0.5	0.5	0.5	250	0.1	0.1000	0.0000
0.2100	1.0	1.0	1.0	225	0.2	0.2105	+0 0005
0.1000	. .	. .	0.5	160	1.0	0.1011	+0 0011
0.2000	...	...	1.0	300	2.0	0.2009	+0 0009

The Iodometric Determination of Telluric Acid.

Gooch and Howland * have shown that telluric acid may be reduced by the action of potassium bromide and sulphuric acid to the condition of tellurous acid and estimated by determining the iodine liberated by the bromine set free in the operation.

According to the method demonstrated, the alkali tellurate is introduced into the apparatus for distillation with 3 grm. of potassium bromide, care being taken to insure in the 50 cm.³ or more of liquid the presence of 10 cm.³ of sulphuric acid of half strength. A current of carbon dioxide is passed through the apparatus, and the solution is boiled to set free the bromine, which is absorbed in potassium iodide and estimated by standard sodium thiosulphate.

The distillation apparatus † consists of a Voit gas-washing flask which is joined by a sealed joint to the inlet tube of a Drexel washing bottle. To the outlet tube of the Drexel bottle is sealed a Will and Varrentrapp absorption apparatus. The

* F. A. Gooch and J. Howland, Am. Jour. Sci., [3], xlviii, 375.

† See Fig. 3, page 4.

washing bottle and attached bulbs contain a solution of 3 grm. of potassium iodide, and the former is kept cool by standing it during the distillation in a vessel of cold water.

The formation of tellurium tetrabromide in the concentrated acid liquid makes it impossible to tell by the color when all the bromine has been distilled, but the evidence of the experiments goes to show that the boiling of the liquid from a volume of 50 cm.³ to 25 cm.³ is sufficient, while concentration from 100 cm.³ to 20 cm.³ apparently does no harm.

The tellurite used in testing this procedure was made from tellurous oxide which had been shown by titration with permanganate to have an equivalent weight of about 159, which corresponds to an atomic weight of 127 for the element tellurium. Results are given in the table.

Reduction by Hydrobromic Acid and Estimation of Bromine Set Free.

Te=127.

Initial volume. cm.³	Final volume. cm.³	TeO_2 taken. grm.	TeO_2 found. grm.	Error grm.
50	20	0.0102	0 0098	−0.0004
50	20	0 0102	0 0099	−0.0003
50	20	0.0102	0 0098	−0 0004
50	20	0.0102	0 0098	−0.0004
100	40	0 1000	0 0994	−0.0006
80	40	0 1001	0 1001	0 0000
100	20	0 1002	0 1001	−0 0001
50	20	0 1000	0.1003	+0 0003
50	25	0 5011	0.5008	−0.0003
50	25	0 5002	0 5006	+0 0004
50	25	0 5000	0 4998	−0.0002
50	20	0 5000	0 4994	−0 0006

The Precipitation of Tellurium Dioxide and the Separation of Tellurium from Selenium.

Browning and Flint * have shown that fairly accurate and concordant estimations of tellurium may be obtained by acting upon the alkaline solution of a tellurite with hydrochloric acid, ammonium hydroxide, and acetic acid, sucessively, and weighing the precipitate as tellurium dioxide, the best of all the forms in which

* Philip E. Browning and William R. Flint, Am. Jour. Sci., [4], xxviii, 112.

tellurium has been weighed. It is unaffected by the air, is anhydrous, is not hydroscopic, and can easily be obtained in pure condition. It can be heated to any temperature under that of low redness without any danger of volatilization.

All processes for the estimation of tellurium in which the tellurium is precipitated and weighed in elementary condition are open to the objections, first, that there is more or less difficulty in securing completeness of precipitation owing to the rapid increase of free acid * in the solution; and, second, that the product is extremely susceptible to oxidation. On the other hand, those methods in which compounds decomposable by heat are transformed to the dioxide by ignition are generally both tedious by reason of the length of time required (as, for example, the basic nitrate process as described by Norris †) and, what is more to the point, liable to errors caused not only by lack of constancy of composition but also by the volatilization of the product to be weighed. The process of Browning and Flint presents therefore special advantages in the determination of tellurium.

According to the preferred procedure set forth, the material is dissolved in hydrochloric acid or in a 10 per cent solution of potassium hydroxide, about 2 $cm.^3$ for 0.2 grm. of dioxide. From the solution acidified with hydrochloric acid and diluted with boiling water to a volume of 200 $cm.^3$ the finely crystalline tellurium dioxide is precipitated by the careful addition of dilute ammonia in faint excess followed by the faintest possible excess of acetic acid.

If these simple operations are properly carried out, the precipitate will become crystalline by the time the alkali hydroxide is in excess; the addition of a few drops of acetic acid causes the precipitation to become entirely quantitative when the solution has cooled, so that no tellurium will be found in the filtrate by stannous chloride. The precipitate can be transferred, and safely and rapidly washed with cold water, and dried to constant weight at about 105° (or even under low redness) in a quarter of an hour. Furthermore, the filtration can be performed at the end of half an hour, or after twenty-four hours, as most convenient.

* Crane, Am. Chem. Jour., xxiii, 409. See also Lenher and Homburger, Jour. Am. Chem. Soc., xxx, 387.

† Jour. Am. Chem. Soc., xxviii, 1675.

Test results are given below:

The Solution in HCl Diluted with Boiling Water, and Treated with Ammonia and Acetic Acid.

TeO_2 taken. grm.	TeO_2 found. grm.	Error. grm.
0.2002	0.2000	−0 0002
0.2019	0.2017	−0 0002
0.2904	0.2002	−0.0002
0.2006	0.2004	−0.0002
0.2011	0 2010	−0 0001
0.2003	0.2003	0.0000

The Solution in KOH Acidified with HCl, Diluted with Boiling Water, and Treated with Ammonia and Acetic Acid.

2 $TeO_2.HNO_3$ taken.* grm.	TeO_2 theory: Te taken as 127.5. grm.	TeO_2 found. grm.	Error. grm.
0.2502	0.2089	0.2083	−0.0006
0.2524	0.2108	0.2110	+0.0002
0 2505	0.2092	0.2089	−0.0003
0.2528	0.2111	0.2106	−0.0005
0.2531	0.2113	0.2106	−0.0007
0.5008	0 4182	0.4182	0.0000
0.5010	0.4183	0.4175	−0.0008
0.5005	0.4179	0.4178	−0.0001

* Dissolved in KOH.

Separation from Selenium.

If hydrochloric acid solutions of tellurium and selenium dioxides be mixed, abundantly diluted with boiling hot water, and the operation of the above described process properly applied, only the tellurium is precipitated, the selenium remaining entirely in solution in the filtrate. This not only provides a simple and rapid preparative process for the purification of tellurium from selenium, but also makes possible the estimation of tellurium directly in the presence of the latter element. The dilution should be made, however, with boiling hot water, as cold water induces a flocky precipitation and inclusion of selenious acid. Results are given in the table.

Separation of Tellurium from Selenium.

TeO_2 taken grm.	SeO_2 taken. grm.	TeO_2 found. grm.	Error. grm.
The solution faintly acid with HCl, diluted cold.			
0 2000	0.1	0.2002	+0 0002
0 2015	0 1	0 2016	+0 0001
0 2038	0 1	0 2040	+0 0002
The solution faintly acid with HCl, diluted hot and treated with NH_4OH and HOC_2H_3O.			
0.2028	0 05	0.2019	−0 0009
0.2024	0 05	0.2024	0 0000
0 2003	0.1	0.1992	−0 0011
0 2009	0 1	0.2003	−0.0006

CHAPTER X.

CHROMIUM; MOLYBDENUM; URANIUM.

CHROMIUM.

The Estimation of Chromium as Silver Chromate.

It has been shown by Autenrieth * that when chromic acid is added to a boiling solution of silver nitrate, or when a soluble chromate or dichromate is added to a solution of silver nitrate previously acidified with nitric acid, or when silver chromate is treated with nitric acid, silver dichromate is formed; and that, on the other hand, it is silver chromate which is precipitated when silver nitrate in excess is added to a solution of a soluble dichromate, cold or hot, the reaction proceeding according to the equation

$$4\,AgNO_3 + K_2Cr_2O_7 + H_2O = 2\,Ag_2CrO_4 + 2\,KNO_3 + 2\,HNO_3.$$

The characteristics of both silver dichromate and silver chromate have recently been summarized and further studied by Margosches.† The solubility of silver dichromate in water and in ordinary solutions is such as to preclude the use of this substance as the final product of a quantitative process depending upon precipitation. The solubility of silver chromate in a moderately large volume of water is also considerable, and the solvent action of free acid, even acetic acid in quantity, is marked. Gooch and Weed ‡ have found, however, that the precipitation of silver chromate is practically complete in a solution only faintly acid with acetic acid and in presence of a large excess of silver nitrate. If such a precipitate is collected in the filtering crucible and washed with a dilute solution of silver nitrate until no other impurities remain, silver chromate does not dissolve, and the excess of silver nitrate may be removed by the cautious use of water without appreciable effect upon the precipitate.

* Ber. Dtsch. chem. Ges., xxxv, 2057.

† Zeit. anorg. Chem., xli, 68; l, 231.

‡ F. A. Gooch and L. H. Weed, Am. Jour. Sci., [4], xxvi, 85.

From the results of test experiments it is apparent that accurate determinations of chromium taken as the chromate or dichromate may be secured by precipitating silver chromate in presence of an excess of silver nitrate, making the solution ammoniacal and then faintly acid with acetic acid, transferring the precipitate, after standing half an hour, to the filtering crucible, washing with a dilute solution of silver nitrate, and, after other soluble impurities have been removed, finishing the washing with small amounts of water applied in successive portions.

The results are given in the table. In no case did the filtrate, with the washings, show by the lead acetate test the presence of a chromate.

Precipitation of Silver Chromate.

$K_2Cr_2O_7$ taken.	$AgNO_3$ used in precipitation.	Volume at precipitation	Ag_2CrO_4.		
			Found.	Theory.	Error.
grm.	grm	cm.³	grm.	grm.	grm.
0 0921	0 4248	100	0 2072	0 2076	−0.0004
0.0921	0 4248	100	0 2073	0.2076	−0 0003
0 0921	0 4248	100	0 2075	0 2076	−0 0001
0 0921	0 4248	100	0 2074	0 2076	−0 0002
0 0921	0 4248	100	0 2075	0 2076	−0 0001
0 0921	0 4248	100	0.2073	0 2076	−0 0003
0 0921	0 4248	100	0.2073	0 2076	−0.0003*
0 0921	0 4248	100	0 2075	0 2076	−0.0001*
0 0921	0 4248	100	0 2080	0 2076	+0.0004†
0.0921	0 4248	100	0.2070	0 2076	−0.0006†
0.5801	3	150	1 3087	1 3082	+0.0005
0.7352	3	200	1.6573	1 6574	−0.0001‡

* The precipitation was made in presence of 5 grm. of NH_4NO_3.
† The precipitation was made in presence of 5 grm. of $NaNO_3$.
‡ An excess of 1 cm.³ of 40 per cent acetic acid was added before filtering.

The Iodometric Determination of Chromic Acid.

Kessler * has shown that arsenious acid may be estimated by treating it, in the presence of hydrochloric acid, with an excess of a chromate solution of known strength, by which treatment the arsenious acid is oxidized and the chromic acid reduced, and determining the excess of chromic acid by adding an excess of a ferrous salt and titrating with chromic acid until a drop taken from the solution fails to give a blue color with a ferricyanide.

* Ann. Phys., xcv, 204.

The amount of the chromate originally used less the excess determined by the ferrous salt gives the amount of the chromate used for the oxidation, from which may be calculated the amount of arsenious acid originally present. Despite the use of a ferrous salt and the numerous steps involved in the manipulation, Kessler claims very satisfactory results for his method. Browning * gives results of experiments showing that Kessler's reaction may be used in the reverse process for the determination of chromic acid, the arsenious acid being used in excess, according to the reaction

$$4\,CrO_3 + 3\,As_2O_3 + (x)As_2O_3 = 2\,Cr_2O_3 + 3\,As_2O_5 + (x)As_2O_3.$$

According to the procedure indicated, an excess of $n/10$ arsenite is added to the cold solution of chromic acid acidulated with 10 cm.^{3} of dilute hydrochloric acid or sulphuric acid [1 : 3], the total volume being less than 100 cm.^{3} After a few minutes, when the solution has taken on the bluish green color characteristic of chromic salts, the solution is treated with acid potassium carbonate or acid sodium carbonate in excess (about 5 grm.). To the alkaline solution $n/10$ iodine is added in excess, the mixture is allowed to stand, with frequent shaking, for about half an hour, residual iodine is taken up with $n/10$ arsenite, and the excess of the last is titrated by $n/10$ iodine in presence of starch.

The long period of standing is made necessary by the tendency of precipitated chromic hydroxide to hold arsenious acid and thus delay the oxidation by iodine. The precipitation of the chromic hydroxide may be obviated by addition of Rochelle salt before the neutralization, but in this event the dark green color taken on by the solution makes the end-reaction of the starch iodide difficult to determine with great accuracy.

In the presence of a ferric salt the brown color of the precipitate also makes the determination of the end-reaction difficult unless the precipitate is allowed to settle after each addition of iodine. Edgar † has shown, however, that this difficulty may be obviated by adding sirupy phosphoric acid (3 cm.^{3} to 5 cm.^{3}) before the neutralization, so that the iron precipitate is white and the starch blue comes out against the pale green of the precipitate.

Results of test determinations are given in the table.

* Philip E. Browning, Am. Jour. Sci., [4], i, 35.

† See page 511.

Reduction by Standard Arsenite.

CrO_3 taken. grm.	CrO_3 found. grm.	Error. grm.	Remarks.
0 1001	0 1004	+0 0003	
0.1005	0 1004	−0 0001	The iodine acted 20 minutes.
0 1006	0 1007	+0 0001	The iodine acted 20 minutes.
0 1004	0 1011	+0 0007	The iodine acted 20 minutes.
0.1009	0.1009	0 0000	The iodine acted 2 hours.
0 1002	0 1003	+0 0001	The iodine acted 2 hours.
0 1011	0 1004	−0 0007	Rochelle salt used.
0.1007	0 1007	0 0000	Rochelle salt used.
0 0401	0 0395	−0 0006	
0 0402	0 0388	−0 0014	0.5 grm. ferric alum present.
0.1001	0.1018	+0 1017	
0 1009	0 1007	−0 0002	
0.1007	0 1011	+0 0004	1 grm. ferric alum present.
0.1005	0 1017	+0 0012	
0.1004	0 1010	+0 0006	
0.1000	0 1032	+0.0032	Rochelle salt used.
0.1005	0.1006	+0 0001	1 grm. ferric alum present.

The Iodometric Estimation of Chromic Acid and Vanadic Acid.

That vanadic acid and chromic acid may be accurately estimated in presence of one another by taking advantage of the differential reducing actions of hydrobromic and hydriodic acids has been shown by Edgar.*

In carrying out the operation, the alkali salts of the chromic and vanadic acid are put in the Voit flask of the distillation apparatus previously described,† one or two grams of potassium bromide are added, the flask is connected with the absorption apparatus containing a solution of potassium iodide made alkaline with sodium carbonate or sodium hydroxide, and the whole apparatus is filled with hydrogen gas. Fifteen to twenty cubic centimeters of concentrated hydrochloric acid are added through the separatory funnel and the solution is boiled for ten minutes, an interval of time found to be enough for the completion of the reduction. A slow current of hydrogen is maintained to avoid back suction of the liquid from the Drexel bottle. The apparatus is disconnected, the Voit flask placed in a beaker containing cold water, and the alkaline solution in the absorption apparatus cooled by running water. The contents of the

* Graham Edgar, Am. Jour. Sci., [4], xxvi, 333.

† See Fig. 3, page 4.

trap are washed into the Drexel bottle and the solution therein is made slightly acid with hydrochloric acid. The liberated iodine is titrated with approximately $n/10$ sodium thiosulphate and the color is brought back by a drop or two of $n/10$ iodine solution, after the addition of starch.

Alkaline potassium iodide is again placed in the absorption apparatus and the latter connected with the Voit flask. The current of hydrogen is turned on and, after the air has been expelled, the apparatus is disconnected momentarily, one or two grams of potassium iodide are added to the solution in the Voit flask, and connections made again. Through the separatory funnel 10 cm.^{3} to 15 cm.^{3} of concentrated hydrochloric acid and 3 cm.^{3} of sirupy phosphoric acid are added and the solution in the reduction flask is boiled to a volume of 10 cm.^{3} to 12 cm.^{3}. The absorption apparatus is removed and cooled, hydrochloric acid is added and the liberated iodine titrated with approximately $n/10$ sodium thiosulphate.

Double Treatment with Hydrobromic Acid and with Hydriodic Acid.

V_2O_5 taken as $NaVO_3$. grm.	CrO_3 taken as $K_2Cr_2O_7$. grm.	I Titration. $Na_2S_2O_3$ $n/10 \times 1.031$. cm.^{3}	II. Titration. $Na_2S_2O_3$ $n/10 \times 1.031$. cm.^{3}	Error on V_2O_5. grm.	Error on CrO_3. grm.
0.1523	......	16.20	16.22	(I) 0.0000 (II) +0.0002	.....
0.1523	......	16.19	16 20	(I) −0.0001 (II) 0.0000	. .
0.2031	......	21.59	21.59	(I) −0.0001 (II) −0.0002	. . .
0.1523	0.0685	36.08	16.22	+0 0002	−0 0001
0.1523	0 0685	36 10	16 20	0.0000	0 0000
0 1523	0 0085	36.12	16.17	−0 0003	+0 0002
0 1523	0 0685	36.07	16 20	0.0000	−0.0001
0 1523	0.1370	56 00	16.17	−0.0003	+0 0001
0 1523	0.1370	56 02	16.22	+0 0002	0.0000
0.1523	0.1370	56.03	16.19	−0.0001	+0 0001
		$n/10 \times 0.992$,	$n/10 \times 0.992$		
0.2031	0 1370	63.82	22.46	0 0000	−0 0001
0.2031	0.1370	63.80	22.48	+0.0002	−0.0003
0.1016	0.0685	32 00	11.25	+0.0002	+0 0001
0.1016	0.0685	31.92	11.24	+0 0001	−0.0001
0.1016	0.0685	31.90	11.25	+0.0002	−0 0002
0.0508	0.0343	15.95	5.63	+0.0002	−0.0001
0 0508	0 0343	15.95	5.62	+0.0001	−0.0001

The iodine determined in the first titration corresponds to a reduction of the chromic and vanadic acids according to the equation

$$V_2O_5 + 2\ CrO_3 + 8\ HBr = V_2O_4 + Cr_2O_3 + 4\ Br_2 + 4\ H_2O,$$

while in the second case the iodine corresponds to a reduction of the vanadium tetroxide to trioxide as indicated in the equation

$$V_2O_4 + 2\ HI = V_2O_3 + I_2 + H_2O.$$

The second titration, therefore, determines the vanadic acid present, and the difference between the first and second furnishes the necessary data for the calculation of the chromium.

Results obtained in the application of this method to the determination of the acidic oxides contained in sodium vanadate and potassium dichromate are given in the table.

The Estimation of Chromic Acid and Vanadic Acid by Reductions and Oxidations.

For the estimation of both vanadic acid and chromic acid, when present together, the following method has been worked out by Palmer:*

Into a measured portion of the solution, made acid with hydrochloric acid, sulphur dioxide is passed until the reduction of the vanadium and the chromium is complete; the solution is then boiled in a current of carbon dioxide until the last traces of sulphur dioxide are expelled. To the cooled solution a sufficient excess of potassium ferricyanide and potassium hydroxide are added in solution. By this process the chromium is oxidized from the condition of Cr_2O_3 to the condition of CrO_3 and the vanadium from the condition of V_2O_4 to the condition of V_2O_5. After allowing the solution to stand a few minutes, a solution of barium hydroxide is added to complete precipitation. The combined precipitates of barium chromate and barium vanadate are filtered off on asbestos and thoroughly washed; the filtrate is made acid with dilute hydrochloric acid and titrated with a known amount of permanganate in excess, and the excess titrated with $n/20$ potassium ferrocyanide to permanent green coloration in presence of a trace of ferric salt.

* Howard E. Palmer, Am. Jour. Sci., [4], xxx, 141.

In another portion of the solution the vanadium is determined as follows: The solution, about 100 cm.^{3} in volume, is made acid with from 10 cm.^{3} to 15 cm.^{3} of glacial acetic acid, and hydrogen peroxide is added. The solution is then heated to boiling and boiled for a few minutes; by this process the perchromic acid and the pervanadic acid, which were first formed in the cold, are decomposed, the chromium being reduced to the condition of Cr_2O_3, while the vanadium appears in the condition of V_2O_5. The solution is diluted somewhat, and a solution of lead acetate is added to complete precipitation of the lead vanadate; the chromium, being in the condition of Cr_2O_3, is not precipitated. The solution is stirred vigorously and heated to boiling to coagulate the precipitate. The precipitate is filtered off on asbestos, washed thoroughly and dissolved in potassium hydroxide, and the solution in potassium hydroxide is made strongly acid with sulphuric acid, whereby the lead is precipitated as the sulphate, while the vanadic acid remains in solution. A current of sulphur dioxide is passed through the solution until the blue color indicates complete reduction of the vanadium to the tetroxide condition; and the sulphur dioxide is expelled by boiling in a current of carbon dioxide. The warm solution is then titrated with permanganate to the appearance of the first permanent pink color, easily recognized in the presence of the white precipitate of lead sulphate.

Determination of Chromic Acid and Vanadic Acid.

V_2O_5 taken. grm.	CrO_3 taken. grm.	V_2O_5 found. grm.	Error. grm.	CrO_3 found. grm.	Error. grm.
0.1139	0.1010	0.1134	−0 0005	0.1010	0.0000
0.1139	0.1010	0.1139	0.0000	0 1017	+0.0007
0.1139	0.1010	0.1134	−0 0000	0 1019	+0.0009
0.1139	0.1010	0.1142	+0.0003	0.1019	+0 0009
0.1139	0 1010	0.1131	−0.0008	0 1015	+0 0005
0.1139	0.1010	0.1134	−0.0005	0.1016	+0 0006
0.1139	0.0505	0.1139	0.0000	0 0507	+0 0002
0.1139	0.0505	0.1134	−0.0005	0.0507	+0 0002
0.0569	0.0505	0.0565	−0 0004	0 0505	0 0000
0.0569	0.0505	0.0563	−0.0006	0.0508	+0.0003

This titration gives a measure of the amount of vanadium present; and by subtracting the number of cubic centimeters of permanganate used in this titration from the number of cubic centimeters used in the preceding titration, the number of cubic centimeters corresponding to the oxidation of the chromium from Cr_2O_3 to CrO_3 is obtained. Experimental results are given in the table.

The Volumetric Estimation of Chromium in the Chromic Condition.

As Bollenbach and Luchmann have shown,* chromium may be quantitatively oxidized from the condition of Cr_2O_3 to the condition of CrO_3 by potassium ferricyanide in alkaline solution, and a measure of the oxidation obtained by titrating with permanganate the ferrocyanide formed.

According to this method an excess of at least four to six times the theoretical amount of potassium ferricyanide and 40 $cm.^3$ to 50 $cm.^3$ of a 2-normal solution of sodium hydroxide are added to the solution containing the chromium to insure complete oxidation. The oxidized chromium is removed by precipitation as barium chromate by means of barium hydroxide in solution, and filtration. The filtrate is acidified with hydrochloric acid and titrated with permanganate according to Bollenbach's modification of De Haen's method,† designed to overcome the difficulty involved in obtaining an exact end point in the titration of large amounts of ferrocyanide, owing, as was first pointed out by Grützner, ‡ to the formation of a precipitate of $K_2MnFeC_6N_6$ during the titration. This modification consists in adding an excess of permanganate to the solution and, after the precipitate has cleared up, titrating back the excess of permanganate with $n/20$ potassium ferrocyanide in the presence of a trace of a ferric salt, the formation of a permanent green coloration due to the ferric ferrocyanide indicating the end point.

According to the experience of Palmer § the proportions of ferricyanide and alkali hydroxide prescribed above are insufficient to bring about complete oxidation of the chromium in chromic condition to that of chromic acid, according to the equation

$$Cr_2O_3 + 6\,K_3FeC_6N_6 + 6\,KOH = 2\,CrO_3 + 6\,K_4FeC_6N_6 + 3\,H_2O.$$

* Zeit. anorg. Chem., lx, 446.

† Zeit. anal. Chem., xlvii, 687.

‡ Chem. Centralblatt, 1902, i, 500.

§ Howard E. Palmer, Am. Jour. Sci., [4], xxx, 141.

But by using about fifteen times the theoretical proportion of potassium ferricyanide and a rather strong solution of potassium hydroxide in a total volume of solution amounting to 100 $cm.^3$ or 125 $cm.^3$ results are obtained in accord with the theory.

Palmer tested the process upon a chromic salt made by treating portions of a standardized solution of potassium chromate, made slightly acid with hydrochloric acid, with a current of sulphur dioxide until the clear green color indicated complete reduction of the chromic acid to the condition of chromic oxide, Cr_2O_3. The sulphur dioxide was then expelled by boiling the solution in a current of carbon dioxide. To determine the chromium in such a solution the following procedure proved effective.

To the cold solution of the chromic salt are added 15 to 20 times the theoretical amount of potassium ferricyanide and potassium hydroxide in rather strong solution, with care to make the final volume 100 $cm.^3$ to 125 $cm.^3$. Barium hydroxide is added to precipitate the chromate formed and the insoluble barium chromate is removed by filtration. The solution is acidified with hydrochloric acid and treated with measured permanganate in excess, and the excess of permanganate is determined by titration to permanent green with potassium ferrocyanide in presence of a ferric salt, with due correction for the amount of permanganate taken up by the same amount of the ferricyanide alone.

Experimental results are given in the table.

Estimation of Chromium in the Chromic Condition.

CrO_3 taken. grm.	$K_3FeC_6N_6$ used. grm.	KOH used. grm.	CrO_3 found. grm.	Error. grm.
0.1010	6	12	0.0979	−0.0031
0.1010	8	16	0.0981	−0.0029
0.1010	8	16	0.0989	−0.0021
0.1010	8	12	0.0997	−0.0013

MOLYBDENUM.

The Gravimetric Estimation of Molybdic Acid by Liberation of Iodine and Absorption of that Element by Silver.

When a soluble molybdate is added to an excess of potassium iodide made acid with hydrochloric acid and shaken with electrolytically prepared silver under an atmosphere of hydrogen,* iodine equivalent to the molybdic acid is set free and then is

* See page 27.

absorbed by the silver. From the increase in weight of the silver the amount of molybdenum trioxide was calculated on the assumption that one molecule of molybdenum trioxide liberates one atom of iodine according to the following equation:

$$2\,MoO_3 + 4\,KI + 4\,HCl = 2\,MoO_2I + 4\,KCl + 2\,H_2O + I_2.$$

Results obtained by Perkins* in the treatment of ammonium molybdate, the composition of which had been determined by fusion with sodium tungstate containing a slight excess of tungstic acid, are given in the table.

Absorption of Iodine by Silver.

Ag taken grm.	MoO_3 taken. grm.	I_2 found. grm.	Calculated MoO_3. grm.	Error grm.
2 0002	0 2127	0 1869	0.2120	−0 0007
2 0006	0 2127	0 1874	0.2126	−0.0001
2.0012	0 2127	0 1870	0 2121	−0 0006
2 0048	0 2127	0 1876	0 2128	+0 0001
2.0000	0 2540	0 2242	0.2543	+0.0003
2 0004	0 2909	0.2571	0.2916	+0 0007

The Iodometric Estimation of Molybdic Acid.

The Digestion Method. Mauro and Danesi † have made use of the reaction which takes place between hydrochloric acid, potassium iodide and a soluble molybdate to determine the amount of molybdic acid from the amount of iodine set free in accordance with the reaction

$$2\,MoO_3 + 4\,HI = 2\,MoO_2I + I_2 + 2\,H_2O.$$

The best results are obtained by acting upon a soluble molybdate containing from 0.1 to 0.5 grm. of molybdic acid, with 1.5 grm. of potassium iodide in 1.5 cm^3. of water and 2.5 cm^3. of strong hydrochloric acid in an atmosphere of carbon dioxide, the whole being heated an hour and a half in a sealed tube. The authors point out that with prolonged heating the action proceeds a little further, and in the cold, under conditions otherwise similar, not quite so far as the theory of the equation would indicate.

* Claude C. Perkins, Am. Jour. Sci., [4], xxix, 540.
† Zeit. anal., Chem., xx, 567.

In studying this method Gooch and Fairbanks * have obtained variable results fairly in accord with the theory of the reduction when the digestion is made in sealed tubes at extremely small volume, and with small amounts of the molybdate, under the exact conditions indicated, but widely deficient for larger amounts, for larger volumes, and for digestions in the cold. This behavior indicates, of course, a tendency on the part of the iodine to reverse the action, and the obvious remedy for the reversal should be found in the removal of the iodine as it is set free. Friedheim and Euler † accomplish this by a process of distillation in the Bunsen apparatus, collecting and determining the iodine in the distillate.

Distillation Process. The experience of Gooch and Fairbanks ‡ confirms in general the utility of the distillation process provided that conditions are exactly defined. It is not sufficient to say that the boiling should be stopped when a clear green color appears and when the steam is no longer colored by iodine; for the green color comes very gradually, and iodine remains in the residue after the green color has developed distinctly. It is safer and more convenient to start the distillation with a definite volume of liquid and boil until the volume is reduced to a definite point. If the initial volume is made about 40 cm^3., no iodine remains in the flask after the liquid has been boiled down to 25 cm^3., and at that degree of concentration the molybdic acid shows the theoretical reduction; but if the concentration is pushed beyond this point, a tendency to further reduction of the molybdic acid becomes evident.

It is necessary to carry on the distillation in an atmosphere of carbon dioxide, inasmuch as the hydriodic acid freed by the action of hydrochloric acid upon the potassium iodide is decomposed by distillation in contact with air, with liberation of iodine. As even a trace of oxygen will immediately set free iodine from boiling hydriodic acid, the carbon dioxide is best evolved from boiled marble by the action of boiled acid to which a little cuprous chloride has been added, and finally passed through solutions of iodine and potassium iodide to free it from any reducing substance.

* F. A. Gooch and Charlotte Fairbanks, Am. Jour. Sci., [4], ii, 156.

† Ber. Dtsch. Chem. Ges., xxviii, 2066.

‡ Loc. cit.

A convenient apparatus, constructed with sealed and ground joints exclusively, is shown in the accompanying figure. The distillation takes place in the first flask, and the iodine collects in the second flask and trap, which hold a solution of potassium iodide kept cool by immersion of the flask in cold water.

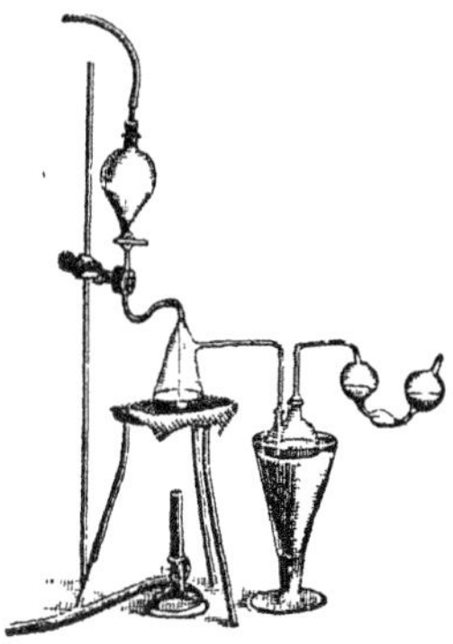

Fig. 27.

In carrying out a determination with this apparatus, the purified carbon dioxide is first passed for some minutes and the stopcock in the funnel is then closed. The molybdate dissolved in 10 cm.³ of boiled water is put in the funnel and almost all of it is allowed to run into the first flask. It is necessary that a few drops be left in the funnel so that the liquid to follow may not carry down bubbles of air. Potassium iodide, never exceeding the theoretical requirement by more than 0.5 grm., is introduced similarly in 10 cm.³ of boiled water, followed by 20 cm.³ of concentrated hydrochloric acid (sp. gr. 1.20). Before the acid is allowed to run in entirely, the funnel is again filled with carbon dioxide and finally left connected with the generator so that carbon dioxide may be passed into the apparatus at pleasure. The liquid in the first flask is boiled until the volume of liquid has decreased to 25 cm.³, indicated by a mark. The iodine collected in the second flask

Distillation Process.

MoO_3 taken as ammonium molybdate. grm.	KI. grm.	MoO_3 found. grm.	Error. grm.
0.2585	0 5	0 2580	−0 0005
0 2995	0.5	0 2991	−0 0004
0.2524	0 5	0 2513	−0 0011
0.2446	0 5	0 2457	+0 0011
0.2903	0.5	0.2914	+0 0011
0 2798	0.5	0.2808	+0 0010
0.2656	0.5	0.2663	+0.0007
MoO_3 dissolved in NaOH. grm.			
0.2273	0.5	0 2281	+0.0008
0.2052	0.5	0 2062	+0 0010
0.3474	0 5	0.3467	−0.0007

and trap is titrated with standardized thiosulphate. Results of test experiments made according to this procedure are given in the table.

Further study of this process has been made by Gooch and Norton,* with a view to testing further the reliability of the process and to determining the progress of the reduction of molybdic acid as the concentration proceeds.

The distillation apparatus employed in this work was constructed with sealed or ground joints of glass wherever contact with iodine was a possibility. It was made by sealing together a separating funnel A, a 100 cm.^{3} Voit flask B, a Drexel wash bottle C, and a bulbed trap *g*, as shown in Figure 3.† Upon the side of the distillation flask B was pasted a graduated scale, by means of which the volume of the liquid within the flask might be known at any time. Carbon dioxide, generated in a Kipp apparatus by the action of dilute hydrochloric acid (carrying in solution cuprous chloride to take up free oxygen) upon marble previously boiled in water, was passed through the apparatus before and during the operation, so that it was possible to interrupt the process of boiling at any point of concentration, to remove the receiver by easy manipulation, to replace the receiver, and to continue the distillation without danger of admitting air to the distillation flask.

The mean error of the indications found for different periods of distillation are given in the accompanying statement.

Concentration. cm.^{3}	Error. grm.
40 to 32	−0.0045
40 to 25	+0.0008
40 to 10	+0.0014

The best results, showing a mean error of +0.0008 grm. between extremes of −0.0001 grm. and +0.0020 grm., were obtained by this process when the distillation was continued until the original volume of 40 cm.^{3} had been diminished to 25 cm.^{3}. Concentration beyond the limit of 25 cm.^{3} plainly develops a

* F. A. Gooch and John T. Norton, Jr., Am. Jour. Sci., [4], vi, 168.

† See page 4.

tendency toward over-reduction, especially when the amount of potassium iodide is increased beyond about 0.5 grm. in excess of that theoretically required.

It is shown, further, that the precaution of conducting the operation in an atmosphere of carbon dioxide does not eliminate all chance of error of this sort unless the liquid of the mixture — the hydrochloric acid as well as the water — is free from air. Thus, 40 cm.[3] of unboiled acid, sp. gr. 1.12, introduced enough air into the apparatus to cause an error of 0.0013 grm. reckoned in terms of molybdenum trioxide, while the iodine set free by the action of the residual acid of this experiment upon another gram of potassium iodide introduced without admission of air corresponded to only 0.0002 grm. in terms of molybdenum trioxide. The use of acid of sp. gr. 1.1, freshly boiled in the air, obviously reduces the error due to the unboiled acid, but even in this case the effect of dissolved oxygen is not wholly obviated.

While it is possible to determine molybdic acid by estimating the iodine set free upon boiling a solution of that acid in hydrochloric acid to which potassium iodide has been added, the operation is beset with difficulties. If due attention be paid to the proportion of acid, the excess of potassium iodide over the amount theoretically necessary to produce the ideal reduction, and the final degree of concentration of the liquid, the change of the molybdic acid to the condition of oxidation of the pentoxide, Mo_2O_5, may, as has been shown, be closely realized. Even when the conditions essential to the ideal reduction obtain, however, the possibility of the interaction of atmospheric oxygen with the hydriodic acid produced from the potassium iodide and hydrochloric acid must be guarded against by conducting the operation in an atmosphere of carbon dioxide. Furthermore, the hydrochloric acid employed must be freed, so far as may be, from dissolved oxygen by previous boiling. The operation is capable of yielding accurate indications (unless by a chance combination of opposite errors) only when the precautions mentioned are scrupulously observed.

On the other hand, the essential principles of the reaction admit of very simple application when the residue of the distillation, instead of the iodine distilled, is made the object of the analytical determination.

Reoxidation of the Residue by Iodine.

According to the method developed by Gooch and Fairbanks * the mixture of molybdic acid, potassium iodide, and hydrochloric acid is boiled between defined limits of volume in an Erlenmeyer flask simply trapped to prevent mechanical loss during the boiling. The residue after the removal of all free iodine is treated with tartaric acid to prevent subsequent precipitation, neutralized with an alkali bicarbonate, and reoxidized by a measured excess of standard iodine, this excess of iodine being finally determined by titration with standard arsenite. The difference between the iodine value of the arsenite used and the amount of standard iodine employed measures the molybdic acid. The action of atmospheric or dissolved oxygen obviously plays no important part in the operation, provided all uncombined iodine, however produced, is finally boiled out. At the outset iodine is liberated by the air present as well as by the molybdic acid, but as the boiling continues the hydriodic acid diminishes in strength, the iodine is driven from the flask kept filled with steam, and if the conditions of the operation have been properly adjusted the molybdic acid undergoes the ideal reduction. When this point is reached dilution with cold water obviates danger of immediate oxidation and gives opportunity for continuing the treatment referred to above.

According to this method, the soluble molybdate in amount not exceeding the equivalent of 0.5 grm. of MoO_3, and at least 20 cm.^{3} of hydrochloric acid (sp. gr. 1.20) with from 0.2 grm. to 0.6 grm. of potassium iodide, according to the amount of molybdate used, are put into a 150 cm.^{3} flask which is then trapped loosely by a short bulbed tube hung in the neck, as shown in Fig. 6.† The solution is boiled until the original volume of 40 cm.^{3} to 60 cm.^{3} has been reduced to exactly 25 cm.^{3}, as determined by a mark upon the flask. The residue is diluted immediately to a volume of 125 cm.^{3}, cooled, and transferred to a reaction bottle of the form shown in Fig. 7,‡ the Will and Varrentrapp absorption tube being charged with a solution of potassium iodide to catch any traces of iodine thrown off mechanically during the process. Through the stoppered funnel is added in

* F. A. Gooch and Charlotte Fairbanks, Am. Jour. Sci., [4], ii, 160.

† See page 6.

‡ See page 6.

solution 1 grm. of tartaric acid to prevent precipitation during the subsequent neutralization. The free acid is neutralized by introduction of acid alkali carbonate (or of alkali hydroxide to partial neutralization, followed by the acid carbonate) and a measured amount of $n/10$ iodine in excess run in. The iodine color begins to fade perceptibly within fifteen minutes, but for complete oxidation the mixture, protected from direct sunlight, should be set aside for an hour and a half or two hours. The iodine remaining is then titrated by standard arsenite.

At the end of this operation it is wise to acidulate the solution with dilute hydrochloric acid and determine by titration with sodium thiosulphate any slight amount of iodine which may have taken the form of iodate in the long digestion.

Results of experimental tests of this procedure are given in the table.

Reoxidation of Residue by Iodine.

MoO_3 taken as ammonium molybdate grm.	KI. grm.	MoO_3 found. grm.	Error. grm.
Neutralized by $NaHCO_3$.			
0 1517	0.5	0.1517	0.0000
0.2530	0.5	0.2537	+0 0007
0 1636	0.5	0.1637	+0 0001
0 1702	0 5	0.1702	0.0000
0.1520	0 5	0.1518	−0 0002
0 1642	0.5	0.1652	+0 0010
0 4560	0 75	0.4560	0 0000
0.1690	0 5	0.1683	−0.0007
Partially neutralized by NaOH: fully neutralized by $NaHCO_3$.			
0 0507	0.5	0.0519	+0 0012
0 1663	0.5	0.1666	+0 0003
0 0101	0.5	0.0095	−0 0006
0 1639	0.5	0.1632	−0 0007
0 1636	0.5	0.1625	−0 0011
0 0507	0.5	0.0510	+0 0003
0.1685	0 5	0.1683	−0 0002
0 1514	0.5	0.1512	−0 0002
0.1649	0 5	0.1646	−0.0003

Reoxidation of the Residue by Permanganate.

The method just described, according to which a soluble molybdate is boiled with hydrochloric acid and a small excess of potassium iodide to a definite concentration, the residue neutralized with an acid alkali carbonate, and the reduced molybdic salt reoxidized by standard iodine, affords an accurate determination of the molybdate but is somewhat tedious on account of the delay necessary before the final titration by standard arsenite may be effected.

Gooch and Pulman * have succeeded in obviating the difficulty of long delay by substituting potassium permanganate for iodine in the process of reoxidation.

When an excess of potassium permanganate is added to a solution containing hydrochloric and hydriodic acids with molybdenum in a condition of oxidation corresponding to the pentoxide, several different effects of oxidation may be expected. There is the immediate liberation of iodine from the hydriodic acid, the production of iodic acid, some slight tendency to liberate chlorine from the hydrochloric acid, the formation of representatives of the higher oxides of manganese, and, lastly, the reproduction of the molybdic acid, which is the object of the operation. All other effects than the last must be prevented or recorded in order that the estimation of the molybdic acid may be accomplished. Of the secondary actions, the liberation of chlorine may be prevented by adding a manganous salt, according to the well-known proposals of Kessler † and Zimmerman; ‡ the iodine set free may be converted to hydriodic acid again by the introduction into the acid solution of a sufficient excess of a standard arsenite solution; the iodic acid may be reduced in the acid solution in the same manner, as was found by experiment, since the value obtained for a solution of iodic acid by adding to it in presence of sulphuric acid a measured amount of standard arsenite, making alkaline, and titrating back with iodine, proved to be the same as that found by acting directly upon the iodic acid with an excess of potassium iodide in presence of dilute sulphuric acid and determining by standard arsenite the iodine liberated. The higher oxides of manganese are likewise reduced in the acid solution by the arsenious acid of the standard arsenite.

Whatever excess of arsenious acid is left over after the reactions described may, obviously, be determined by neutralizing the solution with an alkali bicarbonate and titrating with iodine.

* F. A. Gooch and O. S. Pulman, Jr., Am. Jour. Sci., [4], xii, 449.

† Ann. Phys., cxviii. 48; cxix, 225–226.

‡ Ann. Chem., ccxiii, 302.

Every difficulty introduced by the secondary oxidations may be overcome by adopting the following procedure: the addition of manganous sulphate to the reduced, diluted and cooled solution; the introduction of measured standard potassium permanganate until its characteristic color is evident; the treatment of the solution with a measured amount of standard arsenite in known excess, to destroy the excess of permanganate; the introduction of tartaric acid to prevent subsequent precipitation; neutralization by acid alkali carbonate; and titration of the excess of arsenite by iodine.

The operation requires the use of a standard solution of arsenite, easily made with accuracy, a solution of iodine in potassium iodide standardized directly against the arsenite in the usual manner, and a solution of potassium permanganate the value of which in terms of the arsenite is found by bleaching a measured portion with an excess of arsenite and titrating back with iodine the arsenite remaining. The value (in terms of molybdic acid) of the permanganate used, diminished by that of the arsenite and increased by that of the iodine, gives the amount of molybdic acid present.

According to this procedure the soluble molybdate in amount not exceeding the equivalent of 0.5 grm. of MoO_3, at least 20 $cm.^3$ of hydrochloric acid (sp. gr. 1.20), with 0.2 grm. to 0.6 grm. of potassium iodide, according to the amount of molybdate used, are put into a 150 $cm.^3$ flask trapped by a short bulbed tube hung loosely in the neck, as shown in Fig. 6 on p. 6. The solution, having a volume of 40 $cm.^3$ to 60 $cm.^3$ originally, is boiled to a volume of 25 $cm.^3$, indicated by a mark upon the flask. The residue is diluted immediately to a volume of 125 $cm.^3$, cooled, and transferred to a bottle fitted with a stoppered funnel and trap, as shown in Fig. 6, p. 6. Through the stoppered funnel are added 0.5 grm. of manganese sulphate and a measured amount of $n/10$ potassium permanganate to the characteristic coloration. A measured amount of standard arsenite is then added, enough to correspond approximately to the permanganate used, experience having shown that this amount of arsenite is sufficient to reduce with readiness in the acid solution the iodic acid, the permanganate, the higher oxides of manganese, and nearly all the free iodine, the remainder of the last being taken up immediately after the subsequent neutralization. Next, a solution of about 3 grm. of tartaric acid is run in, and the free acid of the solution is neutralized by acid potassium carbonate. Finally, the liquid adhering to stopper and tubes is washed off into the bottle, the contents of the trap are added, and the residual arsenite is titrated by standard iodine, using the starch indicator.

When many determinations are to be made the process yields results with rapidity. Many operations may be started successively in the Erlenmeyer flasks, and one neutralization bottle may serve for the treatment of all the reduced residues as they come along.

The results of experiments are given in the following table.

Reoxidation of Residue by Permanganate.

Weight of MoO_3 taken as ammonium molybdate.	Weight of KI used.	Weight of MoO_3 found.	Error.
grm.	grm.	grm.	grm.
0 0423	0.2	0.0430	+0 0007
0.0429	0 2	0.0435	+0.0006
0.0420	0.2	0.0427	+0.0007
0.0827	0.3	0.0829	+0 0002
0.0837	0 35	0.0838	+0 0001
0.0826	0 5	0.0832	+0 0006
0.2465	0 6	0.2460	−0 0005
0.2481	0.6	0.2469	−0.0012
0.2470	0.6	0.2465	−0.0005

The Estimation of Molybdic Acid reduced in the Jones Reductor.

Opinions have differed in regard to the degree of reduction obtained when molybdic trioxide, in sulphuric acid solution, is passed through the column of amalgamated zinc as applied in the Jones reductor. Jones,* who first determined molybdenum by this method, considered that the reduction goes to the condition represented by the formula $Mo_{12}O_{19}$, the same degree of reduction that Wernke † obtained with zinc and sulphuric acid in a closed flask. Doolittle ‡ and Eavenson found that, by varying the strength of the acid and the speed at which the molybdenum was passed through, different degrees of reduction might be obtained, but none were lower than that represented by the formula $Mo_{12}O_{19}$. Blair § and Whitfield were unable to

* Am. Inst. Min. Eng., xviii, 705.

† Zeit. anal. Chem., xiv., 1.

‡ Jour. Am. Chem. Soc., xvi, 234.

§ Ibid., xvii, 747.

press the reduction below the condition represented by the symbol $Mo_{24}O_{37}$. Miller* and Frank in repeating the experiments of Blair and Whitfield obtained in general the same results, though by taking extraordinary precautions they were able to get a reduction to a little below the midway point between the conditions represented by the symbols $Mo_{24}O_{37}$ and Mo_2O_3. W. A. Noyes and Frohman,† by taking pains to replace the air in the reductor flask by carbon dioxide, were able to obtain a reduction to the form of Mo_2O_3.

Noting the ease with which the reduced molybdenum solution is oxidized by the air, Randall ‡ has investigated the possibility of charging the reductor flask with an oxidizer unaffected by air, to anticipate the oxidizing effects of the air as the reduced molybdenum compound comes through the reductor, and to register the oxidation. In preliminary experiments an excess of a standard solution of potassium permanganate was used in the receiver. But it was found, in blank tests, that somewhat more than the theoretical amount of permanganate was used up, due either to impurities in the zinc, to small particles of zinc which had worked through the asbestos at the bottom of the reductor, or possibly to the hydrogen formed in the reductor. That the last mentioned possibility is a sufficient cause of the effects obtained was shown by passing hydrogen, formed by the action of hydrochloric acid on zinc in a Kipp generator and washed with water and caustic potash, through 23 cm.^{3} of standard permanganate diluted with 300 cm.^{3} of hot dilute sulphuric acid (2.5 per cent), adding 20 cm.^{3} of a solution of standard ferrous sulphate and titrating back with permanganate. The results obtained by such action of hydrogen during fifteen minutes are shown in the accompanying statement:

Effect of Hydrogen.

$FeSO_4$. cm.^{3}	$KMnO_4$ required. cm.^{3}	$KMnO_4$ theory. cm.^{3}	Reduced by hydrogen. cm.^{3}
20	24.6	23.1	1.5
20	24 7	23.1	1.6
20	23.6	23.1	0.5

* Jour. Am. Chem. Soc., xxv, 919.
† Ibid., xvi, 553.
‡ D. L. Randall, Am. Jour. Sci., [4], xxiv, 313.

The use of ferric alum * in the receiving flask, with phosphoric acid to decolorize it,† Randall found to be unobjectionable and effective. According to the procedure laid down by Randall, the receiving flask‡ is charged with 20 cm.³ to 30 cm.³ of a solution prepared by dissolving 100 grm. of ferric alum in a liter of water; phosphoric acid (4 cm.³ of the sirupy acid) is added to the solution in the receiver; and through the 36 cm. column of amalgamated zinc in the reductor are passed in succession 100 cm.³ of hot dilute sulphuric acid (2.5 per cent), the molybdic acid in the form of ammonium molybdate dissolved in 10 cm.³ of water and acidified with 100 cm.³ of the hot dilute acid (2.5 per cent), 200 cm.³ of the hot dilute acid, and finally 100 cm.³ of hot water. The molybdenum salt is green as it passes through the lower part of the reductor, but on coming in contact with the ferric salt it is changed to a bright red. The solution is titrated while still hot with approximately tenth normal permanganate.

The results in the following table are calculated on the assumption that the molybdic acid is reduced to the form of Mo_2O_3 and the close agreement with theory indicates that the reduction does not stop at an intermediate point.

Collection of Mo_2O_3 *in Ferric Sulphate: Titration of Ferrous Salt by Permanganate.*

Ammonium molybdate taken.	Iron solution.	H_3PO_4.	$KMnO_4$ used.	MoO_3.		Error.
				Found.	Theory.	
grm.	cm.³	cm.³	cm.³	grm.	grm.	grm.
0 2000	20	4	30.38	0.1628	0 1631	−0 0003
0 2000	20	4	30.45	0.1632	0.1631	+0.0001
0 2000	20	4	30.33	0.1626	0.1631	−0.0005
0 3000	30	4	45.65	0 2447	0.2447	0.0000
0.3000	30	4	45.80	0.2455	0.2447	+0 0008
0.3000	30	4	45.73	0.2451	0 2447	+0 0004
0 3000	30	4	45.76	0 2453	0 2447	+0 0006
0 3000	30	4	45.68	0.2448	0.2447	+0 0001
0.3000	30	4	45.64	0.2446	0 2447	−0.0001
0 3000	30	4	45.71	0.2449	0.2447	+0.0002

* Professor Henry Fay kindly gives the information that this mode of treating reduced molybdic oxide was first worked out many years ago by Dr. C. B. Dudley, though never published.

† C. Reinhardt, Chem. Ztg., 13, 33.

‡ See page 347.

The Determination of Molybdic Acid and Vanadic Acid by Reductions and Oxidations.

Molybdic acid and vanadic acid are reduced in a perfectly definite manner by a column of amalgamated zinc, and each by itself, or both together, may be estimated by titration with potassium permanganate if the receiver be charged with a solution of ferric alum to anticipate possible oxidizing action of the air.* On the other hand, of these two acids only vanadic acid is easily reduced in solution by sulphur dioxide. These are characteristics which suggested to Edgar † the search for conditions under which molybdic acid would escape all action by sulphur dioxide, and the development of a method for the determination of the two acids.

Action of Sulphur Dioxide on Molybdic Acid.

Volume of solution. cm.³	MoO_3. grm.	H_2SO_4 (sp. gr. 1.84). cm.³	Time of treatment with SO_2. min	$KMnO_4$ $n/10 \times 1.004$. cm³	Color of solution
25.0	0.200	Faintly acid.	10	0 15	Light blue.
35 0	0.200	0 5	10	0 05	Faint blue.
50 0	0.200	1.0	10	0 0	Colorless.
75.0	0.200	2.0	10	0 0	Colorless.
50.0	0.200	2 0	30	0 0	Colorless.
25.0	0.200	5 0	10	0 0	Colorless.
25.0	0.400	5.0	10	0 0	Colorless.
50.0	0.400	5 0	10	0 0	Colorless.
50.0	0.400	10.0	10	0 0	Colorless.
50.0	0.400	15 0	10	0.0	Colorless.

To determine the conditions under which molybdic acid is unaffected by sulphur dioxide, experiments were made in which solutions of molybdic acid of varying concentrations, acidified with varying amounts of sulphuric acid, were heated to boiling and treated with a current of sulphur dioxide for varying lengths of time. The excess of sulphur dioxide was then removed by boiling the solution, a current of carbon dioxide being meanwhile passed into it, and the degree of reduction was determined by titration with nearly $n/10$ potassium permanganate. The solution of molybdic acid was standardized by the method of Randall ‡ and also by evaporating a portion to dryness and

* See Randall, page 426; Gooch and Edgar, page 349.

† Graham Edgar, Am. Jour. Sci., [4], xxv, 332.

‡ See page 426.

igniting at low red heat. The results of the experiments are given in the table.

The results show that if the concentration be not greater than 0.2 grm. of MoO_3 in 50 cm.³ of solution, and the acidity not less than 1 cm.³ of sulphuric acid (sp. gr. 1.84) in the same volume, the molybdic acid is not reduced, and that if the acidity be increased to 5 cm.³ of sulphuric acid, reduction does not occur at even a concentration of 0.4 grm. in 25 cm.³.

If a solution containing vanadic acid and molybdic acid be treated with sulphur dioxide under suitable conditions, only the vanadic acid will be reduced

$$V_2O_5 + SO_2 = V_2O_4 + SO_3,$$

and the degree of reduction, determined by titration with permanganate, will measure the vanadic acid, according to the equation

$$5\,V_2O_4 + 2\,KMnO_4 + 3\,H_2SO_4 = 5\,V_2O_5 + K_2SO_4 + MnSO_4 + 3\,H_2O.$$

If the solution be passed through the zinc column both the vanadic acid and the molybdic acid will be reduced

$$V_2O_5 + 3\,Zn = V_2O_2 + 3\,ZnO,$$
$$2\,MoO_3 + 3\,Zn = Mo_2O_3 + 3\,ZnO,$$

and oxidation by permanganate will take place according to the equations

$$5\,V_2O_2 + 6\,KMnO_4 + 9\,H_2SO_4 = 5\,V_2O_5 + 3\,K_2SO_4 + 6\,MnSO_4 + 9\,H_2O,$$
$$5\,Mo_2O_3 + 6\,KMnO_4 + 9\,H_2SO_4 = 10\,MoO_3 + 3\,K_2SO_4 + 6\,MnSO_4 + 9\,H_2O.$$

If the number of centimeters of permanganate used in the titration of the vanadium tetroxide be multiplied by three, and the product subtracted from the total number of centimeters of permanganate used in the titration of vanadium dioxide and molybdenum sesquioxide, the result is the number of centimeters used in oxidizing Mo_2O_3 to MoO_3, from which the amount of molybdic acid present may be easily calculated.

According to the procedure recommended by Edgar, the solution containing vanadic and molybdic acids is diluted to 75 cm.³, acidified with 2 cm.³ to 3 cm.³ of strong sulphuric acid, heated to boiling and subjected to a current of sulphur dioxide for a few minutes until the clear blue color indicates the complete reduction of the vanadic acid to the state of tetroxide. The boiling is continued for some time, a current of carbon dioxide being passed into the liquid until the last trace of sulphur dioxide has been removed. Titration is then effected by nearly $n/10$ potassium permanganate and the vanadic acid is calculated.

Next, the solution just titrated, preceded by 100 cm.^{3} of hot water, and 125 cm.^{3} of dilute sulphuric acid (2.5 per cent), and followed by 100 cm.^{3} of the dilute sulphuric acid and then by 200 cm.^{3} of hot water, is passed slowly through a column of amalgamated zinc in a Jones reductor* into the receiver containing a solution of ferric alum, and the hot solution is titrated with nearly $n/10$ potassium permanganate, a little phosphoric acid being added to decolorize the ferric salt. The amount of permanganate used in this titration, diminished by three times the amount used in titrating the vanadium tetroxide reduced by sulphur dioxide, measures the molybdic acid.

The experimental results show that molybdic acid and vanadic acid may be accurately estimated in the presence of one another by two processes of reduction and oxidation, the reduction being made first by sulphur dioxide and last by amalgamated zinc.

Molybdic and Vanadic Acids.

$KMnO_4$ $n/10 \times 1.052$. cm.^{3}	$KMnO_4$ $n/10 \times 1.052$ cm.^{3}	V_2O_5 taken as $NaVO_3$. grm.	MoO_3 taken as $(NH_4)_2MoO_4$. grm.	V_2O_5 found. grm.	MoO_3 found. grm.	Error on V_2O_5. grm.	Error on MoO_3. grm.
11.95*	74.15	0.1144	0.1930	0.1146	0.1934	+0 0002	+0.0004
11 95*	74.00	0.1144	0.1930	0.1146	0.1926	+0 0002	−0.0004
11.94*	74.20	0.1144	0.1930	0.1145	0 1936	+0.0001	+0.0006
5.97†	37.10	0.0572	0.0965	0.0572	0 0965	0 0000	0.0000
5.97†	37.05	0.0572	0.0965	0.0572	0.0962	+0 0000	−0.0003
5.98†	37.12	0.0572	0.0965	0.0573	0.0963	+0 0001	−0.0002
11 95‡	55.0	0 1144	0.0965	0.1146	0.0967	+0 0002	+0.0002
11.95‡	55 0	0.1144	0.0965	0.1146	0.0967	+0 0002	+0.0002
11.96‡	54 86	0.1144	0.0965	0.1147	0.0958	+0.0003	−0.0007
17.92§	92 0	0.1716	0.1930	0.1719	0.1931	+0.0003	+0.0001
17 94§	92.05	0.1716	0.1930	0.1720	0.1931	+0 0004	+0.0001
17.92§	92.02	0.1716	0.1930	0.1719	0.1932	+0.0003	+0.0002

* With 8 cm.^{3} of sirupy phosphoric acid and 50 cm.^{3} of 10 per cent ferric alum.
† With 4 cm.^{3} of sirupy phosphoric acid and 25 cm.^{3} of 10 per cent ferric alum.
‡ With 6 cm.^{3} of sirupy phosphoric acid and 35 cm.^{3} of 10 per cent ferric alum.
§ With 10 cm.^{3} of sirupy phosphoric acid and 65 cm.^{3} of 10 per cent ferric alum.

* See page 426.

URANIUM.

The Determination of Uranium by the Aid of the Jones Reductor.

In studying the behavior of uranyl sulphate in the zinc reductor Pulman * has shown that the degree of reduction varies with conditions, and that when the reduced solution is received in an atmosphere of carbon dioxide the titration with permanganate shows that the uranium salt has been reduced below the uranous stage, represented by the oxide UO_2, but that the over-reduction may be corrected by brief contact with the oxygen of the air.

The details of procedure are as follows: The uranium sulphate solution, in volume 100 cm.³ to 150 cm.³, and containing sulphuric acid in the proportion [1 : 6], is heated nearly to boiling. Preceded by a few cubic centimeters of acid of the same strength, the solution is drawn by gentle suction through the 18-inch reductor charged with 20-mesh amalgamated zinc, and is followed by more of the same acid, used in washing the container, and 250 cm.³ of hot water. The contents of the receiving flask are poured through the air into a porcelain dish, diluted with about 200 cm.³ of hot water, and titrated with $n/10$ potassium permanganate.

The proportion of free sulphuric acid should be kept during the digestion nearly at the ratio [1 : 6], since with less the reduction is delayed, while more produces too rapid evolution of hydrogen. With acid in the ratio [1 : 6], fifteen minutes or more should be taken in passing uranium sulphate equivalent to 0.2 grm. of uranic oxide through the reductor; for 0.3 grm. half an hour or more should be allowed. Care is taken that the liquid in the reductor shall always cover the zinc, lest hydrogen dioxide, formed by contact of nascent hydrogen and air, vitiate the results.

The contents of the receiving flask after the reduction are olive-green, but upon exposure to the air by pouring into the dish the color changes immediately to the sea-green color always possessed by uranous salts, and this change of color is of itself evidence of oxidation. In the titration of the hot solution of uranous sulphate with permanganate the solution gradually becomes more and more yellowish green as the highest condition of oxidation is approached. With small amounts of uranium

* O. S. Pulman, Jr., Am. Jour. Sci., [4], xvi, 229.

the addition of a single drop of permanganate in excess brings out a faint pink color, but with larger amounts the end point is a yellowish pink.

The results obtained in applying this method to uranyl sulphate prepared from the pure nitrate by evaporation with sulphuric acid are shown in the table.

Reduction in the Zinc Column; Exposure to Air; Titration with Permanganate.

Uranyl sulphate taken, in terms of UO_3.	H_2SO_4 (1.84).	Dilution at time of reduction.	Time.	$KMnO_4$.		Error in terms of UO_3.
grm.	cm.³	cm.³	minutes.	cm.³	grm. UO_3.	grm.
0.1336	18	117	15	9.32	0.1334	−0.0002
0.1337	20	120	15	9 37	0.1341	+0.0004
0.1336	25	125	17	9.40	0.1345	+0.0009
0.2005	18	117	20	14.02	0.2006	+0.0001
0.2003	25	125	17	14.01	0.2005	+0 0002
0.2671	23	150	20	18 67	0.2671	0.0000
0.2673	20	140	18	18.65	0.2669	−0.0004
0.1001	25	125	22	7 06	0.1010	+0.0009
0.1002	20	140	17	7.02	0.1004	+0.0002
0.1002	20	140	14	7.01	0 1003	+0.0001
0 0668	18	117	17	4.70	0.0673	+0.0005
0.0994	20	100	16	6.96	0 0995	+0 0001
0.1988	20	120	18	13 90	0.1988	0 0000
0.1988	25	150	18	13.88	0.1985	−0.0003
0.3314	25	150	27	23.14	0.3309	−0.0005
0.3314	30	150	36	23.19	0.3316	+0.0002
0.3314	25	145	33	23.17	0.3313	−0.0001

CHAPTER XI.

FLUORINE; CHLORINE; BROMINE; IODINE.

FLUORINE.

The Detection of Fluorine.

Browning * has shown that small amounts of fluorine may be detected by the converse of the method previously described for the detection of silicates and fluosilicates.† According to the procedure, the fluoride is put, with a suitable amount of silica, in a small lead cup, 1 cm. in diameter and depth a few drops of concentrated sulphuric acid are added; the cup is covered by a flat piece of lead with a small hole in the center; upon the cover is placed a piece of moistened black filter paper and upon this a small pad of moistened filter paper to keep the black paper moist during subsequent heating upon the steam bath. After about ten minutes' heating a white deposit is found on the under side of the black paper, over the opening in the cover, if fluorine is present in appreciable amount.

The results of tests are given in the table.

Tests for Fluorine.

Name and amount of fluoride used. grm.	Approximate per cent of F.	SiO_2 present. grm.	Result.
0.1000 CaF_2	49	None.	Nothing
0.0100 CaF_2	49	0 0500	Very good.
0.0050 CaF_2	49	0 0500	Apparent.
0.0010 CaF_2	49	0 0500	Trace.
0.0100 Na_3AlF_6	54	0.0500	Very good.
0.0050 Na_3AlF_6	54	0 0500	Distinct.

The Acidimetric Estimation of Fluosilicic Acid.

Hileman ‡ has studied methods in use for the determination of fluorine in fluosilicic acid by neutralization with standard alkali. The first set of these methods depends upon the action of fluo-

* Philip E. Browning, Am. Jour. Sci., [4], xxxii. 249.

† See page 241, Fig. 20.

‡ Albert Hileman, Am. Jour. Sci., [4], xxii, 329.

silicic acid upon potassium chloride in alcoholic solution and titration (without removal of the precipitated potassium fluosilicate) of the liberated hydrochloric acid by standard ammonia* or by standard fixed alkali.† The second set of methods depends upon the titration of fluosilicic acid in water solution by standard alkali hydroxide to the complete decomposition of the fluosilicate. Hileman's results are given in the following tables:

Titration in Alcoholic Solution

$$H_2SiF_6 + 2\,KCl = K_2SiF_6 + 2\,HCl$$
$$2\,ROH + 2\,HCl = 2\,RCl + 2\,H_2O.$$

H_2SiF_6. cm.³	Standard NH_4OH. cm.³	Standard KOH. cm.³	Standard NaOH. cm.³	Fluorine found. grm.	Average. grm.
25	7.3	.....	....	0.1433	
25	7.3	....	....	0.1433	
25	7.27	. ...	....	0.1426	0.1428
25	7.23	... .	...	0.1429	
25	7.29	...	....	0 1431	
25	.. .	10 67	..	0.1412	
25	....	10.72	. .	0 1419	
25	....	10.64	.. .	0.1408	0.1411
25	....	10.67	. ..	0.1412	
25	....	10.60	. ..	0.1403	
25	....	.....	9.11	0.1416	
25	....	.....	9.12	0.1418	
25	.. .	.....	9 07	0 1410	0.1415
25	. ..	.....	9 10	0 1414	
25	.	. . .	9 12	0.1418	

The differences between the amounts of fluorine indicated by the individual determinations in any one of these processes are generally slight. The averages of the determinations by potassium hydroxide and sodium hydroxide are very close together, being 0.1411 grm. and 0.1415 grm. of fluorine. The average of the titrations by ammonium hydroxide is a little higher, namely, 0.1428 grm. That the differences between these averages are due to gradual variations in the reading tint is shown by a comparison of three titrations as nearly simultaneous as possible, in which the greatest care was taken to bring all to the same tint at the final reading.

* Penfield, Am. Chem. Jour., i, 27.

† Bullnheimer, Zeit. angew. Chem., 1901, 101.

Comparison of Simultaneous Titrations in Alcoholic Solution.

	Solution used. cm.³	Fluorine found grm.
Titration by NH_4OH	7.2	0.1414
Titration by KOH.	10 71	0.1418
Titration by NaOH.	9 13	0.1419

It appears that the results obtained are practically the same by the three processes of neutralization applied to a solution of fluosilicic acid. But it is to be observed that all are possibly subject to a common and constant error due to the presence of hydrofluoric acid as well as fluosilicic acid. If the former acid is present it tends to raise the apparent value of the latter.

With these results of titrations in alcoholic solution are to be compared the results obtained by the method of titration in water solution (in which the fluosilicate is completely converted to fluoride), recorded in the following table:

Titration of Fluosilicic Acid in Water Solution:

$$H_2SiF_6 + 6\ ROH = 6\ RF + SiO_4H_4 + 2\ H_2O.$$

H_2SiF_6 taken. cm.³	Standard KOH. cm.³	Standard NaOH. cm.³	Fluorine found. grm.	Average. grm.
25	30.9	. . . .	0 1358	
25	30.8	. . . .	0.1353	0.1355
25	30.9	. . .	0.1358	
25	30.79	. . . .	0.1353	
25	. . . . .	26.2	0.1357	
25	. . . . .	26.15	0.1355	
25	. . . . .	26.25	0.1360	0.1358
25	. . . . .	26.2	0.1357	
25	. . . .	26.13	0.1354	
25	. . . . .	26.14	0.1354	

It is obvious that the process of titrating fluosilicic acid in water solution yields uniform indications, both with potassium hydroxide and sodium hydroxide, but that the values for fluorine are very much below those of the titrations in alcoholic solution. And this will be the case if the solution of fluosilicic acid contains hydrofluoric acid, as is probable.

The action of ammonium hydroxide upon fluosilicic acid in water solution proves to be comparable with that of sodium hydroxide, and inferentially with that of potassium hydroxide, though the hydrolysis of the fluosilicate appears to be not quite so complete.

The Iodometric Estimation of Fluosilicic Acid.

It is obvious that the reaction by which fluosilicic acid liberates iodine from a mixture of potassium iodide and potassium iodate may be turned to account in the analysis of fluorides as well as in the determination of fluosilicic acid,* provided the course of action is regular.

Iodometric and Acidimetric Determinations in Water Solutions.

H_2SiF_6.	Standard NaOH.	Standard $Na_2S_2O_3$.	Fluorine found.
Commercial fluosilicic acid.			
	(1 cm.^{3}=0.005137 grm. of fluorine.)	(1 cm.^{3}=0.002335 grm. of fluorine.)	
cm.^{3}	cm.^{3}	cm.^{3}	grm.
25	10.82	....	0 0556
25	10.87	.....	0.0559
25	10.83	.....	0.0557
25	. .	23.54	0 0549
25	...	23.52	0.0549
25	. .	23.48	0.0548
25	. .	23.50	0.0548
25	.	23.45	0.0548
25	.....	23.48	0.0548
Fluosilicic acid made by HF on excess of SiO_2.			
	(1 cm.^{3}=0 005137 grm. of fluorine.)	(1 cm.^{3}=0 002441 grm. of fluorine.)	
25	9.08	.....	0.0466
25	9.05	.....	0.0464
25	9.05	.....	0.0464
25	.....	18.78	0.0458
25	.....	18.80	0.0459

Upon testing the action of fluosilicic acid upon the iodide-iodate mixture, Hileman† has found that, while iodine is liberated

* See page 436.

† Albert Hileman, Am. Jour. Sci., [4], xxii, 383.

freely in the cold, a complete reaction is not obtained in the course of several hours — the amount of iodine liberated indicating that an acid other than fluosilicic acid is acting. It appears, further, that on heating the mixture to the boiling point in a flask closed with a glass stopper and trapped with a solution of potassium iodide, nearly one equivalent of iodine is liberated for every equivalent of fluorine present as fluosilicic acid, according to the reaction

$$5 KI + KIO_3 + H_2SiF_6 = 6 KF + 3 I_2 + SiO_2 + H_2O.$$

The Estimation of Fluorine Evolved as Silicon Fluoride.

As to sources of error in the determination of fluorine by the silicon fluoride processes, due to imperfect elimination and collection of silicon fluoride, there is the testimony of many investigators. The importance of using the fluoride in the finest state of division, of having the sulphuric acid of highest strength, of properly absorbing the vapors of sulphuric acid evolved from the decomposition flask, and of using quartz for the silicon dioxide in the decomposition flask, have all been emphasized. Many forms of apparatus have been employed and the results have varied widely.

Elimination of Silicon Fluoride at High Temperatures. In consequence of difficulty with the silicon fluoride processes in which the decomposition is effected at the usual temperatures, between 150° and 160°, Hileman * has devised a convenient form of apparatus in which the acid mixture may be heated to boiling to facilitate the removal of the silicon fluoride to the absorption system.

A glass stopper, made by drawing out a glass tube 1 cm. in diameter and sealing a small glass tube on each end, is ground into a 70 cm.³ side-neck flask. To one end is sealed a glass stopcock. The other end extends to the bottom of the flask. The side neck is sealed to a Voit flask. The length of the tube between the two flasks is 17 cm., and it is bent at a point about 12 cm. from the Voit flask. The tube leading from the Voit flask is joined to a large empty U-tube and this is connected, through a tube charged with phosphorus pentoxide, with the absorption tube. The absorption apparatus is similar to that

* Albert Hileman, Am. Jour. Sci., [4], xxii, 329.

described by Burk,* and consists of a test tube 34 cm. in length and 2 cm. in diameter, conveniently inclined and containing a few cubic centimeters of mercury into which extends an inlet tube with a capillary opening through which the gas is delivered under the mercury and allowed to bubble into the water with which the tube is charged. The tube is connected with a suction pump by means of a T-tube joined also to an air-trap of mercury to regulate the pressure within the apparatus to any desired degree.

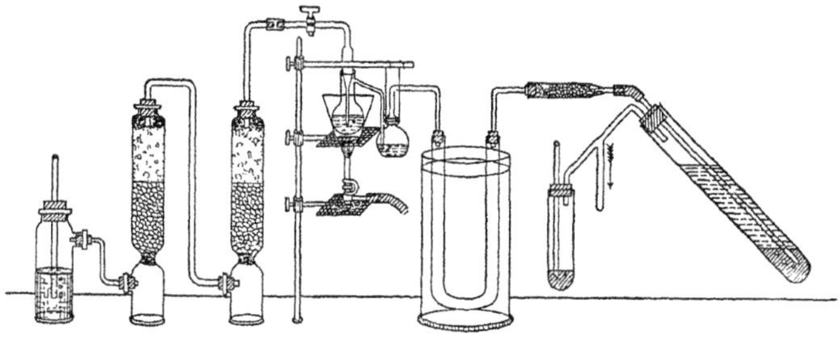

Fig. 28.

Preparatory to making a determination the apparatus is carefully dried. The tip of the inlet tube is then placed beneath the surface of mercury in the bottom of the absorption tube and distilled water is added, care being taken that enough space shall remain to allow for the rise in level when air bubbles through the liquid. The pressure regulator is adjusted so that the suction shall produce slightly diminished pressure in the apparatus. The U-tube is immersed in a vessel of cold water and connected with the system.

Next, the mineral, together with quartz powder to about three times the weight of the fluorine present, is transferred to the decomposition flask, and enough sulphuric acid to seal the delivery tube from the side-neck flask is introduced into the Voit flask. The two flasks are then tilted so that the acid shall moisten the connecting tube to the bend. About 40 $cm.^3$ of sulphuric acid, previously boiled for half an hour and cooled in a current of dry air, and several capillary tubes of the form recommended by Scudder† to prevent bumping, are put in the decom-

* Jour. Am. Chem. Soc., xxiii, 825.
† Ibid., xxv, 113.

position flask. The stopper is quickly replaced and sealed with a drop of sulphuric acid. A thin strip of asbestos is wrapped about the neck of the flask and, the stopcock having been closed, the bulb is heated in a radiator covered over with sheet asbestos.

When the heat is applied, bubbles of gas are given off, the solid material rises to the surface, and during the course of the heating an oily film gathers on the upper part of the flask and in the delivery tube. On boiling, this film is replaced by a white deposit which recedes before the acid vapors. The success of the determination depends, as was found, on the breaking up of this deposit, which is probably a product of the hydrolysis of silicon fluoride. On this account the tube between the two flasks should be as short as is practicable. Unsatisfactory results were obtained when the tube was about one-half longer than the dimensions given above. When the acid vapors have penetrated the length of the tube, leaving it clear or translucent, the decomposition is complete and, the stopcock having been opened, the side-neck flask is cooled to about 75°.

Elimination of Silicon Fluoride; Absorption in Water; Titration of Fluosilicic Acid with Sodium Hydroxide.

CaF_2. grm.	Quartz. grm.	NaOH. cm.³	Theory, fluorine. grm.	Found, fluorine. grm.	Error, fluorine. grm.
0.3000	0.4	28.2	0.1459	0.1444	−0.0015
0.3000	0.4	28.35	0.1459	0.1452	−0.0007
0.3000	0.4	28.34	0.1459	0.1451	−0.0008
0.3000	0.4	23.5	0.1215	0.1203	−0.0012
0.3000	0.4	28.3	0.1459	0.1419	−0.0010
0.3000	0.4	28.37	0.1459	0.1453	−0.0006
NaF.					
0.3000	0.4	26.2	0.1356	0.1342	−0.0012
0.3000	0.4	26.44	0.1356	0.1355	−0.0001
0.3000	0.4	26.46	0.1356	0.1356	0.0000
0.3000	0.4	26.3	0.1356	0.1347	−0.0009
0.3000	0.4	26.34	0.1356	0.1349	−0.0007
0.3000	0.4	26.3	0.1356	0.1347	−0.0009
	Ignited Silicic acid				
0.3000	0.4	26.29	0.1356	0.1346	−0.0010
	Quartz				
0.3000	0.4	26.3	0.1356	0.1347	−0.0009

If the acid tends to suck back from the Voit flask, it is arrested by opening the stopcock to relieve the vacuum, and it is at this point of the experiment that the necessity for the previous adjustment of the pressure regulator, to avoid sudden transfer of acid vapor to the absorption tube, becomes apparent. The decomposition is ended in from fifteen to forty minutes. A current of purified air is drawn through the apparatus, slowly at first and then more rapidly. About six liters are necessary to remove the last trace of silicon fluoride. The delivery tube is washed and the absorption solution transferred to a flask and titrated with sodium hydroxide prepared from sodium *, with phenolphthalein as an indicator, according to the following reaction: †

$$H_2SiF_6 + 6\,NaOH = 6\,NaF + Si(OH)_4 + 2\,H_2O.$$

The results obtained are shown in the preceding table.

With the apparatus described above, in which the sulphuric acid in the decomposition flask may be boiled, the silicon fluoride formed passes rapidly to the absorption system, other products of partial hydrolysis of silicon fluoride formed in the flask or tube are ultimately reconverted to silicon fluoride, and regular results of a fair degree of accuracy are obtained. In all the experiments except two, phosphorus pentoxide (about 2.0 grm.) was introduced into the decomposition flask, the purpose being to retain water formed in the reaction. The results of these two experiments show, however, that phosphorus pentoxide in the flask is not essential.

In three blank experiments, in which the acid in the decomposition flask was heated to boiling, amounts of acid in each case equivalent to 0.0002 grm. of fluorine were found in the absorption solution; the results, therefore, are subject to this trifling error.

Iodometric Determination of Fluorine in Fluorides. Hileman ‡ has also applied the iodometric determination of fluosilicic acid § to the estimation of fluorine in silicon fluoride evolved and absorbed in the manner described above.

Silicon fluoride, evolved from calcium fluoride in the apparatus previously figured and by the method described,|| is absorbed in water, and, after separating the mercury used in the apparatus

* Küster, Zeit. anorg. Chem., xli, 475.

† See page 434.

‡ Am. Jour. Sci., [4], xxii, 383.

§ See page 435.

|| See page 436.

by means of a separating funnel, the iodide-iodate mixture is added to the solution and the iodine liberated is titrated by sodium thiosulphate. The results are given below:

Elimination of Silicon Fluoride; Absorption in Water; Iodometric Estimation of Fluosilicic Acid.

CaF_2. grm.	$Na_2S_2O_3$. cm.³	Theory, fluorine. grm.	Found, fluorine. grm.	Error. fluorine. grm.
0.2500	51.35	0.1216	0 1199	0.0017
0.2300	46.75	0 1119	0 1091	0 0028
0.2300	47 15	0 1119	0 1100	0 0019
NaF.				
0.2000	38.50	0 0903	0 0899	0.0014

The average error of −0.0019 grm. is considerably greater than that of the acidimetric method, −0.0008 grm.

CHLORINE; BROMINE; IODINE.

The Detection of Iodine, Bromine and Chlorine in Presence of One Another.

In the qualitative testing of substances it is not a matter of moment that a portion of a substance looked for escapes the reaction, provided enough is left to furnish the indication sought. In the separation of iodine from bromine and chlorine, Gooch and Brooks * have applied the nitrite reaction † to small amounts of liquid in test tubes and have demonstrated that the losses of chlorine and bromine under the conditions are proportioned to the strength of the solution; or, in other words, that when the amounts of bromine and chlorine are very small they escape volatilization, and when large a sufficient amount remains to give strong tests.

The process elaborated for the rapid qualitative detection of the halogens in presence of one another may be summarized as follows:

To detect iodine, the solution of the substance under examination is acidulated with dilute sulphuric acid and treated with a drop or two of a solution of sodium or potassium nitrite free

* F. A. Gooch and F. T. Brooks, Am. Jour. Sci., [3], xl, 283.

† See page 451.

from chlorine. Unless the amount present is small, the iodine shows itself in the color of the solution and in the vapors which escape. Small amounts may be found by shaking the liquid with carbon disulphide in the usual manner, or, when economy of material is desirable, by gently heating the prepared solution and testing the escaping fumes with red litmus paper, thus utilizing the same portion of material for the detection of the iodine and for its separation preparatory to testing for bromine and chlorine.

To remove the iodine previous to making the tests for bromine and chlorine, a few drops of dilute sulphuric acid and a like amount of a dilute solution of sodium or potassium nitrite (prepared free from chlorine by adding a little silver nitrate, faintly acidulating with nitric acid, and filtering) are added to the solution of the substance in a test tube, and the liquid is boiled with constant agitation. When the color of iodine disappears from the fumes and the solution, a drop or two of sulphuric acid, and of the nitrite, are again added, and the boiling is repeated. When the escaping steam no longer gives to red litmus paper the characteristic gray blue color due to the action of iodine, the process of separation is complete.

A portion of the solution thus prepared is tested for bromine by cautiously adding a dilute solution of sodium hypochlorite and shaking with colorless carbon disulphide.

The test for chlorine is made in a second portion of the solution from which the iodine has been removed. The liquid is neutralized with sodium carbonate or hydroxide free from chloride, evaporated to dryness in a test tube and treated with sulphuric acid and potassium dichromate, the fumes of the chlorochromic anhydride which arise on gentle warming being condensed and converted to chromic acid by a film of moisture upon the interior walls of a trap, such as is shown in Fig. 6,* or in a mushroom trap, shown in Fig. 29. The trap is washed out with a very little distilled water (5 cm.^{3}) and the washings, made slightly ammoniacal to destroy free bromine, if necessary, and after gentle warming again acidified, are tested with lead acetate. If yellow lead chromate is precipitated the presence of chlorine in the original substance is

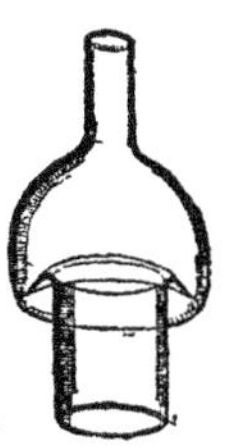

Fig. 29.

* See page 6.

proved. If the precipitate is white, as is very likely to be the case, a few drops of a saturated solution of ammonium acetate are added with caution, and the whole is gently warmed to dissolve the white sulphate. On cooling the solution and shaking (or immediately, if much chromic acid has been formed), the yellow chromate falls, or gives color to the solution, according as the chloride was originally present in large or small amount.

The process is rapid and sufficiently exact for qualitative testing in general. The results of experiments made to determine the delicacy of the tests for bromine and chlorine in a substance treated to remove iodine are shown in the tabular statements:

Hypochlorite Test for Bromine after Treatment to Remove Iodine.

KI taken. grm.	KBr taken. grm.	Total volume. cm.^{3}	Ratio of KBr to solution.	Color in carbon disulphide
0 1000	0.0010	10	1 : 10,000	Pronounced.
0 1000	0 0004	10	1 : 25,000	Pronounced.
0.1000	0.0003	10	1 : 33,000	Pronounced.
0.1000	0.0002	10	1 : 50,000	Faint.
0.1000	0.0002	10	1 : 50,000	Faint.
0 1000	0 0001	10	1 : 100,000	None.
0.0500	0.0005	10	1 : 20,000	Pronounced.
0.0400	0.0004	10	1 : 25,000	Pronounced.
0 0300	0.0003	10	1 : 33,000	Pronounced.
0.0200	0.0002	10	1 : 50,000	Faint.
0.0100	0 0001	10	1 : 100,000	None.
0.1000	0.00007	5	1 : 70,000	Trace.
0.0070	0.00007	5	1 : 70,000	Trace.

Chromate Test for Chlorine after Treatment to Remove Iodine.

KI taken. grm.	KBr taken. grm.	KCl taken. grm.	Final volume. cm.^{3}	Reaction obtained.
...	0.1	0.0030	5	Marked precipitation.
...	0.1	0.0020	5	Distinct precipitation.
...	0.1	0.0010	5	Distinct color.
...	0.1	0.0005	5	Faint color.
...	0.1	0.0005	5	Faint color.
...	0.1	0.0004	5	Faintest color.
...	0.1	0.0003	5	Doubtful color.
...	0.1	0.0002	5	Doubtful color.
...	0.1	0.0001	5	None.
0.1	0.1	0.0010	5	Distinct color.
0.1	0 1	0.0010	5	Distinct color.
0.1	0.1	0.0010	5	Distinct color.

The evolution of considerable amounts of bromine appears to diminish the delicacy of the test in some degree, but 0.0005 grm. of chlorine — the amount in 0.0010 grm. of potassium chloride — is indicated unmistakably in the presence of 0.1 grm. of potassium bromide, and 0.1 grm. of potassium iodide, and the test may probably be relied upon to show half that amount of chlorine.

The Determination of Free Chlorine and Free Bromine by Liberation of Iodine and Absorption of that Element by Silver.

Chlorine and bromine in free condition may be determined by allowing them to act upon potassium iodide in solution made acid with hydrochloric acid and in an atmosphere of hydrogen, shaking the mixture with a weighed amount of specially prepared silver, and determining the increase in weight of the silver due to absorption of iodine set free, according to the method described for the determination of iodine by absorption in metallic silver.* Tests of this process by Perkins † were made upon definite amounts of aqueous solutions of chlorine and bromine determined titrimetrically at the time of the experiments. These definite amounts of the aqueous solutions were put in a flask containing an excess of potassium iodide made acid with hydro-

Liberation of Iodine and Absorption by Silver.

Silver taken. grm.	Halogen taken. grm.	Iodine found. grm.	Calculated amount of halogen. grm	Error. grm.
Determination of bromine.				
3.0000	0.0213	0.0336	0.0211	−0 0002
3.0000	0.0426	0.0678	0.0427	+0 0001
3.0000	0.1065	0.1694	0.1067	+0.0002
Determination of chlorine.				
3.0000	0 0161	0.0574	0.0160	−0 0001
3.0000	0 0322	0.1146	0.0320	−0 0002
3.0000	0 0322	0.1145	0 0320	−0 0002
3.0000	0.0322	0.1141	0.0318	−0 0004
3.0000	0.0483	0.1716	0 0479	−0 0004

* See page 444.

† Claude C. Perkins, Am. Jour. Sci., [4], xxix, 338.

chloric acid in an atmosphere of hydrogen, and the mixture was shaken with a weighed amount of the specially prepared silver. The residue of silver and silver iodide was collected on asbestos in a perforated crucible, washed, dried and weighed. The increase in weight represents the weight of iodine liberated and from this the amount of chlorine or bromine is calculated. Results of this process are given in the table.

The Gravimetric Determination of Iodine by Absorption in Metallic Silver.

Free Iodine. It has been shown by Gooch and Perkins * that free iodine may be determined with accuracy in solution in potassium iodide, either neutral or alkaline with an acid carbonate, by shaking the solution with metallic silver in a closed flask filled with hydrogen and determining the increase in weight of the silver. Silver reduced from a silver salt by zinc, or from silver sulphide by hydrogen, may serve the purpose, provided it is subjected to a preliminary treatment with potassium iodide, and silver reduced from the oxide by hydrogen is also serviceable; but the best form of silver, and the one most easily prepared in the pure state, is that deposited electrolytically upon a small oscillating cathode of platinum from a solution of silver nitrate, the platinum anode being inclosed in a porous cell. The shaking of the silver may be done by hand or by some simple form of mechanical shaker like that described and figured elsewhere.† In the test experiments, detailed in the table, $n/10$ iodine solution was drawn from a burette into a 250 cm.^{3} Erlenmeyer flask containing a weighed amount of finely divided silver. The flask, properly trapped and attached to a mechanical shaker adjusted to give the liquid a rapid rotary motion, was shaken until the iodine color had vanished. The liquid, usually 50 cm.^{3} in volume, was diluted to about 100 cm.^{3}, and the residue of silver and silver iodide, collected in a perforated crucible fitted with asbestos, was washed, dried between 130° and 140°, and weighed. The difference between the weight of silver taken and that of the residue of silver and silver iodide should, according to the

* F. A. Gooch and Claude C. Perkins, Am. Jour. Sci., [4], xxviii, 33.

† See page 9.

The Action of Silver upon n/10 Iodine in an Atmosphere of Hydrogen.

Silver taken	Iodine taken.	Increase in weight of iodine.	Error in iodine.	Average error in iodine.
grm.	grm.	grm.	grm.	grm.
A. Silver reduced from AgCl by zinc and treated with KI.				
3 0000	0 6461	0.6464	+0 0003	......
1.0000	0 6447	0 6448	+0 0001	+0.0002
B. Silver reduced from AgI by zinc and treated with KI.				
3.6293	0 3217	0.3221	+0 0004	.
3.2049	0 3217	0.3225	+0 0008	. ..
3.0000	0.3217	0.3219	+0 0002	. .
3 0068	0.3217	0.3212	−0 0005	.
3.0049	0 3217	0.3221	+0.0004	. .
3.0026	0.6434	0 6441	+0 0007	.... .
2 9990	0 3217	0.3214	−0 0003	......
3.0005	0 3217	0 3214	−0.0003	+0.0002
C. Silver reduced from Ag_2S by hydrogen and treated with KI.				
3.0000	0.6461	0.6463	+0.0002	......
3.0000	0.6461	0.6460	−0.0001	+0.0001
D. Silver reduced from Ag_2O by hydrogen.				
3.0000	0.6434	0.6443	+0.0009	......
3.0000	0.6434	0.6430	−0.0004	+0.0003
E. Silver reduced electrolytically from $AgNO_3$ upon an oscillating cathode.				
4.4189	0.6447	0.6447	0.0000	......
3.0025	0.6447	0.6448	+0.0001	......
3.0009	0.6447	0.6443	−0.0004	......
3.0157	0.6447	0.6445	−0.0002	......
3.0000	0.6447	0 6444	−0.0003	......
3.0000	0.6447	0.6452	+0.0005	......
3.0004	0.6447	0.6443	−0.0004	......
3.0043	0.6447	0 6443	−0.0004	......
3.0000	0.6434	0.6430	−0.0004	......
3 5810	0 3217	0.3221	+0.0004	......
3.0000	0.3217	0.3219	+0.0002	−0.0001
Electrolytic silver in presence of $NaHCO_3$.				
3.0014	0 3217	0 3216	−0.0001	.. ...
3.0169	0 3217	0 3216	−0.0001	......
3.0083	0 6434	0 6433	−0.0001	......
3.0016	0 2500	0 2503	+0.0003	.. . .
3.0069	0.3217	0.3219	+0.0002	+0.0001

theory of action, be the measure of the free iodine. The time required for the absorption of approximately 0.65 grm. of iodine in 50 cm.[3] of liquid was from fifteen to twenty-five minutes. The mean error of sixteen determinations in which electrolytic silver was employed proved to be 0.00004 grm. between extremes of +0.0005 grm. and −0.0004 grm.

This process, in which free iodine is absorbed by specially prepared silver * under hydrogen, either in neutral solution or in a solution made alkaline with an acid carbonate, is applicable in many analytical operations involving liberation of iodine as well as in the gravimetric standardization of the usual iodine solution of volumetric analysis. The adaptation of the process to the determination of iodine in iodides, free chlorine, free bromine, and various oxidizers, has been described by Perkins.†

Iodine in Iodides. In applying the process to the determination of combined iodine a definite amount of the iodide in solution is introduced into a flask with an excess over the calculated amount of an oxidizing reagent (usually potassium nitrite or hydrogen peroxide), and the whole, made acid with hydrochloric acid, is shaken with a weighed amount of silver. The increase in the weight of the silver indicates the amount of iodine liberated, and from this the amount of potassium iodide may be easily calculated. The table shows the results of determinations with potassium iodide in solution previously standardized by the distillation method with sulphuric acid and potassium hydrogen arsenate.‡

Determination of the Iodine of Potassium Iodide.

Silver taken.	KI taken.	Iodine found.	Calculated amount of KI.	Error.
grm.	grm.	grm.	grm.	grm
2.7803	0.1144	0.0872	0.1141	−0.0003
3.0028	0.1346	0.1026	0 1342	−0.0004
2.7800	0.1279	0.0978	0.1281	+0.0002
2.0008	0.1279	0.0974	0.1274	−0.0005
3.0001	0.1346	0.1029	0 1346	0.0000

* See page 27.

† See also pages 443, 444, 361.

‡ See page 457.

The Determination of Halogens in Benzol Derivatives by the Use of Metallic Potassium.

The method of Stephanoff * for the estimation of halogens in aromatic compounds, by reduction with sodium and alcohol, was found by Maryott † to give irregular and low results when applied to chlorbenzol. The action of the sodium is very slow, especially after the liquid has become pasty owing to separation of sodium ethylate, and this difficulty is only partly overcome by increasing the amount of alcohol on account of the unfavorable effect of dilution.

Maryott has shown, however, that a similar method, based on the use of the more active potassium, in place of sodium, as the reducing agent, gives a complete reduction and accurate analytical results when employed upon the halogen substituted benzols.

As the action of potassium upon 98 per cent alcohol is very energetic, the alcohol used is diluted with twice its own volume of thiophene-free benzol. Very little heating is required, so that a plain glass tube, about 50 cm. in length, serves as a return cooler in place of the water cooled condenser used by Stephanoff.

The substance to be analyzed is weighed out in an Erlenmeyer flask, 10 cm.^{3} to 15 cm.^{3} of the 1:2 alcohol-benzol mixture are added, the cooling tube is attached, and the potassium, in small pieces, is gradually dropped in through the tube. The weight of potassium required is about ten times that called for by the equation

$$C_6H_5Cl + 2\,K + C_2H_5OH = C_6H_6 + KCl + C_2H_5OK.$$

The greater efficiency of potassium as a reducer as compared with sodium is shown by the fact that an analysis of chlorbenzol, carried out in exactly the same way as the analyses tabulated below except that sodium (ten times the theoretical amount) was used instead of potassium, gave only 84 per cent of the total chlorine.

A small amount of potassium ethylate usually separates out during the action of the metal, but seems to have no bad effect upon the reduction. After the action has become less vigorous, about 2 cm.^{3} of alcohol are added and the flask is carefully heated and gently shaken from time to time until the potassium is

* Ber. Dtsch. chem. Gesell., xxxix, 4056.

† C. H. Maryott, Am. Jour. Sci., [4], xxx, 378.

dissolved. The contents are then shaken with water, the water layer is acidified with nitric acid, and the halogen precipitated and weighed as the silver salt. The use of Volhard's volumetric method in estimating the halogens is, of course, feasible, though somewhat less accurate. The time required for an analysis, exclusive of weighings, is about twenty-five minutes.

A series of analyses gave the following results:

Chlorbenzol, 31.52 Per Cent Chlorine.

Weight taken. grm.	Per cent of chlorine found.	Per cent error.
0 4036	31.62	+0.1
0.3533	31.46	−0 06
0 4181	31.51	−0.01
0 3278	31.44	−0 08
0.4087	31 48	−0 04
0 4245	31 53	+0 01
0.3324	31.52	0.0

Hexachlorbenzol, 74.71 Per Cent Chlorine.

Weight taken. grm.	Per cent of chlorine found.	Per cent error.
0.1301	75.39	+0.68
0.1190	75.15	+0.44

Brombenzol, 50.92 Per Cent Bromine.

Weight taken. grm.	Per cent of bromine found.	Per cent error.
0.3976	51.29	+0.37
0 3921	51.31	+0.39
0.3928	51.31	+0.39

p-Chloraniline, 27.81 Per Cent Chlorine.

Weight taken. grm.	Per cent of chlorine found.	Per cent error.
0.3081	27.93	+0.12
0.3456	28.00	+0.19

As the hexachlorbenzol, brombenzol, and p-chloraniline used were commercial products only once redistilled or recrystallized, the positive errors observed may be largely due to impurities.

The Direct Determination of Chlorine in Mixtures of Alkali Chlorides and Iodides.

Fleischer justifies his use of hydrochloric acid as a standard in alkalimetric processes by his observation that decinormal solutions of this acid, and even solutions of twice the strength (7.3 grm. to the liter), do not yield after ten minutes' boiling enough acid to redden blue litmus paper held in the steam. Hydriodic acid behaves similarly to hydrochloric acid in the matter of volatilizing from aqueous solutions; but to the decomposing action of oxidizing agents it is far more sensitive. Gooch and Mar * have studied the conditions under which hydriodic acid may be completely broken up and iodine removed from the solution by vaporization while the hydrochloric is retained without appreciable loss. It was found that the volatility of hydrochloric acid from boiling mixtures of potassium chloride and sulphuric acid, dependent upon the concentration of the sulphuric acid as well as upon that of the chloride, is inconsiderable for 1 grm. of potassium chloride and 5 cm.^{3} of concentrated sulphuric acid, or 10 cm.^{3} of the [1 : 1] acid, in 200 cm.^{3} of the boiling solution. At 300 cm.^{3} the dilution is sufficient to guarantee security against volatilization of hydrochloric acid under the conditions named. To remove the iodine of potassium iodide without volatilizing hydrochloric acid use is made, for one method, of a ferric salt according to the reaction of Duflos,

$$2\,HI + Fe_2(SO_4)_3 = 2\,FeSO_4 + H_2SO_4 + I_2,$$

and, for a second method, the action of a nitrite or of nitrogen trioxide,

$$2\,KNO_2 + H_2SO_4 + 2\,HI = K_2SO_4 + 2\,H_2O + 2\,NO + I_2.$$

Use of Ferric Sulphate. In tests by the first method it was found that from a volume of 300 cm.^{3} containing 10 cm.^{3} of sulphuric acid [1 : 1], 5 grm. of ferric alum and 0.005 grm. of potassium iodide, every trace of iodine had disappeared so completely after five minutes' boiling that nitrous acid and chloroform collected no color in the cooled liquid, but when the amount of potassium iodide was increased to 1 grm. iodine was found in considerable amount after boiling for an hour with occasional replacing of water so that the volume should not decrease much

* F. A. Gooch and F. W. Mar, Am. Jour. Sci., [3], xxxix, 293.

below 300 cm.³. The reaction is, therefore, plainly reversible under the conditions. When, however, a sufficient amount of nitric acid is added to restore the iron to the ferric state, boiling brings about complete liberation of the iodine. In dilute solutions the amount of nitric acid necessary to oxidize a fixed quantity of ferrous salt is greater than in concentrated solutions. Thus, while 0.1 cm.³ of strong nitric acid should be more than enough to reoxidize the iron reduced by 1 grm. of potassium iodide, when the full oxidizing action is brought out, it is necessary to add to these dilute solutions about 2 cm.³ of the acid to complete the action satisfactorily. The separation of iodine and the estimation of chlorine according to this process may be summarized as follows: To the solution of the alkali chloride and iodide diluted to about 400 cm.³, in an Erlenmeyer flask capable of containing a liter, are added 10 cm.³ of sulphuric acid of half strength, with 2 grm. of ferric sulphate (either in the form of iron alum, or ferrous sulphate oxidized in concentrated solution by about 0.3 cm.³ of nitric acid) and 3 cm.³ of nitric acid. A trap * is hung in the neck of the flask, and the liquid is boiled until the steam which escapes no longer gives to red litmus paper, after two minutes' exposure, the characteristic gray blue due to traces of iodine. Then 1 cm.³ more of nitric acid is added and the test for iodine again made. When no iodine is found in the

Distillation with Ferric Sulphate: Determination of Chloride in the Residue.

H_2SO_4 [1 : 1].	$Fe_2(SO_4)_3$.	HNO_3.	KCl = HCl.		KI.	Initial volume.	Final volume.	Time in minutes.	AgCl = HCl found.		Error.
cm.³	grm.	cm.³	grm.	grm.	grm.	cm.³	cm.³		grm.	grm.	grm.
10	2*	..	0.4960	0.2425	..	400	300	40	0.9536	0.2425	0.0000
10	2*	..	0.4970	0.2429	..	400	300	40	0.9534	0.2624	−0.0005
10	2*	.	0.4942	0.2416	..	400	300	30	0.9509	0.2418	+0.0002
10	2*	2	0.4969	0.2429	..	400	300	30	0.9559	0.2431	+0.0002
10	2*	3	0.4956	0.2423	1	400	350	30	0.9546	0.2428	+0.0005
10	2*	3	0.4969	0.2429	1	400	350	23	0.9662	0.2432	+0.0003
10	2†	3	0.4949	0.2419	1	400	300	27	0.9523	0.2422	+0.0003
10	2†	5	0.4970	0.2429	1	400	250	55	0.9559	0.2431	+0.0002
10	2†	5	0.4955	0.2422	1	400	300	30	0.9524	0.2422	0.0000
10	2†	5	0.4967	0.2428	1	400	300	33	0.9546	0.2428	0.0000
10	2†	6	0.4964	0.2427	1	400	300	30	0.9550	0.2429	+0.0002

* The iron was added in the form of iron alum.
† The iron was added as $FeSO_4$ oxidized by HNO_3.

* See Fig. 6, page 6.

escaping vapor, silver nitrate is added in excess to the contents of the flask, the precipitate is settled, collected in a perforated crucible on asbestos, washed, dried, and weighed as silver chloride.

Tests of the method, with determinations in blank — that is, experiments from which the iodine was purposely omitted — are detailed in the tabular statement.

The Nitrite Method. With pure sodium nitrite at hand there is probably no serious objection to introducing that substance directly into the solution, but if impurities are present it is desirable to generate the gas outside the solution. For a generator, two straight drying tubes are connected by a rubber tube and set up after the fashion of the von Babo generator, and the rapidity of the current is regulated to a rate of five or six bubbles to the second by changing the relative elevation of the generator tubes. The iodine separates immediately upon the introduction of the nitrous fumes and escapes upon boiling, leaving the solution colorless in a very short time. The litmus test must, however, be relied upon to indicate the removal of the iodine. According to the method, as developed, the solution of the chloride and iodide contained in an Erlenmeyer flask is diluted to 400 cm.^{3}, 10 cm.^{3} of sulphuric acid of half strength are added, and the gas from 2 grm. of sodium nitrite acted upon by dilute sulphuric acid (generated in simple apparatus, such as is described above) is passed with reasonable rapidity into the agitated solution. The liquid is boiled until colorless, and still further until litmus paper placed in the steam gives no reaction for iodine after an exposure of two minutes. The contents of the flask are

Distillation with Nitrite: Determination of Chloride in the Residue.

H_2SO_4 [1 : 1]. cm.^{3}	$NaNO_2$ used in generator. grm.	KCl. = grm.	HCl. grm.	KI. grm.	Initial volume. cm.^{3}	Final volume. cm.^{3}	Time in minutes.	AgCl found. grm.	HCl found. grm.	Error. grm.
10	2	0 4953	0.2421	1	400	350	20	0 9524	0.2422	+0.0001
10	2	0 4975	0.2432	1	400	350	16	0 9573	0.2434	+0.0002
10	2	0 4956	0.2423	1	300	250	15	0.9530	0.2423	0 0000
10	2	0 4973	0.2431	1	300	250	15	0.9550	0.2429	−0 0002
10	2	0 4964	0.2427	1	300	250	15	0.9550	0.2429	+0 0002
10	2	0.4969	0.2429	1	300	250	15	0 9567	0.2433	+0.0004

treated with silver nitrate, and the precipitated chloride is settled, collected on asbestos in the perforated crucible, washed, dried and weighed.

The results of test experiments are given in the preceding table.

The Direct Determination of Bromine (and Chlorine) in Mixtures of Alkali Bromides (and Chlorides) with Iodides.

The methods elaborated by Gooch and Mar * for the direct determination of chlorine in mixtures of alkali chlorides and iodides have been studied by Gooch and Ensign† with a view to similar application in the determination of bromine in mixtures of alkali bromides and iodides. The conditions found suitable in the separation of chlorine from iodine prove to be inappropriate to the separation of bromine from iodine. The use of a ferric salt to eliminate iodine was not found to be practicable under any conditions. Even at the high dilution ranging from 650 cm.^{3} to 500 cm.^{3} bromine was likewise set free even when the concentration of the sulphuric acid present was very low.

On the other hand the nitrite process, only fairly successful when the sulphuric acid present is restricted to 5 cm.^{3} of the [1 : 1] acid in a final volume of 500 cm.^{3}, is established as trustworthy when the sulphuric acid present is held within the limits of 2 cm.^{3} to 4 cm.^{3} of the [1 : 1] mixture. When the quantity of sulphuric acid is still further diminished there is evidently a slight tendency to show an apparent excess of bromide, due in all probability to the retention of a little combined iodine in the solution. The best proportion for practical use is probably 3 cm.^{3} of the half and half acid in an initial volume not less than 600 cm.^{3}, and this proportion proves to be applicable to the separation of iodine from an iodide associated with chloride as well as bromide.

The method may be briefly summarized as follows: The neutral solution containing the bromide and iodide is diluted to 600 cm.^{3} or 700 cm.^{3} (instead of 400 cm.^{3}, which was found to be a sufficient dilution in the case of the separation of chlorine from iodine); 1 cm.^{3} to 1.5 cm.^{3} of strong sulphuric acid, or, better, 2 cm.^{3} to 3 cm.^{3} of the [1:1] mixture (instead of the 10 cm.^{3}

* See pages 449, 451.

† F. A. Gooch and J. R. Ensign, Am. Jour. Sci., [3], xl, 145.

employed in the chlorine separation) are added, a sufficient amount of pure sodium or potassium nitrite is introduced (or, if it is preferred, the gas generated by the action of dilute sulphuric acid upon the ordinary nitrite and introduced from the outside); and the liquid is boiled, after trapping the flask, until the color has vanished and the escaping steam no longer gives to red litmus paper the color characteristic of iodine. The residual liquid is treated with an excess of silver nitrate and the precipitated bromide filtered off, washed, dried and weighed. The process of boiling need not extend beyond half an hour, or a little more, and care should be taken that the volume of the liquid shall never be less than 500 cm.^{3}. The process has been tested for quantities of the potassium bromide and iodide not much larger than 0.5 grm. each. The presence of 0.5 grm. of potassium chloride does not affect the sharpness of the separation. The results are given below:

Separation of Iodine from Bromine.

H_2SO_4 [1 : 1]	KI.	$NaNO_2$ in the liquid.	KBr = HBr taken.		Initial volume.	Final volume.	Time in minutes.	AgBr = HBr found.		Error in HBr.
cm.^{3}	grm.	grm.	grm.	grm.	cm.^{3}	cm.^{3}		grm.	grm.	grm.
3	0.5	0.35	0.5508	0.3745	650	500	30	0.8689	0.3744	−0.0001
3	0.5	0.35	0.5513	0.3747	650	500	30	0.8694	0.3746	−0.0001
3	0.5	0.35	0.5513	0.3747	650	500	30	0.8699	0.3748	+0.0001
3	0.5	0.35	0.3005	0.2042	650	500	30	0.4746	0.2045	+0.0003
3	0.5	0.35	0.2759	0.1875	650	500	30	0.4358	0.1878	+0.0003
3	0.5	1.75	0.5513	0.3747	650	500	30	0.8705	0.3750	+0.0003
3	0.5	1.75	0.5510	0.3746	650	500	30	0.8707	0.3751	+0.0005

H_2SO_4 [1 : 1]	KI.	$NaNO_2$ used in generator.	KBr = HBr taken.		Initial volume.	Final volume.	Time in minutes.	AgBr = HBr found.		Error in HBr.
cm.^{3}	grm.	grm.	grm.	grm.	cm.^{3}	cm.^{3}		grm.	grm.	grm.
2	0.5	2	0.5366	0.3647	650	500	30	0.8478	0.3654	+0.0007
2	0.5	2	0.5369	0.3650	650	500	30	0.8472	0.3651	+0.0001
2	0.5	2	0.5515	0.3747	650	500	30	0.8687	0.3742	−0.0005
3	0.5	2	0.5371	0.3652	650	500	30	0.8459	0.3644	−0.0008
3	0.5	2	0.5365	0.3647	650	500	30	0.8465	0.3647	0.0000
3	0.5	2	0.5368	0.3649	650	500	30	0.8486	0.3656	+0.0007
3	0.5	2	0.5364	0.3646	650	500	30	0.8471	0.3650	+0.0004
3	0.5	2	0.5505	0.3742	650	500	30	0.8690	0.3744	+0.0002
3	0.5	2	0.0576	0.0391	650	500	30	0.0915	0.0394	+0.0003
3	0.5	2	0.0552	0.0375	650	500	30	0.0883	0.0380	+0.0005

Separation of Iodine from Bromine and Chlorine.

H_2SO_4 [1 : 1].	KI.	$NaNO_2$	KBr.	HCl.	Theory for AgCl+ AgBr.	Found AgCl+ AgBr.	Error in silver salt.	Error calculated as HBr.	Error calculated as HCl.
cm.³	grm.	grm.	grm.	grm.	grm.	grm.	grm.	grm.	grm.
3	0.5	0.35	0 5517	0.4981	1.8280	1.8262	−0 0018	−0 0008	−0 0006
3	0 5	0 35	0 5511	0.4980	1.8268	1.8253	−0 0015	−0 0005	−0 0004

The Application of Iodic Acid to the Analysis of Iodides.

Iodic acid may be easily and completely reduced by an excess of hydriodic acid with the liberation of iodine according to the equation:

$$HIO_3 + 5\,HI = 3\,I_2 + 3\,H_2O.$$

To apply this reaction to the quantitative estimation of iodic acid, it is only necessary to add to the free iodic acid or soluble iodate, in suitably concentrated solution, an excess of a soluble iodide, to acidify — best with dilute sulphuric acid — and to titrate with sodium thiosulphate the iodine thus set free, one-sixth of the iodine found being credited to the iodic acid.

It has been shown by Riegler * that this reaction may be also applied to the quantitative estimation of iodides, the iodine set free upon the addition of a known excess of iodic acid to the iodide solution being removed by petroleum ether, and the residual iodic acid determined as described above.

Gooch and Walker † have studied the limit of applicability of this reaction and have developed a direct method for the quantitative estimation of iodides, dependent upon the action of iodic acid or an iodate in the presence of free sulphuric acid, neutralization of the solution by means of an acid carbonate, and titration of the free iodine by arsenious acid — five-sixths of the iodine thus found being credited to the iodide to be estimated. It has been found that by fulfilling certain necessary conditions the proposed method is entirely successful, so far as concerns the estimation of iodine in iodide solutions free from large amounts of chlorides or bromides.

The degree of dilution at the time when the mixture of iodide and iodate is acidified has an important influence upon the com-

* Zeit. anal. Chem., xxxv, 305.

† F. A. Gooch and C. F. Walker, Am. Jour. Sci., [4], iii, 293.

pleteness of the reaction. Thus, the mean error of test determinations in which the volume at the time of the reaction does not exceed 150 cm.^{3} is practically nothing, while the errors at volumes of 300 cm.^{3} and 500 cm.^{3} amount to 0.0016 grm. and 0.0028 grm. respectively. The doubling of the amount of sulphuric acid used in acidifying does not increase the amount of iodine liberated at the highest dilution. The plain inference is that the interaction between the iodide and iodate should be brought about in a volume of liquid not much exceeding 150 cm.^{3}.

The apparatus employed is a reaction bottle * of 500 cm.^{3} or 1000 cm.^{3} capacity, according to requirements, with stopcock and thistle-tube fused to the inlet tube and a Will and Varrentrapp absorption trap sealed to the outlet. The iodide for the test is drawn from a burette into the bottle and carefully washed down, potassium iodate in excess of the amount theoretically necessary is added, and the volume of the liquid is adjusted to 150 cm.^{3}. The stopper with the thistle-tube and trap is put in place and the trap is half filled by means of a pipette with a 5 per cent solution of potassium iodide. Five cubic centimeters of [1 : 3] sulphuric acid are added through the thistle-tube and washed down, the stopcock is closed, and the solution gently agitated, if necessary, to insure a complete separation of iodine. Potassium bicarbonate in saturated solution to an amount about 10 cm.^{3} in excess of that required to neutralize 5 cm.^{3} of dilute [1 : 3] sulphuric acid is poured into the thistle-tube and allowed to flow into the bottle slowly enough to avoid a too violent evolution of gas. The stopcock is closed and the solution agitated by giving to the bottle a rotary motion, at the same time keeping the bottom pressed down upon the work-table, to prevent a possible splashing of the iodide out of the trap into the acid solution. When the neutralization of the solution has been completed, the bottle is shaken until the last trace of violet vapor has been absorbed in the liquid. The greater part of the solution in the trap is then run back into the bottle, the stopper is removed, and the tube and trap are carefully washed, the washings being added to the bulk of the solution. Decinormal arsenious acid is introduced from a burette to the bleaching point, starch emulsion added, and the solution titrated back with decinormal iodine (usually only a few drops) to coloration.

* Shown in Fig. 7, page 6.

The results of experiments made in the manner described upon portions of a solution of potassium iodide standardized by the arsenate method * are given in the accompanying table.

Analysis of Pure Potassium Iodide.

KI taken. grm.	KI found. grm.	Error. grm.
0 0814	0 0816	+0.0002
0.0814	0.0813	−0.0001
0.0814	0 0805	−0 0009
0.0815	0.0809	−0 0006
0 0814	0.0808	−0 0006
0 0814	0.0806	−0 0008
0.0814	0.0812	−0 0002
0.1628	0 1624	−0.0004
0.1628	0 1617	−0 0011
0.1628	0.1621	−0.0007
0.1628	0.1619	−0 0009
0.1628	0.1624	−0 0004
0.1628	0.1621	−0.0007
0.1628	0.1626	−0 0002
0.2442	0.2451	+0 0009
0.2442	0.2442	0.0000
0.2442	0 2439	−0.0003
0.3256	0.3258	+0.0002
0 3256	0.3256	0.0000
0 3256	0.3258	+0.0002
0 3256	0.3272	+0.0016
0.3256	0 3256	0.0000
0.4071	0.4076	+0.0005
0.4071	0.4080	+0.0009
0 4071	0.4073	+0.0002

The presence of any considerable amount of chloride or bromide, resulting no doubt in the formation of iodine chloride or iodine bromide, is prejudicial to the accuracy of the process. This is shown in the table following:

Effects of Chloride and Bromide.

KI taken. grm.	KI found. grm.	Error. grm.	NaCl taken. grm.	KBr taken. grm.
0 0772	0 0795	+0 0023	0 2	...
0 0772	0 0784	+0 0012	0 2	...
0 0771	0.0823	+0 0052	0 5	...
0 0773	0.0819	+0 0046	0.5	...
0.1544	0 1588	+0 0044	0 5	...
0 1544	0 1590	+0 0046	0.5	...
0 0772	0 0802	+0 0030	...	0.2
0 0773	0.0853	+0 0080	...	0 2
0 0772	0.0873	+0 0101	...	0 5
0 0772	0 0861	+0 0089	...	0 5
0 1544	0.1646	+0 0102	...	0.5
0.1543	0 1626	+0.0083	..	0.5

* See page 457.

It is plain that the value of the process in the determination of iodine in an iodide is restricted of necessity to those cases in which it is known that chlorides or bromides are not present to any considerable extent. For determining the standard of a solution of nearly pure potassium iodide, employed in so many laboratory processes, it is useful.

The Iodometric Determination of Iodine in Haloid Salts.

The determination of iodine in a mixture of alkali chloride, bromide and iodide has been made the subject of investigation by Gooch and Browning.* In this work it was shown that the iodine of the iodide may be all liberated, under defined conditions, by the combined action of an arsenate and sulphuric acid, and its amount registered quantitatively by the amount of arsenious oxide produced.† Under similar conditions, the presence of as much as 0.5 grm. of sodium chloride brings about no formation of arsenious oxide, but does induce a loss of that substance by volatilization as arsenic chloride, proportionate to the amounts of both these substances. The effect of potassium bromide is to produce trifling reduction of arsenic acid without volatility. Due correction of the amounts of iodine indicated by determination of the arsenious oxide in the residue may be made by adding to the indicated amount 0.008 of the product of the weight of chlorine present in chlorides by the weight of iodine, and subtracting 0.0024 of the weight of bromine in bromides.

The mode of proceeding in the determination of iodine in a mixture of alkali chlorides, bromides and iodides, according to this method, may be briefly summarized as follows:

The substance (which should not contain of chloride more than an amount corresponding to 0.5 grm. of sodium chloride, nor of bromide more than corresponds to 0.5 grm. of potassium bromide, nor of iodide much more than the equivalent of 0.5 grm. of potassium iodide) is dissolved in water in an Erlenmeyer beaker of 300 cm.^{3} capacity, and to the solution are added 2 grm. of dihydrogen potassium arsenate dissolved in water, 20 cm.^{3} of a mixture of sulphuric acid and water in equal volumes, and enough water to increase the total volume to 100 cm.^{3}, or a little

* F. A. Gooch and P. E. Browning, Am. Jour. Sci., [3], xxxix, 188; xlv, 334.
† For the reaction, see pages 291, 463.

more. A platinum spiral is introduced, a trap made of a straight two-bulb drying-tube cut off short is hung with the larger end downward in the neck of the flask,* and the liquid is boiled until the level reaches the mark put upon the flask to indicate a volume of 35 cm.³. Great care should be taken not to press the concentration beyond this point on account of the double danger of losing arsenious chloride and of setting up reduction of the arsenate by the bromide. On the other hand, though 35 cm.³ is the ideal volume to be attained, failure to concentrate below 40 cm.³ introduces no appreciable error. The liquid remaining is cooled and nearly neutralized by sodium hydroxide (ammonia is not equally good), neutralization is completed by hydrogen potassium carbonate, an excess of 20 cm.³ of the saturated solution of the latter is added, and the arsenious oxide in solution is titrated by standard iodine in the presence of starch.

Reduction of Arsenate and Determination of Arsenious Oxide Produced.

H_2SO_4 [1 : 1]. cm.³	H_2KAsO_4. grm.	NaCl.	KBr.	Final volume. cm.³	Theory for iodine. grm.	Iodine found. grm.	Error found. grm.	Error corrected.
20	2	...	.	35	0 4080	0.4079	−0 0001	−0.0001
20	2	...	...	35	0 4091	0.4086	−0.0005	−0.0005
20	2	.	.	35	0.4083	0.4086	+0.0003	+0.0003
20	2	...	...	35	0.0400	0.0396	−0 0004	−0.0004
20	2	...	...	35	0 0400	0 0391	−0 0009	−0.0009
20	2	..	...	35	0 0400	0 0400	0.0000	0.0000
20	2	...	.	35	0 0400	0 0401	+0.0001	+0.0001
20	2	...	..	35	0 0040	0.0037	−0 0003	−0.0003
20	2	...	..	35	0 0040	0.0038	−0.0002	−0.0002
20	2	0.5	...	35	0 4077	0 4066	−0.0011	−0.0001
20	2	0.5	...	35	0 4082	0.4073	−0.0009	+0.0001
20	2	0.5	...	35	0 4086	0 4073	−0 0013	−0.0003
20	2	0.5	...	35	0 0400	0.0402	+0.0002	+0 0001
20	2	0.5	...	35	0 0400	0.0395	−0 0005	−0.0004
20	2	0.5	...	35	0.0040	0.0037	−0 0003	−0.0002
20	2	0.5	...	35	0 0040	0.0037	−0.0003	−0.0002
20	2	...	0.5	35	0 4082	0.4092	+0 0010	+0.0002
20	2	...	0.5	35	0 4138	0.4136	−0.0002	−0.0010
20	2	...	0.5	35	0.4083	0.4099	+0.0016	+0.0008
20	2	...	0.5	35	0.0400	0.0410	+0.0010	+0.0002
20	2	...	0.5	35	0.0400	0.0404	+0.0004	−0.0004
20	2	...	0.5	35	0.0040	0.0048	+0.0008	0.0000
20	2	...	0.5	35	0.0040	0.0049	+0 0009	+0 0001
20	2	0.5	0.5	35	0 4087	0.4083	−0.0004	−0.0002
20	2	0.5	0.5	35	0 4112	0.4111	−0.0001	+0.0001
20	2	0.5	0.5	35	0.4083	0.4079	−0.0004	−0.0002

* See Fig. 6, page 6.

With ordinary care the method is rapid, reliable and easily executed, and the error is small. In analyses requiring extreme accuracy all but accidental errors may be eliminated from the results by adding (algebraically) to the amount of iodine indicated an amount

$$i = (0.008 \times \text{wt. Cl} \times \text{wt. I}) - (0.0024 \times \text{wt. Br}).$$

Results obtained by this process are given in uncorrected and corrected forms.

Results * obtained in a comparison of titrations of arsenious oxide in the residue with estimations of the iodine expelled under the prescribed conditions and collected in the distillate show close agreement.

Comparison of Arsenious Oxide in Residue and Iodine Expelled.

Iodine taken in form of KI. grm.	Iodine found from As_2O_3 in residue. grm.	Iodine found in distillate by As_2O_3. grm.	Iodine found in distillate. by $Na_2S_2O_3$ grm.	Error in residue. grm.	Error in distillate. grm.
0.4054	0.4052	......	......	−0.0002	......
0.4057	0.4055	......	......	−0.0002	......
0.4054	0.4052	......	......	−0.0002	......
0.4054	0.4052	......	......	−0.0002	......
0.4042	0.4046	0.4046	......	+0.0004	+0.0004
0.4050	0.4052	0.4040	......	+0.0002	−0.0010
0.4050	0.4052	......	0.4039	+0.0002	−0.0011
0.4058	0.4052	......	0.4051	−0.0006	−0.0007
0.4054	0.4046	......	0.4051	−0.0008	−0.0003
0.4042	0.4046	......	0.4039	+0.0004	−0.0003
0.4055	0.4052	......	0.4057	−0.0003	+0.0002

The Determination of the Halogens by the Electrolytic Reduction of Silver in Mixed Silver Salts.

Methods for the estimation of silver in mixed silver salts of the halogens have been based by Gooch and Fairbanks † on the collection of the precipitated salts upon a perforated platinum disk covering the asbestos felt in a perforated crucible for drying and weighing; the fusion of the salts in contact with the platinum disk to give electrical conductivity; reduction of the

* Am. Jour. Sci., [3], xlv, 334.

† F. A. Gooch and Charlotte Fairbanks, Am. Jour. Sci., [3], l, 27.

fused salts by making them the cathode in the electrolysis of a suitable liquid; and the washing, igniting and weighing of the reduced silver.

Silver Chloride and Silver Bromide. Silver chloride and bromide are precipitated; collected, washed, dried at 150° C., and weighed in the filtering crucible provided as usual with a layer of asbestos but in this case covered with the perforated platinum disk. The cap is put in place, the crucible set upon an anvil to keep it cool and prevent soaking of the asbestos with fused silver salts, and the salts are fused with a blowpipe flame carefully directed upon the mass from above. A rubber band is adjusted to cover the junction between cap and crucible. The crucible is nearly filled with a 10 per cent solution of oxalic acid in 25 per cent alcohol and the current passed in the usual manner, the crucible serving as the negative electrode. When the reduction is judged to be complete the band and cap are removed, the crucible set upon the pump, and filtration of the liquid and washing of the residue carried out as usual. Finally the crucible, cap and residue are ignited at a very low red heat and weighed. The entire treatment is repeated until the constant weight of the residue shows that the reduction is complete. Results of this procedure are given in the table.

Electrolytic Reduction of Silver Chloride and Silver Bromide.

AgCl taken. grm.	AgBr taken. grm.	Ag calculated. grm.	Ag found. grm.	Error. grm.
1.0608	...	0.7985	0.7990	+0 0005
1.4380	...	1.0823	1.0823	0 0000
0.9998	....	0.7525	0.7522	−0 0003
...	0.9959	0.5721	0.5723	+0.0002
....	0.9979	0.5731	0 5732	+0 0001
1 0044	0 4988	1.0426	1.0422	−0 0004
0 4933	0.4966	0 6559	0.6568	+0 0009

The manipulation of the method is very easy, and the results show that it is capable of yielding accurate results. In the experiments recorded the current ranged from 0.5 to 0.25 amperes, and for convenience the process was continued over night, though the reduction of amounts such as were treated is usually complete in six or seven hours.

Silver Iodide by Itself and in Mixture with Silver Chloride or Silver Bromide.

This process which works so well with the mixture of chloride and bromide is not applicable to the reduction of silver iodide or to mixtures containing it. Experiment proved that the iodine set free in the electrolysis works over and over again upon the spongy silver, constantly regenerating silver iodide to a greater or less degree. The liberated iodine may, however, be destroyed, without introducing anything objectionable, by conducting the electrolysis in a mixture of ammonium acetate, alcohol, and aldehyde, — made by neutralizing two parts by volume of ordinary (40 per cent) acetic acid with ammonia, adding one part of ammonia, one part of alcohol, and one part of aldehyde (75 per cent). Such a solution works very well on the whole, but as the reduction progresses it frequently happens that a deposit of white ammonium iodate forms upon the anode and introduces too great resistance to the current. This deposit of iodate is, however, easily removed from the electrode by dipping it into hot water. Whenever the solution is so exhausted that free iodine begins to appear the liquid should be carefully decanted and replaced by fresh solution; and before the operation is ended the decanted solutions and the washings of the electrode should be filtered through the crucible, and the residue submitted again to the action of the current, to make it certain that loosened particles of silver or silver salt possibly poured off or removed on the electrode shall not be lost

Electrolytic Reduction of Silver Chloride, Bromide and Iodide.

AgCl taken. grm.	AgBr taken. grm.	AgI taken. grm.	Ag calculated. grm.	Ag found. grm.	Error. grm.
0.4779	.. .	. .	0 3596	0.3591	−0.0005
0 6096	.....	.....	0 4588	0 4591	+0.0003
0 6774	. .	.. .	0 5098	0 5099	+0 0001
.. . .	0 9969	. .	0 5727	0 5726	−0.0001
.....	1 3703	... .	0 7872	0.7875	+0 0003
. . .	.... .	1 0613	0 4878	0 4877	−0.0001
. .	....	1 0621	0 4882	0 4875	−0.0007
...	.....	1.0140	0 4661	0.4662	+0.0001
	....	1.2012	0 5521	0.5530	+0 0009
...	.. ...	1.5031	0.6910	0 6914	+0.0004
0.5035	0.4984	.. ..	0.6653	0 6653	0.0000
1 0020	0 9998	..... .	1 3285	1 3283	−0.0002
0.4939	. ..	0.6561	0 6734	0.6733	−0.0001
.. .	0.5000	0.5304	0 5310	0.5316	+0.0006

finally. The necessity of keeping the process under occasional supervision renders it undesirable to continue the action over night. The formation of gummy carbonaceous matter not easily removed without the application of a degree of heat dangerous to platinum in contact with silver was noted in some cases of prolonged action without attention. Many of the experiments recorded in the preceding table were completed within seven hours with a current not exceeding 0.5 ampere.

These results show that the process affords an accurate reduction of the chloride, bromide, and iodide of silver and mixtures of these salts. When the problem concerns the reduction of the chloride and bromide only, preference is to be given to the simpler process of reduction in alcoholic oxalic acid.

The Estimation of Chlorates by Reduction with Ferrous Sulphate.

Peters and Moody * have pointed out that solutions of ferrous salts which have been allowed to stand until all dissolved oxygen has produced its effect are acted upon with extreme slowness by atmospheric oxygen. Phelps † has made use of this mode of preparing ferrous sulphate to test the process suggested by Corot,‡ as applied to chlorates. The results of experiments in which the chlorate was treated with an excess of approximately $n/5$ ferrous sulphate and 15 cm.^{3} of sulphuric acid [1 : 3] in a flask trapped to prevent mechanical loss are given below.

Reduction by Ferrous Sulphate and Titration of Excess.

$KClO_3$ taken.	Oxygen value of ferrous salt taken.	Oxygen value of ferrous salt found.	Error on $KClO_3$.
grm.	grm.	grm.	grm.
0.0500	0.02756	0 00814	−0.0004
0.0500	0.02739	0.00781	0 0000
0.1000	0.04934	0 01024	−0.0002
0.1000	0.04951	0 01043	−0 0002
0.2000	0.09086	0 01247	+0 0002
0.2000	0.09078	0.01277	−0 0008
0.5000	0.20552	0.00993	−0 0006
0.5000	0.20543	0.00980	−0.0005

* Am. Jour. Sci., [4], xii, 369; see also page 371.

† I. K. Phelps, Am. Jour. Sci., [4], xvii, 201.

‡ Compt. rend., cxxii, 449.

The mixture was brought to the boiling point, cooled to room temperature by running water, diluted to a volume of 600 cm.[3] and titrated with potassium permanganate after addition of 1 grm. to 2 grm. of manganous chloride.

The Iodometric Estimation of Chlorates.

As has been shown,* under conditions properly controlled, arsenic acid in excess is capable of expelling the iodine from hydriodic acid at the boiling temperature of the solution, being itself reduced correspondingly according to the equation

$$H_3AsO_4 + 2\,HI = H_3AsO_3 + H_2O + I_2.$$

On cooling the liquid remaining after such treatment, and neutralizing, the arsenious oxide produced in the reaction may be reoxidized iodometrically in the usual manner, the iodine added to accomplish this purpose being the exact measure of the iodine originally present as hydriodic acid and expelled from the acid solution during the process of boiling.

Gooch and Smith † have found that in a mixture of chloric, hydriodic and arsenic acids the mutual action of the chloric and hydriodic acids takes place according to the equation

$$HClO_3 + 6\,HI = HCl + 3\,H_2O + 3\,I_2,$$

and goes steadily to completion, and that when this effect is accomplished the action of the arsenic acid in liberating iodine from the residual hydriodic acid, and in registering by its own reduction the amount of iodine thus set free, begins. The reaction affords, therefore, a method for the estimation of chlorates which consists in heating the chlorate, in acid solution and under conditions otherwise appropriate, with a known amount of potassium iodide, somewhat in excess of that theoretically equivalent to the chlorate, and in presence of an excess of arsenic acid, the arsenious oxide produced in the process being determined iodometrically and serving to measure the amount of iodide left undecomposed by the chlorate. The difference between the amount of iodide left undecomposed and that originally introduced is the measure of the chlorate entering into the reaction.

* See page 457.

† F. A. Gooch and C. G. Smith, Am. Jour. Sci., [3], xlii, 220.

In the practical application of this process a solution of approximately decinormal potassium iodide is standardized according to the method to which reference has been made. Into an Erlenmeyer flask capable of holding 300 cm.^{3} are put a portion of the iodide solution, 2 grm. of pure potassium arsenate, 20 cm.^{3} of diluted sulphuric acid [1 : 1] and enough water to make the entire volume a little more than 100 cm.^{3} A platinum spiral is introduced to secure quiet boiling, a trap made of a straight two-bulbed drying tube cut short is hung with the larger end in the neck of the flask,* and the liquid is boiled until the level has reached a mark upon the flask indicating a volume of 35 cm.^{3}, experience having shown that this degree of concentration is sufficient and that it is best not to exceed it. The liquid remaining is cooled and nearly neutralized by sodium hydroxide, acid potassium carbonate is added to alkalinity, 20 cm.^{3} of a saturated solution of this salt are added in excess, and the arsenious oxide in solution is titrated by standardized decinormal iodine in presence of starch. The iodine added in the reoxidation of the arsenious oxide is taken as the exact equivalent of the iodine expelled in boiling.

Then a weighed amount of the chlorate is put in an Erlenmeyer flask, a portion of the potassium iodide solution, containing of the iodide a weight amounting at least to twelve and a half times the weight of the chlorate ion (ClO_3), is introduced, and the mixture is treated precisely as is the iodide alone in the process of standardization.

The difference between the iodine added in this titration and the amount similarly added in the titration of the standardization process is the measure of the chlorate according to the reaction

$$KClO_3 + H_2SO_4 + 6\,HI = KHSO_4 + HCl + 3\,H_2O + 3\,I_2.$$

A large excess of iodide over the amount equivalent to the chlorate is unnecessary. The amount of chloride produced from the maximum weight of chlorate which may be handled conveniently in this process is too small to call for any correction † in the indicated amount of residual arsenious oxide. Results are shown in the table.

* See Fig. 6, page 6.

† See page 459.

Reduction by Iodide: Determination of Excess.

H_2SO_4 [1 : 1] taken.	H_2KAsO_4 taken.	KI = I taken.		Iodine corresponding to As_2O_3 reduced.	Difference between iodine taken and iodine added to oxidize As_2O_3.	$KClO_3$ taken.	$KClO_3$ equivalent to difference between I in KI and I added.	Error.
cm.[3]	grm.	grm.	grm.	grm.	grm.	grm.	grm.	grm.
20	2	2.0092	1.5356	0.2962	1.2394	0.2000	0.2000	0.0000
20	2	2.0092	1.5356	0.2973	1.2383	0.2000	0.1999	−0.0001
20	2	1.0380	0.7934	0.0570	0.7364	0.1185	0.1188	+0.0003
20	2	0.8706	0.6654	0.0435	0.6219	0.1000	0.1004	+0.0004
20	2	0.8706	0.6654	0.0429	0.6225	0.1000	0.1005	+0.0005
20	2	0.8706	0.6654	0.0435	0.6219	0.1000	0.1004	+0.0004
20	2	0.8706	0.6654	0.0435	0.6219	0.1000	0.1004	+0.0004
20	2	0.5023	0.3839	0.3208	0.0631	0.0100	0.0102	+0.0002
20	2	0.5032	0.3839	0.3201	0.0638	0.0100	0.0103	+0.0003
20	2	0.2009	0.1536	0.0889	0.0647	0.0100	0.0105	+0.0005
20	2	0.2009	0.1536	0.0903	0.0633	0.0100	0.0102	+0.0002
20	2	0.2009	0.1536	0.0903	0.0633	0.0100	0.0102	+0.0002
20	2	0.1339	0.1024	0.0405	0.0619	0.0100	0.0100	0.0000
20	2	0.1004	0.0768	0.0157	0.0611	0.0100	0.0099	−0.0001
20	2	0.1004	0.0768	0.0182	0.0586	0.0100	0.0095	−0.0005

The Detection of Alkali Perchlorates Associated with Chlorides, Chlorates and Nitrates.

In experimenting at high temperatures with mixtures of alkali perchlorates with salts of the halogens, Gooch and Kreider * have succeeded in developing a simple and delicate method of detecting perchlorates associated with chlorides, chlorates and nitrates. Of the various salts employed preference is given to fused zinc chloride, chiefly because, while sufficiently energetic in its action upon the perchlorate, it does not, like manganese chloride or the double chloride of aluminium and sodium, evolve chlorine under the influence of ordinary air at the high temperature of the reaction.

In making the test the substance to be examined is put in dry condition (or the solution of it is evaporated) in a test tube, a trap made by cutting off a two-bulbed drying tube, or preferably a mushroom trap (Fig. 29), is moistened inside with a solution of potassium iodide and hung with the large end downward in the test tube, anhydrous zinc chloride is added, and the mixture

* F. A. Gooch and D. Albert Kreider, Am. Jour. Sci., [3], xlviii, 38.

heated to fusion. The chlorine evolved during the heating by action of the perchlorate is indicated by the iodine set free from the iodide and subsequently washed with a little water from the trap and tested with starch emulsion. The test for 0.00005 grm. of potassium perchlorate is sure and distinct, as is shown in the experimental results given below:

Tests by Fusion with Zinc Chloride.

$KClO_4$ taken. grm.	Indication by the starch test.
0 00100	Strong.
0.00050	Strong.
0.00020	Strong.
0.00010	Strong.
0.00010	Strong.
0.00005	Distinct.
0.00005	Distinct.
0.00003	Trace.
0.00003	None.
0 00001	None.
0.00000	None.

Obviously, the presence of an alkali chloride in the substance examined cannot interfere with the certainty of action, but all substances which yield chlorine by decomposition or by action of the air must be removed before the test is applied. To break up 0.1 grm. of potassium chlorate it is only necessary to treat with 5 cm.^{3} of the strongest hydrochloric acid and evaporate to dryness. To destroy nitrates the plan of decomposition employed in the quantitative determination by manganous chloride in hydrochloric acid serves best,* the dry substance being treated with 2 cm.^{3} of the saturated solution of manganous chloride in the strongest hydrochloric acid and the liquid evaporated. This method of decomposing the nitrate is peculiarly advantageous, since the decomposing agent is itself an excellent indicator of the completeness of the work of removal. Two or three treatments serve to remove the nitrate entirely; but before proceeding with the test it is necessary to remove the manganese which has been introduced, inasmuch as manganese chloride will of itself evolve chlorine, by exchange for oxygen, when heated in air. Sodium carbonate in solution answers the purpose of

* See page 263.

removing the manganese (together with other interfering substances) and the filtrate from the precipitated manganous carbonate leaves on evaporation a residue, which, when treated with the anhydrous zinc chloride, gives indications for the perchlorate if it is present in appreciable amount. A tenth of a milligram of potassium perchlorate may be found with certainty when associated with 0.1 grm. of the nitrate, or chlorate, or both. The results of a series of tests for potassium perchlorate associated with the chlorate and nitrate of the same element are recorded in the following table:

Tests for Perchlorate after Destruction of Chlorate and Nitrate.

$KClO_4$ taken. grm.	$KClO_3$ taken. grm.	KNO_3 taken. grm.	Indication of the perchlorate.
0.0005	.	...	Strong.*
0.0003	.		Strong.*
0.0002	. .	. .	Good.*
0.0001	. .	.	Good.*
0.0001	...	...	Trace.*
0.0005	0 1	0.1	Strong.†
0.0003	0 1	0.1	Good.†
0.0003	0 1	0.1	Good.†
0 0002	0 1	0.1	Trace.†
0.0001	0 1	0.1	Trace.†
0.0000	0 1	0.1	None.†

* After procedure for the removal of chlorate and nitrate by HCl.
† Chlorate and nitrate removed by $HCl + MnCl_2$.

The Iodometric Determination of Perchlorates.

After attempting without success to apply to the quantitative determination of perchlorates the reaction which has been shown above* to be efficient in detecting perchlorates, Kreider † has succeeded in developing an exact method which is based upon the iodometric determination of the oxygen evolved from a known amount of perchlorate on ignition.

The method is essentially the collection of the oxygen of the perchlorate; its subsequent passage into an atmosphere of nitric oxide over a strong solution of hydriodic acid, and the titration of the iodine thus liberated with decinormal arsenic in alkaline

* See page 465.
† D. Albert Kreider, Am. Jour. Sci., [3], l, 287.

solution. A piece of combustion tubing, 10 or 12 cm. in length, drawn out at one end to a narrow constriction of length sufficient to prevent the action of the heat on the rubber tubing connecting it with a receiver, serves as the heating chamber. The tube must of course be cleansed from all organic materials and cannot be safely employed for more than three fusions. A platinum boat (since porcelain fuses to glass) serves for the introduction of the perchlorate to the combustion tube. As a receiver, two leveling bottles were found vastly superior to a burette and leveling tube, the glass stopcocks of the latter giving continual trouble by the action of the caustic potash upon them. The larger capacity of the bottle is favorable for the reception of the volume of oxygen evolved and its shape offers superior facilities for the absorption of carbon dioxide. Into the neck of the receiver is fitted a rubber stopper carrying a capillary tube just even with the narrower end before the insertion of the stopper into the neck of the bottle. Upon forcing in the stopper with a slight twist, a funnel-shaped orifice is made through which the oxygen may be withdrawn without leaving any residue. The outer end of the capillary tube is connected with the combustion tube by a piece of vacuum tubing fitted with a pinchcock.

In the process of evolving and collecting the oxygen, the perchlorate, weighed in the platinum boat and covered with an equal mixture of sodium and potassium carbonates, is placed in the combustion tube which is connected at the larger end with the carbon dioxide generator. The combustion tube is inclined and carbon dioxide (obtained from a Kipp generator charged with acid and marble previously boiled to expel air, and with cuprous chloride to take up absorbed oxygen, and washed first with a solution of iodine in potassium iodide to oxidize traces of reducing impurities and then with a solution of potassium iodide to absorb volatilized iodine) is sent through to replace the air. Connection is made with the receiver containing caustic potash. Heat is applied to the combustion tube until the mixture in the platinum boat is in quiet fusion and the oxygen is drawn to the receiver under slightly diminished pressure. The tube is carefully cooled and inclined while the current of carbon dioxide carries all oxygen to the receiver.

To bring about action between the oxygen collected and hydriodic acid, through the medium of nitrogen dioxide, a simple piece of special apparatus is employed.* A 100 cm.^{3} pipette is cut off short at both ends and to each end is sealed a glass stopcock. The delivery tube of one of the stopcocks is cut off rather short after being tapered and constricted so as to hold a rubber connector tightly, while the other delivery tube is left long enough to reach to the bottom of an Erlenmeyer beaker. It is a convenience to have these conducting tubes 3 mm. or 4 mm. in diameter rather than capillaries, since for the various connections all air may be expelled from them by displacement with water, which is easily accomplished by using a long-nozzled wash bottle. By attaching the shorter end to an ordinary water pump or to an evacuated flask, the air is partially exhausted. Then, the stopcock is closed and the bulb disconnected and lowered into a solution of hydriodic acid of approximately known strength, obtained by acidifying potassium iodide (3 grm. in 30 cm.^{3}) with hydrochloric acid immediately before use, to avoid liberation of iodine by the action of air. When the desired amount of liquid (30 cm.^{3}) has been drawn in, the stopcock is closed and connection made with the carbon dioxide, by which all residual air is expelled. Then the bulb, held so as to prevent the escape of the liquid, is again exhausted by attachment to the pump. After admitting from a Kipp generator, charged with copper and dilute nitric acid kept hot, about 10 cm.^{3} of nitrogen dioxide washed by passing through acidified potassium iodide and the same reagent made alkaline, attachment is made to the receiver, and the oxygen is allowed to enter slowly under diminished pressure and with continuous shaking. The latter precaution is essential to the process, as otherwise there is imperfect distribution of the hydriodic acid and danger of forming nitric acid. But when the solution of hydriodic acid is kept strong and the shaking continued while the oxygen enters and for a minute or two afterward, depending on the rapidity with which it was admitted, the oxygen may be allowed to enter quite rapidly without any fear of imperfect action. The oxygen being immediately utilized, the partial vacuum is affected only by the heat generated, which is scarcely noticeable.

It is necessary of course to prevent access of air into the bulb until the acid has been neutralized, to accomplish which, without

* See Fig. 25.

loss of iodine, acid potassium carbonate must be used, at least for the end reaction. To remove the contents of the bulb for titration, the two delivery tubes are washed and filled with water, the shorter end connected to a supported funnel containing a saturated solution of the acid carbonate, and the longer one inserted into an Erlenmeyer beaker containing also a saturated solution of the acid carbonate in amount sufficient — as previously determined — to neutralize all the acid taken. On opening the stopcock of the delivery tube which reaches below the liquid in the beaker the bicarbonate is drawn in by the partial vacuum, and sufficient carbon dioxide is liberated to force all the liquid out. Owing to the consequent effervescence as the liquid gains its exit, the flow must be regulated by the stopcock so as to avoid loss of iodine, which is prevented by inclining the beaker so that the bubbles strike quietly against its side. To wash out the bulb, it is raised almost horizontally, so as to prevent the liquid from running through, and the upper stopcock opened to admit the bicarbonate from the funnel. Both stopcocks are then closed, the bulb disconnected and agitated, after which it may be washed with water in presence of air without fear of liberating more iodine. An excess of decinormal arsenic is then run into the beaker and titrated back with iodine.

Results obtained by this method are given in the following table.

Iodometric Determination of Evolved Oxygen.

$KClO_4$ taken. grm.	KI taken. grm.	HCl taken. cm.^{3}	$KClO_4$ found. grm.	Error. grm.
0.1000	3	3	0.1003	+0.0003
0.1000	3	3	0.1006	+0.0006
0.1000	3	3	0 0998	−0.0002
0.1000	4	4	0.1003	+0.0003
0.1000	3	3	0.1003	+0.0003
0.1000	3	4	0.0999	−0.0001
0.1000	3	3	0.1003	+0.0003
0.1000	3	4	0.1001	+0.0001
0.1500	3	4	0.1493	−0.0007
0.2000	6	6	0.1999	−0.0001
0.2000	6	6	0.2009	+0.0009
0.0100	3	3	0.0099	−0.0001
0.0100	3	3	0.0100	0.0000
0.0000	3	3	0.0003	+0.0003

The Estimation of Bromates by Reduction with Ferrous Sulphate.

By treatment of a dissolved bromate with ferrous sulphate standardized iodometrically, and determining the residual ferrous salt similarly, Phelps * has been able to effect the determination of the oxidizing power with considerable accuracy. To the solution of the bromate is added in a trapped flask an excess of approximately *n*/5 ferrous sulphate and 15 cm.^{3} of sulphuric acid [1 : 3]. The mixture is heated to boiling, cooled in running water and nearly neutralized with sodium carbonate. Rochelle salt (2 grm.) is added and an excess of *n*/10 iodine. The mixture is made alkaline with acid potassium carbonate added in excess, bleached with *n*/10 arsenite in presence of starch, and the excess of the arsenite is titrated by iodine.

The following results were obtained:

Reduction by Ferrous Sulphate and Titration of Excess.

$KBrO_3$ taken. grm.	Oxygen value of ferrous salt taken. grm.	Oxygen value of ferrous salt found. grm.	Error on $KBrO_3$. grm.
0.0500	0.01776	0.00357	−0 0006
0.0500	0.01770	0.00336	−0 0001
0.1000	0.03792	0.00942	−0.0008
0.1000	0.03792	0.00922	−0 0001
0.2000	0.06321	0.00576	0 0000
0.2000	0.06321	0.00580	−0.0002
0.5000	0.15670	0.01342	−0 0013
0.5000	0.16212	0.01870	−0.0008

The Iodometric Estimation of Bromates.

Reduction by Hydriodic Acid. In connection with an investigation in respect to other methods for the quantitative determination of bromates, Gooch and Blake † have studied the degree of regularity which may be expected in the method of Kratchmer ‡ which is based upon the action of potassium iodide in presence of acid and titration of the iodine set free according to the equation

$$6\,HI + HBrO_3 = HBr + 3\,H_2O + I_2.$$

* I. K. Phelps, Am. Jour. Sci., [4], xvii, 201.
† F. A. Gooch and J. C. Blake, Am. Jour. Sci., [4], xiv, 285.
‡ Zeit. anal. Chem., xxiv, 546.

The rate at which this reaction proceeds has been investigated by Ostwald,* by Noyes † and by Judson and Walker ‡. Time of action, proportion of iodide to bromate, excess of acid, and dilution are all, within limits, determining factors in the reaction; but in the analytical process it is usually assumed that the reaction goes soon to completion if free acid and a moderate excess of potassium iodide are present.

Gooch and Blake show that the reaction is very incomplete when the excess of potassium iodide is only about 20 per cent over the amount demanded by theory, the time of standing only the few minutes required for the manipulation of the process, and the dilution considerable. When the amount of potassium iodide used is four times that required by theory, the time of standing at least half an hour, and the volume 100 cm.^{3}, variation in the amounts of acid above a reasonable minimum and in the time given to the reaction are without apparent effect.

Reduction by Hydriodic Acid.

$KBrO_3$ taken. grm.	Iodine taken. grm.	KI taken. grm.	H_2SO_4 [1 : 1]. cm.^{3}	Time of standing. hrs.	Approximate volume. cm.^{3}	Iodine found. grm.	Error in terms of $KBrO_3$. grm.
0 1400	0 6378	3	5	½	100	0 6343	−0 0008
0 1400	0.6378	3	5	½	100	0 6329	−0 0011
0 1400	0.6378	3	5	½	100	0.6329	−0 0011
0 1408	0.6416	3	5	22	100	0.6396	−0 0004
0.1408	0.6416	3	5	22	100	0.6364	−0.0011
0 1408	0.6416	3	5	22	100	0.6381	−0 0008
0 1400	0.6378	3	2.5	½	100	0 6340	−0 0008
0.1400	0.6378	3	2.5	½	100	0 6336	−0 0009
0.1400	0.6378	3	2.5	½	100	0 6336	−0 0009
0.1400	0.6378	3	1	½	100	0 6343	−0 0008
0.1400	0.6378	3	0.5	½	100	0.6331	−0.0010
			HCl (sp gr. 1.18) cm.^{3}				
0.1400	0.6378	3	8	½	100	0.6336	−0.0009
0.1400	0.6378	3	8	½	100	0 6329	−0.0011
0.1400	0.6378	3	8	½	100	0.6336	−0.0009
0.1400	0.6378	3	4	½	100	0.6336	−0.0009
0.1400	0.6378	3	4	½	100	0.6333	−0 0010
0.1400	0.6378	3	4	½	100	0 6340	−0.0008
0.1400	0.6378	3	4	1	100	0.6336	−0.0009
0.1400	0 6378	3	4	1	100	0.6333	−0 0010
0 1400	0.6378	3	4	1	100	0 6336	−0.0009

* Zeit. phys. Chem., ii. 127. † *Ibid.*, xix, 599.
‡ Jour. Chem. Soc., lxxiii, 410.

The results of experiments in which measured amounts of standard solutions of potassium bromate (approximately 2.8 grm. to the liter) were thus treated with potassium iodide and hydrochloric or sulphuric acid for definite times in glass-stoppered bottles are given in the table. The iodine liberated was determined by titration with sodium thiosulphate standardized against nearly decinormal iodine the value of which was fixed by comparison with decinormal arsenious oxide dissolved in acid potassium carbonate.

The evolution of iodine by the action of atmospheric oxygen upon the acidified solution of the iodide was found by experiment to vary with the strength of the acid and the time of exposure from 0.0001 grm. to 0.0003 grm. expressed in terms of the bromate, and these values are not greater than the differences observed between parallel determinations of the same sort. Probably, however, all the errors as shown in the table should really be increased a trifle to approximate the truth, notably those of the experiments allowed to stand the longest period, twenty-two hours.

The average apparent error of the process as applied to this particular sample of bromate is −0.0009 grm.; and 2.5 $cm.^3$ of sulphuric acid of half strength [1 : 1] or the equivalent amount of hydrochloric acid, 4 $cm.^3$ of the acid of sp. gr. 1.18, in the presence of about 3 grm. of potassium iodide, complete the action within half an hour, at a dilution of 100 $cm.^3$, as far as it will go under any of the conditions tried. The phenomenon noted by Ostwald,* that small amounts of hydrochloric acid tend to force the reaction more rapidly than equivalent amounts of sulphuric acid, does not appear in these experiments, no doubt because the action was pushed to the limit by the smallest amount of acid employed.

The error of deficiency appears to be due to impurity in the sample of bromate rather than to incompleteness of the reaction. If the reaction were incomplete it would be natural to look for the cause in the possibility of the inhibiting influence of the iodine set free, but it was found in three parallel experiments that the introduction of 0.5 grm. of free iodine dissolved in potassium iodide failed to influence the error appreciably. In a product recrystallized several times, and one in which no chloride

* Loc. cit., page 131.

can be detected, as was the case with the bromate of these experiments, the impurity most natural to look for is potassium chlorate, which might resist removal in the process of purification by recrystallization. The bromate was therefore tested by igniting it and treating the residue with potassium bichromate and sulphuric acid, volatilizing and collecting any chloro-chromic anhydride thus formed, and converting the last into lead chromate.* Traces of chlorine were thus found, which, inasmuch as they were not found in a similar test upon the unignited bromate, must have had their origin in chlorate intercrystallized with the bromate.

Reduction by Arsenious Acid. Use has been made of the fact that iodic acid may be reduced quantitatively by arsenious acid,† and Gooch and Blake ‡ have investigated the similar application of arsenious acid to the quantitative estimation of bromic acid according to the reaction

$$3\,H_3AsO_3 + HBrO_3 = 3\,H_3AsO_4 + HBr.$$

Experiments made to discover the limits within which regularity of action may be expected disclose the fact that for conditions varying within a rather wide range the oxidation of the arsenious oxide reaches a fairly definite limit. The bromate effects practically the same proportionate oxidation of arsenious oxide in a volume of 200 cm.^{3} or less, whether the sulphuric acid of half strength present amounts to 3.5 cm.^{3} or 10 cm.^{3}, and whether the time of digestion is 15 minutes or 30 minutes at the boiling temperature, one and one-half or four hours on the steam bath, or two days at the ordinary temperature. Setting aside experiments in which the addition of acid did not exceed the equivalent of alkali carbonate present by more than 1 cm.^{3}, the average absolute variation from theory of an entire list of forty-two experiments amounts to −0.0007 grm. in terms of bromate, individual variations departing from the average by about the same figure. The meaning of this fact seems to be that the slight error is due to impurity in the potassium bromate employed and not to incomplete oxidizing action on the part of the bromate.§

In making use of this process, the bromate is treated with a considerable excess of arsenious oxide dissolved in acid potassium

* See page 441. † See page 422.

‡ Loc. cit., 471. § See lines 4 to 11, above.

carbonate, the mixture is acidified with 3 cm.[3] to 7 cm.[3] of sulphuric acid [1 : 1], and the liquid, not exceeding 200 cm.[3] in volume, is boiled ten minutes or more, neutralized with acid potassium carbonate and titrated with iodine. Results of this procedure are given in the following table:

Reduction by Arsenite.

$KBrO_3$ weighed. grm.	As_2O_3 taken. grm.	H_2SO_4 [1 : 1]. cm.[3]	Time in minutes.	As_2O_3 unchanged. grm.	As_2O_3 oxidized. grm.	Error in respect to $KBrO_3$ weighed. grm.	Error in respect to $KBrO_3$ by Kratchmer's method. grm.
0.0701	0 1881	5	10	0 0661	0.1220	−0.0014	−0 0003
0 0701	0 1881	5	10	0.0650	0.1231	−0.0009	0.0000
0.0701	0 2475	5	10	0.1232	0 1243	−0.0002	+0 0007
0.0701	0 2475	5	10	0.1236	0.1239	−0.0004	+0.0003
0 0701	0.2475	5	25	0.1234	0.1241	−0 0003	+0.0004
0 0701	0 2475	5	25	0.1234	0.1241	−0 0003	+0.0004
0.1402	0.4950	3	15	0.2479	0 2471	−0 0012	−0 0003
0.1402	0 4950	3	15	0.2476	0.2474	−0 0010	−0.0001
0.1400	0 6188	7	20	0.3708	0.2480	−0.0004	+0.0005
0.1400	0 6188	7	20	0.3710	0 2478	−0 0005	+0.0004
0.1400	0.6188	7	20	0.3706	0.2482	−0.0003	+0.0006
0.1400	0 6188	7	30	0 3708	0.2480	−0.0004	+0.0005
0.1400	0.6188	7	45	0.3711	0.2477	−0.0006	+0 0003

Here again, as in the process of reduction of the bromate by hydriodic acid, the results point to a slight deficiency in the oxidizing power of the bromate. The mean deficiency, 0.0006 grm., differs from that of the hydriodic acid method by about 0.0003 grm.

Reduction by Arsenate-Iodide Mixture.

It has been shown * that a chlorate may be determined by adding to it in solution potassium iodide in known amount, an excess of an arsenate, and sulphuric acid, boiling the mixture between definite limits of concentration, determining by titration with iodine the amount of arsenious oxide produced, and calculating the amount of chlorate present, from the difference between the amount of arsenious oxide thus produced and that which would be produced if the whole amount of iodide added were allowed to act upon the arsenate alone. Gooch and Blake † have shown that a bromate

* See page 463.

† Loc. cit., p. 471.

treated by this process leaves a similar record of its oxidizing power. According to this process, the bromate is treated in an Erlenmeyer 300 cm.³ flask with 2 grm. of potassium arsenate, 20 cm.³ of [1 : 1] sulphuric acid and water to make the entire volume a little more than 100 cm.³ The liquid in the trapped flask * is boiled down to a volume of 35 cm.³, cooled, nearly neutralized with sodium hydroxide, and treated with an excess of 20 cm.³ of acid potassium carbonate. The arsenious oxide in solution is titrated by standardized $n/10$ iodine in presence of starch. The iodine added is taken as the exact equivalent of iodine expelled in boiling and the difference between the iodine thus expelled and that originally added in the form of potassium iodide is the measure of the bromate.

The following table contains the account of experiments made in this manner upon the sample of bromate the action of which in the iodide method and in the arsenious acid method is recorded.

Reduction by Arsenate-Iodide Mixture.

H_2SO_4 [1 : 1] 20 cm.³; initial volume 105 to 170 cm.³; final volume 35 cm.³

$KBrO_3$ taken. grm.	H_2KAsO_4 taken. grm.	I value of KI taken grm.	Iodine corresponding to As_2O_3 produced. grm.	Iodine corresponding to $KBrO_3$. grm.	Error in terms of $KBrO_3$. grm.
0.0700	2	0 4146	0 0948	0 3198	+0 0002
0.0700	2	0.4146	0.0954	0.3192	0.0000
0 0700	2	0 4146	0 0969	0.3177	−0 0003
0.0700	2	0.4146	0 0975	0 3171	−0 0004
0.1400	2	0 7832	0.1458	0 6374	−0 0001
0.1400	2	0 7832	0.1463	0 6369	−0.0002
0.1400	2	0.7832	0.1462	0 6370	−0 0002

The mean error of these determinations, in which all oxidizing material is calculated as bromate, is not far from −0.0001 grm.

Upon comparing these results with those of the iodide process and of the arsenious acid process, it appears that the deficiencies noted above are satisfactorily accounted for by the presence of traces of chlorate in the bromate.†

* See Fig. 6, page 6.

† See page 474.

CHAPTER XII.

MANGANESE; IRON; NICKEL; COBALT.

MANGANESE.

The Determination of Manganese as the Sulphate.

The estimation of manganese by the conversion of salts of that element with volatile acids to the form of the anhydrous sulphate by the action of an excess of sulphuric acid, evaporation, and gentle heating was formerly a recognized procedure. On the authority of Rose,* however, this method was set aside on account of the supposed difficulty of removing the excess of acid without disturbing the composition of the normal salt. Oesten, working under Rose's direction, found losses upon submitting the crystalline hydrous sulphate $MnSO_4.5H_2O$ to gentle ignition several milligrams greater than would have been the case if the salt had been reduced to the normal anhydrous sulphate. At higher temperature the sulphate turned brown and the losses were high. Volhard † studied the process and showed that manganous sulphate may be dehydrated, separated from an excess of sulphuric acid, and brought into definite condition for weighing as the anhydrous salt by careful and protracted heating over a special device of his own — a ring burner enclosed in a sheet-iron casing. Similar results were obtained on evaporating with sulphuric acid and igniting in like manner an aqueous solution of manganous chloride.

That the process of estimating manganese in the form of the sulphate is both simple and accurate has been shown by Gooch and Austin.‡ According to the method, as described, the manganese salt of a volatile acid is treated with sulphuric acid in amount more than equivalent to the manganese, the solution is evaporated on the water bath until the water is removed as far as may be, and then, supported by means of a porcelain ring or triangle, within a larger porcelain crucible used as a radiator,

* Ann. Phys., cx, 125.

† Ann. Chem., cxcviii, 328.

‡ F. A. Gooch and Martha Austin, Am. Jour. Sci., [4], v, 209.

so that the bottom and walls of the one are distant from the bottom and walls of the other by an interval of about 1 cm., the crucible is heated more strongly. The outer porcelain crucible may be heated over a good Bunsen flame to a red heat without risk of overheating the manganese sulphate within the inner crucible, and the ignition may proceed as rapidly as is consistent with the avoidance of mechanical loss by spattering.

Results of this treatment are given in the accompanying statement:

Conversion of Manganese Chloride to Manganese Sulphate.

$MnSO_4$ calculated from AgCl found from 50 cm.³ of solution of $MnCl_2$. grm.	$MnSO_4$ found by treatment of 50 cm.³ of solution with H_2SO_4. grm.	Variation. grm.
0 3515*	0 3513	−0 0002
0 3515*	0 3514	−0 0001
0 3515*	0 3518	+0 0003

* The mean of two determinations — 0.3518 grms. and 0.3512 grms.

$MnSO_4$ *found by Treatment of 50 cm.³ of Each of Various Solutions with* H_2SO_4.

I.	II.	III.	IV.	V.	VI.
(1) 0 3100	(1) 0 3256	(1) 0 3534	(1) 0 3524	(1) 0 3355	(1) 0 5475
(2) 0.3104	(2) 0.3254	(2) 0 3543	(2) 0 3520	(2) 0.3357	(2) 0.5476
(3) 0.3096	.....	....	....	......	....

The Determination of Manganese as Oxide.

The estimation of manganese as mangano-manganic oxide, Mn_3O_4, has been so frequently criticized unfavorably that the method seems to have passed from very general use excepting in certain cases in which the directness of the process is a temptation to incur the risk of some uncertainty. The production of the other oxides of manganese in definite condition is thought to be even more uncertain.

Manganese dioxide, MnO_2, begins, as Wright and Menke * have shown, to lose oxygen at a temperature (about 210° C.) to which the hydrated oxide must be heated to free it from water, and very nearly that at which the nitrate is converted into the

* Jour. Chem. Soc., xxx, 775.

dioxide; so that the chance of producing an undecomposed dioxide by the ignition of the hydrated dioxide (the form in which the dioxide generally appears in analytical processes) or of the nitrate is small.

Manganic oxide, Mn_2O_3, is produced, it is said, from the other oxides by ignition at a low red heat under the ordinary conditions.

The manganoso-manganic oxide, Mn_3O_4, forms, presumably when an oxide of manganese is submitted, under ordinary atmospheric conditions, to the high heat of the blast lamp.

If the proportion of oxygen in the surrounding atmosphere is reduced below the normal, the conversion of Mn_2O_3 to Mn_3O_4 goes on very easily, as Dittmar has shown,* at a temperature between the melting points of silver and aluminum, while if the proportion of oxygen in the surrounding atmosphere is much increased above the normal, the reverse change, from Mn_3O_4 to Mn_2O_3, tends to take place at the same temperature. Gooch and Austin † have pointed out that the condition most favorable to the production of the oxide Mn_3O_4 — a low proportion of oxygen in the surrounding air — can be maintained during the ignition if the products of combustion are made to displace the ordinary air about the crucible, and that this condition is attained when the crucible rests well within the upper part of the flame of a large burner or blast lamp in such manner that the entire wall of the crucible is covered by an oxidizing flame.

If the oxide Mn_3O_4 in finely divided condition within a platinum crucible be moistened with nitric acid and heated gently, the containing crucible being well above a larger porcelain crucible used as a radiator and so heated that only the bottom of the outer crucible shows a faint red glow, the oxide which remains has approximately the composition of manganese dioxide, MnO_2.

If the platinum crucible containing the oxide MnO_2 be now placed in contact with the bottom of the larger porcelain crucible heated to redness the oxide takes a composition approximating more or less closely that of manganic oxide, Mn_2O_3.

If the crucible containing the oxide Mn_2O_3 is now subjected to the large flame of a powerful Bunsen burner or blast lamp in such manner that the oxidizing flame covers the walls of the

* Jour. Chem. Soc., xvii, 294.

† F. A. Gooch and Martha Austin, Am. Jour. Sci., [4], v, 212.

Conversion of Sulphate to Oxides, and of Oxides to Sulphate.

$MnSO_4$ taken.	MnO_2.			Mn_2O_3.			Mn_3O_4.			$MnSO_4$.	
	Found by treatment of Mn_3O_4 with HNO_3 and gentle heating.	Theory.	Error.	Found by heating MnO_2 moderately.	Theory.	Error.	Found by heating the oxide strongly but within the oxidizing flame.	Theory.	Error.	Found by treatment with H_2SO_4 and gentle heating.	Error.
grm.	grm.	grm.	grm.	grm.	grm.	grm.	grm.	grm.	grm.	grm.	grm.
0.3712	....	...	....	....	......	....	I. 0.1880	0.1876	+0.0004		
	II. 0.2114	0.2138	−0.0024	III. 0.1975	0.1954	+0.0021					
	IV. 0.2137		−0.0001	V. 0 1959		+0.0005					
	VI. 0 2147		+0 0009	IX. 0.1986		+0.0032					
	VII. 0.2147		+0 0009	X. 0 1957		+0.0003	XI. 0.1881		+0.0005		
	VIII. 0.2138		0.0000								
	XII. 0.2145		+0.0007	XIII. 0.1948		−0 0006	XIV. 0.1877		+0.0001		
	XV. 0.2115		−0.0023	XVI. 0.1939		−0.0015	XVII. 0.1872		−0.0004		
	XVIII. 0.2158		+0.0020	XIX. 0.1942		−0.0012	XX. 0.1899		+0.0023	XXI. 0.3709	−0.0003
0.6436	....	......	....	....	......	....	I. 0.3247	0.3252	−0.0005		
	II. 0.3443	0.3707	−0.0264								
	III. 0.3704		−0.0003	IV. 0.3372	0.3387	−0.0015	V. 0.3245		−0.0007		
	VI. 0.3705		−0.0002	VII. 0 3367		−0.0020	VIII. 0.3244		−0.0008		
	IX. 0.3597		−0.0110								
	X. 0.3719		+0 0012	XI. 0.3369		−0.0018	XII. 0 3243		−0.0009		
	XIII. 0.3729		+0.0022	XIV. 0.3362		−0.0025	XV. 0.3244		−0.0008		
	XVI. 0.3690		−0.0017	XVII. 0.3366		−0.0021	....	......	....	XVIII. 0.6432	−0 0004
										XIX. 0.6432	−0 0004
										XX. 0.6437	+0 0001

crucible, the composition of the oxide remaining is generally very closely that of the oxide Mn_3O_4. The estimation of manganese in the form of the manganoso-manganic oxide, Mn_3O_4, is by no means to be considered utterly untrustworthy when the process is conducted in the manner described, though it must be recognized that an irregular result may occur occasionally. The danger of accepting such an irregularity as a correct indication may be eliminated to a very considerable extent if the precaution is taken invariably to moisten the ignited oxide with nitric acid, and ignite again. Concordant results thus obtained may be taken with a fair degree of confidence.

Results of this procedure, as well as of the treatment described for producing in succession the oxides MnO_2, Mn_2O_3, Mn_3O_4, are given in the preceding tabular statement.

The Determination of Manganese Separated as the Carbonate.

Austin * has shown that the precipitation as carbonate in presence of large amounts of ammonium chloride, according to Guyard,† and estimation as sulphate ‡ or oxide § offer reliable means for the determination of manganese. The manganese salt dissolved with a considerable amount of ammonium chloride (about 10 grm.) in 200 $cm.^3$ of boiling water is treated with ammonium carbonate in excess. The liquid is kept warm until the precipitate subsides and is then filtered off either upon paper or upon asbestos in a perforated crucible. The crucible and precipitate collected upon asbestos are ignited in the oxidizing flame of a powerful burner ‖ to give the oxide Mn_3O_4 which is weighed as such; the precipitate collected upon paper may be ignited to the oxide and then converted by treatment with a few drops of sulphuric acid and gentle heating to the sulphate.

Weighing as the carbonate is not feasible; for, as Rose has correctly stated, ¶ evolution of carbon dioxide and oxidation of the residue begin before the water is thoroughly removed.

Experimental results are given in the table.

* Martha Austin, Am. Jour. Sci., [4], v, 382.

† Hugo Tamm, Chem. News, xxvi, 37.

‡ See page 477.

§ See page 478.

‖ See page 479.

¶ Ann. Phys., lxxxiv, 52.

Precipitation as Carbonate: Conversion to Oxide and to Sulphate.

NH_4Cl. grm.	Mn_3O_4 taken. grm.	Mn_3O_4 found. grm.	Error. grm.	$MnSO_4$ taken. grm.	$MnSO_4$ found. grm	Error.
10	0.1776	0.1770	−0 0006	. . .		. .
10	0.1776	0.1788	+0 0012	. . .		. . .
10	0 1776	0.1770	−0 0006	. .	.	
10	0.1776	0.1774	−0 0002	. . .	. .	. .
10	0.2478	0 2463	−0 0015	0.4905	0 4903	−0 0002
10	0.1121	0.1110	−0 0011	0.2219	0 2225	+0 0006
10	0.1581	0 1584	+0 0003	0.2128	0.3126	−0 0002
10	0.1699	0 1672	−0 0027	0 3344	0 3355	−0 0009

The Determination of Manganese Precipitated as Ammonium Manganese Phosphate and Weighed as Manganese Pyrophosphate.

By Gibbs' original method manganese phosphate was precipitated by hydrogen disodium phosphate in large excess above the quantity required to cause the precipitation. The flocky white precipitate was dissolved either in sulphuric or hydrochloric acid, and precipitated again at the boiling temperature by ammonia in excess. This semi-gelatinous precipitate, on boiling or long standing even in the cold, becomes crystalline, the crystals forming beautiful talcose scales which have a pearly luster and a pale rose color. The precipitate was filtered off, washed with hot water, dried and ignited. The results obtained by Gibbs' students agree closely with the theory for the pyrophosphate.

When a manganous salt is precipitated in the cold by an excess of an alkaline phosphate, it falls, as Heintz * has shown, in the form of the trimanganous phosphate of the formula $Mn_3P_2O_8$. This same phosphate constitutes the greater part of the precipitate which forms when a manganous salt reacts in the cold, in the presence of ammonium chloride, with microcosmic salt and ammonia in slight excess, but boiling or even subsequent standing may effect a more or less complete conversion of the manganese phosphate to the ammonium manganese phosphate. The success of the analytical process in which manganese is weighed as the pyrophosphate turns, therefore, upon the change of the trimanganous phosphate, $Mn_3P_2O_8$ to the ammonium

* Ann. Phys., lxxiv, 449.

manganese phosphate, NH_4MnPO_4. Gooch and Austin * have shown that the presence of a large amount of ammonium salt is essential to the formation of the precipitate of ideal constitution. Apparently the proportion of ammonium chloride present to ammonium manganese phosphate formed should be at least 40 : 1, or, speaking approximately, 200 molecules of ammonium chloride must be present in the liquid to every molecule of the phosphate formed; and the ammonium chloride may be increased almost to the point of saturation of the liquid without causing more than a trifling solubility of the ammonium manganese phosphate in the presence of an excess of the precipitant. Furthermore, the precipitate may be washed with perfect safety with pure water as well as with slightly ammoniacal water, or with ammoniacal water containing ammonium nitrate, if the filtration is performed rapidly and the precipitate gathered in small space, as is the case when the phosphate is collected on asbestos in the perforated crucible. The finely granular precipitate which may be obtained by slow action of dilute ammonia added gradually to the hot solution of the manganese salt apparently includes a portion of unconverted phosphate which resists the replacement of the manganese by ammonium. On the other hand, the precipitate of flocky condition thrown down in the cold passes easily to the silky and crystalline condition when heated with the proper amount of ammonium salt, and possesses a constitution approaching the ideal under such conditions. The conversion of the flocky manganous phosphate is so rapid that the precipitation may be carried on safely in glass vessels. If the ammonium chloride in the solution were to be included in the precipitate it would volatilize entirely during the ignition, leaving no trace unless, possibly, a portion of its chlorine were to combine with the manganese. Tests for chlorine in the residue of pyrophosphate have resulted negatively, no more than a mere trace being found in any case, so that the contaminating effect of the ammonium chloride proves to be insignificant and the responsibility for the excess of weight above the theory must apparently rest with the included microcosmic salt.

In the practical determination of manganese by the phosphate method of Gibbs, therefore, the presence of large amounts of

* F. A. Gooch and Martha Austin, Am. Jour. Sci., [4], vi, 233.

ammonium chloride is strongly advocated. Good results may be obtained most easily and surely by the following procedure: The slightly acid solution (in platinum or glass), containing in a volume of 200 cm.³ an amount of manganese not more than enough to make 0.4 grm. of the pyrophosphate, 20 grm. of ammonium chloride, and 5 cm.³ to 10 cm.³ of a cold saturated solution of microcosmic salt, is precipitated in the cold by the careful addition of dilute ammonia in only slight excess. The mixture is heated until the precipitate becomes silky and crystalline, the whole is allowed to stand and cool half an hour, the precipitate is collected upon asbestos in a perforated platinum crucible, washed (best with slightly ammoniacal water), dried at gentle heat, and ignited as usual. By this process determinations of the larger amounts of manganese — 0.4 grm. of the pyrophosphate — approximate rather more closely to the theoretical values than do those of the smaller amounts — 0.2 grm. In either case the average error should not exceed 0.0010 grm. in terms of manganese.

Results obtained by this procedure are given in the table.

Determination as Manganese Pyrophosphate.

$Mn_2P_2O_7$ equivalent to MnO_2.		Error in terms of $Mn_2P_2O_7$.	Error in terms of manganese.	Saturated solution of $HNaNH_4PO_4$.	NH_4Cl.	Total volume.	Manganese in the filtrate.
Taken.	Found.						
grm.	grm.	grm.	grm	cm.³	grm.	cm.³	
In platinum.							
0.1885	0.1903	+0.0018	+0.0007	5	20	200	None.
0.1885	0.1910	+0.0025	+0.0010	5	20	200	None.
0.1885	0.1913	+0.0028	+0.0011	5	20	200	None.
0.1885	0.1911	+0.0026	+0.0010	5	20	200	None.
0.3770	0.3776	+0.0006	+0.0002	5	20	200	None.
0.3770	0.3773	+0.0003	+0.0001	5	20	200	None.
0.3770	0.3778	+0.0008	+0.0003	5	20	200	None.
0.3770	0.3783	+0.0013	+0.0005	5	20	200	None.
In glass.							
0.1885	0.1904	+0.0019	+0.0007	5	20	200	None.
0.1885	0.1898	+0.0013	+0.0005	5	20	200	None.
0.3770	0.3767	−0.0003	−0.0001	5	20	200	None.
0.3770	0.3784	+0.0014	+0.0005	5	20	200	None.

The Electrolytic Determination of Manganese.

For the electro-deposition of manganese as the dioxide various processes have been described.* With stationary electrodes solutions containing nitric acid, sulphuric acid, acetic acid, formic acid with or without a formate, or ammonium acetate alone, with chrome alum, or with acetone, have been employed.† For use with the rotating anode, solutions containing ammonium acetate with chrome alum or alcohol have been advocated.‡ In all these processes hydrated manganese dioxide is deposited upon a large anode which is preferably a roughened platinum dish of considerable capacity.

Gooch and Beyer § have described experiments made to test the utility of the electrolytic filtering cell in the determination of manganese as the dioxide. The procedure adopted was the simplest. Portions, 50 cm.^{3} each, of a solution of pure manganous sulphate, standardized by evaporation of measured portions and gentle ignition of the residue over a radiator,|| were treated, in each case, with six drops (0.17 cm.^{3}) of concentrated sulphuric acid, and electrolyzed in the filtering cell with a current of 2 amperes ($N.D._{100} = 5$ amp.) and 20–10 volts, the voltage decreasing as the solution became heated.

In one set of experiments the process of continuous filtration during electrolysis, for which the adjustment of apparatus is shown in Fig. 16, was employed.¶ The time required for the deposition of 0.1860 grm. of the dioxide was one hour and three-quarters. In a second set of experiments the closed cell, shown in Fig. 15, was used during the electrolysis, and the adjustment for filtration made subsequently as previously described.** A period varying from two hours and ten minutes to two hours and fifty minutes is required in the latter process. Tests with hydrogen dioxide and ammonia showed that the deposition was complete in the process of continuous filtration and practically so in the closed-cell process. The closed-cell process naturally

* Smith's Electro-analysis, edition of 1907, page 134 et seq.

† Ibid.

‡ Koester, Zeit. Elektrochem., x, 553.

§ F. A. Gooch and F. B. Beyer, Am. Jour. Sci., [4], xxvii, 62.

|| Am. Jour. Sci., [4], v. 209.

¶ See page 17.

** See page 15.

requires less attention during the electrolysis, and so it is advantageous to run the process for a period, perhaps two hours, with the closed cell, and then to adjust the apparatus for filtration during further electrolytic action, in order that floating particles of the dioxide may be drawn to the felt and completeness of precipitation may be assured. In this way the advantage of the circulating process may be obtained with less attention to manipulation than is required when the filtration is continuous from the start. The deposit is washed with water after interruption of the current, first dried at 200° for ten or fifteen minutes and weighed, and thereafter ignited to low redness in the spreading flame of a large burner.* Results of experiments with the cell arranged for continuous filtration, and of experiments in which the closed cell was used until the electrolysis was nearly over, are given in the accompanying table:

Deposition of Manganese Dioxide.

Solution of $MnSO_4$ taken.	$_2SO_4$ concentrated.	Current.			Theory MnO_2.	MnO_2 weighed as MnO_2.	MnO_2 weighed as Mn_3O_4.	Error.
cm.^{3}	cm.^{3}	Amp.	N. D.$_{100}$.	Volt.	grm.	grm.	grm.	grm.
Electrolysis with continuous filtration.								
50	0.17	2	5	20–12	0.1860	0.1862	.	+0.0002
..	....	..	..	.....	. ...		0.1858	−0.0002
50	0.17	2	5	20–12	0.1860	0.1856	.. .	−0.0004
..	....	..	..	.....	.....	.. ..	0.1856	−0.0004
50	0.17	2	5	20–12	0.1860	0.1843	......	−0.0017
..	....	..	..	.....	.....	......	0.1872	+0.0012
Electrolysis in closed cell with subsequent filtration.								
50	0.17	2	5	20–12	0.1860	0.1860	.. .	0.0000
..	....	..	..	.....	......	. ....	0.1853	−0.0007
50	0.17	2	5	20–10	0.1860	0.1856	.....	−0.0004
..	....	..	.	...	.....	.....	0.1856	−0.0003
50	0.17	2	5	18–10	0.1860	0.1853	.. ..	−0.0007
..	...	..	..	....	......	......	0.1858	−0.0002

The results are evidently as good as could be expected of any process which involves the weighing of a manganese oxide brought to condition by heating. The degree of oxidation of the oxide thrown down under the conditions approximates closely

* Am. Jour. Sci., [4], v, 214.

to that of the ideal oxide represented by the symbol $MnO_2.H_2O$, formerly assigned by Rüdorff * to the electrolytically formed oxide, and differs in that respect from that of the electrolytically deposited oxide which was studied by Groeger.†

The Determination of Manganese Precipitated by the Chlorate Process.

Gooch and Austin ‡ recommend the substitution of sodium chlorate for potassium chlorate in the precipitation of manganese from the nitric acid solution, the greater solubility of the sodium salt and the consequent rapidity with which its decomposition takes place making its use advantageous.

According to the treatment outlined, the manganous nitrate, free from chlorides and sulphates, is dissolved in concentrated nitric acid (85 $cm.^3$), and treated with sodium chlorate (5 grm.). The mixture is boiled five minutes, more nitric acid (15 $cm.^3$) and a few crystals of sodium chlorate are added, and the heating is discontinued as soon as the liquid boils again. The insolubility of the precipitate, if this procedure is followed, is so great that no more than insignificant traces, never exceeding 0.0001 grm., may be recovered from the filtrate after filtering on asbestos upon the perforated cone or crucible and washing with water. On the other hand, prolonged boiling after the last addition of chlorate results in considerable losses of manganese (from 0.0010 grm. to 0.0030 grm.), due to the reducing effect of lower oxides of nitrogen formed (as is always the case in boiling nitric acid) after the chlorine dioxide has been expelled. An excess of chlorate at the end of the operation seems to be essential and a slight yellow color in the solution, due to chlorine dioxide, is a favorable indication. It is best to filter the undiluted solution under pressure upon asbestos on a perforated cone having a filtering surface of about 40 $cm.^2$. Dilution before filtration tends to increase the solubility of the manganese.

While manganese is very completely precipitated in the chlorate process conducted with the precautions indicated, the condition of oxidation cannot be assumed to be that of the dioxide, and the indications of any process which rests upon the assumption

* Zeit. angew. Chem., 1892, 6.

† Zeit. angew. Chem., 1895, 253.

‡ F. A. Gooch and Martha Austin, Am. Jour. Sci., [4], v, 260.

that the oxygen value of the manganese compound precipitated is that of the dioxide is likely to be erroneous. If, therefore, the chlorate method is used to separate manganese, precautions must be taken to secure a definite condition of oxidation before applying a process of determination which depends upon the oxygen value of the precipitated oxide.

Manganese Oxide by Chlorate Process.

Mn taken. grm.	Mn found upon the hypothesis that MnO_2 is the oxide finally obtained. grm.	Error. grm.
By reduction with potassium iodide and titration of free iodine by thiosulphate.		
0 0643	0 0637	−0 0006
0 0643	0 0642	−0 0001
0 0643	0 0642	−0 0001
0.0651	0 0651	0 0000
0.1125	0 1121	−0 0004
0.1125	0.1121	−0.0004
0.1125	0 1120	−0 0005
0.1214	0 1206	−0.0008
0.1214	0 1207	−0 0007
0.1214	0.1223	+0 0009
0.1214	0.1214	0 0000
By reduction with arsenious oxide and titration of the excess by iodine in presence of Rochelle salt.		
0 1213	0 1212	−0.0001
0.1213	0.1201	−0 0012
0.1213	0.1203	−0 0010
0.1213	0 1208	−0.0005

Gooch and Austin show that a definite condition of oxidation may be secured by dissolving the precipitate in hydrochloric acid, diluting a little, adding sulphuric acid, evaporating until no more hydrochloric acid remains, and treating the manganous sulphate as follows, according to the method of Wright and Menke:* The solution of manganous sulphate (not exceeding 0.5 grm.), very nearly neutralized by potassium carbonate, is

* Jour. Chem. Soc., xxxvii, 36.

mixed with a solution of zinc sulphate (2 grm.), and a freshly made and carefully filtered dilute solution of potassium permanganate (1.5 grm.). The liquid now amounting to about 500 cm.^{3} is heated to 80° and acid potassium carbonate is added in quantity a little more than enough to neutralize the free acid. The loose precipitate is collected upon asbestos and after careful washing is returned to the flask in which the precipitation was made. The oxygen value of the oxide thus obtained, now that of the oxide MnO_2, may be determined by any appropriate method of treatment.

The preceding table contains the results of two different methods for determining the oxygen value of manganese dioxide separated by the chlorate method and brought to ideal condition by the procedure described.

NICKEL (COBALT).

The Electrolytic Determination of Nickel with the Rotating Cathode.

Gooch and Medway* determine nickel by deposition upon the rotating cathode,† from 50 cm.^{3} of solution containing one-half its volume of concentrated ammonia and a gram of ammonium sulphate. It should be especially noted that the solution must be kept within the limit of volume indicated above, as further dilution lengthens the time necessary for complete deposition.

Deposition of Nickel.

Nickel taken. grm.	Nickel found. grm.	Error. grm.	Current. amp.	$N.D._{100}$.	Time. min.
0.0954	0.0954	0 0000	1.5	5	30
0.0954	0.0953	−0.0001	3	10	25
0.0954	0.0956	+0.0002	3	10	25
0.0954	0 0953	−0.0001	3.5	11.7	20
0.0954	0 0955	+0 0001	3.5	11.7	20
0.1738	0.1736	−0 0002	3.5	11.7	25
0.1738	0.1740	+0.0002	3.5	11.7	25
0.1738	0 1740	+0 0002	4	13.3	25
0 1738	0 1737	−0.0001	4	13.3	25
0 1738	0 1738	0.0000	4	13.3	25

* F. A. Gooch and H. E. Medway, Am. Jour. Sci., [4], xv, 323.

† See page 11.

The Estimation of Nickel by Precipitation as the Oxalate and Titration with Potassium Permanganate.

Classen * has shown that nickel may be completely precipitated by treating the soluble nickel potassium oxalate with a large amount of acetic acid, and estimated by igniting the oxalate and weighing the oxide. Ward † has adapted this process to the volumetric determination of nickel by titration of the precipitated oxalate with permanganate. Finding that the precipitate formed by precipitation with potassium oxalate tends to include the alkali oxalate and that the nickel oxalate thrown down from the boiling acetic acid solution of a nickel salt falls in finely divided condition and is difficult to filter, Ward makes the first precipitation in water solution and adds the acetic acid afterward. The nickel salt is dissolved in a definite amount of water, crystallized oxalic acid is added to the boiling solution, and an equal volume of acetic acid is added. The precipitate is allowed to settle over night, filtered on asbestos in a perforated crucible, and washed with small amount of water. The crucible is placed in a beaker containing about 25 $cm.^3$ of dilute [1 : 3] sulphuric acid, and heat is applied to dissolve the oxalate. The volume of the solution is made up to about 200 $cm.^3$ with water, and cobalt sulphate is added until the green color of the nickel salt is neutralized and a slight pink tinge appears. The addition of the cobalt salt, recommended by Gibbs, is necessary to secure a definite end point. The solution is heated to boiling and the titration is made in the usual manner with permanganate. Results of this procedure are given in the table.

Precipitation by Oxalic Acid.

Nickel taken as the sulphate. grm.	Volume of water solution at precipitation. $cm.^3$	Oxalic acid. grm.	Acetic acid added. grm.	Nickel found. grm.	Error. grm.
0.0050	100	2	100	0.0054	+0.0004
0.0257	100	2	100	0.0258	+0.0007
0.0503	100	2	100	0.0502	−0.0001
0 1257	100	2	100	0.1271	+0 0014

* Zeit. anal. Chem., xvi, 470.
† H. L. Ward, Am. Jour. Sci., [4], xxxiii, 336.

The Detection of Nickel in Presence of Cobalt.

Browning and Hartwell * have modified advantageously the method of Clarke † for the separation of nickel and cobalt, according to which cobalt precipitated as the ferricyanide remains insoluble in ammonium hydroxide while the nickel salt is dissolved.

The method as modified may be described as follows: To not more than 0.1 grm. of the salts of the two elements in about 5 cm.^{3} of water a few drops of a saturated solution of alum are added, free mineral acid is neutralized with ammonium hydroxide, and the solution is made faintly acid with acetic acid. To this solution is added about 0.5 grm. of solid potassium ferricyanide with shaking to effect the solution of this reagent and the precipitation of the nickel and cobalt salts. The precipitate is treated with about 5 cm.^{3} of strong ammonium hydroxide and the mixture is filtered. To the filtrate, which should have no reddish color, a piece of sodium or potassium hydroxide about the size of a pea is added and the mixture is boiled. The appear-

Detection of Nickel.

$CoSO_4.7H_2O$ grm.	$NiSO_4.7H_2O$ grm.	$KAl(SO_4)_2$ saturated solution. cm.^{3}	$K_3FeC_6N_6$. grm.	NH_4OH (concentrated). cm.^{3}	NaOH, solid. grm.	Result.
....	0.0100	2	0.5	5	1	Heavy precipitate.
....	0.0050	2	0.5	5	1	Heavy precipitate.
....	0.0010	2	0.5	5	1	Heavy precipitate.
....	0.0003	2	0.5	5	1	Distinct.
....	0.0001	2	0.5	5	1	Plain.
0.10	.....	2	0.5	5	1	None.
0.10	0.0100	2	0.5	5	1	Heavy.
0 10	0.0050	2	0.5	5	1	Distinct.
0.10	0.0030	2	0.5	5	1	Very faint.
0.05	......	2	0.5	5	1	None.
0.05	0.0100	2	0.5	5	1	Heavy.
0.05	0.0050	2	0.5	5	1	Distinct.
0.05	0.0030	2	0 5	5	1	Plain.
0.05	0.0010*	2	0.5	5	1	Faint.

* Equivalent to 0.0002 of the metal.

* Philip E. Browning and John B. Hartwell, Am. Jour. Sci., [4], x, 316.
† F. W. Clarke, Am. Jour. Sci., [3], xlviii, 67.

ance of black nickelic hydroxide, very small amounts showing first as a dark coloration, indicates nickel.

The preceding table records the experimental results.

The Separation of Nickel and Cobalt by the Etherial Solution of Hydrochloric Acid.

The work of Havens * shows that very small amounts of nickel may be separated from cobalt, both taken as the chlorides, by treating the dry salts with the least amount of water which will effect solution (1 cm.^{3}), adding ether (10 cm.^{3} to 15 cm.^{3}) and saturating the mixture, cooled in running water, with gaseous hydrochloric acid. When saturation is complete the precipitated nickel chloride is filtered upon asbestos in the perforated crucible and washed with ether previously saturated with hydrochloric acid. From the etherial solution containing the minimum of water the precipitation of nickel chloride is practically complete, but when the amount of cobalt present exceeds a few milligrams errors due to inclusion of cobalt chloride in the mass of precipitated nickel chloride appear.

Results of this method of separation are given below. The nickel in the precipitated chloride and the cobalt in the filtrate were determined by the electrolytic process in the experiments recorded.

Separation by Ether-Hydrochloric Acid.

Nickel taken as the hydrous chloride.	Nickel found.	Error.	Cobalt taken as the hydrous chloride.	Cobalt found.	Error.
0.0068	0.0066	−0.0002	.....	......	......
0.0090	0.0090	0.0000	0.0300	......	......
0 0090	0.0091	+0.0001	0.0123	0.0127	+0.0004
0.0469	0.0490	+0 0021	0.0700	......	......
0.0468	0.0503	+0.0035	0.0700	......	......
0 0472	0.0493	+0 0021	0.0700	......	......

Iron.

The Determination of Iron in the Ferric State by Reduction with Sodium Thiosulphate and Titration of the Excess of the Latter with Iodine.

The action of sodium thiosulphate upon a ferric salt takes place with the formation of a ferrous salt and a tetrathionate, according to the following expression

$$2\,FeCl_3 + 2\,Na_2S_2O_3 = 2\,FeCl_2 + Na_2S_4O_6 + 2NaCl.$$

* Franke Stuart Havens, Am. Jour. Sci., [4], vi, 396.

After many attempts by others * to utilize the reaction in the quantitative estimation of iron, Norton † has succeeded in placing the method upon a successful basis for the determination of moderate amounts of ferric iron by direct titration. Norton has studied carefully the possible sources of error: incompleteness in the reduction of the ferric salt; decomposition of the thiosulphate by the acid, resulting in the subsequent over-run of iodine; the possible tendency of the ferric salt under concentration to oxidize the thiosulphate to the condition of the sulphate rather than to that of the tetrathionate; and finally the oxidizing action of the air, which may tend to keep up progressive oxidation of the iron salt and excessive expenditures of thiosulphate. The first three sources of difficulty tend to produce errors of deficiency; the fourth an error of excess.

It is shown that with quantities of ferric oxide present up to 0.1 grm. the dilution may vary from 400 $cm.^3$ to 1000 $cm.^3$ for each cubic centimeter of strong hydrochloric acid and still give excellent results. At a dilution greater than 1000 $cm.^3$ the action of the thiosulphate is incomplete in presence of 1 $cm.^3$ of the acid, and at a smaller dilution than 400 $cm.^3$ the decomposing action of the acid on the thiosulphate becomes noticeable. When larger quantities of iron oxide are dealt with, the dilution must be increased proportionately to the quantity of ferric oxide. On this account it is necessary, assuming that the quantity of acid present is always kept within the maximum strength mentioned, 1 $cm.^3$ to 400 $cm.^3$, to regulate the dilution by the approximate quantity of the iron so that not less than 400 $cm.^3$ of water shall be used to every 0.1 grm. of iron oxide present. Under properly regulated conditions of dilution as regards acid and the iron salt, the reduction is completed in from five to ten minutes.

Under the conditions of acidity and dilution laid down, the process of reduction is complete at the ordinary room temperature within ten minutes after the introduction of the thiosulphate, and experience shows clearly the danger of submitting mixtures of sodium thiosulphate and acid to temperatures much above the

* Sherer; Gelehrte Anzeigen der könig. Bayrisch. Acad., 1859. Mohr, Ann. Chem., cxiii, 269. Kremer and Landolt, Zeit. anal. Chem., i, 214. Oudemanns, Zeit. anal. Chem., vi, 129; ix, 362. Haswell, Repert. anal. Chem., i, 179. Bruel, Compt. rend., xcvii, 954.

† John T. Norton, Jr., Am. Jour. Sci., [4], viii, 25.

ordinary. On the other hand, artificial reduction of temperature tends to retard the action to an undesirable degree.

To complete the reduction within a reasonable time under the conditions of acidity and concentration there should always be present an excess of thiosulphate in amount from 15 cm.^{3} to 35 cm.^{3} of the $n/10$ solution.

The following procedure is recommended: The dilution must be at least 400 cm.^{3} for each 0.1 of a grm. of iron oxide present; the quantity of acid should never exceed 1 cm.^{3} of the strong acid to each 400 cm.^{3} of water; the time of reduction must be short to avoid progressive oxidation; the temperature of the solution should be kept at the normal temperature of the atmosphere; and the excess of sodium thiosulphate present should never be less than 15 cm.^{3} of the $n/10$ solution. In the case of large dilution freshly boiled water should be used.

Titration by Thiosulphate.

Fe_2O_3 taken. grm.	Fe_2O_3 corrected. grm.	Dilution. cm.^{3}	HCl conc. cm.^{3}	Excess $Na_2S_2O_3$. cm.^{3}	Fe_2O_3 found. grm.	Error. grm.
0.0125	0.0125	200	$\frac{1}{8}$	23.5	0.0125	0.0000
0.0250	0.0250	400	$\frac{1}{4}$	21.98	0.0250	0.0000
0.0250	0.0250	400	$\frac{1}{4}$	17	0.0250	0.0000
0.0250	0.0250	400	$\frac{1}{4}$	17	0.0250	0.0000
0.0500	0.0499	400	$\frac{1}{2}$	24	0.0498	−0.0001
0.0500	0.0499	400	$\frac{1}{2}$	19	0.0498	−0.0001
0.0500	0.0499	400	$\frac{1}{2}$	15.1	0.0497	−0.0002
0.0500	0.0499	400	$\frac{1}{2}$	19	0.0498	−0.0001
0.1001	0.0999	400	1	23.1	0.0993	−0.0006
0.1001	0.0999	400	1	17.93	0.0997	−0.0002
0.1001	0.0999	400	1	22.92	0.0997	−0.0002
0.1001	0.0999	400	1	18	0.0997	−0.0002
0.1001	0.0999	400	1	16	0.0996	−0.0003
0.1498	0.1495	600	$1\frac{1}{2}$	23.26	0.1493	−0.0002
0.1498	0.1495	600	$1\frac{1}{2}$	16.66	0.1493	−0.0002
0.1498	0.1495	600	$1\frac{1}{2}$	26.87	0.1475	−0.0020
0.1996	0.1992	800	2	22.38	0.1990	−0.0002
0.1996	0.1992	800	2	17.29	0.1999	+0.0007
0.1996	0.1992	800	2	22.20	0.1991	−0.0001
0.4045	0.4037	1600	4	16.03	0.4042	+0.0005
0.4045	0.4037	1600	4	16.2	0.4023	−0.0014
0.4018	0.4010	1600	4	16.24	0.4007	−0.0003
0.5051	0.5041	1800	4	15.27	0.5026	−0.0015

In applying the method to an insoluble ferric compound, like the oxide, the material, best not exceeding the equivalent of 0.2 grm. of the oxide, is dissolved in hydrochloric acid; the solution is evaporated to pasty condition and then diluted; a drop of potassium sulphocyanate is introduced with $n/10$ sodium thiosulphate, 15 cm.^{3} to 35 cm.^{3} in excess; the liquid is allowed to stand until perfectly colorless; and the excess of thiosulphate is determined by $n/10$ iodine with the aid of the starch indicator.

Results obtained by this procedure are given in the table.

The Standardization of Potassium Permanganate in Iron Analysis.

Charlotte F. Roberts * has discussed the standardization of potassium permanganate for practical work in iron analysis. The best authorities agree in considering that the standard of the permanganate should finally be referred to a standard solution of ferric chloride, but the difficulty consists in determining with accuracy the amount of iron in this solution. Though the purest iron which can be obtained commercially is used as the basis, the resulting ferric chloride still contains some silica and phosphorus, which must be eliminated or the amount determined gravimetrically. The process of determining the amount of iron in the ferric chloride solution, upon which the potassium permanganate is finally standardized, thus becomes long and tedious.† To obviate this, the potassium permanganate may be compared with a solution containing a known weight of iron and the solution of ferric chloride may then be standardized by reduction and titration with this same potassium permanganate. Since electrolytic iron is undoubtedly the purest form of iron known, it would seem that potassium permanganate titrated against this might be expected to give trustworthy results for the first comparison.

For the production of a definite amount of electrolytic iron two different courses are open. Either a weighed amount of a pure iron salt, as ammonio-ferrous sulphate, may be taken and the iron completely precipitated by electrolysis; or an indefinite quantity of the salt may be taken and subjected to electrolysis for a time, and the amount of iron determined by weighing the electrolytic deposit. This second method is recommended as being much more rapid and free from details of manipulation which render the first difficult. About 10 grm. of ammonio-ferrous sulphate are dissolved in 150 cm.^{3} of water, 5 cm.^{3} of a

* Charlotte F. Roberts, Am. Jour. Sci., [3], xlviii, 290 (1894).

† Blair, The Chemical Analysis of Iron, 7th Ed., page 235.

saturated solution of potassium oxalate are added, and the whole is heated with a considerable quantity of solid ammonium oxalate until a clear solution is obtained. This solution is decomposed in a beaker between two platinum electrodes, the iron being deposited on a piece of platinum foil of a size convenient for insertion in a rather large weighing bottle, in which it is weighed both before and after the electrolysis. An hour and a half, with a current of 2 amperes, is sufficient for the precipitation of 0.4 grm. to 0.5 grm. of pure iron, and it was found unadvisable to use a current much stronger than 2 amperes, since a higher current showed a tendency to render the deposit less smooth and compact. After washing, drying and weighing in the usual way, the iron was dissolved in hydrochloric acid, the weighing bottle being used instead of the small flask ordinarily employed in this operation, the oxidized iron was reduced with zinc, and finally titrated with the solution of potassium permanganate in presence of sulphuric acid and a large amount of water.

The following table shows the results obtained by this procedure, the first column giving the weight of the electrolytic deposit of iron, and the second the weight of iron found by titration with potassium permanganate previously standardized on a special ammonium oxalate which had been shown to give results identical with those obtained by the use of specially prepared anhydrous lead oxalate:

Comparison of Standards.

I.	II.	Difference.	Average.
0.4357	0.4364	+0 0007	
0.3551	0.3559	+0 0008	
0.2552	0.2550	−0 0002	
0 2898	0 2890	−0 0008	
0.1590	0.1599	+0 0009	
0.3528	0.3534	+0 0006	0 0000
0 4498	0 4494	−0 0004	
0 5086	0.5085	−0 0001	
0 4462	0.4457	−0 0005	
0 4226	0.4222	−0 0004	
0 5170	0.5165	−0 0005	

The standard of potassium permanganate as determined from pure iron is identical with that obtained in these experiments with the special ammonium oxalate, but the standard obtained in the former way would under ordinary conditions be more satisfactory for work in iron analysis. A simple and rapid method, then, for standardizing the potassium permanganate solution is to determine its strength, first, by comparison with electrolytic iron in the manner above described. Then by reduction and titration with the permanganate the exact amount of iron in each cubic centimeter of the ferric chloride solution may be determined. This being ascertained, the solution of ferric chloride may be employed at any time for the standardization of potassium permanganate.

The Behavior of Ferric Chloride in the Jones Reductor.

The column of amalgamated zinc as applied in the Jones reductor * has proved very effective in the reduction of ferric sulphate preparatory to the estimation of the ferrous salt by potassium permanganate. The impression has prevailed that the salt of iron acted upon by the amalgamated zinc must be the sulphate and that chlorides and nitrates must not be present even in small amounts. Randall † has shown, however, that it is possible to reduce ferric chloride in the zinc reductor and to determine the iron with success by potassium permanganate, provided the titration is carried on in the presence of manganous sulphate and in solutions sufficiently dilute. A small excess of hydrochloric acid has no influence on the result in dilute solutions, and in a volume of one liter the excess may amount to as much as 25 $cm.^3$ of the strongest acid.

According to the method described, the procedure is to first run 100 $cm.^3$ of warm dilute 2.5 per cent sulphuric acid through the reductor charged with amalgamated 20-mesh zinc, next to pass in the iron solution diluted with 100 $cm.^3$ of warm 2.5 per cent sulphuric acid and then to wash down with 200 $cm.^3$ of the warm dilute acid followed by 100 $cm.^3$ of hot water. The receiving flask of the reductor is kept in a vessel containing running tap water, so that the solution is cooled as fast as it is reduced. Manganous sulphate, 1 grm. to 5 grm., is added before the titration with permanganate. Following are results obtained by this method, and, for comparison, the results obtained upon the same solution of ferric chloride standardized by evaporating

* See page 437.

† D. L. Randall, Am. Jour. Sci., [4], xxi, 128.

measured amounts with 10 cm.^{3} of sulphuric acid to the fuming point of the acid, passing the solution through the reductor and titrating with permanganate.

Reduction After Conversion of Chloride to Sulphate.

$FeCl_3$.	H_2SO_4[1 : 1.]	Volume at titration.	$KMnO_4$.	Fe found.		Variation from average.
cm.^{3}	cm.^{3}	cm.^{3}	cm.^{3}	grm.		grm.
75	25	750	70.75	0.4867		−0.0005
75	25	750	70.83	0.4873	0.4872*	+0.0001
75	25	750	70.83	0.4873		+0.0001
75	25	750	70.88	0.4876		+0.0004

Reduction of the Chloride.

$FeCl_3$.	Fe taken.*	H_2SO_4 2.5 per cent.	HCl conc.	Volume at titration.	$MnSO_4$.	$KMnO_4$.	Fe found.	Error.
cm.^{3}	grm.	cm.^{3}	cm.^{3}	cm.^{3}	grm.	cm.^{3}	grm.	grm.
75	0.4872	100	0	750	1	70.81	0.4871	−0.0001
75	0.4872	100	0	750	1	70.75	0.4867	−0.0005
75	0.4872	100	0	750	1	70.83	0.4873	+0.0001
75	0.4872	100	0	750	1	70.82	0.4872	0.0000
75	0.4872	100	0	750	1	70.83	0.4873	+0.0001
100	0.6497	100	0	750	1	94.43	0.6497	0.0000
100	0.6497	100	0	750	1	94.44	0.6498	+0.0001
100	0.6497	100	10	750	1	94.44	0.6498	+0.0001
100	0.6497	100	20	1000	5	94.53	0.6503	+0.0006
100	0.6497	100	25	1000	5	94.53	0.6503	+0.0006
100	0.6497	100	25	1000	1.25	94.53	0.6503	+0.0006
100	0.6497	100	25	1000	1.25	94.48	0.6500	+0.0003
100	0.6497	100	25	1000	5.00	94.49	0.6501	+0.0004

* Mean of results obtained by titration after the conversion of the chloride to sulphate.

The Effect of Nitric Acid in the Titration of a Ferrous Salt by Potassium Permanganate.

In Schneider's * method for the determination of manganese, permanganic acid is titrated with hydrogen peroxide in the presence of nitric acid, and Ibbotson † and Brearley in their modification of this process recommend the use of standard ferrous ammonium sulphate instead of hydrogen peroxide. Blair ‡ recommends the use of ferrous ammonium sulphate. Obviously it is of interest to know the exact effect of nitric acid upon solu-

* Ding. Pol. Jour., cclxix, 224.

† Chem. News, lxxxiv, 247.

‡ Blair, Jour. Am. Chem. Soc., xxvi, 793. Also, Chemical Analysis of Iron.

tions of the ferrous salt undergoing oxidation by permanganate, and an investigation of this matter by Randall * has shown that when more than 10 per cent by volume (20 cm.^{3} in 200 cm.^{3}) of the concentrated acid is present oxidation of the ferrous salt takes place, as is made plainly evident by the change of color of the solution, low results, and uncertain end reaction. The error due to oxidation of the ferrous salt by nitric acid is, however, in part counterbalanced by the reoxidation by permanganate of any nitrous acid which may have been produced. If the titration is made without unnecessary delay the presence of as much as 5 per cent by volume of nitric acid has no appreciable effect upon the estimation of ferrous iron. Following are results of the titration of two different solutions of ferrous sulphate in presence of varying amounts of nitric acid:

Permanganate Titration in Presence of Nitric Acid.

	$FeSO_4$. cm.^{3}	Dilution. cm.^{3}	H_2SO_4 [1 : 1.] cm.^{3}	HNO_3. cm.^{3}	Approx. $n/10$ $KMnO_4$. cm.^{3}
I	25	200	5	0	13.37
	25	200	5	0	13.39
	25	200	5	0	13.41
	25	200	5	0	13.38
	25	200	5	0	13.38
	25	200	0	3	13.38
	25	200	0	5	13.40
	25	200	0	5	13.41
	25	200	0	5	13.40
	25	200	0	5	13.38
II	25	200	5	0	13.47
	25	200	5	0	13.50
	25	200	5	0	13.49
	25	200	0	5	13.53
	25	200	0	10	13.50
	25	200	0	20	13.51
	20	200	0	30	13.00

The Permanganate Estimation of Iron in Presence of Titanium.

For analytical purposes, a ferric salt in solution is most easily and conveniently reduced to the ferrous condition by the action of zinc; and where many determinations of iron are to be made, the use of the well-known Jones reductor† yields accurate results

* D. L. Randall, Am. Jour. Sci., [4], xxiii, 139.

† See page 347.

very rapidly. The use of zinc, whether in the flask or in the reductor, has, however, been precluded when the ferric salt is accompanied by titanic acid, for this substance is reduced with the iron and subsequently oxidized by the permanganate in the titration process. When, therefore, titanium is present with the iron, it has been customary to have recourse to other methods of reduction. In this event, either hydrogen sulphide or sulphur dioxide is substituted for the zinc to bring about the reduction of the ferric salt, since titanic acid is not reduced by these reagents; but the removal of the excess of hydrogen sulphide or of sulphur dioxide from solution without oxidation of the ferrous salt is not an easy or rapid process.

Gooch and Newton * have studied the problem of adapting the ordinary convenient process of reducing the ferric salt by zinc to the estimation of iron in presence of titanium. It is obvious that to solve this problem it is only necessary to find and employ some reagent which shall be neutral toward the ferrous salt but capable of reoxidizing the titanium compounds formed by the reducing action of the zinc, and shall have no action on the permanganate. Compounds of silver, copper or bismuth oxidize easily the reduced titanium salt; but the use of a compound of silver is precluded by the fact that it oxidizes also the ferrous salt to some extent as well as the titanium salt. Cupric salts and pure bismuth oxide prove, however, to be without action upon the ferrous salt. It is found that the violet color of the solution containing the titanium compound produced by the action of zinc upon the titanium sulphate is discharged upon adding a little cupric sulphate to the solution and heating, and, after filtering, a drop of potassium permanganate gives its characteristic rose tint to the solution. It is found also that when cupric oxide is added to a similarly reduced solution of the titanium salt the characteristic color vanishes on shaking the solution. The following table contains the results obtained in titrating with potassium permanganate the ferrous salt left after reducing by zinc in small flasks carefully measured amounts of ferric sulphate and titanium sulphate, treating the mixture thus obtained with cupric sulphate or with cupric oxide, and filtering off the reduced copper and cuprous salt:

* F. A. Gooch and H. D. Newton, Am. Jour. Sci., [4], xxiii, 365.

Oxidation of the Titanous Sulphate by Cupric Compounds: Titration of Ferrous Sulphate by Permanganate.

Fe_2O_3 taken. grm.	TiO_2 taken. grm.	Fe_2O_3 found. grm.	Error. grm.	
0.1375	0.1	0.1378	+0.0003	Treated with $CuSO_4$.
0.1375	0.1	0.1374	−0.0001	
0.1375	0.1	0.1377	+0.0002	
0.1375	0.1	0.1378	+0.0003	Treated with CuO.
0.1375	0.1	0.1378	+0.0003	
0.1375	0.2	0.1382	+0.0007	

Similar experiments in which bismuth oxide was substituted for the copper oxide are also recorded. To the measured amount of ferric sulphate and titanium sulphate contained in a small flask, provided as usual with the funnel valve, zinc is added and the reduction effected in the ordinary manner. The titanium salt appears to act catalytically in this process, so that reduction goes on more easily and with less expenditure of zinc than in the similar reduction of the ferric salt taken by itself. After the zinc has disappeared, the solution, of characteristic violet color, is cooled in the flask, treated with a little bismuth oxide, gently shaken, filtered from the excess of bismuth oxide and the precipitated bismuth into about a liter of cold water, and titrated with standard potassium permanganate. Results are given in the following table:

Oxidation of Titanous Sulphate by Bismuth Oxide: Titration of Ferrous Sulphate by Permanganate.

Ferric sulphate. $cm.^3$	TiO_2. grm.	$KMnO_4$. $cm.^3$	Fe_2O_3 taken. grm.	Fe_2O_3 found. grm.	Error. grm.
10	0.04	12.84	0.0993	0.0992	−0.0001
10	0.06	12.85	0.0993	0.0993	0.0000
10	0.08	12.90	0.0993	0.0997	+0.0004
10	0.1	12.90	0.0993	0.0997	+0.0004
10	0.1	12.89	0.0993	0.0996	+0.0003
10	0.1	12.85	0.0993	0.0993	0.0000
10	0.1	12.80	0.0993	0.0989	−0.0004
10	0.1	12.90	0.0993	0.0997	+0.0004
10	0.2	12.90	0.0993	0.0997	+0.0004
10	0.2	12.89	0.0993	0.0996	+0.0003
10	0.2	12.90	0.0993	0.0997	+0.0004
20	0.1	25.70	0.1986	0.1986	0.0000

It is evident that either cupric sulphate, cupric oxide, or bismuth oxide may be used to reoxidize the salt of titanium reduced by zinc, without affecting appreciably the ferrous salt in solution.

Similar results are obtained when the ferric sulphate solution is passed through the column of amalgamated zinc in the Jones reductor.* The flask is kept cool in running water, a small amount of bismuth oxide added, the flask shaken and allowed to stand a few minutes, and the mixture filtered with the aid of the suction pump. In the cold solution free from dissolved oxygen there is little danger of reoxidation of ferrous sulphate, as has been shown by Peters and Moody.† The experimental results are given in the table.

The Zinc Reductor: Bismuth Oxide: Permanganate.

Ferric sulphate. cm.^{3}	TiO_2. grm.	$KMnO_4$. cm.^{3}	Fe_2O_3 taken. grm.	Fe_2O_3 found. grm.	Error. grm.
40	0.01	46.80	0.3943	0.3943	0.0000
40	0.02	46.79	0 3943	0.3942	−0.0001
40	0.04	46.80	0.3943	0.3943	0.0000
40	0.06	46.83	0.3943	0.3946	+0.0003
40	0.1	46.75	0.3943	0.3939	−0.0004
40	0.1	46 82	0 3943	0.3945	+0.0002
40	0.1	46.78	0.3943	0.3941	−0.0002
40	0.1	46.80	0 3943	0.3943	0.0000
40	0.1	46.75	0.3943	0.3939	−0.0004
40	0.1	46.80	0.3943	0.3943	0.0000
40	0.2	46.77	0.3943	0 3940	−0.0003
40	0.2	46.81	0.3943	0.3944	+0.0001
40	0.2	46.85	0.3943	0.3947	+0.0004

The Estimation of Iron by Potassium Permanganate after Reduction with Titanous Sulphate.

Knecht ‡ was the first to recommend the use of titanium sesquioxide and its salts in volumetric operations where a rapid and powerful reducing agent is required, and later in collaboration with E. Hibbert § published a method for the direct titration of ferric chloride by a standard solution of titanous chloride,

* See page 347.

† See page 371.

‡ Ber. Dtsch. Chem. Ges. xxvi, 166.

§ Ibid., xl, 3819.

using potassium sulphocyanate as an indicator, the reaction between the two salts taking place according to the following equation:

$$FeCl_3 + TiCl_3 = FeCl_2 + TiCl_4.$$

According to these investigators the method yields excellent results, and rapidly. The only precautions necessary are that the solution of titanous chloride, being naturally very sensitive to the action of atmospheric oxygen, must, after having been boiled with marble to expel occluded oxygen, be kept under a constant pressure of hydrogen. It has been found, however, that even with such precautions the standard of the solution gradually changes and must be checked from time to time against known amounts of ferric iron. The proposal has therefore been made by Newton,* to reduce the iron by titanous sulphate, oxidize the excess of titanous sulphate, and titrate the remaining ferrous salt by permanganate.†

A solution of titanous sulphate of convenient strength may be made up by mixing 20 grm. of commercial titanic acid with three times its own weight of a mixture of sodium and potassium carbonates and fusing in a platinum crucible, treating the melt (after being finely ground) in a platinum dish with hot concentrated sulphuric acid, cooling, diluting a little, filtering through asbestos, treating with zinc until reduction is accomplished, and, while zinc is still left in the flask, filtering the solution quickly through a platinum cone into about two liters of freshly distilled water contained in a small reservoir connected with burette and hydrogen generator.

To determine ferric sulphate in solution it is only necessary to add in the cold an excess of titanous sulphate, prepared as described, destroy this excess by treating with a little bismuth oxide,‡ filter from the excess of bismuth oxide into about a liter of cool distilled water, and titrate with permanganate. If an appreciable amount of hydrochloric acid is present in the solution of the ferric salt it is advisable to evaporate to dryness and convert the chloride to sulphate by use of concentrated sulphuric acid. Upon addition of the sulphuric acid a white pasty mass is formed which rapidly goes into solution on diluting with water

* H. D. Newton, Am. Jour. Sci., [4], xxv, 343.

† See page 500.

‡ See page 501.

and warming. As titanous sulphate prepared in the manner described above always contains some iron, it is necessary to make a correction for this by treating with bismuth oxide an amount of the solution equal to that used, filtering, and running in potassium permanganate to color. This correction should not amount to more than 0.1 cm.[3] when working with 0.3 grm. of ferric oxide. If these simple precautions be taken, ferric iron may be determined with rapidity and exactness by reduction with titanous sulphate, treatment with bismuth oxide, and titration with potassium permanganate. Test results are given in the table.

Permanganate Titration after Reduction with Titanous Sulphate.

$Fe_2(SO_4)_3$. cm.[3]	$KMnO_4$. cm.[3]	Fe_2O_3 taken. grm.	Fe_2O_3 found. grm.	Error. grm.
20	13.46	0 1063	0.1064	+0.0001
20	13 42	0 1063	0 1060	−0.0003
20	13.44	0.1063	0 1062	−0 0001
20	13.44	0 1063	0.1062	−0.0001
20	13.41	0.1063	0 1060	−0 0003
30	20.18	0.1594	0.1594	0.0000
30	20.20	0.1594	0.1596	+0 0002
30	20.20	0.1594	0 1596	+0.0002
30	20.22	0.1594	0 1598	+0 0004
30	20.19	0.1594	0.1595	+0.0001
40	26.92	0.2127	0.2127	0 0000
40	26.90	0.2127	9.2125	−0 0002
40	26.90	0.2127	0.2125	−0 0002
40	26.93	0.2127	0.2128	+0.0001
40	26.90	0.2127	0.2125	−0.0002

Separations of Iron by Volatilization in Gaseous Hydrogen Chloride.

Metallic iron is easily acted upon by an excess of chlorine at moderately elevated temperatures with the formation of ferric chloride and by hydrochloric acid gas with formation of ferrous chloride. Out of contact with air, or moisture, both chlorides may be volatilized at appropriate temperatures — the ferric chloride below 200° C.; the ferrous chloride at a bright red heat. If water vapor, or oxygen, or air, be present during the heating, both chlorides are partially decomposed with the formation of non-volatile residues, ferric oxide or ferric oxychloride. When ferric oxide is submitted to the action of hydrochloric acid gas at about 200° the greater part of the iron sublimes,* as ferric chloride, but a residue remains. At the outset the ferric oxide volatilizes quickly and abundantly in the form of the greenish vapor of ferric chloride, and if the operation is interrupted at this stage the residue which remains is nearly black, insoluble in water, slightly soluble in cold hydrochloric acid, and readily soluble in hot hydrochloric acid with the formation of ferric chloride. It is probably something analogous to the oxychloride which Rousseau† identified as the product of the action of water upon ferric chloride at 275° to 300°. This dark residue yields to the action of the hydrochloric acid at 180° to 200° only slowly; but ultimately the residue is essentially ferrous chloride. Little volatilization occurs within the range of temperature from 200° to 500°.

If the temperature of the oxide is 450° to 500° when the brisk current of acid begins to act, the whole mass of oxide is rapidly converted and volatilizes without residue, the production of the ferrous chloride (formed by dissociation of ferric chloride) being apparently kept at a minimum by the adjustment of equilibrium in the atmosphere of ferric chloride and chlorine resulting from the partial dissociation. If dissociation of ferric chloride to ferrous chloride is the cause of the formation of a residue at 200°, the temperature of slow action, the introduction of chlorine into the atmosphere of hydrochloric acid gas should change the condition of equilibrium and enable the ferric chloride to volatilize without dissociation. Gooch and Havens‡ find, as a matter of fact, that if a little manganese dioxide is added to the contents of the hydrogen chloride generator, so that the gas may carry with it a little chlorine, every trace of ferric oxide is volatilized from the boat at 180° to 200°. The residue of ferrous chloride found at 180° to 200° when the hydrochloric acid is used alone is likewise volatilized at the same temperature when the admixture of chlorine is made.

These facts, that ferric oxide is completely volatile in hydrochloric acid gas applied at once at a temperature of 450° to 500°, and at 180° to 200° if the acid carries a little chlorine, open

* Moyer, Jour. Am. Chem. Soc., xviii, 1029.

† Compt. rend., cxvi, 118.

‡ F. A. Gooch and Franke Stuart Havens, Am. Jour. Sci., [4], vii, 370.

the way to many analytical separations of iron from substances not volatile under these conditions.

The separation of the iron oxide from various oxides proves to be complete at 450° to 500° if the mixture is submitted at once to the action of hydrochloric acid gas, or at 180° to 200° when chlorine is mixed with the hydrochloric acid. The temperature of red heat employed by Deville * is unnecessary if the mixed oxides are submitted at once to the action of hydrochloric acid at 450° to 500° without previous gentle heating in the acid atmosphere. The mixture of chlorine and hydrochloric acid is to be preferred, however, not only because the temperature of the reaction is lower but because it needs no regulation, while the danger of error arising from the liability of ferric chloride to dissociate, or from deficiency of oxidation in the oxide treated, or from mechanical loss due to too rapid volatilization is avoided.

According to the procedure adopted, the mixed oxides, put in a porcelain boat which is placed in a wide combustion tube heated in a small furnace, are submitted to the action of dry hydrochloric acid gas, generated by dropping sulphuric acid upon a mixture of strong hydrochloric acid, common salt, and a small amount of manganese dioxide. The gas is admitted at one end of the combustion tube and passed out at the other through a water trap, while the required temperature, best 200° to 300°, is maintained by regulating the burners of the furnace. The time of action varies somewhat with the condition of the oxide to be volatilized and the temperature. Generally an hour's heating at 200° proves sufficient for the complete removal of 0.1 grm. of iron. At higher temperatures the action is more rapid; but the non-volatile oxide is liable to mechanical loss if the volatilization of the iron is too rapid. It is better, therefore, to make use of a lower temperature until the volatilization of iron is nearly complete, and then to raise the heat for a few minutes to insure the removal of the last traces of the volatile chloride.

Results of this procedure follow.

Iron and Aluminium. Separations of iron oxide and aluminium oxide by the procedure outlined above are shown in the following table: †

* Ann. Chim., [3], xxxviii, 23.

† Gooch and Havens, loc. cit.

Volatilization of Ferric Chloride.

Fe_2O_3 taken. grm.	Al_2O_3 taken. grm.	Al_2O_3 found. grm.	Error. grm.	Time. hours.	Temperature. C°.	Atmosphere
0 1000	. . .	. . .	0 0000	½	450–500	HCl.
0 2000	. .	. .	0.0000	1	450–500	HCl.
0 1020	0.1015	0 1015	0 0000	½	450–500	HCl.
0 2145	0.1006	0 1008	+0 0002	¾	450–500	HCl.
0 1000	. . .	. . . .	0 0000	¾	180–200	$HCl+Cl_2$.
0.1000	0.1032	0 1032	0.0000	¾	180–200	$HCl+Cl_2$.
0 1072	0.1013	0 1015	+0.0002	1	180–200	$HCl+Cl_2$.
0.2045	0.1032	0.1033	+0 0001	1¼	180–200	$HCl+Cl_2$.
0 1050	0.1023	0.1019	−0.0004	½	450–500	$HCl+Cl_2$.
0.2008	0.1007	0 1006	−0 0001	¾	450–500	$HCl+Cl_2$.
. . . .	0.1087	0 1087	0 0000	1¾	450–500	$HCl+Cl_2$.

Iron and Beryllium. The separation of iron and beryllium has been tested by Havens and Way.*

Volatilization of Ferric Chloride.

Fe_2O_3 taken. grm.	BeO taken. grm.	BeO found. grm.	Error. grm.
. . . . . .	0.1309	0 1311	+0.0002
. . . . . .	0.1285	0 1285	0.0000
0.0997	0.0456	0 0457	+0.0001
0.1045	0.1099	0.1099	0.0000
0.1215	0.1080	0 1081	+0.0001
0.1510	0.1305	0.1290	−0.0015
0.2030	0.1081	0 1083	+0.0002

Iron and chromium. Tests of the separation of iron and chromium have also been made by Havens and Way.†

Volatilization of Ferric Chloride.

Fe_2O_3 taken. grm.	Cr_2O_3 taken. grm.	Cr_2O_3 found. grm.	Error. grm.
. . . . . .	0.1008	0 1008	0.0000
0.1007	0.1006	0 1006	0.0000
0.1007	0 1000	0.1002	+0.0002
0 1010	0.1005	0 1003	−0.0002
0 1019	0.1006	0.1005	−0 0001
0 2007	0.1003	0.0999	−0.0004

* Franke Stuart Havens and Arthur Fitch Way, Am. Jour. Sci., [4], viii, 217.
† Loc. cit.

Iron and Zirconium. The following table contains results by Havens and Way * in the separation of iron and zirconium:

Volatilization of Ferric Chloride.

Fe_2O_3 taken. grm.	ZrO_2 taken. grm.	ZrO_2 found. grm.	Error. grm.
.. ..	0.1516	0.1516	0.0000
0.1053	0 1010	0 1010	0 0000
0.1204	0 1519	0 1523	+0 0004
0.1236	0.1516	0 1517	+0.0001
0.2150	0.1517	0 1519	+0.0002

The Estimation of Iron and Vanadium in Presence of Each Other.

In view of the difficulties attendant upon the separation of iron and vanadium, Edgar † has developed a method by which these elements may be estimated in the presence of each other.

If a solution containing vanadic acid and iron be reduced by means of sulphur dioxide the reoxidation by potassium permanganate proceeds according to the equation

$$\text{I.} \quad 5\,V_2O_4 + 10\,FeO + 4\,KMnO_4 = 5\,V_2O_5 + 5\,Fe_2O_3 + 2\,K_2O + 4\,MnO.$$

If this solution, after titration, be passed through a column of amalgamated zinc in the Jones reductor,‡ the receiving flask being charged with a solution of ferric sulphate, the reduction is carried, in the case of vanadic acid, to the condition of V_2O_2 and the reoxidation by permanganate proceeds according to the equation

$$\text{II.} \quad 5\,V_2O_2 + 10\,FeO + 8\,KMnO_4 = 5\,V_2O_5 + 5\,Fe_2O_3 + 4\,K_2O + 8\,MnO.$$

The difference between the amounts of permanganate used in the first and second titrations is evidently used in oxidizing the vanadium from the condition of V_2O_2 to V_2O_4, and measures the amount of vanadic acid present. This being known, the iron present may be calculated from the amount of permanganate used in either titration.

* Loc. cit.

† Graham Edgar, Am. Jour. Sci., [4], xxvi, 79.

‡ See page 349.

Into the slightly acid solution of the ferric salt and vanadic acid, contained in a stoppered flask provided with inlet and outlet tubes, is passed a current of sulphur dioxide until the color changes from red into green and finally into a clear blue. A few cubic centimeters of dilute sulphuric acid are added, and the solution is heated to boiling, the current of sulphur dioxide being replaced by one of air-free carbon dioxide. When the last traces of sulphur dioxide have been removed, the flask is cooled in running water, the atmosphere of carbon dioxide being maintained, and, when thoroughly cool, titrated with potassium permanganate until the color has changed from blue into yellowish green. The solution is then heated to 70° or 80° and the titration is completed at that temperature. The amount of permanganate used indicates the amount of vanadic acid present according to equation I.

Differential Reductions by Sulphur Dioxide and by Zinc.

n/10 × 0.9545.		V_2O_5 taken.	V_2O_5 found.	Error on V_2O_5.	Fe_2O_3 taken.	Fe_2O_3 found.	Error Fe_2O_3.
I. $KMnO_4$.	II. $KMnO_4$.						
cm.³	cm.³	grm.	grm.	grm.	grm.	grm.	grm.
31 90	58 02	0.1136	0.1137	+0 0001	0.1437	0.1436	−0.0001*
31.90	58.04	0.1136	0.1138	+0 0002	0.1437	0.1435	−0.0002*
31.85	58.00	0.1136	0.1138	+0 0002	0.1437	0.1433	−0.0004*
31.90	58.00	0.1136	0.1136	0 0000	0 1437	0.1437	0 0000*
25.30	38.25	0.0568	0.0568	0.0000	0.1437	0.1423	−0 0004†
25.29	38.30	0.0568	0.0566	−0.0002	0.1437	0.1433	−0.0004†
15 98	29.02	0.0568	0.0568	0.0000	0 0719	0.0721	+0.0002†
38.50	77.60	0.1704	0.1702	−0.0002	0.1437	0.1442	+0 0005‡
38.45	77.60	0 1704	0.1704	0 0000	0.1437	0.1438	+0.0001‡
38.45	77.58	0 1704	0.1703	−0.0001	0.1437	0.1439	+0 0002‡
22.50	48.60	0.1136	0.1136	0.0000	0.0719	0.0720	+0.0001*
22.50	48.60	0.1136	0.1136	0.0000	0.0719	0.0720	+0.0001*
22.45	48.58	0.1136	0.1137	+0.0001	0.0719	0.0716	−0.0003*
15.97	29.07	0.0568	0.0570	+0.0002	0.0719	0.0718	−0.0001†

* 35 cm.³ of a 10 per cent solution of ferric alum in the receiver.
† 20 cm.³ of a 10 per cent solution of ferric alum in the receiver.
‡ 50 cm.³ of a 10 per cent solution of ferric alum in the receiver.

The solution just titrated for vanadic acid, having now a volume of 100 cm.³ to 150 cm.³, is passed through a column of amalgamated zinc in a long Jones reductor, being preceded by 150 cm.³ of hot dilute ($2\frac{1}{2}$ per cent) sulphuric acid and followed by 100 cm.³ of the same acid, and finally 200 cm.³ of distilled water. The receiving flask, containing an excess of ferric sulphate, is kept cool by means of running water, and its contents, after the addition of sirupy phosphoric acid to remove the color of the iron, are titrated with permanganate until the color has changed from bluish green to yellow, and the color of the permanganate begins to be persistent and destroyed only by shaking. The flask is then heated to 70° or 80° and the titration completed in the hot solution.

The results given in the table show that iron and vanadium may be readily estimated in the presence of each other by two reductions — the first with sulphur dioxide and the last with amalgamated zinc, under the conditions described above — each followed by an oxidation with permanganate.

The Estimation of Ferric Iron, Vanadic Acid and Chromic Acid in the Presence of One Another.

The estimation of vanadic acid, chromic acid and ferric iron associated with one another has been accomplished by Edgar * by making use of processes of differential reduction.

By the action of hydrobromic acid, vanadic acid and chromic acid are reduced according to the equations

$$V_2O_5 + 2\,HBr = V_2O_4 + H_2O + Br_2,$$
$$V_2O_5 + 2\,CrO_3 + 8\,HBr = V_2O_4 + Cr_2O_3 + 4\,H_2O + 4\,Br_2,$$

the iron being unaffected. By the action of hydriodic acid upon the residue after action of the hydrobromic acid vanadium tetroxide is further reduced and the iron is reduced to the ferrous condition while the chromic salt is unaffected, as shown in the following equations:

$$V_2O_4 + 2\,HI = V_2O_3 + H_2O + I_2,$$
$$Fe_2O_3 + 2\,HI = 2\,FeO + H_2O + I_2.$$

If the halogen liberated in the two reductions be separately determined, two equations result, of which the first gives the sum of the vanadic and chromic acids and the second the sum of the vanadic acid and the iron; the halogen equivalent to the vanadium being the same in each case. If then either the vanadium, iron or chromium be estimated separately, a third equation is obtained, from which, with the aid of the first two, all three constituents may be calculated. Edgar found that the estimation

* Graham Edgar, Am. Jour. Sci., [4], xxvii, 174.

of the chromium in a separate portion by a modification of Browning's method of reduction with arsenious acid * affords a successful solution to this problem.

The distillation flask used by Edgar consists of a 100 $cm.^3$ pipette modified as shown in Fig. 5,† the inlet tube being bent upwards and having a separatory funnel sealed to its end, while the outlet tube is bent upwards and then down to enter the absorption flask, a small bulb being blown in it to prevent mechanical loss during distillation. In carrying out the process a slow current of hydrogen from a Kipp generator is kept up through the apparatus, and this, entering near the bottom of the flask, makes it possible to carry the distillation to very small volume without danger of "bumping."

The entire process in detail is as follows:

(I) The solution, of about 50 $cm.^3$ volume, containing the vanadate, chromate and ferric salt, is divided into two equal parts, one of which is placed in a distillation flask. To this, one to two grams of potassium bromide are added, together with 25 $cm.^3$ of concentrated hydrochloric acid, and the mixture is distilled until a volume of about 25 $cm.^3$ is reached, the reduction having always been found to be complete in that time. The bromine liberated is absorbed in alkaline potassium iodide, and, after cooling and acidifying, the iodine liberated is estimated by titration with approximately tenth normal sodium thiosulphate.

(II) The absorption apparatus is replaced, and, after the addition to the solution in the distillation flask of 1 grm. of potassium iodide, 10 $cm.^3$ of concentrated hydrochloric acid and 3 $cm.^3$ to 5 $cm.^3$ of sirupy phosphoric acid, distillation is again carried on until a volume of 10 $cm.^3$ has been reached; the iodine thus liberated being estimated as before.

(III) In the second half of the original solution the chromic acid is estimated by adding sulphuric acid to slight acidity, 3 $cm.^3$ of sirupy phosphoric acid, and an amount of standard arsenious acid in excess of that required to effect the reduction of the chromic acid. The use of phosphoric acid causes the iron to be precipitated as phosphate, and thus the difficulty mentioned by Browning ‡ in observing the end point, due to the reddish-

* See page 407.

† See page 5.

‡ See page 408.

Differential Reductions: By Hydrobromic Acid, by Hydriodic Acid and by Arsenite.

V_2O_5 taken.	V_2O_5 found.	Error, V_2O_5.	CrO_3 taken.	CrO_3 found.	Error, CrO_3.	Fe_2O_3 taken.	Fe_2O_3 found.	Error. Fe_2O_3.	(I) $n/10 \times$ 1.100 $Na_2S_2O_3$.	(II) $n/10 \times$ 1.100 $Na_2S_2O_3$.	(III) $n/10$ iodine.	(III) $n/10$ As_2O_3.
grm.	grm.	grm.	grm.	grm.	grm.	grm.	grm.	grm.	cm.^{3}	cm.^{3}	cm.^{3}	cm.^{3}
0.1183	0.1185	+0 0002	0 0709	0 0709	0.0000	0.1021	0.1018	−0 0003	31 15	23.40	8 74	30 0
0 1183	0.1179	−0.0004	0.0709	0.0710	+0 0001	0.1021	0.1023	+0 0002	31 12	23 40	8 70	30.0
0 1183	0.1186	+0.0003	0.0709	0 0709	0.0000	0.1021	0.1018	−0 0003	31 16	23.42	8 75	30 0
0.1183	0 1184	+0 0001	0 1064	0 1065	+0 0001	0.1531	0.1533	+0.0002	40.80	29.25	8 10	40.0
0 1183	0.1183	0 0000	0 1064	0 1062	−0 0002	0 1531	0.1534	+0.0003	40.75	29.20	8 15	40 0
0.0591	0 0590	−0.0001	0.1064	0 1065	+0 0001	0.1021	0.1019	−0.0002	34 90	17 50	8 10	40 0
0.0591	0.0590	−0.0001	0 0354	0 0355	+0 0001	0 2042	0.2043	+0 0001	15 58	29.15	9 35	20 0
0.1183	0.1188	+0.0005	0.0354	0 0353	−0.0001	0.1021	0.1018	−0 0003	21 50	23.45	9 40	20.0
0 0591	0.0590	−0.0001	0.1417	0.1417	0.0000	0.1021	0.1024	+0 0003	44 58	17.55	7 45	50 0
0.2366	0.2365	−0 0351	0.0354	0 0352	−0.0002	0.1531	0.1530	−0.0001	33.22	41 00	9 40	20 0
0 2366	0.2361	−0 0005	0.0354	0 0355	+0 0001	0.1021	0.0017	−0.0004	33 25	35.20	9 35	20.0
0.1183	0.1182	−0.0001	0.0709	0.0710	+0.0001	0.0510	0.0511	+0.0001	31.15	17.60	8.70	30.0

brown ferric hydroxide, is in large measure obviated, the blue of the starch iodide being quite clear against the pale green chromic hydroxide. After standing for from fifteen to twenty minutes the solution is made alkaline with sodium bicarbonate and an excess of standard iodine solution added. This is allowed to stand in a stoppered flask for from one-half hour to an hour, the excess of iodine being then removed with arsenious acid and the solution titrated to color with iodine after the addition of starch.

According to Edgar's procedure the first step in the calculation is the reduction to terms of tenth normal solution of the figures of titration obtained in the processes (I), (II) and (III). It is evident that the subtraction of the titration figure of (III) from that of (I) gives the number of cubic centimeters corresponding to reduction of the vanadium pentoxide to tetroxide, while the subtraction of this result from that of (II) gives the number equivalent to the reduction of the ferric salt. By multiplying these figures by the amounts of vanadic acid, chromic acid and ferric oxide equivalent to 1 cm.³ of the *n*/10 reagent the weights of these substances present are obtained. An example of this procedure is given below:

Titration.	cm.³	*n*/10 factor.	cm.³	
(I)	31.15	×1.100	=34.26 (*n*/10)	$\simeq V_2O_5+CrO_3$.
(II)	23.4	×1.100	=25.74 (*n*/10)	$\simeq V_2O_5+Fe_2O_3$.
(III)	30.00 − 8.74	×1.000	=21.26 (*n*/10)	$\simeq CrO_3$.
	34.26 − 21.26		=13.00	$\simeq V_2O_5$.
	25.74 − 13.00		=12.74	$\simeq Fe_2O_3$.

21.26×0.003334 (factor for CrO_3)=0.0709 grm. CrO_3 found.
13.00×0.00912 (factor for V_2O_5)=0.1185 grm. V_2O_5 found.
12.74×0.00799 (factor for Fe_2O_3)=0.1018 grm. Fe_2O_3 found.

Results obtained by this process are given in the table on the preceding page. The calculation given above is that of the first determination of the table.

INDEX OF AUTHORS*

* Workers in the Kent Chemical Laboratory.

INDEX OF SUBJECTS

Printed in Dunstable, United Kingdom

91220874R00312